# Advances in Soil Science

# SOIL MANAGEMENT
## Experimental Basis for Sustainability and Environmental Quality

*Edited by*

**R. Lal**
School of Natural Resources
The Ohio State University
Columbus, Ohio

**B. A. Stewart**
Dryland Agricultural Institute
West Texas A & M University
Canyon, Texas

CRC LEWIS PUBLISHERS
Boca Raton    London    Tokyo

**Library of Congress Cataloging-in-Publication Data**

Soil Management: Experimental Basis for Sustainability and Environmental Quality /
    edited by Rattan Lal and B.A. Stewart.
        p.    cm. -- (Advances in soil science)
        Papers presented at a workshop held June 16-18, 1993 at Ohio State University.
    Includes bibliographical references and index.
    ISBN 1-56670-076-0
    1. Soil management--Congresses.   2.   Soils--Research--Congresses.
I. Lal, R.   II. Stewart, B. A. (Bobby Alton), 1932-
III. Series: Advances in soil science (Boca Raton, Fla.)
S590.2.S88   1995
631.4--dc20                                                                       94-42105
                                                                                        CIP

© 1995 by CRC Press, Inc.
Lewis Publishers is an imprint of CRC Press

No claim to original U.S. Government works
International Standard Book Number 1-56670-076-0
Library of Congress Card Number 94-42105
Printed in the United States of America  1  2  3  4  5  6  7  8  9  0
Printed on acid-free paper

# Foreword

The world community recognizes the importance of sustainable management of soil, water, and nutrient resources. Agenda 21 of the 1992 UN Conference on Environment and Development (UNCED) identified several issues of soil resources management that require global planning and coordinated effort. Critical issues of soil resources management are: (i) Soil degradation caused by over-exploitation of fragile resources and cultivation of marginal soils with no or low level of off-farm inputs, (ii) Shortage of prime agricultural land as indicated by decrease in per capita available land area, and uneven distribution of potentially cultivable land resources, (iii) Declining trends in per capita food production in Sub-Saharan Africa and elsewhere in regions of rapidly growing population, (iv) Lack of or low rate of adaption of improved technologies by subsistence farmers of the tropics and sub-tropics, (v) Non-availability of essential off-farm inputs to resource-poor farmers, and lack of incentives and logistical support to adopt improved technologies, and (vi) Adverse impact of soil-mining and fertility-depletive practices on environment (e.g., global warming, water quality, biodiversity).

Impact of land use and soil management practices on agricultural sustainability and environmental quality can only be assessed by analyses and interpretation of data from well-planned and properly conducted long-term soil management experiments. Long-term experiments are those that are conducted on the same site for 10 years or more. Some examples of classical long-term soil management experiments include: The Rothamsted Experiment in U.K., and Morrow Plots and Sanborne Experiments in USA. Several long-term soil management experiments have been conducted in the tropics and sub-tropics of Africa, Asia, and South America. However, data from such experiments need to be analyzed to establish relationships between soil properties and productivity; identify soil properties and processes that affect soil quality, soil degradation and soil resilience; and establish critical limits of important soil properties in relation to productivity and environmental quality.

The knowledge gap, with regard to trends in soil properties in relation to management systems, has long been recognized, especially in ecologically-sensitive regions of the tropics and sub-tropics. Furthermore, the available research information needs to be compared and contrasted with information from classical experiments conducted in the temperate region climate. It was the recognition of the need for this comparative evaluation of the available research data that led to a workshop on "Long-Term Soil Management Experiments."

The workshop, organized The Ohio State University took place June 16−18, 1993, and was sponsored by the Rockefeller Foundation and US AID. Objectives of the workshop were to: (i) Identify management-induced changes in soil properties and processes and establish temporal trends, (ii) Identify technological options for sustainable use of soil resources in different agro-ecosystems, (iii) Assess limitations of existing long-term experiments, and (iv) Prioritize needs for additional long-term experiments. The workshop was attended by about 50 participants including representatives from 8 International Agricultural Research Centers (IARCs), 12 U.S. universities, USDA, several national research organizations (FAO, The World Bank, etc.).

Successful outcome of the workshop depended on several individuals who worked as chairpersons or rapporteurs (R. Bertram, D. Hansen, R. Herdt, F. Miller, D. Parish, D. Picard, T. Sentz, and T.J. Smyth), and working-group chairpersons or rapporteurs (K. Cassel, J. Duxbury, C. Pieri, M. Singer, G. Uehara, and L. Wilding). In all, 24 research papers were presented, including 18 state-of-the-art review and case studies. These papers were reviewed and revised, and they form the basis of the book. Recommendations drafted in the concluding chapter were based on those prepared by the chairperson and rapporteur of each working group.

The 19-chapter book is divided into four sections. There are seven chapters in the section dealing with the Humid Tropics, four in the Sub-Humid and Semiarid Tropics section, two in the Temperate and Mediterranean Climate section, and five in the concluding section devoted to Synthesis and Future Priorities.

This volume is an important compilation of the data from long-term soil management experiments with particular reference to the tropics and sub-tropics. Technological options identified by the working groups for sustainable management of soil resources of the tropics are outlined, and needs for additional long-term experiments are prioritized. The information contained in the book is of interest to soil scientists, agronomists, soil ecologists, and those interested in agricultural sustainability and environmental quality.

We express our profound appreciation to all contributing authors for their quality manuscripts, to all session chairpersons and rapporteurs, and working-group chairpersons and rapporteurs for their guidance in technical discussions that led to important recommendations. Logistics and financial support provided by The Ohio State University, the Rockefeller Foundation, and US AID made this volume possible.

Rattan Lal                                                              B.A. Stewart
Columbus, Ohio                                                    Amarillo, Texas

# Preface

When Rattan Lal approached the United States Agency for International Development (USAID) seeking support for a workshop on long-term agronomic trials, several other efforts dealing with issues of soil, water, and sustainability were underway. In part, these essentially-related initiatives reflected a general sense that the international agricultural research and development community had not reached a clear consensus on priorities and approaches for soil, water, and nutrient research, despite continuing interest in the question of agricultural sustainability. As we consider long-term agronomic trials, awareness of these related efforts can provide a useful context and background. We would like to note three such efforts (there are probably more).

First, the International Board for Soils Research and Management (IBSRAM) is working on behalf of several donors to develop a position paper of needs and priorities in soil, water, and nutrient management research. This paper was completed in 1994 and presented to the Consultative Group on International Agricultural Research (CGIAR) donors in their May 1994 Mid-Term Meeting in New Delhi, India. Second, in response to the 1992 United Nations Conference on Environment and Development, the CGIAR developed a paper outlining its abilities to contribute to Agenda 21. This paper has been developed with leadership from Stein Bie of the Norwegian delegation to the CGIAR. Third, prior to the Mid-Term Meeting of the CGIAR in May 1993, a meeting of staff of international agricultural research centers and U.S. universities at the University of Florida at Gainesville, focused on the relationship between productivity and sustainability. A report of follow-up activities of the Gainesville group was presented to the CGIAR at the International Centers Week 1994 meeting in October 1994.

How then, do the papers in this volume relate to the considerable interest from donors and researchers noted above? The scientific rigor and experimental results coming from long-term experiments are vital to an informed discussion of sustainability. For example, there is a misconception in some quarters that yield and sustainability are fundamentally antagonistic. Yet the data, and discussion here, if anything, suggest that yield may be the best indicatior of sustainability. Moreover, those making decisions on program emphasis and research approaches need the kind of hard, empirical evidence available from long-term trials. Thus, the initiatives above can benefit directly from the research underpinning provided through sustained research on soil, water, and fertility.

In another sense, the field of long-term research offers a frame of reference within which new research approaches, employing GIS, modeling, and other techniques, can be applied and refined, increasing their predictive value in other situations. The challenge of applying these and other techniques to the problem of site specificity in natural resources management is at the heart of the CGIAR's whole "eco-regional" research thrust. A broader relevance of applications to farm and land-use management questions to help producers is what the development community must seek when committing international research resources.

Finally, USAID is pleased to support and associate itself with the broad participation of the U.S. research community in regards to long-term research. There is frequent reference to the need for research partnerships in the areas of biotechnology and genetics—we are convinced that major advantages will also be associated with natural resources research collaboration. Collaborative efforts involving universities and other centers of excellence in the United States and elsewhere will be key to making the research advances that are needed. Just as in other areas of research, investigation of long-term soil, water, and plant management relationships will prove richly rewarding for all involved.

Robert B. Bertram and Ralph W. Cummings, Jr.
USAID International Agricultural Research Centers Staff

# Preface

Given that agriculture is the production of food and fiber from land, water, carbon dioxide, nutrients, and sunlight, and that the human race will continue to need food for its existence, and that the number of people on the globe has long since passed the point when it could survive on food produced by naturally existing ecosystems, a global sustainable agricultural system is absolutely necessary. If sustainable agriculture means anything, it must at least mean a system of production that continues to be productive over a long period of time.

Long-term agronomic trials are the one type of empirical data that can confirm whether a particular system is sustainable. Long-term agronomic experiments should provide significant insights into such issues as: whether contemporary crop production practices are less sustainable than systems used in the pre-chemical past, whether crop rotations provide protection against productivity decline, whether adequate measures of off-site effects of cropping practices are available to suggest the magnitude of such effects, and whether tropical crop production conditions lead to more rapid resource degradation than do temperate conditions.

Professor Rattan Lal has long experience with considering such questions in the tropics, and when he approached the Rockefeller Foundation with the idea of convening a workshop of professionals working with long-term trials, we were convinced that such a workshop could make a contribution to the evolving discussion on sustainable agriculture. The product, which led to this volume, provides a treasure house of information about long-term trials in the tropics.

The experiments and experiences discussed in this monograph carefully document crop production systems with well-defined system boundaries, over a significant period of time. Although not originally designed to address questions of sustainability, these long-term agronomic trials provide an invaluable data resource that has, up to now, been largely ignored by both the research community and the sustainability gurus. With a rigorous definition of sustainability and this data, the sustainability (or lack thereof) of various cropping systems will be more clearly illustrated than in any previous effort. It will be useful to those who ask whether agriculture is sustainable, to those who want to look at such experiences, to those who wish to initiate new long-term trials, and to those who need to know the limitaitons of long-term trials.

Robert W. Herdt
Agricultural Sciences Division
The Rockefeller Foundation

About the Editors:

*Dr. R. Lal* is a Professor of Soil Science in the School of Natural Resources at The Ohio State University, Columbus, Ohio. Prior to joining Ohio State in 1987, he served as a soil scientist for 18 years at The International Institute of Tropical Agriculture, Ibadan, Nigeria. Prof. Lal is a fellow of the Soil Science Society of America, American Society of Agronomy, and The Third World Academy of Sciences. He is a recipient of both the International Soil Science Award, and the Soil Science Applied Research Award of the Soil Science Society of America.

*Dr. B.A. Stewart* is a Distinguished Professor of Soil Science, and Director of the Dryland Agriculture Institute at West Texas A&M University, Canyon, Texas. Prior to joining West Texas A&M University in 1993, he was Director of the USDA Conservation and Production Research Laboratory, Bushland, Texas. Dr. Stewart is past president of the Soil Science Society of America, and was a member of the 1990-93 Committee of Long Range Soil and Water Policy, National Research Council, National Academy of Sciences. He is a Fellow of the Soil Science Society of America, American Society of Agronomy, Soil and Water Conservation Society, a recipient of the USDA Superior Service Award, and a recipient of the Hugh Hammond Bennett Award by the Soil and Water Conservation Society.

# Contributors

*J. Alcantara*, Division of Agronomy, Plant Physiology, and Agroecology; International Rice Research Institute (IRRI), Los Baños, Philippines.

*M.M. Anders*, International Crops Research Institute for the Semi-Arid Tropics, Patancheru P.O., Andhra Pradesh 502324, India.

*J.B. Aune*, Centre for Sustainable Development, Agricultural University of Norway, N-1432 A, Norway.

*M.A. Ayarza*, Savanna and Forage Programs, Centro Internacional de Agricultural Tropical (CIAT), Cali, Columbia.

*F.H. Beinroth*, Professor of Soil Science, Department of Agronomy and Soils, University of Puerto Rico, Mayagüez, PR 00708.

*Joaquim Braga Bastos*, Center for Agroforestry Research of the Eastern Amazon, Brazilian Enterprise for Agricultural Research, Belém, Brazil.

*D.K. Cassel*, Department of Soil Science, North Carolina State University, Raleigh, North Carolina 27695-7619, U.S.A.

*K.G. Cassman*, Division of Agronomy, Plant Physiology, and Agroecology; International Rice Research Institute (IRRI), Los Baños, Philippines.

*S.K. De Datta*, International Programs, Virginia Polytechnic Institute and State University, Blacksburg, Virginia 24061, U.S.A.

*J. Descalsota*, Division of Agronomy, Plant Physiology, and Agroecology; International Rice Research Institute (IRRI), Los Baños, Philippines.

*M. Dizon*, Division of Agronomy, Plant Physiology, and Agroecology; International Rice Research Institute (IRRI), Los Baños, Philippines.

*S.K. Ehui*, International Livestock Centre for Africa (ILCA), P.O. Box 5689, Addis Ababa, Ethiopia.

*H. Eswaran*, National Leader, World Soil Resources, Soil Conservation Service, P.O. Box 2890, Room 4838-S, Washington, D.C. 20013, U.S.A.

*M.J. Fisher*, Savanna and Forage Programs, Centro Internacional de Agricultural Tropical (CIAT), Cali, Columbia.

*M.J. Glendining*, Rothamsted Experimental Station, Harpenden, Hertfordshire AL5 2JQ, United Kingdom.

*H.C. Goma*, Soil Productivity Research Programme, Misamfu, Regional Research Centre, P.O. Box 410055, Sasama, Zambia.

*Hazel C. Harris*, International Center for Agricultural Research in the Dry Areas (ICARDA), P.O. Box 5466, Aleppo, Syria.

*Peter J. Harris*, Department of Soil Science, University of Reading, London Road, RG1 5AQ, United Kingdom.

*I. Haque*, International Livestock Centre for Africa (ILCA), P.O. Box 5689, Addis Ababa, Ethiopia.

*R. Lal*, School of Natural Resources, Ohio State University, 2021 Coffey Road, Columbus, Ohio 43210-1085, U.S.A.

*K.B. Laryea*, International Crops Research Institute for the Semi-Arid Tropics, Patancheru P.O., Andhra Pradesh 502324, India.

*Raimundo Freire de Oliveria*, Center for Agroforestry Research of the Eastern Amazon, Brazilian Enterprise for Agricultural Research, Belém, Brazil.

*C.S. Ofori*, Land and Water Development Division, Food and Agriculture Organization of the United Nations (FA), Viale delle Terme di Caracalla, 00100, Rome, Italy.

*D.C. Olk*, Division of Agronomy, Plant Physiology, and Agroecology; International Rice Research Institute (IRRI), Los Baños, Philippines.

*P. Pathak*, International Crops Research Institute for the Semi-Arid Tropics, Patancheru P.O., Andhra Pradesh 502324, India.

*J.M. Powell*, International Livestock Centre for Africa (ILCA), P.O. Box 5689, Addis Ababa, Ethiopia.

*E. Pushparajah*, International Board for Soil Research and Management (IBSRAM), P.O. Box 9-109, Bangkhen, Bangkok 10900, Thailand.

*Christian Pieri*, Agriculture and Natural Resources, The World Bank, 1818 H Street, N.W., Washington, D.C. 20433, U.S.A.

*D.S. Powlson*, Rothamsted Experimental Station, Harpenden, Hertfordshire AL5 2JQ, United Kingdom.

*M. Samson*, Division of Agronomy, Plant Physiology, and Agroecology; International Rice Research Institute (IRRI), Los Baños, Philippines.

*J.I. Sanz*, Savanna and Forage Programs, Centro Internacional de Agricultural Tropical (CIAT), Cali, Columbia.

*E. Adilson Serrão*, Center for Agroforestry Research of the Eastern Amazon, Brazilian Enterprise for Agricultural Research, Belém, Brazil.

*B.R. Singh*, Department of Soil and Water Sciences, Agricultural University of Norway, P.O. Box 5028, N-1432, Aas, Norway.

*T.J. Smyth,* Department of Soil Science, North Carolina State University, Raleigh, North Carolina 27695-7619, U.S.A.

*B.A. Stewart,* Dryland Agriculture Institute, West Texas A&M University, Canyon, Texas 79016, U.S.A.

*Leopoldo Brito Teixeira*, Center for Agroforestry Research of the Eastern Amazon, Brazilian Enterprise for Agricultural Research, Belém, Brazil.

*R.J. Thomas*, Savanna and Forage Programs, Centro Internacional de Agricultural Tropical (CIAT), Cali, Columbia.

*Soumaïla Traoré*, Department of Soil Science, University of Reading, London Road, RG1 5AQ, United Kingdom.

*G.Y. Tsuji*, Department of Agronomy and Soil Science, University of Hawaii at Monoa, Honolulu, HI 96822, U.S.A.

*G. Uehara*, Department of Agronomy and Soil Science, University of Hawaii at Manoa, Honolulu, Hawaii 96822, U.S.A.

# Contents

B. Sub-Humid and Semiarid Tropics.

C. Temperate and Mediterranean Climate

D. Synthesis and Future Priorities

# Managing Soils for Enhancing and Sustaining Agricultural Production

R. Lal and B.A. Stewart

## I. Introduction

Enhancement and maintenance of soil productivity is essential to the sustainability of agriculture and to meeting basic food needs of rising population. The world population is projected to rise from 5.3 billion people in 1990 to 8.5 in 2025 and 10 billion by 2050 (Bonhaarts, 1994). Out of the projected increase in world population of 90 million a year, 86 million will occur in the developing countries most of which are in the tropics and sub-tropics. Most of this increase in population will occur in already land-starved regions (e.g., South Asia) or ecologically-sensitive regions (e.g., Sub-Saharan Africa). The population of Sub–Saharan Africa may increase at the rate of 25 million people a year until it reaches 2 billion by the year 2050. This rate of increase in population will double food requirements in developing countries (Blake et al., 1994). It is estimated that in the 1990s more than 700 million people in the developing world, including about 180 million children, do not have access to enough food (Conway et al., 1994). Most of these malnourished people live in rural areas of South Asia and Sub-Saharan Africa. The problem of food shortage is likely to exacerbate in these regions when the population will increase to about 2 billion in South Asia and to 1.2 billion in Sub-Saharan Africa by the year 2025.

In the contest of increasing food demand, several important issues with regard to world soil resources that need to be addressed are:
(i)    Are world soil resources adequate to meet the food demands for a rapid surge in human population?
(ii)   What are the potential and constraints of world soil resources?

ISBN 1-56670-076-0
©1995 by CRC Press, Inc.

1

(iii)     Is generic technology available that can be adopted to soil-specific conditions for sustainable management of soil and water resources?
(iv)     What strategic and applied research is needed to fill in the knowledge gaps that enable sustainable management of soil and water resources?
(v)      How should activities in research and transfer of technology be organized to be cost-effective and avoid duplication and redundancy?

These questions are very pertinent to the developing countries of the tropics and sub-tropics where soil resources are poorly endowed and resource-poor farmers cannot afford the scientific inputs needed to increase food production. In addition to social and political factors, it is the quality of soil resources and their potential carrying capacity that is being questioned.

## II. Soil Resources

World soil resources are finite and non-renewable. Furthermore, most potentially cultivable land has already been brought under cultivation. The remaining land is located in either inaccessible areas or in ecologically-sensitive ecoregions. Therefore, future increases in food production have to come from intensification of cultivation on existing land. Furthermore, cultivable land area may actually decline due to soil degradation and to increasingly non-agricultural uses, e.g., urbanization.

There are severe constraints to intensification of agricultural activities on existing lands. Alfisols, the most predominant soils of the world occupy about $17.1 \times 10^6$ km$^2$ area (Table 1). These soils are widespread in arid and semiarid regions of South Asia and Sub-Saharan Africa. Alfisols have extremely poor soil soil structure, are easily crusted and compacted, and are highly prone to accelerated erosion. In contrast, Oxisols and Ultisols cover $22.4 \times 10^6$ km$^2$ of the land area. These soils have severe nutritional limitations and chemical constraints. And, to a lesser extent, Oxisols are also prone to physical limitations. Vertisols are widespread in arid and semiarid regions. These soils have severe soil physical contraints including poor trafficability during the rainy season and high susceptibilty to erosion. In addition, Vertisols also have major nutritional and chemical constraints. Even with science-based management and high inputs, these soils have low to medium population-carrying capacities (Table 1).

Soil degradation caused by interactive effects of biophysical and socio-economic factors is a severe global problem (Table 2). Worldwide $240 \times 10^6$ ha or 12% of the total land area is estimated to be susceptible to soil chemical degradation. The problem of soil chemical degradation is particularly severe in South America. Soils of Africa prone to chemical constraints include $1296 \times 10^6$ ha (45.6% of land area) with low cation exchange capacity (CEC), 635 million ha (22.3%) with aluminum toxicity, $383 \times 10^6$ ha (13.5%) from high P fixation, and $637 \times 10^6$ ha (22.4%) from low K supply (FAO, 1986). Soils

**Table 1.** Some principle soils of the world and their constraints to intensive and sustainable land use

| Soil order | World area[a] ($10^6$ km$^2$) | Constraints |
|---|---|---|
| Alfisols | 17.1 | Poor soil physical conditions, high soil erosion risks, poor soil tilth, low nutrient-holding capacity, low to medium population-carrying capacity |
| Oxisols | 11.9 | Low nutrient-holding capacity, severe chemical constraints (e.g., P fixation, Al toxicity) low plant-available water reserves, low population-carrying capacity |
| Ultisols | 10.5 | Low nutrient-holding capacity, severe chemical constraints, low plant-available water reserves, poor soil tilth, severe physical constraints, high erodibility, low effective soil volume, low to medium population-carrying capacity |
| Vertisols | 2.9 | Poor soil physical conditions, poor soil tilth water imbalance, poor trafficability, severe soil erosion, severe chemical constraints and nutrient imbalance, low to medium population-carrying capacity |

([a] From Eswaran et al., 1993.)

**Table 2.** Estimates of chemical and physical soil degradation

| | Soil chemical degradation | | Soil physical degradation | |
|---|---|---|---|---|
| | $10^6$ ha | % of total | $10^6$ ha | % of total |
| Africa | 61 | 12 | 19 | 4 |
| Asia | 74 | 10 | 12 | 2 |
| S. America | 70 | 29 | 8 | 3 |
| C. America | 6 | 2 | 5 | 8 |
| N. America | + | + | 1 | 1 |
| Europe | 26 | 12 | 36 | 17 |
| Oceania | 1 | 1 | 2 | 2 |
| World | 240 | 12 | 83 | 4 |

(Modified from Oldeman et al., 1990.)

of Sub-Saharan Africa are particularly low in N, have low water-holding capacity, and are prone to drought. Worldwide, $83 \times 10^6$ ha or 4% of the total land area is presumably susceptible to soil physical degradation. Within the tropical ecosystems, the problem of soil physical degradation is severe in Africa

**Table 3.** Estimates of land degradation in some countries of Asia

| Country | Arable and cropped land area (km$^2$) | Degraded land area (km$^2$) |
|---|---|---|
| China | 96,976 (10) | 280,000 (30) |
| India | 168,990 (57) | 148,100 (50) |
| Indonesia | 21,221 (12) | 43,000 (24) |
| Pakistan | 20,760 (27) | 15,500 (17) |
| Philippines | 7,930 (27) | 5,000 (17) |
| Thailand | 24,050 (39) | 17,200 (34) |

Figures in parentheses represent % of the total land area.
(Modified from Dent, 1990; Eswaran et al., 1983.)

and Asia. Soil erosion is a severe form of physical degradation. It is estimated that about 80% of the agricultural land in the world suffers moderate to severe erosion and 10% suffers slight to moderate erosion (Speth, 1994). Presumably, 17% of the land supporting plant life worldwide has lost agronomic value over the 45 year period ending in 1993 (Bongaarts, 1994).

Land prone to major degradative processes is widespread in Asia (Dent, 1990; Eswaran et al., 1993). Estimates of degraded land area in several countries of Asia show that degraded land covers 50% of the total land area in India, 30% in China, 30% in Thailand, 24% in Indonesia, and 17% each in Pakistan and the Philippines (Table 3).

Management of these degraded lands require in-depth understanding of predominant processes and interactive socio-economic and political factors. Although the conventional plant-breeding approach can be helpful, alleviating soil-related constraints to agronomic intensification and high yields is crucial.

Nutritional constraints can be alleviated by adding chemical fertilizers. However, fertilizers are often prohibitively expensive for resource-poor farmers and they are not readily available. The average fertilizer use is only 30 kg N ha$^{-1}$ in Asia, 15 kg ha$^{-1}$ in Latin America, and 4 kg ha$^{-1}$ in Africa (IFDC, 1987; Conway et al., 1994; Vlex, 1994).

Irrigated agriculture has played an important role in increasing food production in Asia and North Africa. Total irrigated land area is about $170 \times 10^6$ ha in developing countries. An additional $100 \times 10^6$ ha of land is potentially irrigable especially in Asia, e.g., India and China. However, such large scale expansion of irrigation is expensive. Irrigation potential is rather low in Africa. Only 5% ($2.23 \times 10^6$ ha) of arable land is currently irrigated (Wright, 1986).

The potential carrying capacity of land is particularly questionable in Sub-Saharan Africa. Only 30% of the total land area of $2.22 \times 10^6$ ha of Sub-Saharan Africa is cultivable (World Bank, 1989). Of the total harvested land in Sub-Saharan Africa, 64% is considered low potential or problem land (FAO, 1989). The total problem or low potential land in Sub-Saharan Africa is almost four-fold of that in Latin America and two-fold of that in Asia.

## III. Strategies for Generating Technologies

The 50-year period ending in 1990 saw considerable progress regarding the understanding of the basic principles of soil science including nutrient and water cycles and in their assessment of water and nutrient requirements for crop production (Simonson, 1991; Gardner, 1991). Our understanding of potential and constraints of tropical soils of the tropics has also advanced (Sanchez, 1976; Lal, 1987; 1990) and popular myths have been replaced by facts (Lal and Sanchez, 1992). Nonetheless, soil research is site-specific because magnitude and direction of most processes strongly interact with biophysical and socio-economic factors which vary strongly among ecoregions and environments. Furthermore, high rate of soil degradation is driven by socio-economic and political factors, and poverty. Oldeman et al. (1990) estimated that worldwide soil degradation by overgrazing would reach $579 \times 10^6$ ha and $552 \times 10^6$ ha by agricultural activities. Because of the strong effect of these anthropogenic activities, understanding processes that affect agronomic productivity and sustainability is crucial to sustainable use of soil and water resources. An important strategy is to understand the effects of land use, farming/cropping systems, and soil and crop management on soil properties and soil processes. It is the management-induced alterations in soil properties and processes that affect yield, response to inputs, the severity and type of soil degradation, and environmental regulatory capacity of soil. Lal (1994) suggested soil indicators and the frequency that they should be monitored (Table 4).

Interactive effects of anthropogenic perturbations on sustainable use of soil and water resources underlines the importance of studying soils and alterations in their properties and processes under on-farm conditions. This strategy is particularly relevant in developing countries of the tropics and sub-tropics where resource-poor small land holders are forced to mine soil fertility and set in motion land degradative processes. Agriculture in the developing worlds is at crossroads due to issues that necessitate on-farm evaluation of alterations in soil properties and processes. Notable among these factors are:

- Reduction in the fallow period and intensive cultivation for extended period of time,
- Replacement of manual operations of seedbed preparation by animal-driven or motorized equipment,
- Frequent burning of biomass and other crop residues for land clearing and seedbed preparation,
- Increasing use of agricultural chemicals including fertilizers and pesticides,
- Gradual change from subsistence agriculture to semi-commercial and commercial.

Most soils of the tropics have low buffer capacity because of low organic matter content and predominance of low activity clays. Consequently, these soils are highly prone to physical degradation which is set in motion by crusting,

**Table 4.** Suggested frequency of monitoring soil indicators

| Soil indicator | Suggested monitoring frequency |
| --- | --- |
| *Soil physical indicators* | |
| Soil moisture | Every week |
| Bulk density and penetration resistance | Every season |
| Hydraulic conductivity | Yearly |
| Structure | 1 to 2 years |
| Infiltration | 1 to 2 years |
| Available water-holding capacity | 3 to 5 years |
| Texture | 3 to 5 years |
| | |
| *Soil chemical indicators* | |
| pH | Seasonal |
| Total nitrogen | 1 to 2 years |
| Available nutrients | 1 to 2 years |
| Cation-exchange capacity | |
| | |
| *Soil biological indicators* | |
| Earthworm activity | Every season |
| Biomass carbon | 1 to 2 years |
| Soil organic carbon | 1 to 2 years |
| | |
| *Crop indicators* | |
| Yield | Every season |
| Root growth | Every season |
| Nutrient status | 1 to 2 years |
| | |
| *Micro-climate* | |
| Soil temperature | Daily and seasonal |
| Air temperature | Daily |
| Evaporation | Daily |
| Rainfall amount | Seasonal |
| Rainfall intensity | Maximum over 5 to 10 minutes |

(From Lal, 1994.)

compaction, low infiltration, high runoff and accelerated erosion. Physical degradation is accentuated by replacement of manual farm operations by mechanized operations. Applications of inorganic fertilizers can lead to acidification (i.e., decline in pH) of these soils of low-activity clays. Soil application of pesticides has drastic adverse effects on soil fauna, e.g., earthworms and termites. Soil biodiversity and biomass carbon are drastically reduced.

An important strategy for sustainable management of these soils is to maintain an ecological balance between soil-climate-vegetation and yet intensify agricultural production. Maintaining an ecological balance is especially critical in view of the necessity to mechanize farm operations, enhance soil fertility by using off-farm inputs, adopt monoculture or simplified systems, and use pesticides to decrease losses.

## IV. Need for Long-Term Soil Management Research

Agricultural sustainability is defined as a non-negative trend in per capita productivity, and productivity trends can only be assessed over a long-time horizon. Productivity is defined as production per unit consumption of the resource. Total factor productivity is production divided by the summation of the product of all resources and their respective costs. Quantitative assessment of sustainability, therefore, implies evaluation of trends in productivity per unit change in soil characteristics (Lal, 1984; Equations 1 and 2).

$$S_c = f(P_i, P_d, P_m) \tag{1}$$
$$S_c = f(P_i \times S_p \times W_r \times C_f) \, dt \tag{2}$$

when $S_c$ is the coefficient of sustainability, $P_i$ is the productivity per unit input of the limited resource, $P_d$ is productivity per unit decline in soil property, $P_m$ is the maximum assured productivity, $S_p$ is the critical level of soil property, $W_r$ is the soil water regime and its quality, $C_f$ is the climatic factor, and t is time. Data from long-term field experiments are required to:

(i)   assess sustainability and provide detailed information for establishing relationships between crop performance and soil climatic conditions. Because of drastic differences in yield due to seasonal variability in rainfall and climate, yield trends can only be monitored over a long time,

(ii)  evaluate trends in productivity and changes in soil properties,

(iii) establish dynamics of changes in soil properties that accentuate negative impact of global climate change,

(iv)  develop relationship between intensification of agricultural production methods and environmental degradation, e.g., water quality,

(v)   identify linkages of soil conditions to pathogens, weeds, and insects,

(vi)  evaluate the impact of land use and soil management on soil degradative processes,

(vii) establish resilience characteristics of soil systems in relation to degradative processes, and

(viii) assess the impact of land-restorative measures on improvements in soil quality.

# V. Conclusions

- World population is rapidly increasing and the largest increase in population is taking place in developing countries of Asia and Sub-Saharan Africa. The need to drastically increase world food production to feed the increasing population is an important consideration in developing techniques and policies to sustain management of soil, water, and nutrients.
- World soil resources are finite, unevenly distributed in different geographical regions, non-renewable over the human time-scale, fragile, and prone to degradative processes. Enhancement and maintenance of high soil productivity are essential to feed and provide basic necessities of life for the growing population.
- Several critical issues with regard to sustainable management of soil resources are pertaining to (i) adequacy of soil resources, (ii) their potential and constraints, (iii) availability of appropriate technologies, (iv) knowledge gaps, and (v) effective technology transfer.
- Soil degradation is a widespread problem, and the degradative processes are driven by social, economic, and political factors. Resource-poor, subsistence, and low-input agriculture are among major factors that exacerbate soil degradation.
- Data from long-term field experiments are needed to establish yield trends, evaluate sustainability, assess management-induced changes in soil properties, and quantify the impact of land use on global climate change and environmental quality.

# References

Blake, R.O., D.E. Bell, J.T. Mathews, R.S. McNamara, and M.P. McPherson. 1994. Feeding 10 billion people in 2050: The key role of the CGIAR's, IARC's. A report by the action group on food security, World Resources Institute, Washington, D.C. 17 pp.

Bongaarts, J. 1994. Can the growing human population feed itself? *Scientific American*. March 1994: 36-42.

Conway, G., U. Lela, J. Peacock, and M. Pineiro. 1994. Sustainable agriculture for a food secure world: a vision for the Consultative Group on International Agricultural Research, CGIAR, Washington, D.C.

Eswaran, H., S.M. Virmani, and L.D. Spivy, Jr. 1993. Sustainable agriculture in developing countries: constraints, challenges and choices. p. 7-24. In: J. Ragland and R. Lal (eds.), *Technologies for Sustainable Agriculture in the Tropics*. U.S. Agency for International Development Spec. Publ., Washington, D.C.

Food and Agriculture Organization (FAO). 1986. African agriculture: The next 25 years. FAO, Rome, Italy.

Food and Agriculture Organization (FAO). 1986. Irrigation in Africa, south of the Sahara. FAO Investment Centre Tech. Paper 5. FAO, Rome, Italy.

Food and Agriculture Organization (FAO). 1989. The state of food and agriculture. FAO Agric. Ser. 22. FAO, Rome, Italy.

Gardner, W.R. 1991. Soil science as a basic science. *Soil Sci*. 151:2-6.

International Fertilizer Development Center (IFDC). 1987. Africa fertilizer situation. IFDC, Muscle Shoals, AL.

Lal, R. 1987. *Tropical Ecology and Physical Edaphology*. Chichester, U.K.

Lal, R. 1990. *Soil Erosion in the Tropics: Principles and Management*. McGraw Hill, NY.

Lal, R. 1994. Methods and guidelines for assessing sustainable use of soil and water resources in the tropics. SMSS Tech. Bull. 21. USDA Soil Conservation Service, Washington, D.C. 78 pp.

Lal, R. and P.A. Sanchez (eds.). 1992. *Myths and Science of the Soils of the Tropics*. Special Publication, Soil Science Society of America, Madison, WI.

Oldeman, L.R., R.T.A. Hakkeling, and W.G. Sombrock. 1980. World map of the status of human-induced soil degradation: an explanatory note. ISRIC, Wageningen, The Netherlands.

Sanchez, P.A. 1976. *Management and Properties of the Soils of the Tropics*. J. Wiley and Sons, New York, NY.

Simonson, R.W. 1991. Soil science: goals for the next 75 years. *Soil Sci*. 151:7-18.

Speth, J.G. 1994. Towards an effective and operational international convention on desertification. International Negotiating Committee, New York, N.Y.

Vlex, P.G. 1993. Strategies for sustaining agriculture in Sub-Saharan Africa: the fertilizer technology issue. p. 265-277. In: *Technologies for Sustainable Agriculture in the Tropics*. Am. Soc. Agron. Spe. Publ. No. 56. Madison, WI.

World Bank. 1989. Sub-Saharan Africa: from crisis to sustainable growth. World Bank, Washington, D.C.

Wright, E.P. 1986. Water resources for human, livestock, and irrigation. In: L.J. Foster (ed.), *Agricultural Development in Drought-Prone Africa*. Overseas Development Institute, London, United Kingdom.

# A. Humid Tropics

# Synthesis of Long-Term Soil Management Research on Ultisols and Oxisols in the Amazon

T.J. Smyth and D.K. Cassel

ISBN 1-56670-076-0
©1995 by CRC Press, Inc.

# I. Introduction

## A. Agriculture and Deforestation Issues

Three-fourths of the 673 million ha of land under native tropical rainforest and seasonal tropical forest in tropical America are located in the Amazon Basin (Sanchez, 1987). Continued expansion of agricultural frontiers within the Amazon Basin to relieve population growth pressures has brought worldwide attention to the ecological consequences of increased deforestation. Tropical rainforests in Latin America are being cleared at the rate of 2.5 to 2.8 million hectares per year (Melillo et al., 1985). Clearing in the Amazon Basin is estimated to occur at the rate of 1.2 million hectares per year and is associated mainly with subsistence farming by shifting cultivators and pasture establishment by cattle ranchers (Hecht, 1982). Application of land management options which recuperate and/or maintain long-term productivity of deforested and degraded lands was recently proposed as a strategy to reduce further deforestation by eliminating the need to abandon cleared land (Sanchez et al., 1990). This paper summarizes experiences gained from long-term soil management research on continuous use of Oxisols and Ultisols in the Amazon.

## B. Areal Extent of Ultisols and Oxisols, and Their Management Constraints

The land resource base in the Amazon is huge, occupying about 482 million hectares, as reported by Cochrane and Sanchez (1982) in a very thorough summarization of existing data. The following information is extracted from that publication. Three climatic subregions were identified as shown in Table 1. This classification is based on estimates of total wet season potential evapotranspiration (WSPE); length of the wet season, that is, the period when rainfall exceeds potential evapotranspiration; and mean wet season air temperature (WSMT). The tropical rain forest subregion occurs primarily in the western part of the Amazon Basin and occupies 171 million hectares or 35% of the land area. The semi-evergreen seasonal forest subregion occurs primarily in the eastern portion of the Basin and occupies 274 million hectares, 17% of it being poorly drained. The savannas occupy 37 million hectares in the Amazon Basin. The Llanos of Colombia and Venezuela are not included, nor is the Cerrado of Brazil. In general, soils in the Amazon Basin have low fertility although 6% of the total land area has well-drained soils with high native fertility.

Overall, 23% of the land area in the Amazon Basin is flat and poorly drained (Table 1). Of the 171 million hectares of land in the tropical rain forest subregion, 29% is poorly drained. Most of the remainder has a slope between 0 and 30%. Although the degree of erosion susceptibility varies with soil and rainfall characteristics, virtually all of the sloping land in this region of high rainfall is subject to soil erosion if improperly managed.

**Table 1.** Climatic subregions of the Amazon

| Subregion | Climate[1] | Flat, poorly drained | Well drained 0-8 | Well drained 8-30 | Well drained >30 | Total |
|-----------|-----------|---------------------|------|-------|------|-------|
| | | | | million hectares | | |
| Tropical rainforest | WSPE> 1300 mm Wet season > 9 months WSMT > 23.5°C | 50 | 79 | 30 | 12 | 171 |
| Semi-ever-green season-al forest | WSPE: 1061-1300 mm; Wet season: 8-9 months WSMT > 23.5° C | 47 | 142 | 69 | 16 | 274 |
| Savannas | WSPE: 900-1060 mm; Wet season: 6-8 months WSMT> 23.5°C | 12 | 19 | 4 | 2 | 37 |
| Total | | 109 | 240 | 103 | 30 | 482 |
| Percent | | 23 | 50 | 21 | 6 | 100 |

[1] WSPE = total wet season potential evapotranspiration; WSMT = mean wet season air temperature.
(Adapted from Cochrane and Sanchez, 1982.)

Cochrane and Sanchez (1982) estimated that 220 million hectares of Oxisols and 141 million hectares of Ultisols are present in the Amazon Basin. This represents 46% and 29%, respectively, of the total land area. Oxisols occurring in the uplands are well drained, have well-developed soil structure, low effective cation exchange capacity (ECEC), and low native fertility. Ultisols are common in both upland and poorly-drained landscape positions. They are generally acid, infertile, and have weaker soil structure than the Oxisols. Contrary to earlier beliefs, plinthite occurs in only about 4% of the soils in the Amazon Basin.

Under similar conditions of rainfall, slope, soil cover, and soil texture, Oxisols are less subject to soil erosion that Ultisols. The stronger soil structure of Oxisols leads to higher infiltration rates. Infiltration rates are high for native soils even when the clay percentage is high. When the surfaces of Oxisols are left bare, the strong soil structure helps resist erosion, but some soil erosion is likely depending upon rainfall characteristics. The soil structure of Ultisols, being weaker than Oxisols, is more easily broken down by raindrop impact. Thus soil erosion losses for Ultisols tend to be greater than for Oxisols. While Cochrane and Sanchez (1982) estimated that only 6 and 10% of the soils in the tropical rainforest and semi-evergreen seasonal forest, respectively, are highly erodible, we believe the erosion hazard is greater. Our belief is that major erosion hazards exist for soils in the 8-30% slope category and on a large fraction of the soils in the 0-8% slope category (Table 1).

Another limitation to soils in the Amazon Basin is soil moisture stress (Cassel and Lal, 1992; Cochrane and Sanchez, 1982). Even though mean total rainfall exceeds mean total potential evapotranspiration for 8 or more months each year, there often are one or more short periods during the year when plants undergo moisture stress severe enough to reduce crop yields.

The main chemical soil constraints to crop production in the Amazon are soil acidity, P deficiency, micronutrient deficiency, and low ECEC (Cochrane and Sanchez, 1982). Of the 482 million hectares in the Amazon Basin, 81% of the area had native pH values in the topsoil of less than 5.3, and 82% had native pH values less than 5.3 in the subsoil. Associated with these low pH values is aluminum toxicity. Assuming that 60% aluminum saturation in the top 50 cm of soil is toxic to aluminum-sensitive plant species, Cochrane and Sanchez (1982) reported that 73% of the soils in the Amazon have this problem.

Ninety percent of the soils in the Amazon have topsoil P levels less than 7 mg/kg (Cochrane and Sanchez, 1982). If one assumes a critical P level of 10 mg/kg, these soils will not support crops without additions of P. Fortunately, only 16% of the soils are estimated to be strong P fixers, that is, they have over 35% clay and a high percentage of iron oxides. The remaining soils can be managed by prescribing small P applications on a crop by crop basis.

The low ECEC is considered to be a soil constraint (Cassel and Lal, 1992). The susceptibility of leaching of mobile nutrients increases as ECEC decreases. This is of major importance in an environment where rainfall exceeds potential evapotranspiration most of the year and where nutrients are in short supply to begin with. It is critical that mobile nutrients added to the soil remain in the soil as long as possible giving the plant adequate opportunity to utilize them. For example, potassium is considered to be a constraint on 56% of the land area (Cochrane and Sanchez, 1982). It is critical that K applied to the soil be retained by the soil ECEC until it is absorbed by plant roots.

## II. Characterization of Soils and Traditional Agricultural Practices in Yurimaguas and Manaus

### A. Physical and Chemical Properties of Profiles for a Yurimaguas Ultisol and a Manaus Oxisol

Selected soil chemical and physical property data for the Yurimaguas soil series (Typic Paleudult) in Yurimaguas, Peru is given in Table 2. This soil was covered with a secondary forest when the first experiments began in 1972. Similar data are presented in Table 3 for the Xanthic Hapludox under a secondary forest in Manaus, Brazil. Comparison of the data in these two tables show that the soils have both similarities and dissimilarities. Mean monthly

**Table 2.** Properties of the Typic Paleudults (fine-loamy, siliceous, isohyperthermic) at Yurimaguas, Peru

| Depth | Clay | Sand | Org. C | pH | Exchangeable | | | | Effective CEC | Al Saturation | Bulk Density | Porosity | |
|---|---|---|---|---|---|---|---|---|---|---|---|---|---|
| | | | | | Ca | Mg | K | Al | | | | Micro | Macro |
| | ----%---- | | | | ----------cmol kg⁻¹---------- | | | | cmol kg⁻¹ | % | Mg m⁻³ | -----m³ m⁻³----- | |
| 0-5 | 6 | 80 | 1.25 | 3.8 | 0.84 | 0.20 | 0.20 | 2.05 | 3.49 | 59 | 1.16 | 0.28 | 0.19 |
| 5-13 | 10 | 70 | 0.84 | 3.7 | 0.05 | 0.04 | 0.04 | 2.63 | 2.76 | 95 | 1.16 | 0.28 | 0.19 |
| 13-43 | 15 | 61 | 0.42 | 3.9 | 0.05 | 0.03 | 0.03 | 3.11 | 3.24 | 96 | 1.39 | 0.14 | 0.14 |
| 43-77 | 17 | 57 | 0.29 | 4.0 | 0.03 | 0.02 | 0.02 | 3.12 | 3.20 | 98 | -- | -- | -- |
| 77-140 | 25 | 50 | 0.18 | 4.1 | 0.03 | 0.01 | 0.03 | 4.48 | 4.58 | 98 | -- | -- | -- |
| 140-200 | 24 | 54 | 0.17 | 4.4 | 0.06 | 0.03 | 0.04 | 3.80 | 3.94 | 96 | -- | -- | -- |

(From Cochrane and Sanchez , 1982; and Alegre et al., 1986a.)

**Table 3.** Properties of the Xanthic Hapludox (clayey, kaolinitic, isohyperthermin) at Manaus, Brazil

| Depth | Clay | Sand | Org. C | pH | Exchangeable | | | Effective CEC | Al Saturation | Bulk Density | Sat. Hyd. Conductivity |
|---|---|---|---|---|---|---|---|---|---|---|---|
| | | | | | Ca+Mg | K | Al | | | | |
| cm | ----%---- | | | | ----------cmol kg⁻¹---------- | | | cmol kg⁻¹ | % | Mg m⁻³ | cm h⁻¹ |
| 0-8 | 76 | 15 | 3.0 | 4.6 | 2.00 | 0.19 | 1.1 | 3.29 | 33 | 1.04 | 28.1 |
| 8-22 | 80 | 12 | 0.9 | 4.4 | 0.20 | 0.09 | 1.1 | 1.39 | 79 | 1.12 | 8.1 |
| 22-50 | 84 | 8 | 0.7 | 4.3 | 0.20 | 0.07 | 1.2 | 1.47 | 82 | 1.11 | 9.6 |
| 50-125 | 88 | 7 | 0.3 | 4.6 | 0.10 | 0.04 | 1.0 | 1.14 | 88 | -- | -- |
| 125-265 | 89 | 5 | 0.2 | 4.9 | 0.10 | 0.01 | 0.2 | 0.31 | 65 | -- | -- |

(From Camargo and Rodrigues, 1979; and Melgar et al., 1992.)

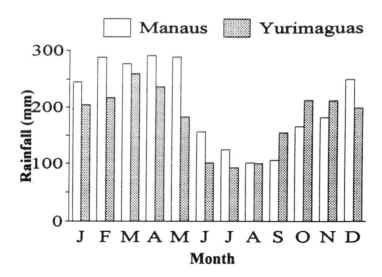

**Figure 1.** Mean monthly rainfall for research sites near Yurimaguas, Peru and Manaus, Brazil. (Adapted from Alegre et al., 1991; and Cravo and Smyth, 1991.)

rainfall for both sites are shown in Figure 1. Mean annual rainfall in Manaus exceeds that of Yurimaguas by about 300 mm.

Sand content of the Yurimaguas soil is 80% at the soil surface and decreases with depth; the soil has a well-developed B horizon with the clay content reaching 25%. In the rooting zone for most crops (0-43 cm) pH is 4.0 or less. Soil organic carbon content exceeds 1% at the soil surface, but decreases to less than 0.5% below the 13-cm depth. The low amounts of exchangeable Ca, Mg, and K coupled with the ECEC of about 4 cmol kg$^{-1}$ below the 5-cm depth create aluminum saturations of 95% or greater.

The Oxisol at Manaus, on the other hand, has a clay content of 76% at the soil surface. The clay content increases with depth. Soil pH is about one-half unit higher than the Ultisol and is quite uniform with depth. The exchangeable Ca plus Mg is greater at all depths for the Oxisol, but exchangeable K is similar to the Ultisol. The consistently lower ECEC for the Oxisol coupled with similar or only slightly higher concentrations of exchangeable cations lead to lower aluminum saturations. However, the exchangeable aluminum values below 8 cm limit root growth to acid-tolerant crops. The bulk density values near 1.1 Mg m$^{-3}$ coupled with the high saturated hydraulic conductivities indicate that infiltration, water retention, and aeration are adequate for this soil.

## B. Shifting Cultivation Practices

Shifting cultivation practices may vary around the world, but a common denominator in all cases is a longer fallow period than cropping period (Moran, 1981; Sanchez, 1976). For tropical America it was described as (1) the annual slash and burn clearing of less than 5 ha, (2) subsequent planting of corn (*Zea mays*), upland rice (*Oryza sativa*), and cassava (*Manihot esculenta*) or plantain (*Musa paradisiaca*), and (3) land abandonment to forest regrowth or pasture establishment after one or two years of cropping (Sanchez and Cochrane, 1980). After clearing a 10-year-old secondary forest in Yurimaguas, a farmer was contracted to simulate a shifting cultivation treatment on three replicated plots of 75 m by 25 m. After cropping the land to rice, cassava, and plantains for one year, he abandoned the land to forest regrowth. Crop yields were 1.7, 8.7, and 10 t ha$^{-1}$ for rice, cassava, and plantain, respectively (Alegre et al., 1989).

Two major factors contributing to land abandonment and forest regrowth after a few years of cropping are weed encroachment and the gradual decline in soil nutrient availability with time after cutting and burning (Valverde and Bandy, 1982; Sanchez and Cochrane, 1980). Mt Pleasant and co-workers (1992) assessed weed population dynamics in an Ultisol at Yurimaguas for five consecutive crops of a rice-cowpea (*Vigna unguiculata*) rotation following secondary forest clearing. In the absence of tillage, weed dry weight increased from 48 to 280 g m$^{-2}$ between the initial and final crops. Weed population composition also changed with time. Sedges comprised 84% of weeds in the first crop, whereas grasses dominated 79% in the fifth crop.

The nutrient content of ash, upon felling and burning rainforests of different ages, has been measured in both Yurimaguas and Manaus (Table 4). Differences in soil properties, clearing techniques, forest biomass, and the proportion of forest biomass actually burned lead to considerable variation among sites in the quantity of ash and its nutrient composition. Smyth and co-workers (1991a) estimated that only 29% of the aboveground biomass (77 t ha$^{-1}$) in an 11-year-old secondary forest at Yurimaguas was converted to ash. The beneficial effects of ash on soil chemical properties are exemplified in Figure 2 by a treatment without tillage, lime, or fertilization in the Ultisol cleared from an 11-year-old secondary forest in Yurimaguas, Peru. The ash markedly increased the soil nutrient supply, providing a twofold increase in exchangeable Ca and Mg and a threefold increase in Modified Olsen-extractable P. The liming effect of the ash was evidenced by a decrease in Al saturation from 76% before burning to 47% immediately after burning. The gradual increase in Al saturation with time and re-equilibration to pre-clearing levels by 12 months after burning also indicated that the residual effects of the ash on soil acidity were short-term. By the third year after burning, availability of most nutrients had declined to levels similar to those prior to clearing.

Lopes and co-workers (1987) compared for a 2.5-year-long period after clearing the soil nutrient dynamics of a 25-year secondary forest site on an Ultisol at Yurimaguas and a primary rainforest site on an Oxisol at Manaus.

**Table 4.** Nutrient contribution of ash upon burning rainforests of different ages on Oxisols at Manaus, Brazil and Ultisols at Yurimaguas, Peru

| Location and Soil | Vegetation | Ash Dry wt. | Nutrient Additions | | | | | | | | |
|---|---|---|---|---|---|---|---|---|---|---|---|
| | | t ha⁻¹ | N | Ca | Mg | K | P | Zn | Cu | Fe | Mn |
| | | | -----------------------------------------kg ha⁻¹----------------------------------------- | | | | | | | | |
| Manaus, Brazil Xanthic Hapludox | Primary forest | 9.2 | 80 | 82 | 22 | 19 | 6 | 0.2 | 0.2 | 58 | 2.3 |
| | Secondary forest (12 years) | 4.8 | 41 | 76 | 26 | 83 | 8 | 0.3 | 0.1 | 22 | 1.3 |
| | Abandoned pasture (5 years regrwoth) | 2.2 | 18 | 58 | 14 | 40 | 3 | -- | -- | -- | -- |
| Yurimaguas, Peru Typic Paleudult | Secondary forest (25 years) | 12.1 | 127 | 174 | 42 | 131 | 17 | 0.5 | 0.2 | 4 | 11.1 |
| | Secondary forest (17 years) | 4.0 | 67 | 75 | 16 | 38 | 6 | 0.5 | 0.3 | 8 | 7.3 |
| | Secondary forest (11 years) | 1.1 | 10 | 217 | 51 | 81 | 8 | 0.7 | 0.1 | 2.7 | 3.4 |

(From Smyth and Bastos, 1984; McKerrow, 1992; Seubert et al., 1977; Sanchez, 1987; and Smyth et al., 1991a.)

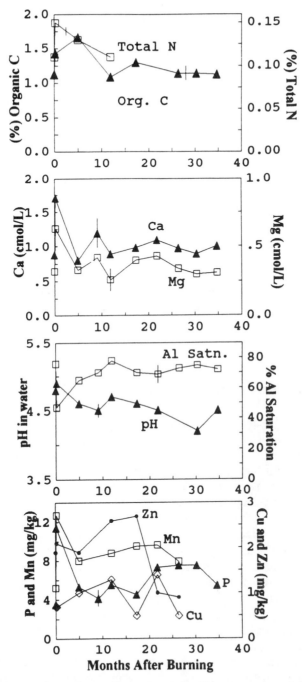

**Figure 2.** Changes in surface soil (0-15 cm) fertility parameters for an Ultisol in Yurimaguas, Peru during 3 years of cultivation following slash and burn clearing of an 11-year-old secondary rainforest. (From Smyth et al., 1991.)

Some of the differences in nutrient depletion patterns reflected differences in native soil properties. A marked increase in soil P availability after burning for the Yurimaguas site, as opposed to no change in the Manaus site, was attributed to differences in P sorption capacity between the loamy Ultisol and clayey Oxisol. Initial topsoil organic C and total N levels in the Oxisol were higher than in the Ultisol. Whereas the Oxisol showed a continual decline in both parameters across 2.5 years, there was little change in organic C and total N after clearing the Ultisol. Trends for soil acidity and exchangeable bases over time appeared to be associated with the amount of ash rather than particular soil properties. Initial increases in soil pH and exchangeable Ca and Mg were larger in the Oxisol and were sustained for a longer time than in the Ultisol. Consequently, Al saturation remained lower in the Hapludox than in the Paleudult throughout the two initial years of cultivation.

Soil management practices are often evaluated by whether soil conditions are improved or degraded with time. Long-term experiments are of major importance to the development of improved soil management practices, given the diversity of soils in the Amazon and the potential variation in soil properties associated with differences in forest biomass, clearing techniques, quality of the burn and rainfall distribution. Continuous monitoring of selected soil characteristics over time at the same site reduces the risk of confounding effects of time with spatial differences.

## III. Land Clearing and Reclamation

### A. Land Clearing to Initiate Continuous Cropping Experiments

The major thrust of the initial research begining in 1972 at Yurimaguas was to determine if continuous cultivation of basic food crops in acid, infertile soils of the Amazon Basin was possible. Introduction of one or more continuous cropping system technologies might reduce the ever increasing requirement to clear more and more land for slash and burn agriculture. Excellent foresight by Sanchez and co-workers (Seubert et al., 1977) led to the establishment of an experiment to evaluate two land clearing methodologies. Both manual and mechanical methods were employed. The manual method was the traditional slash and burn procedure indigenous to the area, whereas mechanical clearing involved a small bulldozer with a straight blade. Straight blades are not desirable for clearing land, but they were sometimes available in the area for road construction and maintenance operations, and were used whenever mechanical clearing was practiced.

The experimental design for this first experiment was a split-split plot with four replications (Seubert et al., 1977). Main plot treatments were clearing methods (slash/burn or bulldozer with straight blade); several long-term cropping systems were subplots; and several fertility regimes were sub-subplots. The land was cleared in August 1972. Vegetation, except for the tree trunks which were cut into pieces and manually removed for firewood, was burned three weeks after slashing. The bulldozer removed all vegetation from the land and, although care was exercised in an attempt to minimize the removal of topsoil, some of it was removed during clearing. Crops utilized in the various cropping systems were upland rice, maize, soybean (*Glycine max*), cassava, plantain, and guinea grass (*Panicum maximum*). During the experiment, soil samples were collected periodically to follow the fertility status of the soil; adjustments in fertilizer and lime application rates as time progressed were based on analysis of these samples. In addition, periodic measurement of bulk density and infiltration rate were made.

Bulk density of the topsoil 10 months after clearing was 1.24 g cm$^{-3}$ for manual clearing compared to 1.46 for the bulldozer-cleared land. The infiltration rate one month after clearing was less than 1 cm hr$^{-1}$ for the bulldozer cleared soil compared to greater than 10 cm hr$^{-1}$ for the manually cleared soil. The infiltration rate for the bulldozed soil did not improve as time progressed.

The continuous cropping systems experiment on the manually cleared land continued until 1989 during which time 37 consecutive crops were grown (see section IIIC). Crop production on the bulldozer-cleared land was usually less than that for the manually-cleared land and decreased with time. Liming improved yields on the bulldozed land, but when the yields of the unfertilized plots in 1974 plunged to 33% of the yield of the unfertilized plots on the manually-cleared land, the bulldozer-cleared land was abandoned (Alegre et al.,1986a).

## B. Physical Reclamation

Maintaining the permanent research site at Yurimaguas allowed us to solve a second generation problem, that is, the reclamation of the bulldozer-cleared land that was abandoned in 1974. Six years after abandonment of the bulldozer-cleared land, a reclamation experiment was designed. The land had been covered by guinea grass during the 6-year-long "fallow" period.

The first step in the reclamation process was to evaluate the soil chemical and physical properties as a function of depth to at least 45 cm. Bulk density and mechanical impedance measurements in 1980 revealed a compacted zone in the 15 to 45-cm depth. Bulk density and mechanical impedance in the top 60 cm of soil for both the bulldozer- and manually-cleared land are shown in Figure 3. Both areas were covered with guinea grass. The bulldozed soil had a pH of 4.5, Al saturation of 65%, and 6.5 mg kg$^{-1}$ of available P. The chemical properties could be modified by liming and fertilization, but the major limitation was identified to be subsoil compaction. Clearly, reclamation of this soil would require some management technique to alleviate subsurface compaction.

The resulting experiment had eight reclamation treatments: (1) control (no till), plant by stick; (2) hand rototiller, stick planted; (3) tractor rototiller; (4) chisel plow to 25 cm with 48-kW tractor; (5) controlled foot traffic; (6) bedded

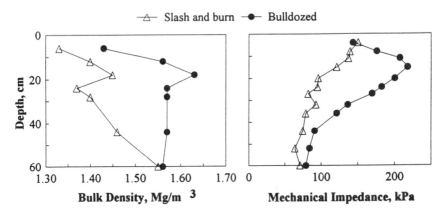

**Figure 3.** Bulk density and mechanical impedance form the 0 to 60-cm depth prior to land reclamation of the bulldozer-cleared land. (From Alegre et al., 1986a.)

with hoe, stick planted; (7) simulated subsoiling to 25 cm; and (8) mulching. Chisel plowing and simulated subsoiling were the two deep tillage operations imposed. Having no access to a tractor-operated subsoiler, the subsoil operation was simulated by manually loosening soil to the 25-cm depth with a tile spade. Each crop was fertilized according to soil tests; all treatments were fertilized and limed identically.

Infiltration, a process that integrates the soil physical properties of total porosity and pore size distribution, was greatest for the chisel plow and subsoil treatments when measured two years after reclamation began (Figure 4). Cumulative infiltration for the no till control treatment in 1982 remained essentially unchanged from the value measured by Seubert et al. (1977) nine years earlier.

Other measurements taken periodically during the 2-year-long reclamation study showed that mechanical impedance in the 12 to 21-cm depth was lowest for the chisel plow treatment, and bulk density below the 15-cm depth decreased only for the two deep tillage treatments, indicating that subsurface compaction was alleviated. The two deep tillage treatments produced the highest yields (Table 5). This reclaimed land is still under continuous crop production.

## C. Chemical Reclamation and Maintenance

The complete fertilization and control (no lime or fertilizers) treatments initiated by Seubert et al. (1977) were planted to a rotation of rice, corn and soybean for 17 consecutive years. Individual crop yields are compared between treatments in Figure 5. Whereas crop yields for the control treatment seldom deviated from zero since the third crop in the experiment, average yields for rice, corn, and

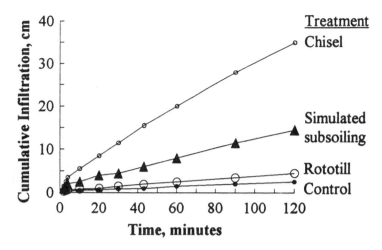

**Figure 4.** Cumulative infiltration for selected treatments after growing six consecutive crops after reclamation began. (From Alegre et al., 1986a.)

**Table 5.** Relative crop yield as affected by reclamation treatment

| Treatment | Rice | Soybean | Corn | Mean |
|-----------|------|---------|------|------|
| | ------------------------ % ---------------------- | | | |
| Control | 69[1] | 23 | 20 | 37 |
| Rototill (12 kW) | 93 | 69 | 76 | 79 |
| Rototill (48 kW) | 87 | 59 | 73 | 73 |
| Chisel plow | 100 | 100 | 100 | 100 |
| Controlled traffic | 85 | 54 | 64 | 68 |
| Bedding | 80 | 71 | 49 | 67 |
| Simulated subsoiling | 92 | 83 | 94 | 90 |
| Mulching | 84 | 76 | 88 | 83 |

[1] Each value is the mean of the relative yield for two crops. Relative yield of 100% is 2.56, 1.72, and 3.34 t ha$^{-1}$ for rice, soybean, and corn, respectively. (From Alegre et al., 1986.)

soybean in the complete treatment were 2.8, 2.9, and 2.3 t ha$^{-1}$, respectively. Monocultures without rotations do not maintain high yields because of the buildup of pathogens.

Fertilizer and lime requirements to maintain these crop yields were frequently adjusted on the basis of soil test analyses after each crop and findings from adjacent satellite experiments (Sanchez et al., 1983; Alegre et al., 1991). The initial fertilization in crop 2 consisted of 80, 100, 80, 0.5 and 0.1 kg ha$^{-1}$ of N,

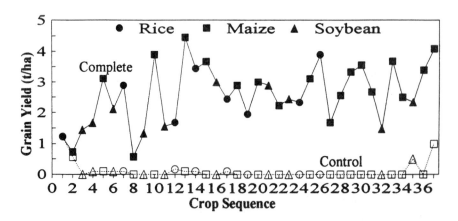

**Figure 5.** Grain yields for 37 consecutive crops over the period of 1972 to 1989 in an Ultisol at Yurimaguas, Peru. (From Alegre et al., 1991.)

P, K, B, and Zn, respectively. Thereafter N inputs for rice and corn increased to 100 kg ha$^{-1}$ in crops 8-10, 160 kg ha$^{-1}$ in crops 11-24, and 100-120 kg ha$^{-1}$ in crops 25-36. Phosphorus fertilization was 20-30 kg ha$^{-1}$ in crops 3-10, and 70 kg ha$^{-1}$ in crops 11-24. Thereafter P was only applied to corn crops at rates of 30-45 kg ha$^{-1}$. The initial rate of K was maintained through crop 10 before increasing to 125 kg ha$^{-1}$ in crops 11-24. Refinements in soil test K interpretations from adjacent field trials (Cox and Uribe, 1992b) led to the application of 50 kg K ha$^{-1}$ in the final corn crops. Magnesium was applied to crop 9 and crops 11-31 at rates of 9 and 30 kg ha$^{-1}$, respectively. Initial B and Mo applications were repeated in crops 11-24 along with Cu and Zn. Lime was applied at rates equivalent to 1.0-1.5 times the exchangeable Al in crops 2, 11, 14, 20, 29, 32, and 35.

Rice was excluded from the rotation after crop 26 because weed-management experiments indicated that its poor competition with weeds restricted potential weed-control options for the continuous cropping system (Mt Pleasant et al., 1990). Prior to crop 29, land preparation was performed with a hand tractor which tilled the soil to a 7.5-cm depth. Thereafter tillage was performed to a 20-25 cm depth with tractor-drawn implements. Exclusion of rice from the rotation and difficulties with mechanized cultivation during high rainfall periods led to the inclusion of a *Mucuna cochinchinensis* cover crop after crop 36 for use in weed control and as a green manure to the succeeding corn crop.

After 10 years of continuous cropping, comparison of chemical soil profile data between the control and complete fertilization treatments revealed that there was a significant movement of Ca and Mg and a reduction in Al saturation to the 45-cm depth (Alegre et al., 1991) (Figure 6). The subsoil chemical environment for root development was, thus, improved with continuous cultivation. This experiment clearly established that continuous cropping was possible in an

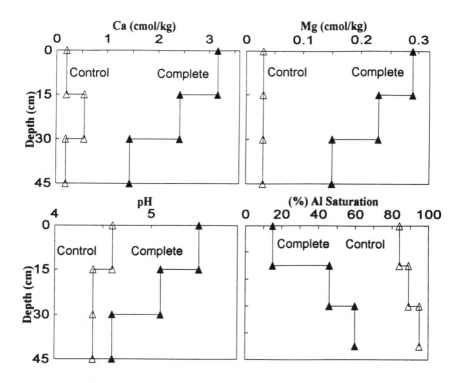

**Figure 6.** Improvement in subsoil chemical properties in an Ultisol at Yuri-maguas, Peru after 10 years of continuous cropping with adequate lime and fertilizer. (From Alegre et al., 1991.)

Ultisol of the Amazon when lime and fertilizer applications were based on continuous monitoring of soil nutrients by routine soil testing procedures.

Investigations were also conducted on the use of organic nutrients as an alternative or supplement to inorganic fertilizers. In the absence of inorganic fertilization green manuring with kudzu (*Pueraria phaseoloides*) from adjacent fields in a soybean-cowpea-corn-peanut (*Arachis hypogea*)-rice rotation provided 90% of the yields obtained with a reference fertilization treatment which received 120 kg N ha[-1] to corn, 100 kg K ha[-1] crop[-1], 48 kg P ha[-1] yr[-1], 10 kg S ha[-1], and 2 t lime ha[-1] (Wade and Sanchez, 1983). Compost from crop residues was compared to complete inorganic fertilization in a field where fertility had been improved by prior continuous cultivation. Yields for the four initial crops with compost were within 80% of the yields for the complete fertilization treatment, but declined to less than 50% in the fourth and fifth crops. After supplementing the compost with 100 kg K ha[-1] yields in six subsequent crops approximated those obtained with complete fertilization (Bandy and Sanchez, 1986). Disadvantages to the use of these organic fertilization alternatives were the labor requirements for harvesting organic materials and the detrimental effects of the continued export of a finite nutrient reserve from adjacent land areas where materials were produced.

## D. Land Clearing/Soil Management Interactions

Deforestation and subsequent changes in land use in the tropics accounted for about 18% of the global warming during the 1980s (Sanchez et al., 1990). It is of interest to stop or at least reduce the rate of clearing in the immediate future. If land is going to be cleared, however, it should be cleared using procedures that minimize soil degradation. The success in ameliorating the compacted subsoil in the land reclamation study discussed in section IIIA (Alegre et al.,1986a) led to the development of a third experiment (Alegre et al., 1986b, 1986c, 1988, 1990). Since the major problem arising from bulldozer clearing of the Yurimaguas soil was subsurface compaction, possibly the land could be cleared in such a manner to eliminate or minimize compaction. This new land-clearing experiment integrated various combinations of land clearing, burning of vegetation, post-clearing tillage, and soil management practices. No detailed information was available on the temporal effects of these various practices on soil properties and crop yields.

The salient features for the six "package" treatments used as main plots in the land clearing experiment were: (1) slash and burn; (2) bulldozer clearing with straight blade; (3) straight-blade clearing followed by chisel plowing 25 cm deep; (4) bulldozer clearing with shear blade, vegetation burned, followed by disking 30 cm deep; (5) shear blade only; and (6) shear blade followed by disking 30 cm deep. The three subplot treatments were post-clearing soil management practice packages: (1) no chemical amendments; (2) lime and fertilizer amendments; and (3) lime, fertilizer, and bedding.

Of major importance to this study was the inclusion of a bulldozer-mounted shear blade to clear forest vegetation. A sharp point, called a "stinger", is mounted at the base at one side of the shear blade. The stinger is designed to pierce and splinter large tree trunks, thus facilitating the shearing action by the base of the blade. When used correctly, the shear blade does not disturb the soil surface.

All six treatments compacted the surface soil and all mechanical clearing operations increased bulk density below 15 cm, but the disking and chisel plow operations following clearing alleviated this compaction. The optimum post-clearing management was with fertilization and bedding. Planting the crop on raised beds gave an unexpected advantage: the beds forced the laborers who hand weeded the plots to walk in the valleys between the beds thus reducing compaction near the plant row.

Few chemical properties were affected by clearing method or post-clearing management. In general, burning the dried vegetation for treatments 1 and 4 increased the initial amount of Ca, Mg, P, and K in the topsoil for the first few crops. Liming was necessary to produce an acceptable yield for the first crop for the remaining treatment which had no fly ash.

Relative grain yields for five of the highest yielding combinations of land-clearing soil management treatments and the straight-blade cleared treatment are presented in Table 6. The traditional slash and burn method, regardless of whether the soil surface configuration was flat or bedded, produced the highest yields of rice, soybean, and corn compared to the mechanized clearing packages. Comparison of straight-blade entries in the table shows the benefit of the 25-cm deep chisel plow operation to alleviate soil compaction incurred during land clearing. The high relative crop yields for shear-blade clearing occurred when the vegetation was burned to provide nutrients for the first crop.

## IV. Tillage/Soil Management Interactions

From 1972 to 1984 the long-term continuous cropping experiment at Yurimaguas had been tilled with a small hand rototiller. While rototilling is an improvement over hand hoeing, rototilling was slow, it failed to incorporate crop residues and soil amendments deeper than 10 cm, and over time it developed a compacted zone below the depth of tillage. The acquirement of a large tractor, tool bar, and assorted basic tillage equipment in 1984 allowed the establishment of studies to evaluate various power machinery tillage systems. Studies discussed below were conducted from 1986 to 1989 and include soil and crop responses to subsoiling, conservation tillage, and compaction.

### A. Deep Tillage

Despite the high annual rainfall, crops often suffer from severe moisture stress during rainless periods of two or more weeks. Having obtained responses to simulated subsoiling in the land reclamation study (Alegre et al., 1986a), and having a powerful tractor, tool bar, and subsoil shanks available, the capability existed to evaluate the practice of subsoiling on crop production on medium to coarse textured soils in the humid tropics.

The experiment had the following six treatments: (1) subsoil plus bedded, P broadcast; (2) bedded only, P broadcast (3) subsoil, flat planted (no bed), P broadcast; (4) flat planted only, P broadcast; (5) subsoil, bedded, P application in bands; and (6) bedded, P application in bands (Alegre et al., 1991). Subsoiling was done one time per year with subsoil shanks spaced 0.8 m apart. When subsoiling and bedding operations were imposed, both were done simultaneously with the same tractor pass. A portion of the bed was removed during the planting operation on a second tractor pass. Application of a herbicide

**Table 6.** Relative crop grain yields for selected land-clearing/soil management combinations. All treatments received fertilizer and lime

| Treatment | | Rice | Soybean | Rice | Corn | |
| Main Plot | Subplot | Crop 1 | Crop 2 | Crop 4 | Crop 5 | Mean |
|---|---|---|---|---|---|---|
| | | | | ---% yield[1]--- | | |
| Slash/burn | flat | 89 | 100 | 100 | 87 | 94 |
| Slash/burn | bed | 100 | 80 | 81 | 100 | 90 |
| Straight blade | flat | 69 | 44 | 45 | 45 | 51 |
| Straight blade/chisel | bed | 72 | 88 | 84 | 87 | 83 |
| Shear blade/burn/disk | flat | 77 | 94 | 99 | 74 | 86 |
| Shear blade/burn/disk | bed | 92 | 88 | 85 | 85 | 88 |

[1] Maximum grain yields for crops 1, 2, 4, and 5 were 3.98, 2.32, 3.48, and 3.30 t ha$^{-1}$, respectively. (From Alegre et al., 1990.)

**Table 7.** Bulk density in the corn row at two depths prior to harvesting the third crop as affected by tillage

| Treatment | 0-15 | 15-30 |
|---|---|---|
| | ------------Mg m$^{-3}$-------------- | |
| Bed, broadcast P, subsoil | 1.43 | 1.50 |
| Bed, broadcast P | 1.47 | 1.67 |
| Flat plant, broadcast P, subsoil | 1.45 | 1.52 |
| Flat plant, broadcast P | 1.51 | 1.65 |
| Bed, band P, subsoil | 1.42 | 1.55 |
| Bed, band P | 1.50 | 1.64 |
| LSD 0.05 | 0.09 | 0.09 |

(From Alegre et al., 1991.)

was made during a third tractor pass. Treatments not subsoiled were disk plowed prior to planting. Six corn crops were grown during the period from late 1986 through October 1989. All treatments received the same amount of lime and fertilizers.

Bulk density, measured in the corn row prior to harvesting the third corn crop, was not affected by tillage in the topsoil, but at the lower depth it was reduced by 0.09 to 0.17 Mg m$^{-3}$ for the three subsoiled treatments (Table 7). Soil strength in the 15- to 30-cm depth was also less for the subsoiled plots. While subsoiling did not increase corn grain yield for the first crop, yields did increase for all remaining crops in the bedded, broadcast P treatments, and for the last four crops that were flat planted (Table 8). Yield response to subsoiling for the treatments with banded P were more variable.

The major reason for the positive yield responses to subsoiling is related to the corn root distribution patterns (Figure 7). Roots grew deeper in the subsoiled treatments and occupied a greater volume of soil. Water uptake from this larger volume of soil during the dryer portions of the growing season allowed the plant to avoid the degree of water stress encountered by the non-subsoiled treatments.

## B. Conservation Tillage

Conservation tillage practices continue to gain importance in many parts of the world. These types of practices, which typically utilize plant residues to protect the soil surface, alter the water balance and soil moisture regime. Earlier research on this site (Wade and Sanchez, 1983) found that, compared to bare soil, surface mulches of guinea grass and kudzu decreased soil temperature, decreased mechanical impedance, and increased soil water content in the topsoil. Do conservation tillage practices have a role in the humid tropics where annual rainfall often exceeds 2000 mm per year? One might hypothesize that the presence of plant residues on the soil surface would reduce surface runoff and

**Table 8.** Corn grain yield as affected by tillage and phosphorus management systems conducted from 1986 to 1989

| Treatment | Corn Crop | | | | | | Mean |
|---|---|---|---|---|---|---|---|
| | 1 | 2 | 3 | 4 | 5 | 6 | |
| | -------grain yield, t ha$^{-1}$------- | | | | | | |
| Bed, broadcast P, subsoil | 3.5 | 3.9 | 3.5 | 4.2 | 4.1 | 4.5 | 4.6 |
| Bed, broadcast P | 3.4 | 3.2 | 2.5 | 2.5 | 2.3 | 2.6 | 2.8 |
| Flat plant, broadcast P, subsoil | 3.4 | 3.5 | 3.7 | 3.8 | 3.9 | 3.7 | 3.7 |
| Flat plant, broadcast P | 3.4 | 3.3 | 2.1 | 2.0 | 2.2 | 2.0 | 2.5 |
| Bed, band P, subsoil | 4.1 | 3.3 | 3.4 | 3.6 | 2.8 | 4.3 | 3.7 |
| Bed, band P | 3.8 | 3.4 | 3.2 | 2.4 | 2.8 | 3.2 | 3.1 |
| LSD 0.05 | 0.05 | 0.7 | 0.5 | 0.8 | 0.6 | 0.4 | |

(From Alegre et al., 1991.)

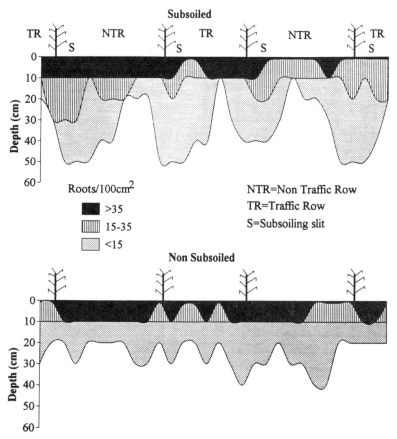

**Figure 7.** Root distribution of corn for subsoil and non-subsoil flat planted treatments. (From Alegre et al., 1991.)

soil erosion. On the other hand, the plant residues might also reduce evaporation and keep the soil so wet that soil trafficability would be affected. The soil might not support the conservation tillage machinery during the planting season. With these questions in mind the following experiment was designed to determine the feasibility of alternative minimum and no-till planting practices for continuous crop production systems at Yurimaguas.

This experiment began in late 1984, continued until early 1989, and was conducted on land having a 1 to 2% slope that was cleared in 1972 and was previously planted to 31 successive crops (Alegre et al., 1991). Definitions of the six tillage treatments are given in Table 9. All tillage operations were performed using a large tractor. Treatments 1 and 3 were subsoiled to the 30-cm depth during the planting operations for each crop. Treatment 5 was subsoiled only one time per year. Runoff plots were constructed to measure surface runoff

**Table 9.** Description of treatments in the conservation tillage study, Yurimaguas, Peru

| Treatment | Description |
| --- | --- |
| No-tillage/subsoil | Mow, kill vegetation with herbicide, simultaneous subsoil and no-till plant, P and lime broadcast without incorporation. |
| No-tillage | Same as above, except no subsoiling. |
| Minimum-tillage/subsoil | Mow, kill vegetation with herbicide, simultaneous subsoil and no-till plant, lime and P incorporated with disk every two years. |
| Minimum-tillage | Same as above, except no subsoiling. |
| Conventional tillage/subsoil | Mow, disk plow, S-tine field cultivator, subsoil and bed, shape bed and plant, band P. |
| Conventional | Same as above, except no subsoiling. |

and soil loss from all three treatments involving subsoiling. A corn-soybean rotation was used.

Although grain yields were significantly affected by tillage for three of the five crops, no decidedly clear advantage occurred for any one of the six tillage systems (Table 10). For those crops where a difference did occur, the highest yield was associated with a subsoiled treatment. Overall, there was a consistent trend for yields to be greater when subsoiling was incorporated into the tillage treatment.

Surface runoff and soil loss for this nearly level soil for a 15-month period beginning in November 1987 were greater than anticipated (Table 11). Clearly the amount of runoff decreased significantly for the two conservation tillage systems. Soil loss was least for the no-tillage system. We believe the amounts of surface runoff and soil loss for the three systems without subsoiling would be even greater.

Based on this limited information it is still valid to ask if the extra effort required to properly implement these conservation tillage practices in the humid tropics is worth it. Depending upon the rainfall patterns, conservation tillage systems may delay planting operations for weeks at a time. Much additional work on conservation tillage with heavy power machinery in the humid tropics needs to be conducted.

**Table 10.** Corn and soybean yield as affected by conservation tillage practices

| Treatment | Corn | | | | Soybean | | |
|---|---|---|---|---|---|---|---|
| | 1 | 2 | 3 | Mean | 1 | 2 | Mean |
| | ------t ha$^{-1}$------ | | | | | | |
| No-tillage/subsoil | 1.90 | 2.83 | 3.27 | 2.33 | 2.25 | 1.25 | 1.75 |
| No-tillage | 1.39 | 2.58 | 2.90 | 2.29 | 2.18 | 0.90 | 1.54 |
| Minimum-tillage/subsoil | 2.20 | 2.41 | 3.40 | 2.67 | 1.63 | 1.40 | 1.52 |
| Minimum-tillage | 1.44 | 2.66 | 2.58 | 2.23 | 1.79 | 1.20 | 1.50 |
| Conventional-tillage/subsoil | 2.00 | 2.39 | 2.68 | 2.69 | 1.99 | 1.18 | 1.59 |
| Conventional-tillage | 1.90 | 2.35 | 2.91 | 2.39 | 1.80 | 0.80 | 1.30 |
| LSD 0.05 | 0.63 | 0.96 | NS | | 0.52 | NS | |

(From Alegre et al., 1991.)

**Table 11.** Surface runoff and soil loss for three tillage systems during a 15-month period in Yurimaguas

| Treatment | Runoff | Soil loss |
|---|---|---|
| | mm | t ha$^{-1}$ |
| No-tillage, subsoil | 165 | 14.9 |
| Minimum tillage, subsoil | 120 | 19.4 |
| Conventional tillage, subsoil | 249 | 22.4 |
| LSD 0.05 | 43 | 5.5 |

(From Alegre et al., 1991.)

### C. Compaction

The frequency, amount and duration of rainfall can limit the effectiveness of heavy power machinery to perform tillage operations. It is conceivable that in some years, rainfall conditions would be such that the soil would remain so wet for extended periods of time that a crop could not be planted at the appropriate time. Tilling the soil when it is too wet either puddles or severely compacts it. A study was designed to evaluate the effects of soil water content and the number of compaction events by tractor wheels on crop yield and soil compaction on sandy loam and loamy sand Yurimaguas soils.

The greatest amount of surface compaction at all three soil water contents occurred in response to compaction imposed by the first pass of the tractor wheel (Table 12). Compaction decreased as water content decreased. At the highest soil water content no compaction occurred after the first tractor pass. For the remaining two water contents the degree of compaction was not as great for the first pass as for the wettest soil, but the second tractor pass continued to significantly increase bulk density. Wheel track ruts as deep as 10 cm developed under the wettest soil. Corn yield decreased with number of tractor passes and was inversely related to water content at the time of compaction (Table 13). Comparison of Tables 12 and 13 shows that corn yield and bulk density are inversely related.

## V. Land-Use Options for Acid-Tolerant Crops

Investigations on fertilizer-based and mechanized continuous crop production served to quantify the production potential and management requirements to overcome the major constraints of Ultisols and Oxisols in the Amazon Basin. It is intended, however, to be used by farmers near large centers which have favorable access to capital, fertilizers, lime, machinery, and stable markets for their products. A second major research thrust in our program was to develop alternatives suitable for farmers on acid, infertile soils in rural areas with limited market infrastructure and capital. This section describes investigations on the use

**Table 12.** Bulk density at the 0- to 10-cm depth after tractor passes at different water contents in a sandy loam Ultisol, Yurimaguas, Peru

| Soil water content | Number of tractor passes | | | | Mean |
|---|---|---|---|---|---|
| | 0 | 1 | 2 | 3 | |
| $m^3m^{-3}$ | ------------------------------Mg $m^{-3}$------------------------------ | | | | |
| 0.36 | 1.33 | 1.57 | 1.58 | 1.59 | 1.52 |
| 0.27 | 1.35 | 1.46 | 1.52 | 1.56 | 1.47 |
| 0.20 | 1.36 | 1.42 | 1.49 | 1.50 | 1.45 |
| Mean | 1.35 | 1.48 | 1.55 | 1.55 | |

LSD 0.05: Soil water content = 0.04; tractor passes = 0.04; soil water within the same tractor pass = 0.07; tractor pass within the same soil water content = 0.07.
(From Alegre et al., 1991.)

**Table 13.** Corn yield as a function of soil water content and number of tractor passes in a sandy loam Ultisol, Yurimaguas, Peru

| Soil water content | Number of tractor passes | | | | Mean |
|---|---|---|---|---|---|
| | 0 | 1 | 2 | 3 | |
| $m^3m^{-3}$ | ------------------------------Mg $ha^{-1}$------------------------------ | | | | |
| 0.36 | 2.28 | 1.13 | 0.93 | 0.81 | 1.29 |
| 0.27 | 2.61 | 2.14 | 2.33 | 1.91 | 2.25 |
| 0.20 | 2.48 | 2.83 | 2.12 | 2.44 | 2.47 |
| Mean | 2.46 | 2.03 | 1.79 | 1.72 | |

LSD 0.05: Soil water content = 0.64; tractor passes = 0.38; soil water within the same tractor pass = 0.66; tractor pass within the same soil water content = 0.85.
(From Alegre et al., 1991.)

of acid-tolerant germplasm in low-input cropping, alley cropping, and legume-grass pasture management options.

## A. Low-Input Cropping

The first step toward developing a viable land use option which minimizes lime and fertilizer inputs in Ultisols and Oxisols was to select species and cultivars which produced acceptable yields under conditions of high Al saturation. Germplasm considered to have high yield potential under humid tropical conditions was supplied by various national and international institutions for field testing in plots with and without lime in Ultisols at Yurimaguas. Rice, cowpea, soybean, corn, winged bean (*Psophocarpus tetraglobulus*), sweet potato

**Table 14.** Germplasm tested for acid tolerance in Yurimaguas, Peru from 1979 to 1982

| Species | Acessions tested | Very tolerant (RY>85%)[1] | Tolerant (RY=65-85%) | Sensity (RY<65%) |
|---|---|---|---|---|
| | | ---------------number of acessions--------------- | | |
| Rice | 32 | 8 | 16 | 8 |
| Cowpea | 30 | 12 | 7 | 11 |
| Soybean | 22 | 0 | 0 | 22 |
| Corn | 20 | 0 | 0 | 20 |
| Winged bean | 16 | 0 | 0 | 16 |
| Sweet Potato | 10 | 0 | 2 | 8 |
| Peanut | 10 | 0 | 1 | 9 |

[1] RY (relative yield) = {(yield without lime)/(yield with lime)} x 100.
(Adapted from Nicholaides and Piha (unpublished) by Bandy and Sanchez, 1986.)

(*Ipomoea batatas*), and peanut germplasm were tested in six field trials from 1979 to 1982. Surface soil Al saturation ranged from 63 to 82% in treatments without lime and 7 to 34% with lime. All nutrients were maintained under non-limiting conditions (Nicholaides and Piha, 1987). Germplasms were considered very tolerant if yields without lime were >85% of those obtained in limed treatments, and sensitive if relative yields were <65%. Results, summarized in Table 14, indicate several rice and cowpea cultivars rank as very tolerant with yields up to 3.5 t ha$^{-1}$ for rice and 2.5 t ha$^{-1}$ for cowpea. None of the tested germplasm for other species met the rigorous criteria for high tolerance to Al.

In 1982 Sanchez and Benites (1987) began a field evaluation of low cost practices to extend the cropping cycle in shifting cultivation which included the use of Al tolerant rice and cowpea cultivars. A secondary forest was cleared by slash and burn, planted with a local upland rice variety by traditional shifting cultivation methods, followed by a rice-cowpea rotation for a total of four rice and two cowpea crops during 2.5 years. A fertilized treatment, where each rice crop received 30, 22, and 48 kg ha$^{-1}$ of N, P, and K, respectively, was compared to a treatment without fertilization. There were no yield differences between treatments with or without fertilizer until the final crop of rice. The authors attributed the yield response to fertilizer in the seventh crop to decreased availability of P and K in the soil. Use of Al-tolerant cultivars, maintenance of crop residues, and zero tillage were considered to be the major factors which enabled this system to maintain favorable yields over 2.5 years and six crops without lime or fertilizers.

The most important constraint to extending the no-till low input system beyond six crops was weed control (Sanchez and Benites, 1987). A one-year kudzu fallow and subsequent slash and burn suppressed weeds and increased surface soil pH, exchangeable bases, and available P. Two rice crops were harvested in a second cropping cycle before obtaining a significant yield reduction in cowpea (third crop) without fertilization (Alegre et al., unpublished). A second cropping cycle without external replenishment of nutrients removed in harvested grain from this system was, therefore, limited to two rice crops.

A separate experiment evaluated the role of crop residue management, tillage method, and K fertilization on soil K availability to a rice-cowpea rotation during 12 consecutive crops (Cox and Uribe, 1992a). Throughout the study there was no difference in yields among conventional tillage, strip tillage (30% of land area), and no-till systems. Yields were lower with crop residue removal than with residue maintenance, and the detrimental effect of residue removal increased with time. Whereas K fertilization (0 to 120 kg K ha$^{-1}$) of the six initial crops always increased grain yields when residues were removed, responses were seldom observed with residue maintenance. Residue maintenance with no K fertilization maintained soil K above the critical level of 0.10 cmol L$^{-1}$ for 11 consecutive crops. Residual effects of fertilizer K were prolonged by residue maintenance.

Weed control by tillage, herbicides, mulches, planting density, and manual practices was compared during five consecutive crops of a rice-cowpea rotation after slash and burn clearing of a secondary forest (Mt Pleasant et al., 1992). The no-till treatment had more weeds and lower relative yields than the tilled treatment in all but the initial crop. Mulching crop residues had little effect on weed control, and yields were consistently higher when residues were incorporated. This suggested that residues were more closely related to nutrient supply than to weed suppression. Weed infestation increased with time of cropping as did the costs for their control. The authors suggested that after 5 to 6 crops weed control in the low-input system would require either tillage or a fallow period.

Low-input cropping is considered a transitional management option because of nutrient depletion and weed encroachment constraints (Sanchez et al., 1990). Nevertheless, extension of land use by resource-limited farmers from one to three years increases short-term crop production and provides time to prepare land for long-term land use alternatives.

## B. Alley Cropping

Trees in this agroforestry system are grown in hedgerows and pruned periodically to provide mulch and nutrients to crops growing in the alleys. A major research concern for the acid, infertile soils is the extent to which trees can improve nutrient cycling when soil nutrient availability is inherently low (Szott et al., 1991). Selection of leguminous trees included both exotic and native species.

Seedling survival, growth rate, and biomass production of the promising species are shown in Table 15. Noticeable absences are *Leucaena leucocephala*,

**Table 15.** Survival, growth rate, pruning yield, and composition of selected tree species during the first year after transplanting in an Ultisol without lime and fertilizer at Yurimaguas, Peru

| Species | Survival | Growth rate | Pruning yield[1] | Leaves in pruning |
|---------|----------|-------------|------------------|-------------------|
| | % | cm mo$^{-1}$ | kg m$^{-1}$yr$^{-1}$ | % |
| *Cassia reticulata* | 100 | 40.9 | 3.5 | 55 |
| *Inga felulei* | 100 | 30.9 | 4.1 | 59 |
| *Calliandra calothyrus*[2] | 80 | 23.0 | 1.4 | 54 |
| *Samamea saman* | 100 | 20.3 | 1.9 | 59 |
| *Gliricidia sepium* | 85 | 15.9 | 2.7 | 54 |
| *Cassia* sp. | 100 | 12.4 | -- | -- |
| *Schilozobium amazonicum* | 70 | 10.1 | -- | -- |
| *Pithecellobium dulce* | 100 | 9.5 | -- | -- |
| *Acacia auriculiformis*[2] | 70 | 7.6 | -- | -- |
| *Flemingia congesta*[2] | 0 | 6.0 | -- | -- |
| *Albizia procera*[2] | 0 | 2.3 | -- | -- |

[1] Average of three prunings; [2] not native to Tropical America.
(From Salazar and Palm, 1987.)

*Leucaena diversifolia*, and *Cajanus cajan* due to Al toxicity or short period of survival when subjected to pruning. Native species had the best overall performance in establishment, growth, and biomass production. Pruning yields would correspond to 5 t ha$^{-1}$ every 3 months for trees planted in double-row hedges separated by 4-m alleys. Seedling mortality resulted from susceptibility to leaf-cutter ants (Salazar and Palm, 1987).

Subsequent evaluations under acid soil conditions in Yurimaguas of *Gliricidia sepium* accessions from the Oxford Forestry Institute revealed marked differences among provenances in growth and biomass production (Figure 8). Provenance 14/84 from Retalhuleu, Guatemala had superior performance to the other Central American accessions. Similar evaluations with other species may provide significant improvements (Fernandes, 1991).

For alley cropping studies conducted in Yurimaguas during the last 10 years, Szott and co-workers (1991) recently compared responses to pruning applications for crops with and without direct competition of tree hedges. When *Inga edulis* and *Erythrina* sp. mulches were applied to plots without tree hedges, rice yields were similar to that obtained with inorganic N, P, and K fertilization (Table 16). In the presence of *Inga edulis* hedges crop yields in the 4.5 m alleys increased with distance from the trees but seldom exceeded yields for the unfertilized control without trees (Table 17). In both situations crop yields for the treatments without hedges, mulch, and fertilization decreased with time of cultivation. The authors suggested that differences in crop performance between the two

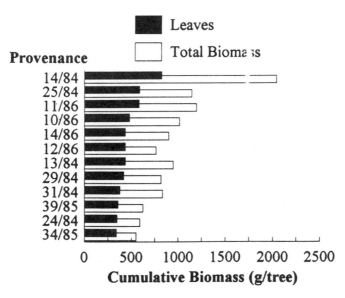

**Figure 8.** Cumulative leaf and total biomass for three prunings of *Gliricidia sepium* provenances during the 20 initial months after outplanting to an Ultisol at Yurimaguas, Peru. (From Fernades, 1991.)

**Table 16.** Grain yields for four consecutive rice crops in a Yurimaguas Ultisol as a function of hedgerow mulch and inorganic fertilizer inputs

| Treatment | Mulch rate | Rice crop | | | |
|---|---|---|---|---|---|
| | | 1 | 2 | 3 | 4 |
| | t ha$^{-1}$ crop$^{-1}$ | ---------grain yield, t ha$^{-1}$--------- | | | |
| *Inga edulis* mulch | 6.7 | 1.3 | 2.4 | 1.1 | 0.9 |
| *Erythrina* sp. mulch | 6.7 | 2.8 | 1.7 | 1.2 | 1.3 |
| NPK fertilizer[1] | -- | 1.8 | 2.1 | 1.2 | 1.2 |
| No mulch or fertilizer | -- | 1.9 | 0.7 | 0.2 | 0.5 |

[1] Each crop received 100, 25, and 100 kg ha$^{-1}$ of N (urea), P (simple superphosphate), and K (KCl), respectively. All treatments received 25 kg P and 50 kg K ha$^{-1}$ in the final crop.
(From Palm, 1988.)

experiments were related primarily to belowground competition of trees and crops for water and nutrients.

Salazar and Palm (1987) measured tree root biomass in the 0-30-cm depth along 4-m transects perpendicular to the hedgerow. Although root biomass for both *Inga edulis* and *Erythrina* sp. declined with increasing distance into the alleys, the former species had consistently higher root production for the first

**Table 17.** Crop grain yields as a function of distance from *Inga edulis* hedges in an Ultisol at Yurimaguas, Peru

| Distance from hedge | Crop sequence | | | | |
|---|---|---|---|---|---|
| | Cowpea | Rice | Rice | Cowpea | Rice |
| m | ------------------------------grain yield, t ha$^{-1}$------------------------------ | | | | |
| 0.75 | 0.43 | 0.43 | 0.15 | 0.34 | 0.26 |
| 1.25 | 0.52 | 0.29 | 0.17 | 0.38 | 0.46 |
| 1.75 | 0.58 | 0.21 | 0.17 | 0.39 | 0.49 |
| 2.25 | 0.59 | 0.21 | 0.24 | 0.40 | 0.43 |
| No hedge or mulch | 1.01 | 0.49 | 0.31 | 0.53 | 0.38 |

(From Szott, 1987.)

**Table 18.** Comparison between nutrients contained in hedgerow prunings[1] from selected tree species and nutrient uptake for a 2 t ha$^{-1}$ rice crop

| Component | N | P | K | Ca | Mg |
|---|---|---|---|---|---|
| | ------------------------------kg ha$^{-1}$------------------------------ | | | | |
| *Cassia reticulata* pruning | 72 | 7 | 37 | 25 | 6 |
| *Gliricidia sepium* pruning | 64 | 5 | 37 | 22 | 8 |
| *Erythrina* sp. pruning | 67 | 6 | 36 | 16 | 7 |
| *Inga edulis* pruning | 62 | 5 | 24 | 15 | 4 |
| Rice grain and straw[2] | 55 | 10 | 37 | 5 | 3 |

[1] Based on a dry matter yield of 2.5 t ha$^{-1}$ and one pruning per crop.
[2] Calculated from Sanchez, 1976.
(From Szott, 1987; Palm, 1988; and A. Salazar, unpublished data.)

3 m from the hedgerow. Fernandes (1990) evaluated the effects of pruning *Inga edulis* roots to a 20-cm depth on rice yields in the alleys. Root pruning had no effect on rice yields at six months after hedgerow establishment, but prunings at 12 and 24 months increased associated rice crop yields by about 20%.

Decreasing yields with increasing time of alley cropping on acid, infertile soils may also be related to declining nutrient availability (Szott et al., 1991). When nutrients contained in prunings of the most promising tree species were compared with nutrient uptake for a 2 t ha$^{-1}$ rice crop, excess nutrient supply was only apparent for Ca and Mg (Table 18). Potassium content of prunings was equal or inferior to crop demands and P requirements were not balanced by pruning additions. The nutrient deficit from prunings is increased upon considering their rate of release during decomposition. When decomposition of field-applied pruned mulches were monitored with litterbag techniques the proportions of N, P, and K remaining in the leaves after 20 weeks were in the ranges of 10-58, 18-66, and 3-22%, respectively. Mineralization was faster from

*Erythrina* leaves than from *Inga edulis* and *Cajanus cajan* leaves (Palm and Sanchez, 1990). Nutrient balance deficits may be compensated by increasing pruning applications from denser hedgerows, but belowground tree-crop competition implies the need for increased alley widths. Based on current findings Szott et al. (1990) concluded that continuous alley cropping could not be sustained on acid, infertile soils without additions of chemical fertilizers.

## C. Legume-Grass Pastures

There are about 7 to 10 million cattle and 4.2 to 6.0 million hectares of pasture in the Amazon Basin. Predominant grasses are *Axonopus scoparius* and *Axonopus micay* in the Eastern Amazon and *Panicum maximum* in the rest (Toledo and Serrão, 1982). Without legumes or fertilization, these pastures have a carrying capacity of one animal per hectare and 100 kg ha$^{-1}$ of annual liveweight gain. After three or four years productivity declines and pastures revert to a secondary forest fallow (Sanchez, 1987). Phosphorus deficiency, grazing management, and poor adaptation of these grasses to the humid tropical environment were main factors associated with the declining yields (Serrão et al., 1979).

Pasture research in Ultisols at Yurimaguas sought to develop management practices and minimum fertilizer requirements for legume and grass species adapted to acid soil conditions which would increase the animal carrying capacity and beef and milk production. Associations of promising selections of grass and legume ecotypes from CIAT's Tropical Pastures Program have been evaluated in replicated long-term grazing trials since 1980. Results for the eight initial years of the study were recently summarized by Lara et al. (1991). Management history of the associations are described in Table 19. Pastures established during 1980 were subjected to continuous grazing which led to high proportions of legumes. Alternate grazing was, therefore, used in subsequent years and the grazing-rest periods were gradually adjusted from 42 days to the current 28-day rotation. Fertilization at pasture establishment consisted of 250, 22, 42, and 10 kg ha$^{-1}$ of lime, P as triple superphosphate, K as KCl, and Mg as MgSO$_4$, respectively. Annual maintenance fertilization consisted of the same quantities of inputs except for lime. Fertilization was suspended after 1987.

Annual liveweight gains for most pastures have remained between 500 and 600 kg ha$^{-1}$ yr$^{-1}$ (Figure 9). These are favorable productivity levels when compared with 100 kg ha$^{-1}$ yr$^{-1}$ for unimproved pastures. There has been considerable fluctuation in the grass/legume proportions of the associations during the grazing period (Table 20). The pure stand of *Centrosema pubescens* 438 was initially established in association with *Andropogon gayanus*, but after the initial year of continuous grazing the legume dominated. Nevertheless, this legume monoculture maintains good animal production and presents no problems with weed infestation. In 1988 the *Brachiaria decumbens/Desmodium ovalifolium* pasture had a poor legume stand and low animal weight gains that were

**Table 19.** Composition and management history of the grass-legume associations evaluated in the grazing trial at Yurimaguas, Peru

| Number | Association Grass/Legume | Grazing Continuous | Grazing Alternating | Time grazed years | Animal load Min. | Animal load Max. |
|---|---|---|---|---|---|---|
| | | ----initiation date---- | | | ----animals ha$^{-1}$---- | |
| 1 | *Brachiaria decumbens/Desmodium ovalifolium* (Bh-Do) | Nov. 15, 1980 | Oct. 6, 1982 | 9 | 4.4 | 5.5 |
| 2 | *Brachiaria humidicola/Desmodium ovalifolium* (Bh-Do) | | Oct. 10, 1982 | 7 | 4.4 | 5.5-6.6 |
| 3 | *Centrosema pubescens* 438 (Cp) | Nov. 15, 1980 | Oct. 6, 1981 | 9 | 3.3 | 4.4 |
| 4 | *Andropogon gayanus/Stylosanthes guianensis* (Ag-Sg) | Nov. 15, 1980 | Oct. 6, 1981 | 9 | 3.3 | 4.4-5.5 |
| 5 | *Andropogon gayanus/Centrosema macrocarpum* (Ag-Cm) | | May 1, 1985 | 5 | 3.3 | 4.4 |

(From Lara et al., 1991.)

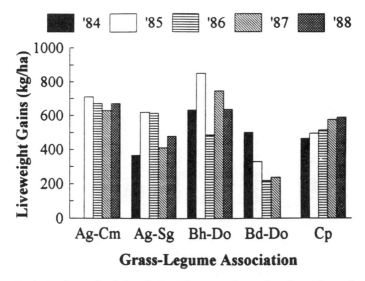

**Figure 9.** Animal productivity during 6 years of rotational grazing of pasture associations in an Ultisol at Yurimaguas, Peru. (From Lara et al., 1991.)

**Table 20.** Changes in the proportion of legumes in pasture associations with time of grazing

| Year | Association[1] | | | | |
| | Bd-Do | Bh-Do | Ag-Sg | Cp | Ag-Cm |
|------|-------|-------|-------|-----|-------|
| | ----------------------------% legume---------------------------- | | | | |
| 1981 | 54 | -- | 25 | 42 | -- |
| 1982 | 56 | -- | 28 | 91 | -- |
| 1983 | 66 | 50 | 40 | 100 | -- |
| 1984 | -- | 29 | 8 | 100 | -- |
| 1985 | 27 | 31 | 50 | 100 | 14 |
| 1986 | 23 | 44 | 41 | 100 | 34 |
| 1987 | 13 | 61 | 9 | 100 | 32 |
| 1988 | 5 | 52 | 9 | 100 | 36 |

[1] Grass-legume species abbreviations are defined in Table 19.
(From Lara et al., 1991.)

attributed to photosensitivity. In the succeeding year, however, the legume increased to 15% of the total dry matter and liveweight gains corresponded to 329 kg ha$^{-1}$ yr$^{-1}$. Two associations, *P. maximum/P. phaseoloides* and *B. dictyoneura/D. ovalifolium*, were discontinued after the third year due to low animal productivity and difficulties in maintaining proper balances of the grass-legume mixtures.

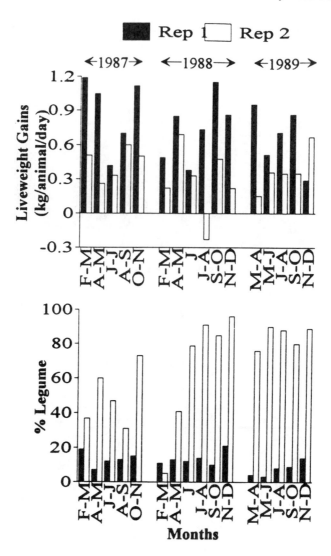

**Figure 10.** Relation between liveweight gains and legume content for replications of the *A gayanus/C. macrocarpum* pasture association during 3 years of grazing. (From Lara et al., 1991.)

The interactions between soil chemical properties and pasture composition, and their potential consequences on grass-legume imbalances in the associations are illustrated with liveweight gains for replicates of the *A. gayanus/C. macrocarpum* association (Figure 10). Dominance of the legume in replicate 2 was associated with a lower Al saturation; 38 to 60% during 1987-1989 as opposed to 74 to 76% for the same period in replicate 1. In this pasture there is an animal preference for the grass which results in superior liveweight gains for replicate 1 which is dominated by the grass. Animal preference for the grass and soil conditions favoring legume dominance may eventually lead to a *C. macrocarpum* monoculture in replicate 2.

This long-term grazing trial provides several grass and legume options for the acid, infertile Ultisols. With minimum fertilizer inputs and proper grazing management, annual liveweight gains relative to unimproved pastures can be increased fourfold. Difficulties remain, however, in maintaining a proper grass-legume balance over time. Favorable results from grazing legume monocultures could be implemented in agroforestry systems where the legume is used as a ground cover.

Castilla (1992) evaluated stocking rates of 0, 3.3, 6.6, and 8.3 animals ha$^{-1}$ on a 4-year-old *B. humidicola/D. ovalifolium* pasture. The three latter stocking rates represented low, adequate, and overgrazed conditions, respectively. Analysis of soil samples several times during the study showed that most of the exchangeable bases, total C, and available P were concentrated in the 0- to 5-cm depth. Bulk density in the 0- to 10-cm depth decreased with time for the 0 and 3.3 animal ha$^{-1}$ stocking rates (Figure 11). Within an 18-month period bulk density decreased 0.10 Mg m$^{-3}$ when animals were absent, and 0.05 Mg m$^{-3}$ at the low stocking rate. These reductions are attributed to the absence or a reduced rate of animal trampling, increased root growth, and greater earthworm populations. Sorptivity, a measure of the ability of water to infiltrate the soil early in the infiltration process, was two times greater for the treatment without animals compared to the three treatments with animals present. Available soil water-storage capacity decreased as stocking rate increased and, at the highest stocking rate, the soil tended to dry out faster during dry periods. Grass in all stocking rate treatments extracted water from the 50-cm depth during dry periods. These results were unexpected and suggest that the resiliency of some soil-pasture associations in the Amazon is sufficient to recover from degradation resulting from overgrazing.

## VI. Extrapolation to Oxisols

Research at Yurimaguas showed that effective crop rotations and judicious application of lime and fertilizers were keys to mechanized continuous cropping. With good crop growth and adequate lime, fertilizers, and mechanization chemical soil properties improved with time. The main objective of research in Manaus, Brazil was to determine the extent to which lime and fertilizer recommendations developed for Yurimaguas Ultisols would change when continuous cultivation was practiced in the clayey Oxisols, which are predominant in the Central Amazon.

We anticipated that the nature of the crop rotation and site-specific soil properties would influence the quantity, frequency, and timing of lime and fertilizer inputs. A combined evaluation of these factors implied an experimental

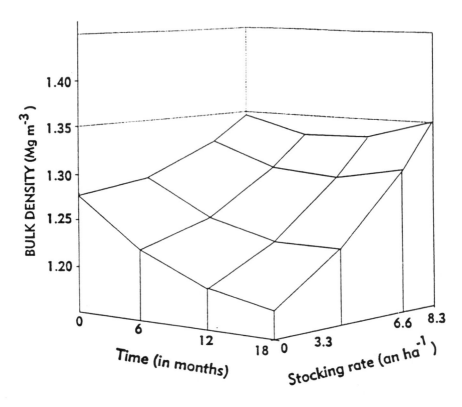

**Figure 11.** Changes in bulk density of a Yurimaguas Ultisol over time as a function of stocking rates for a *B. humidicola/Desmodium ovalifolium* pasture. (From Castilla, 1991.)

design which anticipated changes in required inputs across time without confounding spatial variation in soil properties resulting from clearing the forest by different methods or different years.

A long-term experiment was initiated at Manaus in 1981 (Cravo and Smyth, 1991). The primary forest vegetation was cut, allowed to dry for three months, and burned at the end of annual dry season ($<100$ mm rainfall month$^{-1}$). Unburnt vegetation was manually removed and the field was plowed and disked prior to planting the initial crop. Nutrient content of the ash is shown in Table 4. Changes in soil chemical properties upon incorporating the ash are shown in Table 21 by comparing soil analyses for surface soil samples taken before and after burning.

The experiment contained 35 treatments with four replications in a randomized complete block design. In addition to an absolute check (no fertilizers or lime), crop responses to N, P, K, S, Mg, Cu, B, Mn, Zn, and lime were evaluated individually in three or four treatments with increasing rates of each nutrient.

**Table 21.** Surface soil (0-15 cm) chemical properties for the Oxisol in Manaus, Brazil before and after burning the primary forest vegetation

| Sampling time | pH | Ca | Mg | K | Al | Al Sat. | Org. C | P | Cu | Fe | Mn | Zn |
|---|---|---|---|---|---|---|---|---|---|---|---|---|
| | | | | | | | | | | Mehlich 1 extractable | | |
| | | -------cmol $L^{-1}$------- | | | | -------%------- | | -------mg $L^{-1}$------- | | | | |
| Before burning | 4.2 | 0.1 | 0.3 | 0.06 | 1.8 | 78 | 3.30 | 2 | 0.1 | 175 | 2 | 1 |
| After burning | 5.3 | 2.1 | 0.8 | 0.27 | 0.6 | 16 | 3.09 | 6 | 2.0 | 99 | 10 | 1 |
| Difference | 1.1 | 2.0 | 0.5 | 0.21 | -1.2 | -62 | -0.21 | 4 | 1.9 | -76 | 8 | 0 |

(From Smyth and Bastos, 1984.)

**Table 22.** Sequence of crops and time when fertilizer and lime treatments were initiated

| Crop | Time between planting and harvesting | Treatments initiated |
|------|-------------------------------|----------------------|
|      | months after burning          |                      |
| Rice | 3.0-7.4 | N and P |
| Soybean | 8.9-12.6 | K |
| Soybean | 18.5-22.3 | Lime and Cu |
| Cowpea | 22.9-25.2 | |
| Corn | 27.6-31.3 | S |
| Cowpea | 32.7-34.9 | |
| Corn | 37.2-41.8 | B and Zn |
| Soybean | 42.3-46.2 | |
| Cowpea | 46.9-48.9 | Mn |
| Corn | 52.1-56.2 | |
| Cowpea | 58.1-61.0 | |
| Corn | 63.1-67.3 | Mg |
| Cowpea | 69.2-71.5 | |
| Corn | 76.1-80.1 | |
| Cowpea | 82.2-84.4 | |
| Corn | 87.0-91.2 | |
| Cowpea | 93.9-96.2 | |

(From Cravo and Smyth, 1991.)

Treatments for each nutrient were not initiated until soil and plant analyses from preceding crops and the yields suggested that a particular nutrient deficiency was likely to occur. Once a yield response to a given nutrient or to lime was obtained, it was added to all other treatments (except the absolute check) to avoid confounding crop performance in individual treatments with limitations of other nutrients. Fertilizer rates in the blanket applications were adjusted for each crop, based on previous crop performance in treatments for each nutrient.

Crop sequence and fertilizer and lime treatments initiated during the eight years of continuous cropping are shown in Table 22. Three treatments each for N and P were initiated with the first crop, because (1) N needs are difficult to assess without local crop response data, and (2) P sorption data suggested an immediate crop response to this element. Phosphorus was applied uniformly to all remaining plots before planting the subsequent soybean crop. Similarly, K and lime were uniformly applied to all treatments after obtaining significant yield responses in Crops 2 and 3, respectively. Responses to Cu have been sporadic and are not associated with any specific crop in the rotation. Nevertheless, Cu was included in blanket fertilizer applications for the first corn crop to avoid potential interaction with other nutrients under evaluation. Potential Mo

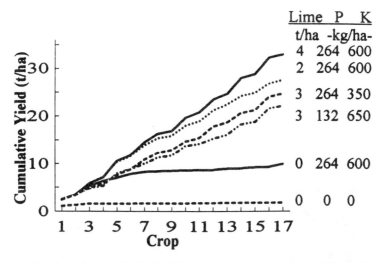

**Figure 12.** Cumulative yields for 17 consecutive crops of rice (1), soybean (3), corn (6), and cowpea (7) as a function of lime, P, and K inputs during 8 years of continuous cropping in an Oxisol at Manaus, Brazil. (From Cravo and Smyth, 1991.)

constraints were avoided with applications of 20 g ha$^{-1}$ to the first soybean crop and fifth cowpea crop.

Potential S losses during forest clearing by burning and the avoidance of S-carrying fertilizers when applying N, P, and K suggested a possible early depletion of soil reserves for this nutrient. Although treatments for S and all remaining nutrients were initiated after the second year of cultivation, no significant yield responses were obtained in subsequent crops. During the two final cowpea crops, yield responses to Mg were traced to imbalances caused by a buildup of soil K. Likewise, Mn deficiencies in legumes have occurred only with overliming at the highest lime rate.

Cumulative grain yields for 17 consecutive crops with variable rates of P, K, and lime are compared in Figure 12 to yields without lime and fertilizers. Additional constant inputs to all fertilized treatments were 634 kg N ha$^{-1}$ and 1 kg Cu ha$^{-1}$. Maximum yields in Manaus were similar to those reported for the complete lime and fertilization treatment in the Ultisol at Yurimaguas. Without fertilizers and lime, total yield for 17 crops was 1.7 t ha$^{-1}$. The importance of lime is exemplified by yield comparisons among 0, 2, and 4 t lime ha$^{-1}$ at fixed levels of P and K. Across these lime rates cumulative yield increased from 8.5 to 32.5 t ha$^{-1}$. Treatments with 3 t lime ha$^{-1}$ received an initial lime application of 2 t ha$^{-1}$ in the first corn crop, followed by 1 t ha$^{-1}$ in the fifth corn crop when residual effects from the first application began to decline.

The magnitude of the yield response to lime depended on both the crop and time after lime was applied. As the Al saturation in the unlimed treatment

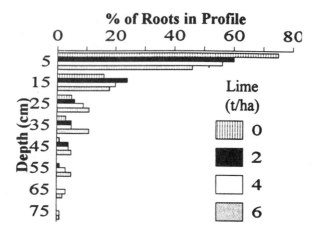

**Figure 13.** Corn root distribution in the Manaus Oxisol as a function of lime applied 6 years earlier. (From Smyth and Cravo, unpublished data.)

increased with time of cultivation to levels of 82 to 87% in the final crops, corn yield responses were detected up to 4 and, eventually, 6 t lime ha$^{-1}$. Cowpea yields were increased by liming when Al saturation was >58%, but there was never a yield response above 2 t lime ha$^{-1}$. In a satellite lime experiment with smaller lime increments across the range of 0 to 4.6 t ha$^{-1}$ the critical Al saturation for corn and soybean was established at 27% as opposed to 54% for peanut (Smyth and Cravo, 1992).

Based on changes in soil acidity with time after liming, we estimated that 2 t lime ha$^{-1}$ would maintain Al saturation below the critical value for corn and soybean for three years. In comparison, average annual lime inputs for continuous cropping in the Ultisol at Yurimaguas was 1.5 t ha$^{-1}$ (Sanchez et al., 1983). The smaller lime requirement for the Oxisol is in agreement with a lower ECEC and exchangeable Al level than in the Ultisol.

Subsoil chemical restrictions to root growth were alleviated with time after liming by significant increases in exchangeable bases and a decrease in Al saturation to a depth of 70 cm. Improvements in the subsoil chemical environment were apparent when corn root distribution was assessed by the trench-profile method six years after the lime treatments were initiated (Figure 13). Lime increased the quantity of roots in the profile by a factor of 2.5 and increased the proportion of total roots in the profile which extended into the subsoil. With 4 and 6 t lime ha$^{-1}$ corn roots extended to a depth of 80 cm. The observed improvements in subsoil chemical properties with increasing time under appropriately-managed continuous cropping are similar to the findings reported for Ultisols in Yurimaguas (section III.C).

Phosphorus was the first nutrient deficiency detected in the long-term experiment and yields for all crops were increased by applied P. Reductions in

P inputs from an average of 16 to 18 kg ha$^{-1}$ crop$^{-1}$ decreased total yields by 6.3 t ha$^{-1}$ (Figure 12). A separate field experiment was conducted to evaluate long-term P requirements of a corn-cowpea annual rotation, when fertilizer was applied in band and broadcast placement (Smyth and Cravo, 1990). Treatments were arranged in a split-plot factorial combination. Main plots consisted of five broadcast P rates (0, 22, 44, 88, and 176 kg ha$^{-1}$) applied once at the beginning of the experiment. Subplots contained four banded P rates (0, 11, 22, and 44 kg ha$^{-1}$) applied to each crop. Banded P was discontinued once total P applied by this placement method reached 176 kg P ha$^{-1}$. Lime and all other nutrients were applied to all plots according to periodic soil analyses. A total of six crops of corn and five crops of cowpea were evaluated in the experiment during five consecutive years.

At equal amounts of applied P, yields were greater with banded than with broadcast P during the initial crops. Although cowpea yields were increased up to 44 kg banded P ha$^{-1}$ no response was observed for corn above 22 kg banded P ha$^{-1}$. A higher yield response and soil P critical level for cowpea than for corn was attributed to superior P requirements for plants depending on symbiotic N$_2$ fixation for their N supply. After 11 crops of corn and cowpea, total yields at a fixed level of total applied P were similar for all broadcast and band combinations. This suggested that cumulative yields for long-term continuous cropping depended more on the amount of applied P than on the placement method.

Net cumulative profit for three P fertilization strategies were compared for the eight initial crops (Figure 14). The increasing deficit for the treatment without applied P resulted from both declining yields, from a maximum of 0.2 t ha$^{-1}$ in the initial corn crop, and the costs of lime and other added nutrients. Cumulative yields with a single broadcast application of 176 kg P ha$^{-1}$ or with eight banded applications of 44 kg P ha$^{-1}$ reached a similar value of 17 t ha$^{-1}$. Despite similarities in final net profit for both P treatments, banded P was more advantageous during the four initial crops because fertilizer P costs were distributed equally across each crop. Banded applications of 22 to 44 kg P ha$^{-1}$ crop$^{-1}$ would maintain near-optimum yields and would be more conducive to the capital constraints and manual tillage practices predominant among small farmers in the region. It is indeed interesting that these levels of fertilizer P inputs to a clayey Oxisol with appreciable P sorption capacity are similar to the average of 45 kg P ha$^{-1}$ crop$^{-1}$ which was applied to the 19 initial crops in the loamy Ultisol at Yurimaguas (Sanchez et al., 1983).

Yield response to fertilizer N in the long-term continuous cropping experiment was only observed with corn, and suggested the need for a gradual increase in N inputs with time of cultivation. Individual corn crop yields with N treatments are shown in Table 23. During the two initial crops, there was no significant yield increase to rates of 40 to 120 kg N ha$^{-1}$. With a subsequent reduction in N rates, yields for the third and fourth crops increased significantly with 80 kg N ha$^{-1}$. Nitrogen requirements for the two final crops were 120 kg ha$^{-1}$. Nitrogen was applied to this study in three equal split-applications at planting and sidedressed at 25 and 55 days after planting. A separate field

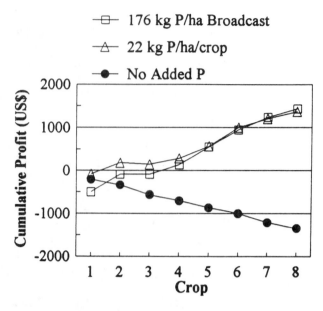

**Figure 14.** Cumulative net profit of P fertilization strategies for an annual rotation of corn and cowpea in a clayey Oxisol at Manaus, Brazil. (From Cravo and Smyth, 1991.)

experiment which compared timing and distribution of fertilizer N on corn yields indicated that there was no advantage in partitioning applied N into two or three applications among the above growth stages, relative to a single application at 25 days after planting. Yields were reduced, however, with a single application at planting (Melgar et al., 1991).

The total N input of 634 kg ha[-1] during the 17 crops in the Manaus experiment translates into an average of 91 kg N ha[-1] for each crop of rice and corn. Based on N responses obtained with each crop (Table 23) the average rate for all rice and corn crops could be reduced to 69 kg N ha[-1]. In the long-term trial in the Yurimaguas Ultisol total fertilizer N applied during the 19 initial crops would correspond to 114 kg N ha[-1] per crop of rice and corn (Sanchez et al., 1983). The lower N requirement for continuous cropping in Manaus may be related to a higher proportion of legume crops in the rotation than in Yurimaguas. In a satellite experiment in the Manaus Oxisol, we estimated that the fertilizer-N substitution value of cowpea residues for a subsequent corn crop, during three consecutive years, averaged 26 kg ha[-1] (Smyth et al,. 1991b). We also observed that delayed N movement beyond a 60-cm depth in this Oxisol corresponded to net positively charged subsoil layers where $NO_3^-$ could be adsorbed (Melgar et al., 1992). Corn crops may have access to this N reserve provided there were no subsoil chemical barriers to root growth.

**Table 23.** Corn yield as a function of applied N in six consecutive years of cultivation

| Applied N | Corn crop | | | | | |
|---|---|---|---|---|---|---|
| | 1984 | 1985 | 1986 | 1987 | 1988 | 1989 |
| kg ha$^{-1}$ | -----------------------------grain yield, t ha$^{-1}$--------------------------- | | | | | |
| 20 | -- | -- | 1.5 | 1.5 | -- | -- |
| 40 | 2.7 | 2.2 | 1.7 | 1.7 | -- | -- |
| 80 | 2.5 | 1.8 | 2.7 | 2.2 | 2.0 | 2.8 |
| 120 | 2.9 | 2.3 | -- | -- | 2.2 | 3.1 |
| 160 | -- | -- | -- | -- | 2.0 | 3.2 |

(From Cravo and Smyth, 1991.)

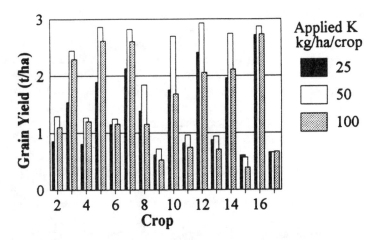

**Figure 15.** Soybean, corn, and cowpea yields as a function of K rates applied to an Oxisol in Manaus, Brazil. (From Cravo and Smyth, 1991.)

Reductions in K inputs to the Oxisol from an average of 35 to 21 kg ha$^{-1}$ crop$^{-1}$, during 17 consecutive crops, reduced cumulative yields by 9.2 t ha$^{-1}$ (Figure 12). Individual crop responses to applications of 25, 50, and 100 kg K ha$^{-1}$ are shown in Figure 15. Potassium was applied to all crops except numbers 4 and 16. There was no yield advantage in any crop to rates greater than 50 kg K ha$^{-1}$. In general, the optimum rate for corn and soybean was 50 kg K ha$^{-1}$; the optimum rate for cowpea was 25 kg K ha$^{-1}$. Soil analysis data indicated that yield depressions with the highest K rate were associated with reductions in the Mg saturation to less than 5% of the ECEC. The absence of a buildup in soil K levels during the 17 crops suggested that K inputs should be made to individual crops.

In a separate K fertilizer trial for corn and cowpea, downward K movement in this Oxisol was only detected in treatments with K rates of 66 and 132 kg ha$^{-1}$ (Smyth and Bastos, 1990). Maintenance of crop residues in the field was an important management practice for minimizing external K inputs.

# VII. Conclusions

The "long-term" nature of research with collaborators in Yurimaguas and Manaus was to address soil management problems as they evolved over time of continued land use, rather than to make comparisons over time among a pre-determined set of treatments. Results from 20 contiguous years of research in the Amazon indicated that favorable crop yields and soil properties were maintained in mechanized continuous cropping provided that crops are grown in rotation, lime and fertilizer inputs are based on frequent soil analyses, and tillage practices are established based on soil properties. There is no particular lime and fertilizer recommendation which can be readily applied to Ultisols and Oxisols throughout the Amazon without adjustments for site-specific conditions. Soils can be severely damaged by improper use of mechanized land clearing techniques. Management requirements for maintaining good physical properties in sandy Ultisols are similar to those reported for other regions of the world.

A central component for the development of improved land use practices which minimize lime and fertilizer inputs in Ultisols and Oxisols was the selection of acid-tolerant germplasm. Although crop rotations with such germplasm can prolong land use by shifting cultivators, long-term cropping is limited by increasing problems with weed management and the need to replenish nutrients exported in grain harvests from a limited soil reserve. External nutrient inputs will also be needed to sustain yields under continuous alley cropping because trees and crops compete for limited soil reserves. Additional studies are needed on selection of trees able to tap soil nutrient reserves that are not accessible to crops, thus, enriching the available nutrient balance for the system in acid soils. Despite the existing limitations, alley cropping may be particularly useful in reducing runoff and erosion hazards on sloping lands. Although animal productivity was significantly increased in long-term grazing trials with selected grass-legume associations, questions remain as to the appropriate proportions of the grass and legume in the pasture, and the management practices which will favor persistence of the legumes.

Knowledge acquired through long-term research in Ultisols and Oxisols of the Amazon has provided the basic concepts for several land use alternatives to shifting cultivation. Suitability of these land management options to a particular set of soil constraints, and ecological and socio-economical conditions would entail a systematic comparison of their agronomic and economic performances under conditions where temporal effects are not confounded with spatial variability in soil properties. Given the limited number of such studies, this would be a logical step toward developing readily-accessible land management alternatives.

# References

Alegre, J.C., D.K. Cassel, and D.E. Bandy. 1986a. Reclamation of an Ultisol damaged by mechanical land clearing. *Soil Sci. Soc. Am. J.* 50:1026-1031.

Alegre, J.C., D.K. Cassel, D.E. Bandy, and P.A. Sanchez. 1986b. Effect of land clearing on soil properties of an Ultisol and subsequent crop production in Yurimaguas, Peru. p. 167-177. In: R. Lal, P.A. Sanchez, and R.W. Cummings, Jr. (eds.), *Land Clearing and Development in the Tropics*. A.A. Balkema Press, Boston, MA.

Alegre, J.C., D.K. Cassel, and D.E. Bandy. 1986c. Effect of land clearing and subsequent management on soil physical properties. *Soil Sci. Soc. Am. J.* 50:1379-1384.

Alegre, J.C., D.K. Cassel, and D.E. Bandy. 1988. Effect of land clearing methods on soil chemical properties of an Ultisol in the Amazon. *Soil Sci. Soc. Am. J.* 52:1283-1288.

Alegre, J.C., D.K. Cassel, and D.E. Bandy. 1990. Effects of land clearing method and soil management on crop production in the Amazon. *Field Crops Res.* 24:131-141.

Alegre, J.C., P.A. Sanchez, C.A. Palm, and J.M. Perez. 1989. Comparative soil dynamics under different management options. p. 102-108. In: N. Caudle (ed.), *TropSoils Technical Report, 1986-1987*. North Carolina State Univ., Raleigh, NC.

Alegre, J.C., P.A. Sanchez, and T.J. Smyth. 1991. Manejo de suelos con cultivos continuos en los trópicos húmedos del Perú. p. 157-168. In: T.J. Smyth, W.R. Raun, and F. Bertsch (eds.), *Manejo de Suelos Tropicales en Latinoamerica*. North Carolina State Univ., Raleigh, NC.

Bandy, D.E. and P.A. Sanchez. 1986. Post-clearing soil management alternatives for sustained production in the Amazon. p. 347-361. In: R. Lal, P.A. Sanchez, and R.W. Cummings, Jr. (eds.) *Land Clearing and Development in the Tropics*. A.A. Balkema Press, Boston, MA.

Camargo, M.N. and T.E. Rodrigues. 1979. *Guia de Excursão XVII Congreso Brasileiro de Ciência do Solo (Manaus)*. Seviço Nacional de Levantamento e Conservação de Solos, Rio de Janeiro, Brazil. 71 pp.

Cassel, D.K. and R. Lal. 1992. Soil physical properties of the tropics: common beliefs and management restraints. *Soil Sci. Soc. Am. Spec. Publ.* 29:61-89.

Castilla, C.E. 1992. Carbon dynamics in managed tropical pastures: the effect of stocking rate on soil properties and on above- and below-ground carbon inputs. Ph.D. Dissertation. North Carolina State Univ., Raleigh, NC. 175 pp.

Cochrane, T.T. and P.A. Sanchez. 1982. Land resources, soils and their management in the Amazon region: a state of knowledge report. p. 137-209. In: S.B. Hecht (ed.), *Amazonia: Agriculture and Land Use Research.* CIAT, Cali, Colombia.

Cox, F.R. and E. Uribe. 1992a. Management and dynamics of potassium in a humid tropical Ultisol under a rice-cowpea rotation. *Agron. J.* 84:655-660.

Cox, F.R. and E. Uribe. 1992b. Potassium in two humid tropical Ultisols under a corn and soybean cropping system. *Agron. J.* 84:480-489.

Cravo, M.S. and T.J. Smyth. 1991. Sistema de cultivo com altos insumos na Amazonia Brasileira. p. 144-156. In: T.J. Smyth, W.R. Raun, and F. Bertsch (eds.), *Manejo de Suelos Tropicales en Latinoamerica.* North Carolina State Univ., Raleigh, NC.

Fernandes, E.C.M. 1991. Ensayo de proveniencias de *Gliricidia sepium* (Jacq.) Walp. en un Ultisol de la Amazonia Peruana. pp. 275-281. In: T.J. Smyth, W.R. Raun, and F. Bertsch (eds.), *Manejo de suelos tropicales en Latinoamerica.* North Carolina State Univ., Raleigh, NC.

Fernandes, E.C.M. 1990. Alley cropping on acid soils. Ph.D. Dissertation. North Carolina State Univ., Raleigh, NC. 157 pp.

Hecht, S.B. 1982. Agroforestry in the Amazon Basin: practice, theory and limits of a promising land use. p. 331-371. In: S.B. Hecht (ed.), *Amazonia: Agriculture and Land Use Research.* CIAT, Cali, Colombia.

Lara, D., C. Castilla, and P.A. Sanchez. 1991. Productividad y persistencia de pasturas asociadas bajo pastoreo en un Ultisol de Yurimaguas. pp. 86-89. In: T.J. Smyth, W.R. Raun, and F. Bertsch (eds.), *Manejo de Suelos Tropicales en Latinoamerica.* North Carolina State Univ., Raleigh, NC.

Lopes, A.S., T.J. Smyth, and N. Curi. 1987. The need for a soil fertility reference base and nutrient dynamics studies. p. 147-166. In: P.A. Sanchez, E.R. Stoner and E. Pushparajah (eds.), *Management of acid tropical soils for sustainable agriculture.* International Board for Soil Research and Management, Bangkok, Thailand.

McKerrow, A.J. 1992. Nutrient stocks in abandoned pastures of the Central Amazon Basin prior to and following cutting and burning. M.Sc. Thesis. North Carolina State Univ., Raleigh, NC. 116 pp.

Melgar, R.J., T.J. Smyth, M.S. Cravo, and P.A. Sanchez. 1991. Doses e épocas de aplicação de fertilizante nitrogenado para milho em Latossolo da Amazônia Central. *R. Bras. Ci. Solo* 15:289-296.

Melgar, R.J., T.J. Smyth, P.A. Sanchez, and M.S. Cravo. 1992. Fertilizer nitrogen movement in a Central Amazon Oxisol and Entisol cropped to corn. *Fertilizer Res.* 31:241-252.

Melillo, J.M., C.A. Palm, R.A. Houghton, G.M. Woodwell, and N. Myers. 1985. A comparison of two recent estimates of disturbance in tropical forests. *Environmental Conservation* 12:37-40.

Moran, E.F. 1981. *Developing the Amazon.* Indiana University Press, Bloomington, IN.

Mt Pleasant, J., R.E. McCollum, and H.D. Coble. 1990. Weed population dynamics and weed control in the Peruvian Amazon. *Agron. J.* 82:102-112.

Mt Pleasant, J., R.E. McCollum, and H.D. Coble. 1992. Weed management in a low-input cropping system in the Peruvian Amazon region. *Trop. Agric.* 69:250-259.

Nicholaides, J.J. and M.J. Piha. 1987. A new methodology to select cultivars tolerant to aluminum and with high yield potential. p. 103-116. In: L.M. Gourley and J.G. Salinas (eds.), *Sorghum for acid soils: Proceedings of a workshop on evaluating sorghum for tolerance to Al-toxic tropical soils in Latin America, Cali, Colombia, 28 May to 2 June, 1984.* CIAT, Cali, Colombia.

Palm, C.A. 1988. Mulch quality and nitrogen dynamics in an alley cropping system in the Peruvian Amazon. Ph.D. Dissertation. North Carolina State Univ., Raleigh, NC. 112 pp.

Palm, C.A. and P.A. Sanchez. 1990. Decomposition and nutrient release patterns of the leaves of three tropical legumes. *Biotropica* 22:330-338.

Salazar, A. and C.A. Palm. 1987. Screening of leguminous trees for alley cropping on acid soils of the humid tropics. pp. 61-67. In: *Gliricidia sepium*: management and improvement. Spec. Publ. 87-01. Nitrogen Fixing Tree Assoc., Honolulu, HI.

Sanchez, P.A. 1976. *Properties and management of soils in the tropics.* John Wiley and Sons, New York.

Sanchez, P.A. 1987. Management of acid soils in the humid tropics of Latin America. p. 63-107. In: P.A. Sanchez, E.R. Stoner, and E. Pushparajah (eds.) *Management of acid tropical soils for sustainable agriculture.* International Board for Soil Research and Management, Bangkok, Thailand.

Sanchez, P.A. and J.R. Benites. 1987. Low-input cropping for acid soils of the humid tropics. *Science* 238:1521-1527.

Sanchez, P.A. and T.T. Cochrane. 1980. Soil constraints in relation to major farming systems of tropical America. p. 107-139. In: M. Drosdoff, H. Zandstra, and W.G. Rockwood (eds.), *Priorities for alleviating soil-related constraints to food production in the tropics.* IRRI, Los Baños, Philippines.

Sanchez, P.A., C.A. Palm, and T.J. Smyth. 1990. Approaches to mitigate tropical deforestation by sustainable soil management practices. p. 211-220. In: H.W. Scharpenseel, M. Schomaker, and A. Ayoub (eds.), *Soils on a warmer earth* (Developments in Soil Sci. 20). Elsevier, NY.

Sanchez, P.A., J.H. Villachica, and D.E. Bandy. 1983. Soil fertility dynamics after clearing a tropical rainforest in Peru. *Soil Sci. Soc. Am. J.* 47:1171-1178.

Serrão, E.A.S., I.C. Falesi, J.B. DaVeiga, and J.F. Teixeira. 1979. Productivity of cultivated pastures on low fertility soils of the Amazon of Brazil. pp. 195-225. In: P.A. Sanchez and L.E. Tergas (eds.), *Pasture Production in Acid Soils of the Tropics.* CIAT, Cali, Colombia.

Seubert, C.E., P.A. Sanchez, and C. Valverde. 1977. Effects of land clearing methods on soil properties of an Ultisol and crop performance in the Amazon jungle of Peru. *Trop. Agric.* 54:307-321.

Smyth, T.J., J.C. Alegre, and C.A. Palm. 1991a. Dinámica de nutrientes del suelo durante tres años de cultivos de bajos insumos en un Ultisol de la Amazonia Peruana. p. 39-47. In: T.J. Smyth, W.R. Raun, and F. Bertsch (eds.), *Manejo de Suelos Tropicales en Latinoamerica*. North Carolina State Univ., Raleigh, NC.

Smyth, T.J. and J.B. Bastos. 1984. Alterações na fertilidade de um Latossolo Amarelo àlico pela queima de vegetação. *Rev. Bras. Ci. Solo* 8:127-132.

Smyth, T.J. and J.B. Bastos. 1990. Adubação potássica para milho e caupi em Latossolo Amarelo àlico do Estado do Amazonas. pp. 173-181. In: *Anais 1º Simpósio do Tropico Umido*. EMBRAPA, Belem, Brazil.

Smyth, T.J. and M.S. Cravo. 1990. Phosphorus management for continuous corn-cowpea production in a Brazilian Amazon Oxisol. *Agron. J.* 82:305-309.

Smyth, T.J. and M.S. Cravo. 1992. Aluminum and calcium constraints to continuous crop production in a Brazilian Amazon Oxisol. *Agron. J.* 84:843-850.

Smyth, T.J., M.S. Cravo, and R.J. Melgar. 1991b. Nitrogen supplied to corn by legumes in a Central Amazon Oxisol. *Trop. Agric.* 68:366-372.

Szott, L.T. 1987. Improving the productivity of shifting cultivation in the Amazon Basin of Peru through the use of leguminous vegetation. Ph.D. Dissertation. North Carolina State Univ., Raleigh, NC. 168p.

Szott, L.T., C.A. Palm, and P.A. Sanchez. 1991. Agroforestry in acid soils of the humid tropics. *Adv. Agron.* 45:275-301.

Toledo, J.M. and E.A.S. Serrão. 1982. Pasture and animal production in Amazonia. pp. 281-309. In: S.B. Hecht (ed.), *Amazonia: Agriculture and Land Use Research*. CIAT, Cali, Colombia.

Valverde, C. and D.E. Bandy. 1982. Production of annual food crops in the Amazon. p. 243-280. In: S.B. Hecht (ed.) *Amazonia: agriculture and land use research*. CIAT, Cali, Colombia.

Wade, M.K. and P.A. Sanchez. 1983. Mulching and green manure applications for continuous crop production in the Amazon Basin. *Agron. J.* 73:39-45.

# The Role of Forage Grasses and Legumes in Maintaining the Productivity of Acid Soils in Latin America

R.J. Thomas, M.J. Fisher, M.A. Ayarza, and J.I. Sanz

## I. Introduction

Sustainable agricultural productivity is now the key goal of both developed and developing countries (Edwards et al., 1990). In the latter, the option of increasing productivity solely by means of external inputs such as lime, fertilizer, and herbicides, remains beyond the reach of most farmers. Good stewardship of the soil then becomes of paramount concern as ideally, the nutrients needed for agricultural production should come predominantly from fluxes into and out of the soil organic matter rather than from fertilizer additions to the soil nutrient solution (Harwood, 1990). Any net loss of soil organic matter represents a loss of the farmer's capital, especially the poorer farmer as, in terms of a farmer's lifetime, the soil is a non-renewable resource.

Latin American population growth rate is exceeding the rate of increase in agricultural output (Vera et al., 1993) and therefore productivity must increase with a concurrent maintenance or improvement of the soil. Various options to

ISBN 1-56670-076-0

achieve such sustainable increases in productivity have been discussed by
Sánchez and Salinas (1981), Toledo and Nores (1986), Sánchez (1987), and
Goedert (1987) for both the humid tropics (mainly forest) and the acid soil
savannas. Grass/legume pastures feature prominently in the proposed options
discussed in the above articles. However, in a summary of the research policies
on soil management for sustainability, Bentley (1991) targeted research on the
use of fodder grasses and legumes for sustainable agriculture and soil conservation as an "outrageously and consistently neglected" area. In this article we
examine the role of forage grasses and legumes in production systems of the
acid-soil savannas of Latin America. Evidence is presented from long-term
experiments with grass/legume pastures, which demonstrates both the soil-
improving qualities and the increases in productivity obtainable from the use of
acid-tolerant forage germplasm with minimum external inputs.

## II. Land Use in Latin America

Two ecosystems dominate the land in tropical Latin America, the rain forests,
and the savannas. Here we will address the main problems associated with
agricultural production and soil management pertaining mainly to the savanna
ecosystems in the region. Aspects of soil management in the rain forest regions
are discussed elsewhere (e.g., Sánchez, 1987; Villachica et al., 1990; Smith et al.,
this meeting).

Generally, the lands of tropical Latin America are used for extensive cattle
ranching with savannas covering about 45% of the land area or 243 million
hectares (Huntley and Walker, 1982; Vera and Seré, 1985). The savannas
represent the major under-utilized resource in the continent and are found on the
Cerrados of Brazil (180 million hectares), the Llanos of Colombia (17 million),
Venezuela (28 million), and Guyana (4 million), and the savanna of Bolivia (4
million) (Cole, 1986). In Brazil and Venezuela these lands are being rapidly
exploited. Table 1 shows the trends in the areas under crop and pasture in Latin
America.

The Latin American tropics carry some 250 m head of cattle, with an annual
production of some 7 million metric tonnes or megatonnes (Mt) of beef and veal,
and 31 Mt of milk representing 13.8 and 6.6 % of the world total, respectively
(Table 2). The region, however, is not self sufficient in these commodities as
production increases lag behind population growth and potential demand for beef
(Vera et al., 1993).

There is, in addition, substantial crop production in tropical Latin America,
with the major production areas occurring in Brazil. Crop production includes 44
Mt of maize, of which Brazil (55 percent) and Mexico (27 %) produce 82 % of
the total. In 1990, 17 Mt of rice was produced, 58 % of which was produced in
Brazil, and 26 % in the remaining tropical South American countries, and 23 Mt
of soybean over 88 % of which was produced in Brazil. Soybean production has
expanded by over 12 % on an annual rate in the 24 years 1966 to 1990

**Table 1.** Trends in the land areas in crop and pasture in Latin America

| Regional group | Permanent pastures | | Annual and permanent crops | |
|---|---|---|---|---|
| | 1973/80 | 1989 | 1973/80 | 1989 |
| | -------------------millions of ha---------------------- | | | |
| Brazil and Mexico | 213.47 | 244.50 | 87.98 | 103.36 |
| Tropical South America | 126.87 | 137.62 | 19.23 | 21.33 |
| Central America | 11.86 | 13.77 | 6.33 | 6.86 |
| Caribbean | 6.59 | 7.00 | 6.16 | 6.60 |
| Tropical Latin America | 376.79 | 402.89 | 119.70 | 138.09 |

(From CIAT, 1992.)

**Table 2.** Livestock commodity data for tropical Latin America, including the Caribbean

| Commodity | --------------Annual production------------- | | | | --------Stock-------- | |
|---|---|---|---|---|---|---|
| | 1974/81 | 1990 | PWT[1] | PC[2] | 1974/81 | 1990 |
| Beef and veal | 4.61 | 7.05 | 13.8 | 18 | 201.5 | 250.0 |
| Cow milk | 25.15 | 31.12 | 6.6 | 81 | 26.9 | 32.5 |
| Pig meat | 2.46 | 2.54 | 3.7 | 7 | 67.8 | 71.8 |
| Poultry meat | 2.00 | 4.30 | 11.6 | 11 | 747.0 | 1211.0 |

[1] Proportion of world total; [2] per capita.
(From CIAT, 1992.)

(Table 3). Although integrated crop-pasture systems exist in the region, over the last twelve years there have been substantial structural changes away from integrated crop-pasture systems particularly in the Brazilian Cerrados. In this region upland rice yields are low at around 1.5 t/ha, largely because the traditional varieties lodge easily when supplied with more than modest levels of fertilizer (Sanint et al., 1992). Moreover, grain quality is poor and prices are discounted by 20-30% when compared with the irrigated sector prices (Teixeira and Sanint 1988). With the restructuring of the economic support to crop production in Brazil, in many areas rice has given way to soybeans, and in many parts soybean monocropping is now the norm. In the newly pioneered areas particularly, where beef still remains profitable, there have been large swings away from cropping to pastures (Sanint et al., 1992).

## III. Constraints to Production

The main soil chemical constraints in Latin America are low nutrient reserves, soil acidity and phosphorus fixation (Table 4). Around 75% of the soils in the

**Table 3.** Data for the principal crop commodities in tropical Latin America, including the Caribbean.

| Commodity | --------Annual production-------- | | | Annual growth ----rate 1966/90---- | |
|---|---|---|---|---|---|
|  | 1974/81 | 1990 | per capita | area sown | yield ha[-1] |
|  | millons metric tons | | kg | % | % |
| Wheat | 3.05 | 9.62 | 25 | 3.00 | 2.10 |
| Maize | 25.96 | 44.38 | 115 | 0.60 | 2.04 |
| Potatoes | 5.58 | 9.06 | 23 | 0.17 | 2.00 |
| Cassava | 31.32 | 30.94 | 80 | 0.18 | -0.61 |
| Rice, paddy | 9.50 | 17.34 | 45 | 1.09 | 1.69 |
| Dry beans | 3.81 | 3.97 | 10 | 1.33 | -1.07 |
| Soybeans | 0.92 | 23.32 | 67 | 12.23 | 1.61 |
| Sorghum | 2.31 | 7.89 | 20 | 4.15 | 1.49 |

(From Ciat, 1992.)

savanna regions are either Oxisols (43%) or Ultisols (32%). Further details of the climatic and topographic parameters for the savannas have been summarized by Goedert (1987) and Vera et al. (1993). Eswaran et al. (1992) have estimated that oxisols can lose 50% of their sustainability under low-input agriculture within 20 years, a rate which is much more rapid than alfisols, mollisols or vertisols. Thus even though these soils are endowed with relatively good physical conditions and great depth, they must be managed carefully for sustained production.

The pastures of the Cerrados of Brazil are mainly pure grass with about 40 million ha planted to a single genotype of *Brachiaria decumbens* which is usually established after a pioneer crop of upland rice (Zeigler et al., 1993). However, without maintenance fertilizer these pastures are degrading mainly due to phosphorus and nitrogen deficiencies and attacks from spittlebugs (*Cercopids*), (Fenster and Leon, 1979; Spain, 1993; Salinas and Saif, 1990; Valerio J.R. and Nakano, 1988).

Similarly, on areas under monocropping of soybeans and other annual crops in Brazil, yields are not sustainable due to weed competition, depletion of low nutrient reserves and soil physical problems such as compaction and erosion (EMPA, 1987; Sánchez, 1976; Seguy et al., 1988; Spain, 1993; Stoner et al., 1991; Zeigler et al., 1993). In addition, cropping in the cerrados usually requires liming at rates of at least 2 t/ha to adjust the low soil pH (Goedert, 1983;1987).

In unopened savanna regions such as the Llanos or plains of Colombia, cultivation of crops is limited by deficiencies in infrastructure, such as roads (Table 5) and a lack of a local source of lime to ameliorate soil acidity. These areas remain as extensive low-input low-output ranching systems utilising mainly native savanna grasses. The latter have a very low nutritional quality for grazing animals due predominantly to the low reserves of plant available nutrients (Lascano, 1992; Fisher et al., 1992).

**Table 4.** Main constraints in terms of soil chemistry in Latin America

| Constraints | Area x $10^6$ ha | % of total area |
|---|---|---|
| Low nutrient status | 941 | 43 |
| Aluminum toxicity | 821 | 38 |
| High P fixation by iron oxides | 615 | 28 |
| Acidity without Al toxicity | 313 | 14 |
| Low CEC | 118 | 5 |
| High P fixation by allophane | 44 | 2 |
| Others (e.g. salinity, alkalinity) | 246 | 10 |
| Total area | 2172 | |

(Data adapted from Sanchez and Logan, 1992.)

**Table 5.** Road length density in savanna regions

| Region | Paved roads (km/100 $km^2$) |
|---|---|
| Venezuelan llanos | 50.9 |
| Brazilian cerrados | 5.7 |
| Colombian llanos | 0.1 |

(Data from Vera and Seré, 1985.)

# IV. Approaches to Alleviating the Constraints

## A. Selection of Acid-Tolerant Germplasm

### 1. Forage Grasses and Legumes

For areas of acid infertile soils of the lowland tropics, which have little prospect of utilizing even moderate levels of inputs of lime and fertilizer, the Tropical Forage Program of CIAT adopted the strategy of selecting acid-tolerant forage grass and legume species and developing low-input management techniques to establish the selected materials. Using a regional germplasm evaluation approach, a portfolio of acid-tolerant germplasm options for the major ecosystems of tropical America now exists (Toledo et al., 1989; Miles and Lapointe, 1992).

The results of this technology in terms of animal live weight gains (LWG) have been spectacular especially with a legume in the pasture. For example, in the Colombian Llanos grass/legume pastures have more than doubled animal LWG and shown a 10-fold increase in productivity per ha compared with a managed native savanna (Figure 1). The grass/legume pasture increased LWG per head by 50% and LWG/ha increased by 20-30% compared with the grass only pasture. Impressive improvements in reproductive performance and milk production have also been documented (Thomas et al., 1992).

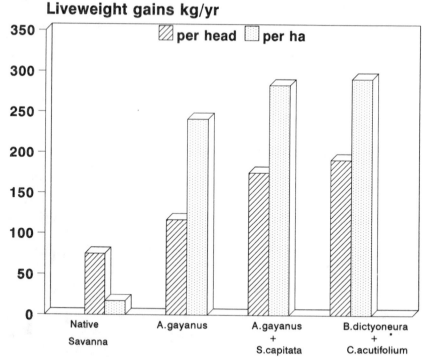

**Figure 1.** Productivity of savanna and improved pastures on an Oxisol. (Data from CIAT, 1990.)

## 2. Upland Rice

CIAT's rice program has developed lines of upland rice that are adapted to savanna acid soils with aluminum saturation levels > 75% (Sarkarung and Zeigler, 1990). These lines are tolerant to moderate drought stress and prevalent diseases such as rice blast and the "hoja blanca" virus. The yield potentials are high, in the order of 3.5-5.0 t/ha and lodging problems are considerably reduced compared with other upland varieties for the savannas (Zeigler et al., 1993; Vera et al., 1993).

## B. The Role of Legumes — Long-Term Effects

The fixation of nitrogen by forage legumes is a key component of the low-input technology for tropical pastures as grasses are usually N-deficient (Toledo and Nores, 1986). The input of N via biological fixation improves the nutritive value of the forage (protein and minerals) and hence animal production. In addition, nutrient cycling is thought to be improved via increased litter quality and a faster transformation of nutrients into plant-available forms via animal ingestion and excretion (Floate, 1981). However, for the Latin American tropics there is little evidence for this or indeed for the expected build up of organic matter under legumes (Gethin Jones, 1942) or pastures (Tate, 1987).

A grazing experiment on an Oxisol in Colombia with grass and grass/legume pastures has been used to examine the beneficial long-term effects of a forage legume in terms of soil improvement and increased animal productivity.

The pastures were sown in 1978 at Carimagua 4°30'N, 71°19'W, 150 masl. This area receives around 2200 mm rainfall annually, mainly during May through November with a distinct dry season between December and March. Soils are Oxisols (tropeptic haplustox isohyperthermic) with $pH(H_2O)$ 4.7-4.9, aluminum saturation > 84%, 3 cmol(+)/kg Al, and low amounts of available nutrients in the order of 0.1 cmol/kg Ca and Mg; 0.1 cmol/kg K and P (Bray II).

The grass *Brachiaria decumbens cv. Basilisk* (CIAT 606) was sown in 1978 alone or with 6 m strips of *Pueraria phaseoloides* CIAT 9900 (kudzu) in duplicated 2 ha paddocks (Tergas et al, 1984). Rates of fertilization at establishment were (kg/ha), 33 P on the grass and 44 P, 40 K, 14 Mg and 22 S on the strips of legume. In 1979, the strips received 18 K, 10 Mg and 22 S as a maintenance fertilizer and thereafter all pasture treatments received 10 P, 9 K, 92.5 Ca, 8 Mg and 11 S every two years up until 1987. After this date, fertilization of the pastures was discontinued.

Continuous grazing started in December 1978 and details of the stocking rates used are given by Lascano and Estrada (1989). Basically the stocking rates varied between 1 and 2 animals/ha depending on the season. Liveweight gains were measured every 56 days and forage availability and botanical composition 4 times per year using quadrat cuts.

## 1. Animal Production

These pastures have now persisted for 14 years and increases in animal production due to the legume *P. phaseoloides* were first observed in the rainy season after 3 years compared with the *B. decumbens* grass only pasture (Figure 2). Differences in animal LWG during the dry seasons however, were noted during the first year (Figure 3). The legume content in this pasture has fluctuated around a mean of 37% (Lascano and Estrada, 1989).

## 2. Soil Chemical Parameters

After 10 years the soil organic matter levels were highest in the 0-2 cm profiles of the grass and grass/legume pasture compared with the native savanna but there were little or no differences in soil organic nitrogen levels between the pastures (Table 6). In these soils the contribution of the legume carbon to soil organic matter was estimated by $^{13}C$ isotope measurements using the difference in natural levels of $^{13}C$ between the $C_4$ grasses and $C_3$ legumes (e.g. Cerri et al., 1991). The

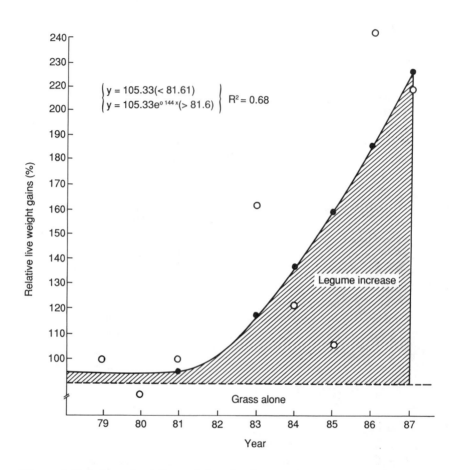

**Figure 2.** Relative animal live weight gains from B. decumbens/P. phaseoloides pastures compared with B. decumbens alone during the rainy season on an Oxisol in the Colombian llanos. ○ - measured values; ● - fitted regression line. (Data from Lascano and Estrada, 1989.)

results (Table 7) show that the legume contributed 29% of the total soil carbon in the top 2 cm of soil and this contribution decreased gradually down the soil profile (Rao et al., 1992).

## 3. Soil Physical Parameters

There were little or no significant differences between the treatments in terms of soil bulk density, resistance to penetration, water infiltration rates or total

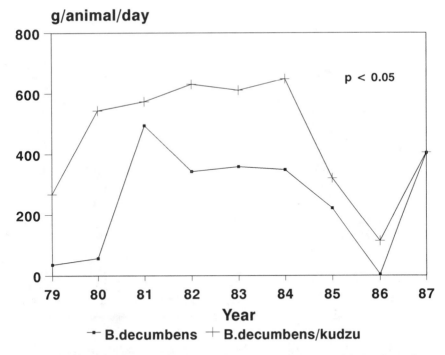

**Figure 3.** Animal live weight gains during the dry season with B. *decumbens* alone or in association with P. *phaseoloides*. Mean production values (1979-87) were significantly different between treatments at P<0.048.
(Data from Lascano and Estrada, 1989.)

**Table 6.** Soil OM and total N in 10-year old pastures compared with a native savanna

| Pasture | Depth (cm) | O.M. (%) | N (%) |
|---|---|---|---|
| Native savanna | 0-2 | 4.07 ± 0.13 | 0.160 ± 0.007 |
| | 2-10 | 3.50 ± 0.17 | 0.096 ± 0.007 |
| B. *decumbens* | 0-2 | 4.82 ± 0.27 | 0.189 ± 0.016 |
| | 2-10 | 3.56 ± 0.13 | 0.110 ± 0.008 |
| B. *decumbens*/ | 0-2 | 4.90 ± 0.22 | 0.164 ± 0.009 |
| P. *phaseoloides* | 2-10 | 3.62 ± 0.10 | 0.101 ± 0.006 |

(Data from Thomas unpublished, means of 10 samples ± S.E.

**Table 7.** $\delta$ [13]C values of soil organic matter under grass, legume and grass/legume pastures.

| Depth (cm) | $\delta$ [13]C |  |  |
|---|---|---|---|
|  | Pasture |  |  |
|  | B. decumbens | Kudzu | B. decumbens + kudzu |
| 0 - 2 | - 13.83 | - 18.68 | - 18.28 (29.4) |
| 2 - 4 | - 12.97 | - 17.02 | - 18.12 (20.8) |
| 4 - 60 | - 11.81 | - 12.51 | - 12.39 (3.8) |

Numbers in brackets represent the % of soil C derived from the legume in the grass/legume pasture and is calculated from:

$\delta$ [13]C of soil under grass/legume - $\delta$ [13]C of soil under grass x 100

$\delta$ [13]C of legume tissue - $\delta$ [13]C value of grass tissue

(Data from Rao et al., 1992.)

porosity (Ayarza unpublished). However there was a better distribution of soil aggregates under the grass and especially the grass/legume pasture compared with the undisturbed savanna (Figure 4). Over 33% of the aggregates from the grass/legume pasture were larger than 0.5mm. For the grass only pasture this value was 25% while only 10% of the savanna soil aggregates were greater than 0.5mm.

## 4. Soil Fertility

N mineralization: The qualitative differences in soil organic matter and soil physical properties are accompanied by differences in potential rates of N mineralization between soil from the grass only and grass/legume pastures as measured by two incubation methods (Table 8). Potential rates of mineralization were consistently higher with soil from grass/legume pastures compared with grass only pastures irrespective of the methods used.

Yields of a subsequent upland rice crop: In 1989 one half of the experiment described above was sown to one of the new acid-tolerant upland rice varieties (CIAT line 3) with differing levels of fertilization (Sanz et al., 1993). The experiment was repeated with a second rice crop in 1990. The benefit of the legume to the rice crop can be seen in the data from the treatment receiving 25 kg/ha P but no N (Table 9). For the first rice crop an extra 1.7 t rice/ha was obtained after the grass/legume pasture compared with the grass only pasture representing about 28 kg N/ha or approximately the equivalent of 86 kg N fertilizer/ha (assuming a nitrogen harvest index of 0.65 and a fertilizer recovery rate of 50%). For the second rice crop the yield advantage from the grass/legume pasture was 0.8 t/ha.

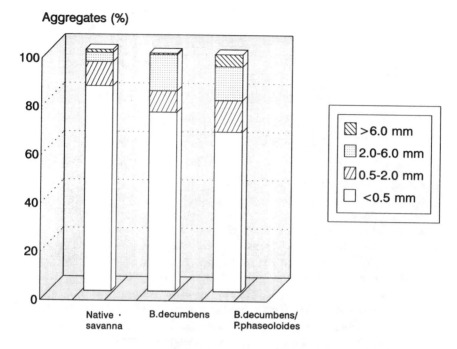

**Figure 4.** Aggregate size distribution of soil particles in the 0-10 cm layer from a clay loam Oxisol under a long term B. *decumbens* pasture with and without P. *phaseoloides compared with the native savanna.*
(From Ayarza, unpublished.)

**Table 8.** Potential N mineralization rates of soil under B. *decumbens* or B. *decumbens* + kudzu pasture measured by two methods

|  | μg N g soil$^{-1}$ day$^{-1}$ | |
| --- | --- | --- |
|  | B. decumbens | B. decumbens + kudzu |
| Incubation in pots in glasshouse for 4 weeks | 0.519 ± 0.085 | 0.913 ± 0.116 |
| Anaerobic incubation at 40°C for 7 days | 3.86 ± 1.33 | 6.84 ± 1.69 |

Data from Thomas (unpublished), means of 15 samples ± S.E.
Soil depth 0-20 cm.

**Table 9.** Yields of a rice crop sown after a 10-year-old grass or grass/legume pasture

| | tonnes rice ha$^{-1}$ | |
| Year | B. decumbens | B. decumbens + kudzu |
| --- | --- | --- |
| 1989 | 1.36 ± 0.17 | 3.07 ± 0.26 |
| 1990 | 1.40 ± 0.21 | 2.22 ± 0.21 |

The rice crop received 0 N and 25 kg P ha$^{-1}$ ± S.E.
(From Sanz et al., 1993; Fisher unpublished.)


## 5. Period to Attain 80% of the Benefits

It is relevant to ask how long is it necessary to maintain a grass/legume pasture on an Oxisol in order to obtain the yield benefits noted above. One approach is to use a model of the expected levels of N mineralization from the grass/legume residues using data weighted for both grass and legume annual dry matter production and their respective rates of litter decomposition during the wet season (Thomas and Asakawa, 1993). During the dry season decomposition is about 2-3 times slower than it is during the wet season which is ignored in that approach. Annual pasture DM production was estimated to be about 15 t DM/ha/yr (212-day wet season) with animals consuming 40% of the pasture. This leaves a total of 9.0 t DM/ha/yr as above ground material which turns over each year (assuming production is at steady state on an annual basis). Then using a 37% legume content and concentrations of N in grass litter of 0.6% N and 1.91% N in legume litter (Lascano and Estrada, 1989; Thomas and Asakawa, 1993), we estimate a return to the soil of 63.6 kg legume N (3.33 t x 1.91% N) and 34.0 kg grass N (5.67 t x 0.6% N). For the contribution from roots we assume a shoot:root ratio of 1.4 (Rao unpublished) and a N concentration of 0.6% N giving 36 kg root N (6 t x O.6% N). Thus the total amount of N available for recycling via decomposition is about 133.6 kg N /ha (63.6+34+36). Decomposition of tropical forage grass and legume litter can be described by a single exponential decay model (Thomas and Asakawa, 1993). From this study the weighted mean decomposition rate constant for grass and legume N was around 0.0015 d$^{-1}$. Using this rate constant we can estimate the cumulative amounts of N available from mineralized plant residues for each year of a 10 year period. These data are presented in Figure 5. We estimate that after 2 years we would obtain around 80% of the benefits of a 10-year-old pasture in terms of N mineralized per year, and 90% after 3 years. The actual amounts of N available may vary depending on the validity of some of the assumptions made above, but the pattern of N availability over time is unlikely to differ greatly. Thus we might expect significant benefits in terms of N supply to a subsequent crop from a moderately grazed grass/legume pasture, with a legume content of 37%, after 3-4 years if year 1 is considered only for establishment. Changes in soil physical conditions, however, are likely to require longer periods. The

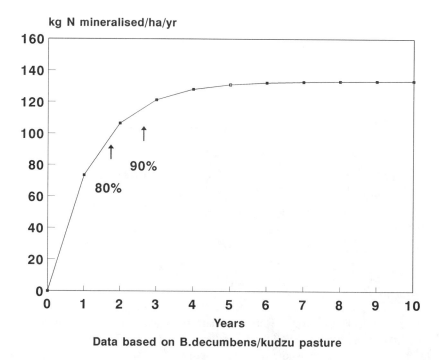

Data based on B.decumbens/kudzu pasture

**Figure 5.** Estimated kg N mineralized ha$^{-1}$ yr$^{-1}$ for a 10 year old B. *decumbens* / P. *phaseoloides* pasture on an oxisol.

expected input of N into the soil pools with varying legume contents and degrees of pasture utilization have been discussed elsewhere (Thomas, 1992).

## C. The Use of Inputs

### 1. Forage Grasses and Legumes

The establishment of acid-tolerant grasses and legumes requires moderate levels of fertilization (Sánchez and Salinas, 1981). Ranges in kg/ha for the major nutrients are: 10-20 P; 10-20 K; 50-400 Ca; 10-15 Mg. The legume component is expected to supply the N input although there is some evidence that small doses of starter N may aid the establishment of certain legumes e.g., *Arachis pintoi* (Thomas, 1993). These levels can be further reduced if the fertilizer is applied in bands e.g., 5 kg P/ha can be used instead of 20 kg (Ayarza and Spain, 1991). Where micronutrient deficiencies occur with Zn, Cu, and B, applications of between 1-3 kg element/ha are recommended (Ayarza, 1991).

**Table 10.** Range of nutrient contents per tonne of upland rice

| Nutrient | kg nutrient/ tonne rice | | | |
| | Rice variety[1] | | | |
| | CIAT 3 | CIAT 23 | IAC 47 | IR8 |
|---|---|---|---|---|
| P | 2-5 | 2 | 10-15 | 5 |
| K | 10-26 | 13-22 | 58-66 | 36 |
| Ca | 0.2-3 | 2.3 | 16-19 | 3 |
| Mg | 1.5-6 | 1.5-3 | 10-13 | 4 |
| S | 0.9-3 | 1.2 | 6-20 | 2 |
| Zn | 0.004-0.110 | 0.004-0.10 | 0.1-0.15 | 0.04 |

[1]CIAT lines are acid-tolerant upland rice varieties.
[1]IAC47 is a Brazilian variety.
[1]IR8 is an irrigated rice variety.
(Data from J.I. Sanz, unpublished.)

## 2. Upland Rice

The new CIAT upland rice varieties planted in the plains of Colombia are currently receiving in kg/ha, 300 dolomitic lime, 50 P, 100 K, 80 N, and 5 Zn. Of major importance is the fact that some of these lines e.g., CIAT 23 (Oryzicasabana 6), do not require the amounts of liming needed to raise the soil pH and they are much less nutrient demanding than either an irrigated rice variety or the other upland varieties currently in use in the region (Table 10).

The low input technology described by Toledo and Nores (1986) was aimed at production from very marginal lands but as these soils are improved under grass/legume pastures and crops are grown, possibilities will arise for intensification of this integrated crop-pasture system. However, from a nutrient balance standpoint, it is obvious that as more nutrients are removed in crop products and as the nutrient cycles become more "leaky" as a result of this intensification, there will be a need to replace the losses from the system. Fertilizers supplemented with other on-farm resources such as crop residues, animal and green manures will then become increasingly important.

It is also evident that although the technologies developed for the region are called "low-input," as pointed out by Sánchez and Salinas (1981), this does not imply the elimination of fertilizer. The aim is to maximize the output per unit of fertilizer added and minimize nutrient losses from the soil-plant system by improving nutrient cycling. Phosphorus is probably the nutrient requiring most attention as the acid-tolerant pasture and rice germplasm cannot be established without additions of purchased phosphorus and there are insufficient amounts available in the unamended soils to improve recycling. Improvements in the efficiency of fertilizer use are possible as it has been demonstrated, for example, in long-term experiments on the Brazilian Cerrados, that the recovery of phosphorus fertilizer is high even with high phosphorus fixing soils, and this, combined with the residual value of the fertilizer, makes phosphorus fertilization economically worthwhile (Goedert, 1983). Further, Le Mare et al. (1987) have shown with Cerrado soil, that phosphorus fertilizer could be used more efficiently when combined with green manures or crop residues. More studies of this sort are needed to improve fertilizer efficiency.

Another important aspect of the use of inputs relates to the persistence of the legume component, which can be a problem in grazed pastures. As discussed by Sprent and Raven (1985), if the major soil nutrients other than N are limiting, then, from an evolutionary viewpoint, it is better that legumes invest energy and photosynthate into producing a greater volume of root tissue than nodule tissue so that they can scavenge the soil for nutrients including N. Legumes are often quoted as requiring greater amounts of nutrients, such as phosphorus, than grasses but this remains controversial and unproven. In addition, some tropical forage legumes require less phosphorus than certain grasses (Sánchez and Salinas, 1981). Nevertheless, if we can extend Sprent and Raven's evolutionary perspective to legume persistence, then the legume-based systems will have greater chances of success if all nutrient deficiencies other than N are alleviated. This will also increase the competitiveness of the legumes in grass/legume pastures. In acid soils, external inputs and the rapid recycling of the limiting nutrients will play a major role in this scenario.

The use of acid-tolerant germplasm which are less nutrient demanding than unadapted varieties, with improved nutrient use efficiency and the strategic use of external inputs are key factors in the successful management of the savanna acid soils for increased production. Much needs to be done to improve our knowledge of this area and especially, the integration of fertilizer inputs with other on-farm resources including crop residues, animal and green manures.

## V. Integrated Rice-Pasture Systems

Since 1989 the technologies associated with the use of acid-tolerant forage and upland rice germplasm has been combined and tested in rice-pasture systems (Vera et al., 1993; Sanz et al. 1993; Zeigler et al., 1993). This has usually taken the form of the simultaneous sowing of forage grasses and legumes with rice to establish an improved grass/legume pasture with a pioneer crop of rice in (1), previously unopened areas of native savanna and (2), already opened savannas with degraded pastures. The latter has been used successfully on limed soils or soils with pH(H$_2$O) 5.0-5.9 in Brazil since 1987 in a system known locally as the "sistema Barreirâo" where upland rice is sown with *Brachiaria spp.* but without forage legumes (Kluthcouski et al., 1991). A third possible use is the rotation of cereals and pastures in a classic legume-based ley. In all of these systems the rice crop is harvested between 105-120 days, the pastures establish much faster than the traditional low-input technology by utilizing the residual fertilizer not removed by the rice crop. The improved grass/legume pastures are then ready for grazing after the rice harvest, whereas, traditionally the pasture requires one year to establish. The system is more efficient in land preparation and fertilization and reduces erosion and nutrient leaching by establishing a ground cover faster and more completely than the separate establishment of rice or pasture. In addition, the overall nutrient status of the acid soils is improved as a result of the application of fertilizers (Vera et al., 1993) in a manner similar to that reported for the humid tropics (Sánchez, 1987).

Further, as we reported above (Table 8), when rice is grown after a grass/legume pasture, increased yields can be expected compared with cropping after a grass only or native savanna pasture. Thus there is the potential for intensifying the system in repeated pasture/crop sequences.

The rice/pasture systems are proving to be successful in the Llanos of Colombia where the break-even yield for the cost of the rice/pasture system is around 1.6 t/ha of rice. Yields in excess of 2 t/ha have been obtained in on-farm trials over 3 consecutive years (Vera et al., 1993). In 1992 between 5-6,000 ha were sown to either rice or rice/pastures. Prior to 1990 these areas were used exclusively for extensive ranching. In Brazil it has been estimated that about 50,000 ha of degraded pastures have been renovated by sowing rice/pastures using the "sistema Barreirâo" (Kluthcouski pers. comm.). Thus the systems are both agronomically feasible and economically attractive (Rivas et al., 1991; Sanint et al., 1990).

In financial terms the rice crop dramatically improves the cash flow of pasture improvement by providing substantial revenues within the year of establishment (Sanint et al., 1990). This improvement in cash flow can allow further cropping and reduce the number of years a pasture is used. Thus the problem of pasture management and particularly legume persistence, becomes less critical as the pasture can be converted to crops with or without undersowing with improved grass and legume species. The system will, however, require crop production expertise either from the grazier or via contract arrangements. The latter is already occurring in the Colombian Llanos.

## A. Potential of Rice/Pasture Systems

In Brazil where upland rice has been a traditional pioneer crop it has been usual to use large amounts of lime (at least 2 t/ha) and the traditional rice lines have been prone to lodging where N supply is also high (Teixeira et al., 1989). The new acid-tolerant lines require less lime (currently 300 kg/ha are used) and are less prone to lodging. The root systems are more able to withstand stresses from aluminum and water compared with the traditional lines. Yields of 3-4 t/ha are possible compared with average yields of 1.2 t/ha in the Cerrado. In addition they attract prices equal to those of the irrigated sector.

It has been estimated that by the year 2025 there is likely to be 50 million ha of improved pasture in the Cerrado (currently there are 30 m ha). If half of the area is renovated every 6 years with rice/pasture, an additional 4.2 m ha of rice would be established yielding 7.6 million tons of rice (at 1.8 t/ha). This would meet half of the additional rice needs for Brazil by 2025 (Ernstberger, 1989).

During the 1980's Brazil's economic difficulties resulted in a drastic reduction in subsidies to agriculture in the Cerrados (Sanint et al., 1992). As mentioned earlier there is a trend away from cropping to pasture in some of this area, which will require a more input-efficient system to reverse. The rice/pasture system is an example of such an improved system, but it is recognised that policies will undoubtedly play a key role in the adoption of such systems. In particular the concept of ley farming in the tropics, i.e., rotating rice and other grain crops with pastures, is relatively new and will need more information on opportunity costs and policy requirements as well as further evidence for its advantages over alternative land uses (Saleem and Fisher, 1993).

Although yields of monocropped rice using the new CIAT lines have been maintained for 3 successive years in Colombia, it is not known how long monocropping can be sustained on the Oxisols of the Llanos with the present technology. It is possible that the problems of soil compaction and erosion encountered in the Brazilian Cerrados will also occur in the Llanos.

## VI. Future Need for Long-Term Research

Although the successful use of grass/legume pastures has been demonstrated for enhanced animal and subsequent crop production on an Oxisol, there remains the need to repeat these findings with other environments and soil types and with different crop/pasture sequences. Lack of legume persistence remains an unsolved problem in many environments but this aspect assumes lesser importance if pastures can be renovated economically with the rice-pasture system described above. An alternative currently under study is the sowing of a cocktail of legumes combining rapid establishment of one legume (e.g., *Stylosanthes spp.* or *Centrosema spp.)* with the longer-term persistence of a legume such as *Arachis pintoi.*

Studies of factors affecting legume persistence and pasture degradation are ongoing in a 5 year experiment at the ICA-CIAT experimental station in Colombia (Fisher et al., 1993). The testing of crop/pasture rotations with combinations of rice, maize, soybeans, sorghum, green manures and grass/legume pastures is currently underway both in Colombia (ICA-CIAT) and in Brazil at EMBRAPA-CPAC, Planaltina. Both of these efforts involve long term rotation experiments using 5 year cycles. The main hypothesis being tested is that integrated crop/pasture systems are more efficient in the use of inputs and that they are more productive and sustainable than either continuous cropping or pasture alone. The role of tillage and other crop management practices currently ongoing in Brazil (e.g., Seguy et al., 1993), which mainly involve crops and green manures, are complementary to the activities described above.

By comparing native savanna with improved grass and grass/legume pastures with known differences in soil quality, soil chemical and physical factors are being identified which may serve as indices of sustainability for the soil resource base. These will need to be tested in other agroecosystems.

The on-station experiments are planned for a minimum of two rotation cycles and are being complemented with on-farm monitoring of similar systems together with on-farm participatory research. This approach will expose the hypothesis and technologies mentioned above to the rigors of actual farm conditions and will provide an historical perspective of land use in existing integrated crop-pasture systems.

## VII. Conclusions

1 - Data from a long-term grazing experiment confirm the value of forage grasses, and especially legumes, for the improvement of both soil quality and animal productivity with only minimum inputs of maintenance fertilization.

2 - On acid soils, the improved soil conditions can result in increased yields of a subsequent acid-tolerant rice crop. The combining of rice and pastures in pasture-crop rotations has been demonstrated to be agronomically and economically viable on acid soils that are representative of large areas of Latin American savannas. This technology is an example of enhanced productivity while contributing to the maintenance and/or improvement of soil quality. The technology has the potential for a large impact with a minimum financial and ecological cost (Sanint et al., 1990).

The two key factors in the success of this technology are (1), acid-tolerant germplasm and (2), high levels of fertilizer use efficiency by both pasture and rice plants.

3 - To avoid problems associated with monocropping on savanna soils e.g., soil compaction and erosion, an option would be the establishment of pasture-crop rotations with a pasture phase of 3-4 years rather like a classical ley farming system. The likelihood of the adoption of ley farming systems will depend on technological factors such as the availability of a viable forage seed industry, fertilizers, improved extension activities and infrastructure. In addition, various policy requirements such as access to credit and subsidy incentives should be in place in order to stimulate interest from farmers (e.g. see Saleem and Fisher, 1993).

4 - There is now an urgent need to explore additional alternatives to rice/pasture, which should include a wider set of acid-tolerant germplasm (e.g., maize, sorghum, soybeans) in order to achieve sustainable crop/crop and crop/pasture systems.

5 - Although the research reported here has been exclusively concerned with acid soil savannas, the principle of the integrated use of grass/legume pastures with

crops for soil improvement and production increases at relatively low costs should be applicable to other agroecosystems in the region, including the forest margins and hillsides, where pastures are ubiquitous (CIAT, 1991). The technology can increase production from existing agricultural lands thereby relieving pressure on areas such as the tropical rain forests and it can also be used to recuperate degraded lands. A key factor is the ability to establish, at low or economical cost, productive grass/legume pastures as a means of overcoming the problems of decreasing fallow periods in cropping systems and the beginning of soil degradation.

## Acknowledgments

The data reported in this review are the result of multidisciplinary research activities conducted by the rice, forage, and savanna programs of CIAT together with counterparts in the national research programs of Colombia (ICA) and Brazil (EMBRAPA). We acknowledge these contributions and thank our colleagues for their permission to use unpublished data.

## References

Ayarza, M.A. 1991. Efecto de las propiedades químicas de los suelos ácidos en el establecimiento de las especies forrajeras. p. 161-185. In: C.E. Lascano and J.M. Spain (eds.), *Establecimiento y renovación de pasturas*. CIAT, Cali, Colombia.

Ayarza, M.A. and J.M. Spain. 1991. Manejo del ambiente físico y químico en el establecimiento de pasturas mejoradas. p. 189-208. In: C.E. Lascano and J.M. Spain (eds.), *Establecimiento y Renovación de Pasturas*. CIAT, Cali, Colombia.

Bentley, C.F. 1991. Soil management research in the search for sustainable agriculture. p. 167-173. In: R. Lal, and F.J. Pierce (eds.), *Soil Management for Sustainability*. Soil and Water Conservation Society, Ankeny, Iowa.

Cerri, C.C., B.P. Eduardo, and M.C. Piccolo. 1991. Use of stable isotopes in soil organic matter studies. p. 247-259. In: *Stable Isotopes in Plant Nutrition, Soil Fertility and Environmental Studies*, IAEA, Vienna.

CIAT. 1990. *CIAT Tropical Pastures Annual report 1989*. Working Document No. 70. CIAT, Cali, Colombia.

CIAT, 1991. *CIAT in the 1990s and beyond: A Strategic Plan, Part 2 (supplement)*, CIAT, Cali, Colombia. pp.125.

CIAT. 1992. *Trends in CIAT Commodities 1992*. Working document No. 111. CIAT, Cali, Colombia.

Cole, M.M. 1986. *The Savannas Biography and Geobotany*. Academic Press, London. 438 pp.

Edwards, C.A., R. Lal, P. Madden, R.H. Miller, and G. House (eds.) 1990. *Sustainable Agricultural Systems*. Soil and Water Conservation Society, Ankeny, Iowa, 696 pp.

EMPA, 1987. *Diagnostico e prioridades para a agropecuaria do Estado de Mato Groso*. Empresa de Pesquisa Agropecuaria do Estado de Mato Groso, Cuiaba, MT, Brazil. 147 pp.

Ernstberger, J. 1989. Study on the social benefits of rice research in Brazil. GTZ report, Weikenstephan Project No. 857860.101.100. 229 pp.

Eswaran, H., J. Kimble, T. Cook, and F.H. Beinroth. 1992. Soil diversity in the tropics: implications for agricultural development. p. 1-16. In: R. Lal and P.A. Sanchez (eds.), *Myths and Science of Soils of the Tropics*. SSSA Spec. Publ. No. 29. SSSA, Madison, Wisconsin.

Fenster, W. and L. Leon. 1979. Management of phosphorus fertilization in establishing and maintaining improved pastures on acid, infertile soils of tropical America. p. 109-123. In: P.A. Sanchez and L.E. Tergas (eds.), *Pasture Production in Acid Soils of the Tropics*. CIAT, Cali, Colombia.

Fisher, M.J., C.E. Lascano, R.R. Vera, and G. Rippstein. 1992. Integrating the native savanna resource with improved pastures. p. 75-99. In: *Pastures for the Tropical Lowlands: CIAT's contribution*. CIAT, Cali, Colombia.

Fisher, M.J., C.E. Lascano, R.J. Thomas, M.A. Ayarza, I.M. Rao, G. Rippstein, and J.H.M. Thornley. 1993. An integrated approach to soil-plant-animal interactions on grazed legume-based pastures on tropical acid soils. In: *Proc. XVII Intl. Grass. Cong. New Zealand/Australia*, Feb 8-21, 1993. in press.

Floate, M.J.S. 1981. Effects of grazing by large herbivores on nitrogen cycling in agricultural ecosystems. p. 585-601. In: F.E. Clark and T. Rosswall (eds.), *Terrestrial Nitrogen Cycles*. Ecol. Bull. No. 33. Swedish Natural Science Research Council, Sweden.

Gethin Jones, G.H. 1942. The effect of a leguminous cover crop in building up soil fertility. *E. Afric. Agric. J.* 8:48-52.

Goedert, W.J. 1983. Management of the Cerrado soils of Brazil: a review. *J. Soil Sci.* 34:405-428.

Goedert, W.J. 1987. Management of acid tropical soils in the savannas of South America. p. 109-127. In: P.A. Sanchez, E.R. Stoner, and E. Pushparajah (eds.), *Management of acid tropical soils for sustainable agriculture: proc. of an IBSRAM inaugural workshop*. IBSRAM, Bangkok, Thailand.

Harwood, R.R. 1990. A history of sustainable agriculture. p. 3-19. In: C.A. Edwards, R. Lal, P. Madden, R.H. Miller and G. House (eds.) *Sustainable Agricultural Systems*. Soil and Water Conservation Society, Ankeny, Iowa.

Huntley, B.J. and B.H. Walker. 1982. Introduction. p. 1-2, In: B.J. Huntley and B.H. Walker (eds.), *Ecology of Tropical Savannas*, Ecological Studies Series, Vol. 42. Springer-Verlag, Berlin.

Kluthcouski, J., A.R. Pacheco, S.M. Teixeira, and E.T. de Oliveira. 1991. *Renovacion de pastagens de cerrado com arroz: I. Sistema barreirão*. Documento 33, EMBRAPA-CNPAF, Goiania, Go. Brasil. 20pp.

Lascano, C.E. 1992. Managing the grazing resource for animal production in savannas of tropical America. *Trop. Grass.* 25:66-72.

Lascano, C.E. and Estrada, J. 1989. Long-term productivity of legume-based and pure grass pastures in the eastern plains of Colombia. p. 1179-1180. In: *Proc. XVI Int. Grass. Cong.*, Nice, France.

Le Mare, P.H., J. Pereira, and W.J. Goedert. 1987. Effects of green manure on isotopically exchangeable phosphate in a dark-red latosol in Brazil. *J. Soil Sci.* 38:199-209.

Miles, J.W. and S.L. Lapointe. 1992. Regional germplasm evaluation: A portfolio of germplasm options for the major ecosystems of tropical America. p. 9-28. In: *Pastures for the Tropical Lowlands: CIAT's contribution.* CIAT, Cali, Colombia.

Rao, I., M.A. Ayarza, R.J. Thomas, M.J. Fisher, J.I. Sanz, J.M. Spain, and C.E. Lascano. 1992. Soil-plant factors and processes affecting productivity in ley farming. p. 145-175. In: *Pastures for the Tropical Lowlands: CIAT's contribution.* CIAT, Cali, Colombia.

Rivas, L., J.M. Toledo, and L.R. Sanint. 1991. Potential impact of the use of pastures associated with crops in the tropical savannas of Latin America. p. 78-91. In: G. Henry, (ed.), *Trends in CIAT Commodities 1991*, CIAT Working Document No. 93. CIAT, Cali, Colombia.

Saleem, M.M.A. and M.J. Fisher. 1993. Role of ley farming in crop rotations in the tropics. *Proc XVII Inter. Grass. Cong. New Zealand/Australia*, Feb. 8-21, 1993. in press.

Salinas, J.G. and S.R. Saif. 1990. Nutritional requirements of *Andropogon gayanus.* p. 99-156. In: J.M. Toledo, R. Vera, and J.M. Lenné (eds.), *Andropogon gayanus Kunth: A Grass for Tropical Acid Soils.* CIAT, Cali, Colombia.

Sánchez, P.A. 1976. *Properties and management of soils in the tropics.* J. Wiley & Sons, NY. 618 pp.

Sánchez, P.A. 1987. Management of acid soils in the humid tropics of Latin America. p. 63-107. In: P.A. Sanchez, E.R. Stoner, and E. Pushparajah (eds.), *Management of acid tropical soils for sustainable agriculture: proceedings of an IBSRAM inaugural workshop.* IBSRAM, Bangkok, Thailand.

Sánchez, P.A. and J.G. Salinas. 1981. Low-input technology for managing Oxisols and Ultisols in tropical America. *Adv. Soil Sci.* 34: 279-406.

Sánchez, P.A. and T.J. Logan. 1992. Myths and science about the chemistry and fertility of soils in the tropics. p. 35-46. In: R. Lal and P.A. Sanchez (eds.), *Myths and Science of Soils of the Ttropics.* SSSA Spec. Publ. No. 29, SSSA, Madison, Wisconsin.

Sanint, L.R., L. Rivas, and C.O. Seré. 1990. Improved technologies for Latin America's new economic reality: Rice-pasture systems for the acid savannas. p.117-131. In: *Trends in CIAT Commodities 1990.* Working Document No. 74, CIAT, Cali, Colombia.

Sanint, L.R., C.O. Seré, L. Rivas, and A. Ramírez. 1992. The savannas of South America: Towards a research agenda for CIAT. p. 149-177. In: *Trends in CIAT Commodities 1992*. Working Document No. 111, CIAT, Cali, Colombia.

Sanz, J.I., R.S. Zeigler, S. Sarkarung, .D.L. Molina, M. Rivera. 1993. *Rice-Pasture Systems for Acid Soil Savannas in S. America*. in press.

Sarkarung, S. and R. Zeigler. 1990. Developing rice varieties for sustainable cropping systems for high rainfall acid upland soils of tropical America. Paper presented at the Rice Symposium for Acid Soils of the Tropics, Sri Lanka, 1990.

Seguy, L., S. Bouzinac, A. Pacheco, V. Carpenedo, and V. da Silva. 1988. *Perspectiva de fixacao da agricultura na regiao centro-norte do Mato Groso*. EMPA-MT, EMBRAPA-CNPAF, CIRAD-IRAT. 52 pp.

Seguy, L., S. Bouzinac, and C. Pieri. 1993. Tillage and crop management practices. In: *Evaluation for sustainable land management in the developing world. Proc. Intl. Workshop, Sept 15-20, 1992*, Chiang Rai, Thailand. in press.

Spain, J.M. 1993. Neotropical savannas: prospects for economically and ecologically sustainable crop-livestock production systems. In: *Manejo de los Recursos Naturales en Ecosistemas Tropicales para una Agricultura Sostenible*. Instituto Colombiano Agropecuaria, Bogota, Colombia. in press.

Sprent, J.I. and J.A. Raven. 1985. Evolution of nitrogen-fixing symbioses. In: *Proc. Roy. Soc. Edin.* 85B:215-237.

Stoner, E.R., E. de Freitas Jr., J. Macedo, A. Mendes, I.M. Cardoso, R.F. Amabile, R.B. Bryant, and D.J. Lathwell. 1991. Physical constraints to root growth in savanna oxisols. *TropSoils Bull.* No. 91-01. North Carolina State Univ. Raleigh, NC. 28pp.

Tate III, R.L. 1987. *Soil Organic Matter, Biological and Ecological Effects*. J. Wiley & Sons, NY. 291 pp.

Teixeira, S.M. and L.R. Sanint. 1988. Arroz de sequeiro. *Agroanalysis* 12:11.16.

Teixeira, S.M., L. Yokoyama, and L. Seguy. 1989. Technical change in agriculture: the impact of alternative agricultural systems for the central west Brazil. In: *Proc 10th Seminaire d'economie et sociologie*, CIRAD, Montpellier, France.

Toledo, J.M. and G.A. Nores. 1986. Tropical pasture technology for marginal lands of Tropical America. *Outlook Agri*. 15:2-9.

Toledo, J.M., J.M. Lenné, and R. Schultze-Kraft. 1989. Effective utilization of tropical pasture germplasm. p. 27-57. In: *Utilization of genetic resources: suitable approaches, agronomic evaluation and use*. FAO Plant production and protection paper No. 94, FAO, Rome, Italy.

Tergas, L.E., O. Paladines, I. Kleinheisterkamp, and J. Velásquez. 1984. Productividad de *Brachiaria decumbens* sola y con pastoreo complementario en *Pueraria phaseoloides* en los llanos orientales de Colombia. *Producción Animal Tropical* 9: 1-11.

Thomas, R.J. 1992. The role of the legume in the nitrogen cycle of productive and sustainable pastures. *Grass. For. Sci.* 47:133-142.

Thomas, R.J., C.E. Lascano, J.I. Sanz, M.A. Ara, J.M. Spain, R.R. Vera, and M.J. Fisher. 1992. The role of pastures in production systems. p. 121-144. In: *Pastures for the Tropical Lowlands: CIAT's Contribution.* CIAT, Cali, Colombia.

Thomas, R.J. 1993. Rhizobium requirements, nitrogen fixation and nutrient cycling. In: *The Biology and Agronomy of forage Arachis*, May 25-28, 1993, CIAT, Colombia. in press.

Thomas, R.J. and N.M. Asakawa. 1993. Decomposition of leaf litter from tropical forage grasses and legumes. *Soil Biol. Biochem.* in press.

Valerio, J.R. and O. Nakano. 1988. Danos causadas pelo adulto da Cigarrhina *Zulia entreriana* na producao e qualidade de *Brachiaria decumbens. Pesq. agropec. bras. Brasilia.* 23: 447-453.

Vera, R.R. and C.O. Seré (eds.) 1985. Sistemas de producción pecuaria extensiva; Brasil, Colombia, Venezuela. *Informe final del Proyecto ETES*, 1978-1982. CIAT, Cali. 538 pp.

Vera, R.R., R.J. Thomas, L.R. Sanint, and J.I. Sanz. 1993. Development of sustainable ley-farming systems for the acid-soil savannas of tropical America. *Symposium on the ecology and sustainable Agriculture in tropical biomes*, Rio de Janeiro, Brazil, 2-9 February, 1992. in press.

Villachica, H., J.E. Silva, J.R. Peres, and C. Magno, and C. da Rocha. 1990. Sustainable agricultural systems in the humid tropics of South America. p. 391-437. In: C.A. Edwards, R. Lal, P. Madden, R.H. Miller, and G. House (eds.), *Sustainable Agricultural Systems.* Soil and Water Conservation Society, Ankeny, Iowa.

Zeigler, R.S., J.M. Toledo, and J.I. Sanz. 1993. Developing sustainable agricultural production systems for the acid soil savannas of Latin America. In: *International symposium on agroecology and conservation in temperate and tropical regions.* Sept. 26-29, 1991, Padova, Italy. in press.

# Soil Alterations in Perennial Pasture and Agroforestry Systems in the Brazilian Amazon

E. Adilson Serrão, Leopoldo Brito Teixeira,
Raimundo Freire de Oliveria, and Joaquim Braga Bastos

## I. Introduction

The Brazilian Amazon has been, in the past two decades, particularly, in the past decade, the center of world attention due to present and potential ecological implications of man's utilization of natural resources for development purposes. In this context, agricultural development has been the most important factor of environmental disturbances in the past three decades. At least 40 million hectares of forest land (an area corresponding to the size of eight Costa Ricas) have been altered for development of land use systems that, in general, have shown low levels of sustainability from a biophysical (agrotechnical and environmental) and socioeconomic point of view.

The search for sustainable agricultural and forestry development in the Amazon is a great challenge for farmers and governmental and non-governmental institutions involved in the process. Within this context, sustainable manage-

ISBN 1-56670-076-0

ment of soil resources in the Amazon is a very important issue for agronomic, ecologic and socioeconomic considerations.

This paper presents research information on productivity and physical and chemical soil alterations in land use systems with present and potential importance for sustainable land development in the Amazon.

## II. The Basis for Agricultural Sustainability in the Amazon

Any proposal for agricultural and forestry development in the Amazon must take into account the need to promote sustainable land use systems. The possibility of developing a sustainable land use system in the region depends on its permanence in an area and on increasing land and labor productivity standards, thereby reducing the pressure for more deforestation. This concept of sustainability necessarily implies an equilibrium in time among agrotechnical, economic, ecological and social feasibility. This equilibrium is frequently fragile in Amazonian agricultural systems and no agricultural land use system in this region meets all these pre-requisites of sustainability at highly satisfactory levels (Serrão and Homma 1993).

## III. Land Use Systems and Their Present Sustainability in the Amazon

Serrão and Homma (1993) made an analysis of the sustainability (as defined above) of selected land use systems because of their present and potential importance due to scale of utilization, types of farmers involved, economic importance, possibilities for future markets, environmental implications, and technological and productivity patterns. In general, these authors suggest that agrotechnical, environmental, and socioeconomic sustainability equilibrium is low for most land use systems and that it is lower for those systems that involve forest conversion at varying levels (mainly ranching, shifting agriculture, upland perennial and semi-perennial crop production, and timber extraction) as compared to those developed on non-forest ecosystems (crop and cattle production on *varzea* floodplains and well- and poorly-drained savannas).

Of major importance, due to their environmental and socioeconomic implications, are those land use systems that demand deforestation for their development. In the frontier opening process in the region, and due to prevailing socioeconomic environment, these land use systems have involved large-scale slash-and-burn activities. Typically, land use intensity decreases from road side areas to areas further into the forest. Negative environmental implications (in terms of carbon emission and loss of biodiversity, biomass, soil, water, nutrients and fire resistence) increase with extensive agriculture activities.

## IV. The Need to Increase Sustainability of Agricultural Activities on Already Altered/Deforested Lands

More than enough land has already been deforested for unsustainable agricultural development in the Amazon. From an agrotechnical point of view, it is possible to produce sufficient amounts of food and fiber to meet the demand of the region's population for the next couple of decades by properly utilizing (increasing sustainability) already altered/deforested lands with no further need to use primary forests, except for selective timber extraction and extraction of non-timber products.

On those lands, agricultural land use systems must have higher levels of sustainability than the unsustainable systems that have been practiced, especially in the past 30 years. Alternative sustainable agricultural land use systems will require efficiency of resources (soil nutrients, biomass, biodiversity, genetic resources, climate) whether they are monoculture, polyculture, or integrated systems.

## V. The Importance of Soil Resource Management for Securing Agricultural Sustainability in Deforested Lands in the Amazon

Agrotechnical sustainability—the capacity of a land use system to maintain its productivity in the same area for as long a period of time as possible—depends on prevailing climatic, biotic (pests, deseases, weeds, etc.), and edaphic (chemical, physical, and biological) conditions. Climatic and biotic factors are of the utmost importance, but they are not within the scope of this paper.

Soil resource management is critical for agricultural development in deforested land in the Amazon. On the one hand, most of the area that has been deforested for agricultural development has soils that are acid and nutrient-deficient, although adequate from the point of view of their physical features. On the other hand, the socioeconomic environment in the region does not allow for intensive fertilizer and liming utilization in agriculture. Therefore, soil resources conservation, especially soil nutrient conservation, is essential for sustainable agricultural development from the point of view of agrotechnical and economic sustainability. For example, it has been estimated (Uhl et al., in press; Serrão, in press) that, in the process of deforestation for pasture formation and during the pasture-use period, a significant portion of the nutrients embodied in the slashed forest biomass is leached from the ecosystem. This loss is valued at close to US$3,500 per hectare, given current prices of NPK fertilizers in Brazil. This does not mean to suggest that this much fertilizer would be necessary to restore soil fertility in degraded pastures. Nor does it imply that these pastures are closed ecosystems in regard to nutrients. Clearly, nutrients enter those ecosystems via atmospheric deposition and weathering. The point is that measurable quantities of nutrients are lost in the conversion of forest to pasture, (there is a similar situation in shifting agriculture) and it is necessary to consider about how to estimate the cost of these losses and search for and develop alternative agricultural production systems that will minimize them and thus increase agrotechnical and economic sustainability with the consequent positive effect on social and ecological sustainability.

From the above considerations, it becomes clear that it is absolutely necessary to understand the soil processes and modifications which lead to soil improvement or degradation due to land utilization for agricultural and forestry development activities in the Amazon.

## VI. Long-Term Soil Management Experiments in the Amazon

Very few research attempts have been made to evaluate chemical and physical soil modifications under agricultural and forestry activities in the Amazon. This is explained mainly due to the fact that the majority of the region's most important transformations in the agricultural production sector started in the 1960s, with an aggressive expansion of the agricultural frontier which has been characterized by accelerated large-scale and aggressive exploration of natural resources. This type of development began to raise concern within the local and foreign research community in searching for ways to evaluate soil-plant and soil-plant-animal relations in the region. The following are probably the few most relevant research attempts to evaluate physical and chemical soil modifications under prevailing land use systems in the Amazon. These research attempts have been made mainly in two types of land use systems which demand deforestation for their development: large-scale extensive pasture-based ranching and small-scale agroforestry systems.

### A. Soil Transformation in Large-Scale Pasture Experiments Systems

It is estimated that at least 60% of the present total deforestation (around 40 million hectares) in the Amazon have been for cattle-raising activities, most of which has occurred in the last three decades.

For a better understanding of this segment, a few definitions and explanations are needed.

## 1. Definitions

lst-cycle pastures (or pioneer pastures): those formed from the moment the primary forest is cleared.

2nd-cycle pastures (or 2nd generation pastures): those resulting from the recuperation (or renovation) of degraded pioneer pastures.

Degraded pastures: those in which the biomass of the weed community predominates in relation to that of sown forage plants.

## 2. Explanations

Typically, lst-cycle pastures (presently still the majority of planted pastures in the region) have been formed by sowing soil nutrient-demanding pioneer grasses, such as "guineagrass" (*Panicum maximum*), which has been the most common species, after the clearing and burning of the forest. In general, during the first 3 or 4 years after establishment, productivity in these pioneer pastures is relatively high. However, after this period, there is a gradual (but fairly rapid) decline in the productivity of the planted grass, which is accompanied by an increasing presence of weeds, the pasture reaching advanced stages of degradation on the average after 5 to 7 years after establishment. This pasture-life period varies with prevailing environmental and management factors. More recently, productivity in lst-cycle pastures has improved with the planting of less soil-fertility-demanding grasses such as *Brachiaria humidicola, Andropogon gayanus* and *Brachiaria brizantha.*

When lst-cycle pastures reach degradation, ranchers tend either to abandon degraded pastures to fallow for an intermediate period or re-establish new pasture systems—the 2nd-cycle pastures. Establishment of 2nd-cycle pastures involves more intensive technological practices, such as mechanization for land preparation and planting and fertilization (especially P fertilization, as will be seen later in this paper) for better pasture establishment. Details of pasture establishment, degradation, and reclamation in the Amazon can be found in Serrão et al. (1979), Serrão and Toledo (1982), Serrão and Toledo (1992), Serrão and Dias Filho (1991).

## 3. "Long-Term" Pasture Experiments

There are no long-term experiments (with at least 10-year duration) to evaluate soil changes under pasture in the Amazon. However, two relevant research attempts have been made with important information resulting from them. Two studies were carried out simultaneously in northern State of Mato Grosso and eastern State of Pará between 1968 and 1973 in extensive commercial lst-cycle pastures of "guineagrass" (Falesi, 1976; Baena, 1978; Serrão et al., 1979). These studies had the objective of "accompanying" and comparing physical and chemical soil changes in pastures of different ages (years after slashing and burning of the forest biomass and sowing of guineagrass) with soil conditions in the adjacent unexploited primary forest. This was done by selecting two private ranches which had pastures with ages varying from 1 to 11 years (in northern Mato Grosso) and 1 to 10 years (in Eastern Pará) after establishment.

The eastern Pará site soil is an Oxisol (a medium-texture Yellow Latosol) area, originally covered by dense rain forest, the climate being a transition of Am to Aw climatic type of the Koppen climate classification.

The northern Mato Grosso site soil is also an Oxisol (a medium-texture Dark Yellow Latosol) area originally covered by open forest with the predominant climate being the Aw type of the Koppen classification. For each age, a representative average pasture (in terms of management history) was chosen to be sampled for determining the soil physical and chemical attributes from the soil surface to a 20-cm depth.

In this traditional process of pasture establishment and utilization in the region, after the clearing and burning of the forest, substantial amounts of nutrients are incorporated in the soil through ash deposition, significantly increasing soil fertility, raising pH in at least one unit and practically neutralizing soil Al. Most importantly, these results indicate that with time, in those soils under guineagrass pastures, nutrients such as Ca and Mg are maintained at fairly satisfactory levels, pH values are maintained between 5.5 and 6.5 and Al is kept practically neutralized. As a consequence, BS seldom is smaller than 50% and Al is pratically nil. K is maintained fairly stable at fairly satisfactory levels for maintaining pasture productivity. A similar trend can be observed for OM and N, in spite of periodic pasture burning, a typical 1st-cycle pasture management practice in the region.

Phosphorus, in its available form, increases considerably in the soil right after burning of the slashed forest to levels sufficient to propitiate fairly high pasture productivity during the first 4 to 5 years. After these first years, available soil P declines rapidly to levels which are incompatible with pasture growth as can frequently be seen in degraded commercial 10-year-old, 1st-cycle pastures.

Although inadequate grazing systems and grazing pressures undoubtedly contribute to a more or less rapid process of 1st-cycle pasture degradation in the region, these results indicate that P is the most limiting soil nutrient for maintaining pasture productivity in forest-replacing pastures in the Amazon. On the other hand, it has been observed by the authors of this paper that pasture productivity decline tends to be faster and more accentuated in soils with higher clay content, which is aggravated by low levels of soil P and high grazing pressures.

Although this experiment cannot be referred to as a true long-term soil resource management experiment and did not include measurement of pasture productivity, its results have become very valuable for a better understanding of the trends in soil nutrients under extensive pasture development in the region.

Following these findings, a series of small-plot and grazing experiments in the region have confirmed the importance of P as the main limiting nutrient for increasing pasture agrotechnical sustainability in the region (Serrão et al., 1979). In the last decade, at least 1.5 million hectares of degraded 1st-cycle pastures have been reclaimed into more sustainable 2nd-cycle pastures. In this process, P fertilization has been imperative. Abandoned degraded pastures seem to be able to conserve most of the other nutrients at relatively satisfactory levels, imitating, in some measure, the forest nutrient-conserving mechanisms which allow for satisfactory regeneration of a forest ecosystem which may be similar to the primary forest ecosystem (Bushbacher et al., 1988; Uhl et al., 1988).

A similar study was conducted, at the Distrito Agropecuário, on the Br-174 highway, about 50km east of Manaus, in Central Amazon (Teixeira, 1977; Teixeira & Bastos, 1989). In this region the area of forest transformed into pasture is small when compared with the areas in the State of Pará where the experiments described above were conducted.

In this case, the experimental area was of a typical primary forest ecosystem. The predominant soil in the region of Manaus is clayey Oxisol (Yellow Latosol), acidic and of low fertility. The climate is of the Am type in the Koppen classification system. In this experiment a forest area and five adjacent *Brachiaria humidicola* pastures, of one, two, six, seven, and eight years of age were studied.

The objective of the study was to identify the alterations in the soil in the pastures of different ages in relation to the soil of the primary forest. The pastures were established following the usual procedure of cutting (May 1977) and burning (October 1977) of the forest biomass and planting of *B. humidicola* (March 1978).

Table 1 shows the parameters of fertility of the soil under forest and pastures. The burning of the forest biomass, substantially increased, in the layer from 0-10cm, the total base and base saturation values. The soil pH also increased. On the other hand, the values for exchange capacity were not altered. The values of exchangeable aluminum diminished from levels considered toxic to less harmful levels, and the values for aluminum saturation had a very accentuated initial reduction, stabilizing in the soil under pasture at about 30%.

The increase in total base and base saturation, as well as in soil pH, occurred due to the increase in exchange bases in the soil. The cation exchange capacity (CEC) value found in the soil under pasture is similar to that determined for the soil under forest. According to Bittencourt (1977), the CEC in the soil of the Manaus region, is due mainly to the organic part of the clay fraction.

The exchangeable aluminum of the soil was reduced as a function of the increase in the pH due to the addition of existing bases in the ash from the burned forest biomass and the polimerizing effect of heat on the soil aluminum. Similar information was obtained by Falesi (1976) and Baena (1978).

The P values in the soil under pasture of various ages were similar to those encountered in the forest soil.

**Table 1.** Average values of total exchange bases (S), cation exchange capacity (CAC), aluminum (Al), base saturation (V), aluminum saturation (Sat. Al), pH, and phosphorus (P) in soils (0-10 cm depth) of primary forest and *Brachiaria humidicola* pastures of various ages in a cleye Oxisol in Central Amazon

| Ecosystem | S (meq/100 g) | CEC (meq/100 g) | Al (meq/100 g) | V (%) | Sat. Al (%) | pH ($H_2O$) | P (ppm) |
|---|---|---|---|---|---|---|---|
| Primary forest | 0.42b | 10.69a | 1.54a | 3.96b | 79a | 4.38b | 2.3a |
| Slashed and burned primary forest | 1.63a | 9.78a | 0.38b | 16.66a | 19b | 5.25 | – |
| 1-year-old pasture | 1.62a | 10.18a | 0.50b | 16.00a | 25b | 5.18a | 3.3a |
| 2-year-old pasture | 1.48a | a0.05a | 0.67b | 14.68a | 34b | 4.85ab | 2.0a |
| 6-year-old pasture | 1.39a | a0.13a | 0.55b | 13.78a | 31b | 4.97ab | 2.8a |
| 7-year-old pasture | 1.36a | 10.37a | 0.38b | 13.04b | 23b | 5.20a | 2.3a |
| 8-year-old pasture | 1.34a | 9.32a | 0.62b | 14.37a | 35b | 5.08a | 2.0a |
| Coeficient of variation | 23.41 | 10.48 | 31.24 | 21.55 | 27.47 | 5.41 | 30.7% |

Mean values followed by the same letter in the same column are not statistically different (Tukey test as 0.05).
(Adapted from Teixeira and Bastos, 1989.)

**Table 2.** Average values for organic C, OM, organic N and C/N relation in the soil (0-20 cm depth) of primary forest and pasture of *Brachiaria humidicola* of various ages in a claye Oxisol in Central Amazon

| Ecosystem | C (%) | OM (%) | N (%) | C/N |
|---|---|---|---|---|
| Primary forest | 2.62 | 4.55 | 0.18 | 15 |
| 1-year-old pasture | 2.40 | 3.90 | 0.18 | 15 |
| 2-year-old pasture | 2.10 | 3.45 | 0.16 | 13 |
| 6-year-old pasture | 2.32 | 4.00 | 0.16 | 14 |
| 7-year-old pasture | 2.48 | 4.25 | 0.19 | 13 |
| 8-year-old pasture | 2.40 | 4.15 | 0.18 | 13 |

(Adapted from Teixeira and Bastos, 1989.)

In Table 2 the mean values for organic carbon, OM, organic nitrogen, and C/N in the 0-20cm layer in the ecosystems of primary forest and pasture of 1, 2, 6, 7 and 8 years found by Teixeira and Bastos (1989) are presented. The levels of organic C and for OM do not show any marked difference, mainly between the forest and the 1, 7 and 8 year-old pastures. These authors, while studying the soil to a depth of 1m, found 148 t/ha of organic C in the primary forest and 160 t/h in the pasture, attributing the higher values in the pasture to the deposit of some of the C that had been stocked in the forest biomass that stayed on the soil as residue.

The levels of N in the soil also show similar values in the ecosystems studied, scarcely altered by the burning of the forest vegetation.

The values for C/N were similar in the various ecosystems, and are found within the range of stability for arable soils. According to this relation, the pasture ecosystems do not appear to have immobilized the N already existing in the environment, probably due to the gradual deposit to the soil of new material from the pasture.

The values of apparent density (which is the most direct measure of compaction) in the soil under pastures of different ages did not show any significant difference when compared to the values for the forest soil.

Teixeira and Schubart (1988) studied the soil fauna that play an important role in the process of degrading organic material, by increasing the area exposed to the action of bacteria and fungus. The authors show that burning forest plant biomass causes the death of a great number of these micro organisms, thus reducing the faunal diversity of the soil. There was a great reduction in the number of faunistic groups in the recently burned soil. However, in the soil under the six year-old pasture, the number of animal groups doubled, but still remained below that found in the forest soil. In both the forest soil and soils under pasture, the groups of *Acari* and *Collembola* predominate in the number of individuals per $m^2$, corresponding to 78.06% and 84.77% for *Acari* and 9.79% and 7.72% for *Collembola*, respectively in the forests and pasture soils.

Although these experiments fail in relating the soil's physical and chemical changes to pasture productivity, they serve to indicate that adequately managed forest-replacing pastures in the Amazon may be appropriate land use systems for efficiently conserving soil nutrients and maintaining satisfactory physical conditions for plant growth.

## B. Soil Transformation in Agroforestry Systems

Agroforestry systems have recently been promoted as alternative land use systems that will use land resources in the Amazon more efficiently (in a more sustained manner). They should, in some measure, gradually replace or be associated with present extensive low-sustainability land use systems such as open monoculture pasture-based cattle raising and shifting agriculture and may also be important in improving social and economic sustainability of extractive reserves.

Interesting forms of agroforestry systems are being developed in the Amazon, the most important and typical being those developed by Nippo-Brazilian farmers in the eastern part of the State of Pará (Subler and Uhl, 1990; Serrão and Homma, 1993). There are also an infinity of back-yard home-garden type systems carried out by small subsistence farmers all over the region (Fernandes and Serrão 1992).

Due to their present and potential importance for increasing sustainability of agricultural and forestry development in the region (Serrão and Homma, 1993), there is an increasing interest in research for developing sustainable agroforestry systems through domestication and introduction of high-value, multi-purpose native and exotic trees and food crops for development and management of integrated crops and trees.

Unfortunately, very few experiments have been designed to evaluate the effect of these systems on modifications of soil resources. Next, the most important research attempts which may be characterized as long-term experiments will be briefly described.

For the purpose of this paper, the following five agroforestry combinations were selected (because of their present and potential importance) from a large-scale experiment being carried out at an EMBRAPA-CPATU experimental station in the eastern part of the State of Pará, where dense rain forest is the dominant natural vegetation, medium-texture Oxisol is the predominant soil type and the Am climate of the Köppen classification prevails. The agroforestry combinations were planted in 1977 after the usual slashing and burning of the forest in the second half of 1976.

1. Agroforestry System 1: Traditional Cocoa (*Theobroma caco*)/Erythrina (*Erythrina glauca*) Association in Capitão Poço, Pará.

In this system, temporary (banana) and permanent (erithryna) shading plants were planted in 1977, spaced at 2.5 m x 2.5 m and 10.0 m x 10.0 m respectively, and cocoa seedlings were planted in 1977.

In a 12-year period (1978-1989), the system received fertilization which corresponded to approximately 70 kg N, 40 kg P, 84 kg K, 25 kg Ca, and 12 kg Mg per hectare annually. Table 3 shows that there were changes in soil fertility due to fertilizer application. Initially, a decline in soil OM was observed, but, after the seventh year of the cropping system, the trend was one of increase in soil OM content. Although there was no definite trend in soil pH values with time, they were always higher under the crop system than under the forest. The relatively high value of Ca + Mg in the planting year (1977) is attributed to the liming effect of slashing and burning in 1976 (the same positive effect can be said in relation to soil pH and Al). Two years after planting, a decline in soil Ca + Mg was observed, but this trend moved toward higher levels after the application of lime in 1984, which also resulted in maintaining Al at low levels, almost to the point of neutralization.

The level of P in the soil did not suffer significant alterations after cutting and burning the forest biomass. With the application of fertilizers in 1978, the level of P in the soil had a considerable increase in relation to the levels found after burning, and since 1982, the levels of P have remained at constant levels. The P fertilizer application was only to replace those nutrients lost due to cocoa fruit production and harvest. On the other hand, even with annual applications, K was present in the soil at low levels. This is explained by the fact that K is required for fruit production and is also lost through leaching of the soil (EMPRESA...1990).

Table 3 shows that the production of cocoa was high in the first three years, but dropped significantly in the years 1985 and 1986. From 1987, the cocoa production reached higher levels, corresponding to higher levels of soil K.

2. Agroforestry System 2: Cocoa/Peach Palm (*Bactris gasipaes*) Association in Capitão Poço, Pará

In this system, cocoa seedlings (spaced at 2.50m x 2.50m) and the permanent shading plants of peach palm (spaced at 10m x 10m) were planted in early 1978 following planting of the temporary shading plants (banana) in 1977.

In this system, in the 1978-1989 period, annual fertilization corresponded to approximately 76 kg N, 50 kg P, 102 kg K, 30 kg Ca, and 15 kg Mg, somewhat higher than in Agroforestry System 1, due to additional fertilization to peach palm trees.

Phosphorus was the soil nutrient that was most altered under the agroforestry system in relation to the forest soil. Soil P content was maintained at medium

**Table 3.** Productivity under five agroforestry systems

| Year | Cocoa production in cocoa/erythrina association | Cocoa/peach palm system | | Cocoa/rubber system | | Rubber production in tree system | Rubber production in kudzu system | |
| --- | --- | --- | --- | --- | --- | --- | --- | --- |
| | | Cocoa | Peach palm | Cocoa | Rubber | | with fertilizer | without fertilizer |
| 1982 | 1548 | 1473 | ---- | ---- | ---- | ---- | ---- | ---- |
| 1983 | 1797 | 1415 | ---- | ---- | ---- | ---- | ---- | ---- |
| 1984 | 2002 | 2241 | 7075 | ---- | ---- | ---- | ---- | ---- |
| 1985 | 974 | 1048 | 8400 | 1705 | ---- | ---- | ---- | ---- |
| 1986 | 939 | 1788 | 6920 | 1648 | ---- | ---- | 420 | 140 |
| 1987 | 1167 | 1308 | 7060 | 1185 | 350 | 505 | 1020 | 220 |
| 1988 | 1216 | 1559 | 5809 | 1819 | 321 | 510 | 810 | 405 |
| 1989 | 1271 | 1691 | 6400 | 914 | 237 | 423 | 795 | 380 |
| 1990 | 1361 | ---- | ---- | 1514 | 300 | 420 | ---- | ---- |
| 1991 | ---- | ---- | ---- | ---- | 414 | 427 | ---- | ---- |

to high levels in the crop system over the years. Soil K was low in the system during the first 6 years, but increases to medium levels thereafter.

The soil content of Ca + Mg increased after the clearing and burning of the forest biomass, then showed some decline in the first four years of the system and increased beginning in the fifth year. It can be observed that Al content in the soil was considerably lower in the agroforestry system, initially due to the liming effect of burning of the forest biomass and, after 1987, due to lime application that year. As a result of the initial clearing and burning, pH values also increased and were maintained always higher than in the forest soil.

The contents of the soil OM also showed decreases after cutting and burning, falling to 50% from that observed in the forest soil in 1981. However, from the seventh year the system showed a gradual increase in the OM content and by 1990 had reached the same levels as had been observed in the forest soil.

Average yields of cocoa and peach palm can be considered high in relation to those observed in average commercial production systems in the region. In this case, there seems to be no significant relation between cocoa and peach palm productivity and soil fertility parameters. According to Silva & Dias (1987), in general, there is no interference in the yield of both cocoa and peach palm in mixed systems, as long as peach palm tree density is not superior to 130 plants per hectare.


## 3. Agroforestry System 3: Cocoa/Rubber Tree (*Hevea* sp) Association in Capitão Poço, Pará

Similar to the agroforestry systems 1 and 2, banana trees were used as temporary shade and rubber trees (5m X 15m spacing) were planted in 1977; followed by cocoa (2.5m X 2.5m spacing) in 1978. Annual fertilization of the system during the 1978 through 1990 period corresponded to an annual application of 64 kg N, 36 kg P, 78 kg K, 25 kg Ca, and 11 kg Mg per hectare.

The low soil P content observed under the forest increased considerably in the soil of the agroforestry system during the first few years after slash-and-burning and then declined to somewhat lower levels which were maintained over the years.

Soil K was maintained at low levels in the agroforestry system through the years after a slight increase in the first couple of years following slash-and-burning. Toward the end of the period, soil K levels in the system became as low as those found in the forest soil.

The drop in the levels of P and K in the soil of the agroforestry system was associated with the beginning of cocoa production, a tropical crop with high nutrient demands, requiring large quantities, mainly of K, as shown by Morais (1988) and Relatório (1992).

The levels of Ca+Mg in the soil of the agroforestry system were well above those observed in the forest soil, with higher levels in the years from 1983 to 1985. Afterward they showed a decline, and then stabilized at values found in the first years of the system. Soil Al was reduced with the liming effect of the forest biomass burning and was maintained at low levels in the system through the years with a slight increase toward somewhat toxic levels at the end of the period. Conversely, pH values raised with slash-and-burning and were maintained over the years in the soil under the system.

Soil organic matter content was reduced (in relation to forest soil) with biomass burning, maintained at low levels during the first few years in the system, and increased after the 6th year, being maintained at levels close to those found in the soil under the forest.

In relation to the physical characteristics of the forest soils and the agro-forestry system, Costa and Teixeira (1992) found apparent densities in the soil under primary forest varying from $1.53g/cm^3$ in the 0-20-cm layer to $1.45g/cm^3$ in the 100- to 150-cm layer. In the soil under the agroforestry system the variation of apparent density was $1.55g/cm^3$ in the 0-20cm layer and $1.47g/cm^3$ in 100- to 150-cm layer , while the water retention capacity in the soils under forest and under the agroforestry system were similar. This data shows that after 14 years of substituting the forest with an agroforestry system, there has been practically no physical change in the soil.

Cocoa production was variable but within expected limits, except in 1990 which has some coincidence with the observed low soil content of P, K, and Ca + Mg. But a similar trend was not observed in rubber production.

## 4. Agroforestry System 4: Rubber Tree as a Monocrop in Capitão Poço, Pará

In this system, rubber tree seedlings were planted in 1977, 2.5m apart within rows 7.5m apart from each other. Annual fertilization in the 13-year period (1978-1990) corresponded to approximately 38 kg N, 16 kg P, 77 kg K, 8 kg Ca, and 5 kg Mg per hectare.

All soil parameters were significantly altered after slash-and-burning. Soil P was maintained in the system at adequate levels after the third year, but showed a reduction toward the end of the period. The same pattern was observed for soil K.

Soil content of Ca + Mg was initially low in the system but tended to increase with time. Conversely, soil Al was reduced with the liming effect of burning and was maintained at very low levels in the system over the years. Soil pH had similar patterns to that of Ca + Mg and, over the years in the system, was maintained at satisfactory levels, acidity being considerably lower than in the forest soil. As for soil OM, its tendency was one of reduction in the first few years in the system with a tendency to increase with time, in general, with higher values than in the forest soil. In spite of some reduction in the soil P and K content towards the end of the period, apparently these reductions were not sufficient to reduce rubber yield in the system.

5. Agroforestry System 5: Rubber Tree with Soil Cover of Tropical Kudzu (*Pueraria phaseoloides*) with and without Fertilization in Capitão Poço, Pará

This system was carried to try to separate fertilizer effects on rubber production and on the soil system in a rubber tree/kudzu association, a common production system in the region. Rubber tree seedling (grafted with clones IAN 717, IAN 710, and FX 3899) and kudzu seeds were planted in 1977/78, rubber seedlings being spaced at 3 m apart in rows 7 m apart from each other. The fertilized plots received an annual fertilization which corresponded to 41 kg N, 17 kg P, 24 kg K, 9 kg Ca, and 5 kg Mg per hectare in the 1978-1988 period.

Results of rubber production are shown in Table 3. As in other systems, considerable changes occurred in the soil due to slash-and-burning of the forest biomass. Also, the general pattern observed in the previously described systems can be observed in this system in the fertilized treatment.

However, important differences in the soil chemical composition were observed between the fertilized and unfertilized systems. This was noticed mostly in terms of Ca + Mg, pH, Al, and P. Fertilization had a positive effect on soil P which was maintained at adquate levels in the system through the years. The same positive effect was not observed for soil Ca + Mg, Al, and pH which maintained more favorable levels in the soil with time in the agroforestry system without fertilization.

The agroforestry system of rubber trees in association with cocoa, 15 years after its implantation, had stocks of organic material of 228 t/ha., with 57.5% concentrated in the soil down to a depth of 100cm and 38.8% at depths to down 30cm (Teixeira et al., in press), while Kling (1976) found for primary forest 612 t/ha of organic material, with 18.6% in the soil depth down to 30cm.

The OM levels in the soil of this agroforestry system of rubber with cocoa were 1.78% in the layer of 0-10cm depth and 0.6% in the layer of 50-100cm depth. The stock of organic material in the soil at a depth up to 100 cm was 131 t/ha, with 63% in the superficial layers and 37% in the 50-100cm layer. The greater concentration of organic material in the superficial layer is due to the addition of organic residues incorporated to the soil by the action of microbes and to climatic factors.

Forests, according to Pearce (1993) are the greatest absorbers of carbon ($CO_2$) on land. For each ton of wood burned, the exchange is nil in relation to the green house effect, if new trees replace those burned. These trees re-absorb carbon through photosynthesis. The author cites that to absorb 1 ton of carbon per year, one hectare of forest is required. In the 15-year-old agroforestry system, the accumulations of phytomass was 86.74 t/ha. This corresponds to an annual average increase of 5.80 t/ha., resulting from the capture of carbon from the atmosphere through photosynthesis, making evident that besides the socio-economic aspects, there are also ecological advantages.

Again, as in the previous systems, soil OM content was reduced to about 50% after forest cutting and burning, but about 11 years later in the agroforestry system the OM content was comparable to that of the original forest. Table 3 shows that from 1986 through 1989, average annual dry rubber production in the fertilized system was almost three times higher than in the unfertilized system.

The levels of K are lower in the systems with rubber trees than those of the forest soil, but, from the third year on, the values for K have been above those observed in the forest.

The transformation of primary forests to cultivated agroecosystems, hardly alters the physical parameters of the soil. Chemical parameters are positively affected in relation to the forest soils, mainly in relation to P, K, and Ca+Mg. The sustainability of these systems in terms of nutrients can only be achieved with the application of fertilizers, mainly P and K, to replace the quantities used by production and lost through leaching.

Of the five agroforestry systems presented, the agroforestry systems 2 (cocoa in association with peach palm) and 3 (cocoa in association with rubber trees), are shown as the most viable alternatives for small and medium producers that, traditionally, plant subsistence crops. The use of these systems will provide capital which will give year-long income from the marketing of rubber, cocoa and peach palm, besides promoting social forestry which are agro-technically and economically sustainable.

## C. Soil Transformation in Silvopastoral Systems

Although in their initial stages of development in the Amazon and still mostly concentrated in the eastern Amazon in small- and medium-size properties, where Veiga and Serrão (1990) found several associations of fruit and timber trees with a wide strata of grasses and legumes for cattle raising, silvopastoral systems are promising land use systems for improving agricultural sustainability in the Amazon.

These systems have only recently attracted the attention of researchers because there is potential for increased agrotechnical, environmental, and socioeconomic sustainability. Only one known research attempt has been made to study silvopastoral systems in the Amazon, and is summarized as follows.

**Silvopastoral systems: An association of forest species with forage grasses for reclamation of degraded pasture lands in the Eastern Amazon.**

This research was developed in a degraded commercial pasture area on a large private ranch in the county of Paragominas, State of Pará. The original vegetation of the area was a dense rain forest, the predominant climate is a transition between the Am to the Aw climate type of the Köppen classification. The soil of the area is predominantly heavy-clay Oxisol (Yellow Latosol) with naturally high acidity and low fertililty (Falesi, 1976). The land was cleared and burned for pasture establishment with guineagrass in the early 1960s. After a few years of relatively high productivity, the pasture went through the process of degradation as briefly described before in this paper.

The degraded pasture area was mechanically prepared (windrowed and disked) in 1984, and in early 1985 the experiment was planted with the following design: The experimental area was divided into three blocks of approximately 3 ha each. In each block, one grass (out of three grasses selected, namely: *Brachiaria humidicola, B. brizantha*, and *B. dyctioneura*) was planted in association (actually, in 12m-wide strips between the three-line rows of tree species) with each of three forest species, namely: *Eucalyptus teriticornis, Schyzolobium amazonicum* and *Bagassa guianensis*.

In 1985, the forest species were planted together with corn which was harvested the same year. In 1986, corn was again planted between the tree rows. This operation was again repeated in 1987, this time together with the forage grasses. Steers were used to rotationally graze the plots at the stocking rate of one head per hectare to introduce the grazing effect on the system

The growth of the trees was considered to be very satisfactory, especially that of *S. amazonicum*, and *E. teriticornis*. As for the associated grasses, excellent establishment and growth was observed in *B. brizantha,* and, secondarily but still with adequate performance, in *B. humidicola* and *B. dyctioneura*.

Soil samples (0-20 cm depth) for determining the chemical and physical composition of the soil in the different pasture/tree combinations were taken twice a year in the wet and in the dry seasons from 1986 through 1990. The following comments on soil change with time are based on average (of wet and dry season) annual soil composition.

The experiment was interrupted in the second semester of 1991 because of an accidental fire which damaged the forest species mainly.

Ca + Mg in the *B humidicola* plots had a tendency to decline the same trend was observed in the *B. dyctineura* area. A peak was observed in *B. humidicola* area in 1989, possibly due to the mineral supplementation which was supplied to the grazing cattle in all plots, but which may have been localized at sampling time, thus influencing the soil analysis, in comparison with the other treatment combination.

Soil K tended to decline in the soil under all grass species after four years of grazing. Probably, most of this is due to the K going out of the system in the consumed grass forage and some is probably leached from the soil surface.

Apparently, the tree-grass combinations were not able to retain enough carbon in the soil through the years, a reduction of about 20% being observed from 1986 to 1990.

Soil P seemed to increase in the pasture, except for the *B. humidicola* pasture. This increase seemed to be related to the mineral supplementation offered to the cattle in all pastures, which is probably higher than the P removed from the pasture through the consumed grass forage. Besides, P is considered a nutrient practically immobile in the soil.

Soil Ca + Mg contents were similar (around 5 mg/100g) under the trees through time until 1989, but showed some differentiated reduction in the 1990 sampling. Loss of K under the trees with time was more accentuated than that of Ca + Mg, probably because K is a more leachable mineral in the soil and is removed in higher quantities in the consumed grass than Ca and Mg. Reduction of soil K was about 30 % under the trees.

Soil C losses were practically the same under all three timber trees (about 20%), similar to those in the soil under the grasses. However, soil C content was above 2% (3.4% OM) which can be considered satisfactory for kaolinitic soils in the Amazon in which OM is responsible for 70% of the cation exchange capacity (CEC).

Similar to the soil under the pastures, P content under the trees had a considerable increase with time (explanation can be the same for the soil under grass strips) but the levels are still low as usual under these land use conditions.

In general, the soil under the tree-grass combination had a decline in its content of most nutrients with time except for small increases in P content but still at low, deficient levels. However, Ca, Mg, K and C were still at satisfactory levels, except for K in the grass strips which has probably started to become limiting for tree and grass growth.

As far as the soil moisture content was concerned, a general trend with time was of slight increases in the soil under the grasses and under the timber trees, and were more noticeable under the trees, as expected.

Although the experiment discussed give some indication of the trends in soil chemical and physical transformations in agrosilvopastoral systems, much needs to be learned from long-term, well-planned agrosilvopastoral experiments.

# References

Baena, A.R.C. 1978. O efeito de pastagens (*Panicum maximum*) na composição química do solo em floresta tropical de terra firme. p. 355-77. In: *Encontro Nacional de Pesquisa Sobre Conservação do Solo*. 2, Passo Fundo, RS, 1978. Anais. Passo Fundo, EMBRAPA–CNPT.

Bittencourt, V.C. 1977. Solos tropicais. p.59-62. In: *Seminário de Uso E manejo das Terras. Manaus, AM,* 1977. Resumo das Palestras. Manaus, EMATER-AM/SUFRAMA.     (SUFRAMA. Distrito Agropecuário da SUFRAMA, 6).

Brienza Junior, S. and J.A.G. Yared. 1991. Agroforestry systems as an ecological approach in the Brazilian Amazon development. *Forest Ecology and Managment*: special issue on agroforestry. 45:1-4.

Buschbacher, R., C. Uhl, and E.A.S. Serrão. 1988. Abandoned pasture in eastern Amazonia. II Nutrient stocks in the soil and vegetation. *J. Ecol.*. 76:682-699.

Costa, M.P. da and L.B. Teixeira. 1992. *Caracterização físico-hídrica de Latossolo Amarelo da região de Capitão Poço, Pará*. Belém, Brazil. EMBRAPA–CPATU. (EMBRAPA–CPATU. Boletim de Pesquisa, 133). 23 pp.

Empresa Brasileira De Pesquisa Agropecuária. 1991. Centro de Pesquisa Agropecuária do Trópico Úmido, Belém – Pará. *Associação de espécies florestais com forrageiras para recuperação de áreas degradadas.* Belém, Brazil. (Projeto de Pesquisa, FORM 13 – Relatório). 10 pp.

Empresa Brasileira De Pesquisa Agropecuária. 1990. Centro de Pesquisa Agropecuária do Trópico Úmido, Belém – Pará. *Sistema de produção com plantas perenes em consórcio.* Belém, Brazil. (Projeto de Pesquisa. FORM 13 – Relatório). 49 pp.

Empresa Brasileira De Pesquisa Agropecuária. 1992. *Centro de Pesquisa Agropecuária do Trópico Úmido,* Belém – Pará. S Belém, Brazil. (Projeto de Pesquisa. FORM 13 – Relatório). 49 pp.

Falesi, I.C. 1976. *Ecossistema de pastagem cultivada na Amazônia Brasileira.* Belém, Brazil, EMBRAPA–CPATU. 1(EMBRAPA–CPATU. Boletim Técnico, 1). 193 pp.

Fernandes, E. and E.A.S. Serrão. 1992. Protótipos de modelos agrossilvipastoris sustentáveis. p. 245-304. In: *Seminário internacional Sobre Meio Ambiente, Pobreza E Desenvolvimento Da Amazônia (SIMDAMAZÔNIA).* Belém, Brazil. Anais. PRODEPA.

Klinge, H. 1976. Bilanzierung von hanptnhrstoffen in okosysten tropischer regenwold (Manaus) – vorlänfige date. *Biogeographica* 7:59-77.

Moraes, F.I. de O. 1988. O cultivo do cacaueiro na Amazônia brasileira – potencialidade e limitações. p. 41-55. In: *Simpósio sobre a Produtividade Agroflorestal da Amazônia: Problemas e Perspectivas.* Belém, Brazil. FCAP.

Pearce, F. 1993. Sequestro de carbono vira prioridade mundical. 15 de agosto, Sào Paulo, folha de São Paulo, Argentina. Caderno CIÊNCIA.

Serrão, E.A.S. and M.B. Dias Filho. 1991. Establecimiento y recuperación de pasturas entre los productores del Tropico Húmedo Brasileño. p. 347-383. In: E.C. Lascano and J.M. Spain. (eds.), E*stablecimiento y Renovación de Pasturas. Conceptos.* Experiencias y Enfoque de la Investigacion. CIAT. Cali, Colombia.

Serrão, E.A.S., I.C. Falesi, J.B. Veiga, and J.F. Teixeira Neto. 1979. Productivity of cultivated pastures on low fertility soils in the Amazon of Brazil. p.195-225. In: P.A. Sanchez and L.E. Tergas. (eds.), *Pasture Production in Acid Soil of the Tropics.* Cali, Colombia. CIAT.

Serrão, E.A.S. 1990. Pature Development and carbon emission/accumulation in the Amazon: Topics for discussion. p. 210-222. In: *Tropical Forestry Response Options to Global Climate Change.* São Paulo Conf. Proc. U.S. Environmental Protection Agency. Washington, D.C.

Serrão, E.A.S. and A.K.O. Homma. 1993. Brazil. p. 265-351. In: *Sustainable Agriculture and the Environment in the Humid Tropics.* National Research Council, National Academy Press, Washington, D.C.

Serrão, E.A.S. and J.M. Toledo. 1990. The search for sustainability in Amazonian pastures. p. 195-214. In: A.B. Anderson (ed.), *Alternatives to Deforestation. Steps Toward Sustainable Use of Amazon Rain Forest.* Columbia University Press. New York..

Serrão, E.A.S. and J.M. Toledo. 1992. Sustaining pasture-based prodution systems in the humid tropics. p.257-280. In: S.B. Hecht (ed.), *Development or Destruction: The Conversion of Tropical Forest to Pasture in Latim America*. Westview Press, Bolder, Colorado.

Silva, I.C. and A.C. da C.P. Dias. 1987. Intercultivo de pupunheira com cacauiro na Amazônia brasileira, resultados parciais. *Rev. Theobroma* 17(2):93-100. CEPLAC, Ilhéus, Bahia, Brasil.

Teixeira, L.B., J.B. Bastos, and R.F. de Oliveira. 1993. Biomassa vegetal em agroecossistema de seringueira consorciada com cacaueiro no nordeste paraense. Belém, Brazil. EMBRAPA–CPATU, 1993. (EMBRAPA–CPATU. Boletim de Pesquisa). in press.

Teixeira, L.B. 1987. *Dinâmica do ecossistema de pastagem cultivada em área de floresta na Amazônia Central*. Tese de Doutorado, Manaus, INDA/FUA. 100 pp.

Teixeira, L.B. and J.B. BASTOS. 1989. *Matéria orgânica nos ecossistemas de floresta primária e pastagens na Amazônia Central*. Belém, Brazil. EMBRAPA–CPATU. (EMBRAPA–CPATU. Boletim de Pesquisa). 26 pp.

Teixeira, L.B. and J.B. BASTOS 1989. *Nutruentes nos solos de floresta primária e pastagem de Brachiaria humidicola na Amazônia Central*. Belém, Brazil. EMBRAPA–CPATU. (EMBRAPA–CPATU. Boletim de Pesquisa, 98). 31 pp.

Teixeira, L.B. and H.O.R. Schubart. 1988. *Mesofauna do solo em áreas de floresta e pastagem na Amazônia Oriental*. Belém, Brazil. EMBRAPA–CPATU. (EMBRAPA–CPATU. Boletim de Pesquisa, 95). 16 pp.

Uhl, C., R.J. Buschbacher, and E.A.S. Serrão. 1988. Abandoned pasture in eastern Amazonia. I. Patterns of plant succession. *J. Ecol.* 76:663-681.

Uhl, C., O. Bezerra, and A. Martini. *An ecosystem perspectives on threats to biodiversity in Exstern Amazonia, Pará State*. Amer. Assoc. Adv. Sci. Press. Washington, D.C. in press.

Veiga, J.B. and E.A.S. Serrão. 1990. Sistemas silvopastoris e produção animal nos trópicos úmidos: a experiência da Amazônia Brasileira. p. 37-68. In: *Pastagens*. Sociedade Brasileira de Zootecnia. Piracicaba.

# Extrapolating Results of Long-Term Experiments

G. Uehara, G.Y. Tsuji, and F.H. Beinroth

## I. Introduction

In May 1974, a contract was signed between the U.S. Agency for International Development and the University of Hawaii to initiate a project entitled "Crop Production and Land Capabilities of a Network of Tropical Soil Families". A parallel project entitled "Crop Production and Land Potential of Benchmark Soils of Latin America" was also signed between AID and the University of Puerto Rico in early 1975. Both projects used identical methods to achieve their objectives and were jointly referred to as the Benchmark Soils Project or simply BSP.

The purpose of the BSP was to test the hypothesis that long-term behavior and performance of a soil could be predicted on the basis of its membership in a specific soil family, provided the behavior and performance of another member of the same soil family were known. This hypothesis was based on the implicit assumption contained in the then new Soil Taxonomy (Soil Survey Staff, 1975) and its predecessor, the "7th Approximation" (Soil Survey Staff, 1960). The hypothesis was that soils belonging to the same soil family as defined in Soil Taxonomy would behave and perform similarly irrespective of their location in the world. It should be noted, however, that soils belonging to a particular soil family cannot occur in widely separated latitudes because Soil Taxonomy

ISBN 1-56670-076-0

stratifies soils on the basis of soil-climate parameters including soil temperature and soil moisture regimes. Thus, while it is possible for two soils separated halfway around the world to belong to the same soil family, it is highly unlikely for a soil that occurs in the tropics to be in the same family as one found more than 25 degrees north or south of the equator.

The notion of the transferability of soil knowledge and experience to other similar soils on the basis of their membership in a common taxon was considered important because much of what was known about named kinds of soil was being rediscovered elsewhere in the world at great expense to agricultural development projects. Thus, the phrase "agrotechnology transfer" was coined to mean the taking of knowledge or technology from its site of origin to other locations where it was likely to succeed. In the BSP, the rational basis for knowledge and technology transfer became the soil family.

## II. The Soil Family

Soil Taxonomy (Soil Survey Staff, 1975) is a hierarchal system of Soil Classification consisting of six categories or levels, starting with the order at the highest level, followed by the suborder, great group, subgroup, family, and series as indicated by the example in Table 1. As in most multi-categorical classification systems, the information content of the object being classified increases as one moves down the taxonomic ladder. The unique and innovative feature of Soil Taxonomy is that names of the higher categories are retained in taxa of all lower categories except the series. As illustrated in Table 1, knowledge of the Family name guarantees knowledge of taxa of all higher categories to which the soil belongs. In the binomial system of classification, for example, the genus gives no hint of the higher categories to which the organism belongs, and requires the user to commit that knowledge to memory.

According to Soil Taxonomy, the intent of the soil family category is to group soils within a subgroup having similar physical and chemical properties that affect their response to management and manipulation for use. Soil Taxonomy further assumes that the responses of comparable phases of all soils in a family are sufficiently alike to meet most of human needs for practical interpretation of such responses.

To be a member of a family, Soil Taxonomy requires that a soil possesses restricted ranges in the following properties:

1.    Particle-size distribution in a control section of major biological activity below plow depth.
2.    Mineral composition in the same control section.
3.    Soil temperature measured at 50-cm depth.
4.    Thickness of the soil penetrable by roots.
5.    A few other properties that are used in defining some families to produce the needed homogeneity. In addition, the soils in a family have all of the

**Table 1.** Example of relationships among category subdivisions in Soil Taxonomy

| Category name | Basis for differentiation | Example of class name | Main features of the class |
|---|---|---|---|
| Order | Dominant soil process that developed soil | *Ult*isol | Clay accumulation; depletion of bases |
| Suborder | Major control of current processes | Ud*ult* | Soil moist most of time; humid (udic) climate |
| Great group | Additional control of current process | Trop*udult* | Fairly constant soil temperature all year; tropical environment |
| Subgroup | Blending of processes (integrades) or extragrades | Aquic *Tropudult* | Temporary wetness in rooting zone |
| Family | Internal features that influence soil-water-air relationships | Fine loamy, mixed isothermic *Aquic Tropudult* | Texture and mineralogy in a control section, and soil temperature |
| Series | Nature of materials that affect homogeneity of composition and morphology | Cerrada | Soil forming in weathering diabase |

properties that are diagnostic for the order, suborder, great group, and subgroup to which the family belongs.

Soil Taxonomy, like most classification systems, is a means of organizing what is known about the objects being classified so that their behavior and performance can be predicted by the names assigned to them. The designers of the BSP chose soils of a particular phase of a soil family to test the underlying assumptions of Soil Taxonomy.

## III. Relationship to Long-Term Experiments

The BSP was not designed to conduct long-term experiments. Its aim was to see if signs of long-term soil behavior and performance observed in Hawaii and Puerto Rico could be duplicated in similar soils elsewhere in the world. The justification for doing so arose from the conflicting opinions of agricultural development experts on the productivity, stability and resi,liency of soils of the tropics. While some tropical soils, particularly those in the Mollisol, Alfisol, Vertisol, and Entisol (alluvial members) orders were judged to be productive, soils in the Ultisol and Oxisol orders were treated as the irreversibly impover-ished, end-products of weathering.

Many of the impressions about tropical soils were promulgated by pedologists such as Buchanan (1807), Mohr and van Baren (1954), and Prescott and Pendleton (1952) who were intrigued by, and focused on, the most highly weathered soils of the tropics. While it was true that lateritic soils and latosols, as they were later to be called, fit the definition of Oxisols or Ultisols, it was also true that not all Oxisols and Ultisols conformed to the definition of lateritic soils. As more systematic surveys of tropical regions were undertaken, it became increasingly evident that laterites and lateritic soils, although unique to the tropics, were not as common as the public had been led to believe. Even so, tropical soils classified as Ultisols or Oxisols continued to be viewed as lateritic soils. But in Hawaii and Puerto Rico, productive soils that had been in continuous cultivation for over a century turned out to be Oxisols and Ultisols. In Hawaii, the Oxisols, although not as productive initially as the more fertile Vertisols, now outperforms the latter by a significant margin. As illustrated in Figure 1, yields have increased in both soils, but the rate of increase has been greater in the Oxisols.

The higher performance of Oxisols today can be attributed to at least two factors. First, the build up of residual phosphorus from decades of fertilizer application has transformed P-deficient Oxisols to some of the most productive soils in the world. But the more critical factor has been the superior physical characteristics of the Oxisol. Oxisols by definition must be composed of low-activity clays (Uehara and Gillman, 1981) with low cation exchange capacity and low shrink-well potential. Although declining soil fertility is frequently cited as a factor contributing to unsustainable agriculture, fertility is one of the easiest soil constraints to overcome. In fact it is a rare soil that does not require some input of nutrients to sustain high productivity. What is less frequently cited, but is more critical to sustainable land management, is the physical characteristics of a soil, which are largely determined by clay mineralogy, particle size distribution, and stability and arrangement of aggregates in each horizon of a soil. In the final analysis, it was soil physics and not soil fertility that enabled the Oxisol to surpass the Vertisol in productivity. But were the Oxisols of Hawaii and Puerto Rico pedological oddities that occurred only in the two island ecosystems? If not, could similar Oxisols be found elsewhere in the world and would they behave and perform similarly to those in Hawaii and Puerto Rico? Such questions had

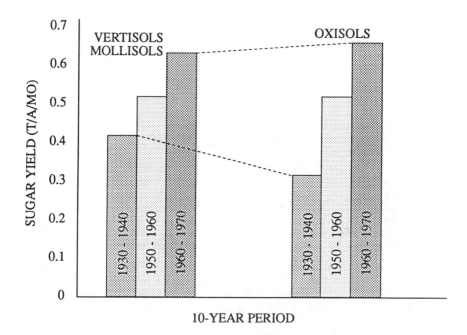

**Figure 1.** Sugar yield over 40-year period on some soils in Hawaii.

long occupied the minds of applied pedologists, but the lack of a suitable quantitative international system of soil classification prevented them from validating this critical pedological principle.

The publication of Soil Taxonomy in 1975 provided opportunities for answering compelling questions about the utilitarian goals of pedology. Soil scientists in Hawaii and Puerto Rico, cognizant of the diversity of their soils and the limited number of scientists assigned to study them, wondered if the global soil knowledge and experience could not be made accessible to all on the basis of local soil resource inventories based on an international system of soil classification. Could Soil Taxonomy serve this purpose and become the universal language for communicating soil data, information and knowledge? Long-term knowledge and experience gained through many years of land use in one location might now be transferred to similar soils elsewhere in the world. When one realizes that there are 11 soil orders consisting of an estimated 57 suborders, 300 Great Groups, 2500 subgroups, 7000 soil families, and 19,000 soil series in the U.S. alone, it soon becomes evident that only a small fraction of all distinct kinds of soil will ever be subjected to long-term studies. Can we then extrapolate what we learn from a limited number of long-term experiments to all other soils? If not how many more long-term experiments do we need to conduct?

While long-term studies can provide invaluable knowledge of soil resiliency and sustained productivity, it would be physically impossible to conduct the

| Practice / SOIL | Practice A | Practice B | Practice C | Practice — |
|---|---|---|---|---|
| SOIL I | Performance Index | Performance Index | Performance Index | Performance Index |
| SOIL II | Performance Index | Performance Index | Performance Index | Performance Index |
| SOIL III | Performance Index | Performance Index | Performance Index | Performance Index |
| SOIL N | Performance Index | Performance Index | Performance Index | Performance Index |

**Figure 2.** Soil by practice matrix to indicate how different soils respond to a given practice and how a given soil performs under alternative practices.

critical number of genotype x environment x management experiments on even a small fraction of key soils to satisfy future information needs. One alternative might be to study existing soil management practices on land which has been under intensive cultivation for 50 years or more, and to create a soils x practice matrix for assessing long-term performance. A skeleton matrix is shown in Figure 2. The matrix is designed to document how a particular type of soil performs under different land use practices, and how different soils respond to a specific management practice.

In this regard, the BSP was able to install a global network of "benchmark" experiments to examine just one dimension of Figure 2, namely to see how different soils responded to a particular treatment. The project, however, added a third dimension to the matrix. The prescribed practice was repeated on the "same" soil at three "primary" sites and as many as nine "secondary" sites. Secondary sites were generally located within a few kilometers of the primary site, but the three primary sites for a particular soil family were located in different countries or continents. As stated earlier, the aim was not to conduct long-term experiments, but to test whether knowledge and experience gained through long-term land use practices in one location could be transferred to other locations with similar soils. This concept would eventually be called "agrotechnology transfer by analogy."

## IV. The Benchmark Soils Project

In 1973, efforts were initiated to seek support from the U.S. Agency for International Development to test the hypothesis of agrotechnology transfer by analogy. The proposal would hinge on the all-important premise upon which Soil Taxonomy is based, namely, that soils of the same family behave and perform sufficiently alike to enable experience to be transferred among them. To test this hypothesis, a plan was designed to conduct common "benchmark" experiments (BSP, 1982) on soils belonging to three families in the Inceptisol, Oxisol, and Ultisol soil orders. The three soil families selected were:

1.     thixotropic, isothermic Hydric Dystrandepts,
2.     clayey, kaolinitic, isohyperthermic Tropeptic Eutrustox, and
3.     clayey, kaolinitic, isohyperthermic Typic Paleudults.

All three families are well-drained, upland soils of the tropics occurring in "moist and cool", "dry and warm", and "moist and warm" agroenvironments, respectively. The first soil family (henceforth referred to by its great group name, Dystrandept) was selected because it was easy to locate in the tropics and project personnel had long experience with its management, behavior, and performance. Dystandept experimental sites were located in Hawaii, Indonesia, and the Philippines.

The second family (Eutrustox great group) was selected to compare results on a soil common to Hawaii and Puerto Rico. A third site located in Central Brazil completed the Eutrustox network.

The third soil family (Paleudult great group) was included at the suggestion of the collaborating host-countries. This and closely related families occur extensively in the humid tropics, and the role they might play in national agriculture development was the center of a major international debate in the 1970's. Benchmark sites for Paleudults were established in Cameroon, Indonesia and the Philippines.

The experimental locations for all families are shown in Table 2. It should be noted that although the three families differed greatly in appearance, behavior, and performance, they still possess common features that enabled them to accommodate a common set of agronomic treatments. These features included small differences between mean winter and summer soil temperatures indicated by the "iso" prefix in the soil temperature regime, high clay content implied by the thixotropic and clayey designation, and low effective cation exchange capacity (ECEC) of the clay fraction and high P-fixation inferred by the thixotropic and   kaolinitic modifiers. An example of the key soil properties associated with each taxon is shown in Table 2. An experienced soil scientist can

**Table 2.** Soil family network of experimental sites

| Location | Area (nearest city) | Site |
|---|---|---|
| **Thixotropic, Isothermic Hydric Dystrandepts** | | |
| Hawaii | Hawi, Hawaii | IOLE |
| | Honokaa, Hawaii | KUK |
| | Hawi, Hawaii | HAL |
| Philippines | Panicuason, Camarines Sur, Luzon | PUC |
| | Pili, Camarines Sur, Luzon | PAL |
| | Calabanga, Camarines Sur, Luzon | BUR |
| Indonesia | Cisarua, Bandung, Java | ITKA |
| | Lembang, Bandung, Java | PLP |
| | Cipanas, West Java, Java | LPH |
| **Clayey, Kaolinitic, Isohyperthermic Tropeptic Eutrustox** | | |
| Puerto Rico | Isabela | ISA |
| | Isabela | ISA-2 |
| | Isabela | ISA-3 |
| Brazil | Jaiba, Minas Gerais | PAR |
| | Jaiba, Minas Gerais | BAH |
| | Jaiba, Minas Gerais | CEA |
| Hawaii | Maunaloa, Molokai | MOL |
| | Waipio, Oahu | WAI |
| **Clayey, Kaolinitic, Isohyperthermic Typic Paleudults** | | |
| Philippines | Davao City, Davao, Mindanao | DAV |
| | Sorsogon, Camarines Sur, Luzon | SOR |
| Indonesia | Kotabumi, Lampung, Sumatra | NAK |
| | Sukanegeri, Lampung, Sumatra | BUK |
| | Kotabumi, Lampung, Sumatra | BPMD |
| Cameroon | Kumba, Southwest | CAM |
| | Kumba, Southwest | BAK |

(From Silva, 1985.)

infer much more about the management requirements of soils from the soil family name than is indicated by Table 2. For consistency with cited reference, the original classification is used in this paper. The reader should know that the name and definitions and keys to Soil Taxonomy have undergone several revisions.

Soil Taxonomy, like all utilitarian classification systems, was prepared to organize knowledge so that knowledge could be used by others to predict behavior and performance of the objects being classified. The BSP scientists were well aware that soils conforming to the three chosen families were considered problem soils elsewhere in the tropics. But they also realized that these and closely related soils could become the future breadbasket of the tropics, and based on long-term experiences in Hawaii and Puerto Rico, were optimistic that that would be the case. The opportunity was at hand to test the reality of what was then mere speculation and hope.

## V. Research Design

The BSP staff believed that in the case of the three soil families, the principal factor constraining productivity of these low ECEC, high-clay soils was P-deficiency. This would normally be a trivial soil fertility problem, but what was unique about the soils of the three families was their high P-fixing capacity. As illustrated in Figure 3, P-application rates that would normally correct P-deficiency result in marginal benefits to high P-fixing soils. The high P-requirement of such soils was used by development experts to discourage their inclusion in development projects on economic grounds. But what was not generally known to the development experts was that once corrected, the P-requirement of high P-fixing soils would approach those of low P-fixing soils. As a one-time investment, the initial expense of raising P to an optimum level could be treated as a capital cost, just as one would treat an irrigation system as capital investment.

## VI. Experimental Design

The treatment design consisted of the $5^2$ partial factorial modification by Escobar described by Laird and Turrent (1981). It was chosen on the basis of (1) appropriateness for use in a graphic estimation of economic optima, (2) magnitude of bias error, (3) number of treatment combinations, (4) flexibility in number of factors and number of levels of each factor, and (5) magnitude of variance error. Three replications were considered adequate for estimating the site-specific coefficients of the yield-response relationship to applied P and N. A diagram of the design is shown in Figure 4. For convenience, the treatment

G. Uehara, G.Y. Tsuji, and F.H. Beinroth

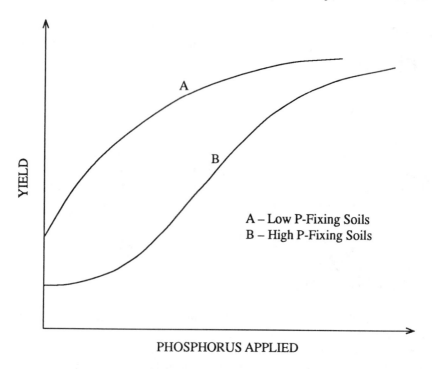

PHOSPHORUS APPLIED

**Figure 3.** A common occurrence in the Tropics is high P-fixing soils.

range was coded as -1 to +1 with the 0 level representing the middle rate (Silva, 1985).

The test crop was maize, and two hybrids, Pioneer X304C and H-610, adapted to tropical environments were selected. All experiments were drip-irrigated and kept as free of weeds, insects, and pathogens, as possible. The project staff had been warned by consultants invited to review the research plan that heavy and frequent downy mildew infestation would render maize unsuitable as a test crop. Accordingly, the maize experiments installed at sites of the first soil family were carefully monitored for the disease. It was later learned that the "isothermic" designation of the Dystrandept precluded downy mildew from flourishing in that environment. Soil Taxonomy had successfully stratified soils into performance classes which neither the consultants nor the BSP staff had anticipated. The second soil family network also failed to encounter problems with the disease, but by then the environmental requirements of the pathogen was known and the staff was able to predict that the ustic (semi-arid) moisture regime of the Eutrustox would also preclude heavy infestation by downy mildew. What Soil Taxonomy was allowing its users to do was to match the biological requirements of organisms to the physical characteristics of the land. The performance of the organisms, whether they be pests or cultivated crops depended on the goodness of the match. The mismatch for maize growing on the three soil family was

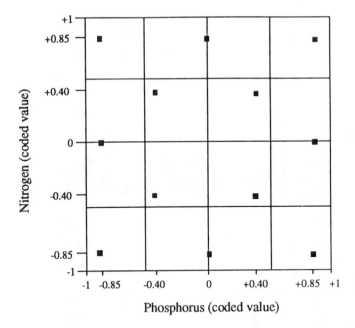

**Figure 4.** Thirteen treatment combinations of applied phosphorus and applied nitrogen. The actual applied levels are coded on a scale from -0.85 to +0.85. (From Silva, 1985.)

P–deficiency, and the mismatch for downy mildew was low temperatures in the Dystrandepts and dry condition in the Eutrustox.

This realization led to detailed characterization of the land including position on the landscape, slope, aspect, soil, hydrology, and weather. Each site was thoroughly characterized for soil and continuously monitored for soil and air temperature, solar radiation, rainfall, maximum and minimum temperature, relative humidity, and wind direction and velocity. Summary weather data for the Dystandept network is provided in Figure 5.

For the crop, date of planting, date of emergence, date of tasselling (50%), and date of physiological maturity were recorded. Biomass and nutrient content of specified tissue were sampled and analyzed at specified growth stages.

## VII. Results

What was learned from the BSP far exceeded the expectations of its designers. That P-fixation and P-deficiency were major production constraints for these soils did not come as a surprise. But because P was selected as the project focus, an inordinate amount of time and effort was invested to analyze the results and

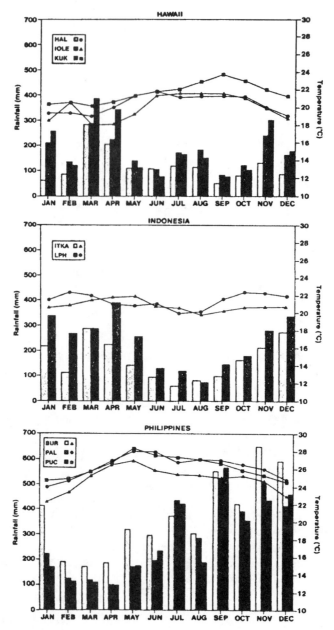

**Figure 5.** Mean monthly rainfall and air temperatures for the Hydric Dystrandept network. (From Silva, 1985.)

**Table 5.** Persistent weeds at Hawaii sites by soil family

| Tropeptic Eutrustox | Hydric Dystrandepts | Common to both families |
|---|---|---|
| *Convolvulus arvensis* L. | *Ageratum conyzoides* L. | *Bidens pilosa* L. |
| *Amaranthus spinosus* L. | *Erechtites hieracifolia* (L.) Raf | *Emilia sonchifolia* L. |
| *Ipomea triloba* L. | | |
| *Momordica charantia* var. pavel Crantz | *Kyllinga brevifolia* Rottb. | *Sonchus oleraceus* L. |
| *Cyperus rotundus* L. | *Cyperus hypochlorus* Hilleb | *Digitaria sanguinalis* (L.) Scop |
| *Cenchrus echinatus* L. | *Setaria glauca* (L.) Beaur. | |
| *Chloris barbata* Swartz | | |

develop a statistical basis for extending project findings to other soils of the same family. The result of this effort is summarized by Cady et al. (1985).

The unplanned and probably more important outcome of the project was in demonstrating the power of Soil Taxonomy to stratify soils according to soil physics, nutrient requirements, crop protection needs, and other factors which heretofore have been based on anecdotal evidence. As the real evidence began accumulating it was easy to fault the project for proving the obvious.

Encounters with insects, weeds, and disease by soil families are summarized in Tables 3, 4, 5, and 6. The fact that organisms occupy well-defined environmental niches was not new, but that these niches often corresponded to soil families as defined by Soil Taxonomy was revealing. Not many users of soil surveys think of soil boundaries as more than lines marking the limits between two recognizably different soils. We have since learned that nematodes and symbiotic nitrogen fixing organisms behave and perform differently in different soils. What we do not know is what fraction of all production factors Soil Taxonomy manages to stratify.

Another impressive feature of the soils in the three families is their robust physical attributes. Although heavy textured, they possess high infiltration rates and allow workers to re-enter a field within one or two hours after a heavy downpour. Their open soil structure is attributable to the high water-stability of the soil aggregates. Thixotropic Dystrandepts and high clay, kaolinitic Eutrustox and Paleudults not only infer high P-fixation and P-deficiency, but point to a

**Table 6.** Maize disease incidence and severity rating for six Indonesian sites

| Disease | Hydric Dystrandepts | | | Typic Paleudults | | |
|---|---|---|---|---|---|---|
|  | ITKA | PLP | LPH | NAK | BUK | BPMD |
| Downy mildews (*Peronosclerospora maydis* [RAC} C.G. Shaw) (*P. philippinensis* [Weston] C.G. Shaw) | O | O | M | S | S | S |
| Common maize rust (*Puccinia sorghi* Schw.) | O | O | O | M | M | M |
| Common smut (*Ustilago maydis* [DC] Cda.) | O | O | O | M | M | M |
| Southern leaf blight (*Helminthosporium maydis* Nisik) | S | S | S | M | M | M |
| Stalk rots (*Gibberalla zeae* [Schw.] Petch) (*Diplodia maydis* [Berk.] Sacc.) | M | M | M | O | O | O |

Key: "O" = pest not observed at site; "M" = pest of minor importance; "S" = pest of major importance.

physically resilient soil. Phosphorus deficiency may be expensive, but physical resiliency is priceless because it cannot be purchased. The three soil families and their relatives have too long been treated with disdain for their nutritionally impoverished state; they should instead have been respected and valued for their physical robustness.

Lastly, an important contribution not documented in any BSP publication is the concept of minimum data sets (MDS) initiated by project personnel. Although remembered as a "soils" project, the BSP staff was diligent in its care to collect a balanced set of crop, weather, soil, pest, landscape, and management data throughout the life of the project. The aim of the MDS was to use it to explain large anticipated yield variances resulting from genotype-by-environment-by-management interactions spanning a decade of field trials in Africa, Asia, and South America. As expected, experimental outcomes were often governed more by uncontrollable factors such as weather than by the treatment variables.

The uncontrolled, random environmental variables, somehow had to be factored into the yield equation, and were eventually accommodated in soil family-specific, regression equations. Because soil families are stratified according to soil climates, including soil temperature and moisture regimes, it was possible to predict maize response to P for soils occurring in different continents, provided the soils belonged to the same family. Although all soils in all three families responded in about the same manner to P applications, the final yield and the growth and development habits of the maize cultivar differed among soil families owing to differences in weather and climate.

Just about the time the BSP regression models were being finalized, word came from the Agricultural Research Service crop modelling group in Temple, Texas that a maize simulation model had been developed and was ready for testing. The Temple group asked if it could obtain crop, soil, weather, and management data for each benchmark site as input data to enable the maize model to predict grain yield and total biomass production. Figures 6 and 7 illustrate how well the model output compares with observed results. What was remarkable about the maize simulation model was its ability to predict the growth, development, and final yield of a maize cultivar in locations never before visited by the model or model developers. What was required by the model to predict outcomes of genotype-by-environment-by-management interactions was a site-specific minimum data set of the soil-plant-atmosphere continuum.

In 1983 a new project was initiated to develop and validate simulation models for eleven food crops including four cereals (rice, wheat, barley, sorghum, and millet), three grain legumes (groundnut, dry beans, and soybean), and three root crops (potato, cassava, and taro). The purpose of the new project known as the International Benchmark Sites Network for Agrotechnology Transfer (IBSNAT) project, was to enable users to match the biological requirements of crops to the physical characteristics of land by conducting agronomic experiments in the computer (Uehara and Tsuji, 1991).

## VIII. Transferring Results of Long-term Trials

There are three general ways of learning about the long-term behavior and performance of agroecosystems. They are (1) by trial-and-error, (2) by analogy, and (3) by systems analysis and simulation. The BSP was an attempt to circumvent the high cost and slowness of learning by trial-and-error. It tried to do so by enabling users of Soil Taxonomy to take the hard-earned lessons gained through long-term trial-and-error experience to other similar agroenvironments by analogy. Soil Taxonomy specified the attributes that make soils behave and perform alike.

The weakness of Soil Taxonomy, and other classification systems like it, is that attributes employed to define an object are more often than not a range of values of one of its properties. The mean annual soil temperature is an example of average characteristics used as differentiating criteria in Soil Taxonomy. What is pertinent to agroecosystems analysis is not only annual or monthly means but daily and even hourly means of variables that affect crop development and overall systems performance.

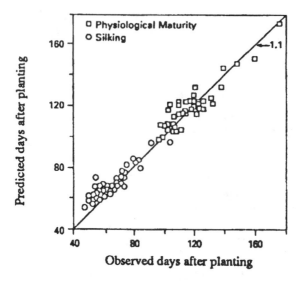

**Figure 6.** Comparison of predicted and observed maize yield. The CERES-Maize model was used to simulate grain yields of experiments conducted in seven locations on three soil types in Hawaii, Indonesia, and the Philippines over a period of eight years. (From Uehara and Tsuji, 1991.)

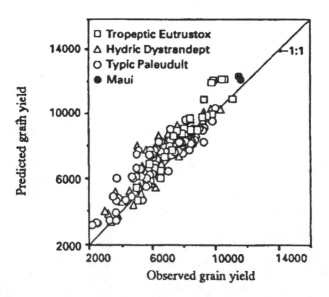

**Figure 7.** Comparison of predicted and observed days to silking and physiological maturity of maize. The CERES-Maize model was used to simulate the phenological events of maize grown in seven sites, on three soil types in Hawaii, Indonesia, and Philippines over a period of eight years. (From Uehara and Tsuji, 1991.)

A crop that performs optimally at 20° C but suffers irreversibly when exposed to a few hours of temperatures at 10° C or 30° C needs more than average soil temperatures or even a range of average values to enable the crop's temperature requirement to be matched to highly variable air temperatures. Dynamic, process-based simulation models can be used to supplement learning by trial-and-error and analogy and should increasingly replace much of the tedious and expensive trial-and-error field experiments.

## IX. Conclusion

Long-term trials concern themselves with temporal changes in agroecosystems. The BSP concerned itself with the transfer of long-term experience from one location to other locations with similar agroenvironments as defined by the soil family. Thus the results of the BSP can be employed to extend results of long-term experiments to similar environments elsewhere in the world. The BSP, however, was not designed to extend results of long-term experiments conducted in one agroenvironment to other sites with dissimilar soil environments.

Application of systems analysis and simulations approaches by the IBSNAT project offers hope that long-term agroecosystems performance can be predicted, provided the underlying processes affecting long-term outcomes are known and a minimum data set of the key driving variables is known. The existing long-term trials can serve as primary sites for investigating processes and defining the minimum data set needed to predict long-term performance in dissimilar agroenvironments.

## References

Buchanan, F. 1807. *A journey from Madras through the Countries of Mysore, Canara and Malabar*. Vol. 2. East India Company, London.

Benchmark Soils Project. 1982. Procedures and guidelines for agrotechnology transfer experiments with maize in a network of benchmark soils. HITAHR Resource Extension Series 015 (BSP Technical Report 3). Hawaii Inst. of Trop. Agr. and Human Resources, College of Trop. Agr. and Human Resources, University of Hawaii. 64 pp.

Cady, F.B., C.P.Y. Chan, J.A. Silva, and C.L. Wood. 1985. Transfer of Yield responses to phosphorus and nitrogen fertilizer. p. 55-73. In: I.A. Silva (ed.) *Soil-Based Agrotechnology Transfer*. Benchmark Soils Project. Hawaii Inst. for Trop. Agr. and Human Resources. College of Trop. Agr. and Human Resources, University of Hawaii.

Laird, R.J. and A. Turrent. 1981. Key elements in field experimentation for generating crop production technology. In: J. A. Silva (ed.), *Experimental designs for predicting crop productivity with environmental and economic inputs for agrotechnology transfer.* Dept. Paper 49. Hawaii Inst. of Trop. Agr. and Human Resources, College of Trop. Agr. and Human Resources, University of Hawaii.

Mohr, E.C., F.A. vanBaren., and J. vanSchuylenborgh. *Tropical Soils: A Critical Study of Soil Genesis as Related to Climate, Rock, and Vegetation.* Interscience, New York.

Prescott, J.A. and R.L. Pendleton. 1952. Laterite and lateritic soils. *Comm. Bur. Soil Sci. Techn. Commun.* 47:1-51.

Silva, J.A. 1985. *Soil-Based Agrotechnology Transfer.* Benchmarks Soils Project. Hawaii Institute of Trop. Agr. and Human Resources, College of Trop. Agr. and Human Resources, University of Hawaii.

Soil Survey Staff. 1975. *Soil Taxonomy. A Basic System for Making and Interpreting Soil Surveys.* Agricultural Handbook No. 436. U.S. Government Printing Office, Washington, D.C.

Soil Survey Staff. 1960. *Soil Classification. A Comprehensive System, 7th Approximation.* U.S. Department of Agriculture, Soil Conservation Service, Washington, D.C.

Uehara, G. and G.Y. Tsuji. 1991. Progress in Crop Modelling in the IBSNAT Project. In: R.C. Muchow and J.A. Bellamy (eds.), *Climate Risk in Crop Production.* CAB International. London, United Kingdom.

Uehara, G. and G.P. Gillam. 1981. *The Mineralogy, Chemistry, and Physics of Soils with Variable Charge Clays.* Westview Press, Boulder, CO. 170 pp.

# Technological Options for Sustainable Management of Alfisols and Ultisols in Nigeria

R. Lal

## I. Introduction

The population of Sub-Saharan Africa was 308 million in 1975, 418 million in 1985, 490 million in 1990, and 506 million in 1991 (FAO, 1991). Compared with 1975, the population of Sub-Saharan Africa in 1991 increased by 64%. The region's population is expected to grow by an average of 25 million a year up to 2050 from a total population of 490 million in 1990 to 1.38 billion in 2025 and 2.1 billion in 2050 (Blake et al., 1994). It is also estimated that one-third of all Africans, about 170 million people, had inadequate diet in 1991. For the decade ending in 1991, the per capita agricultural production in Sub-Saharan Africa declined by 8.2%   (FAO, 1991).

This imbalance between food production and population increase in Sub-Saharan Africa is widely recognized. Consequently, there has been profusion of suggestions about improved technologies that may bring about an increase in agricultural production without degrading soils and the environment (Ragland and Lal, 1993). While basic principles underlying the conservation-effective and highly productive technologies may be known and well established, specificity of soils' properties and environmental factors limit the generalization of these technologies and necessitate on-site adaptation and validation in diverse soils and ecological conditions.

ISBN 1-56670-076-0

It is also the diversity of soils, environments, and socio-economic factors that have led to an apparent contradiction among results of some technological interventions. Apparent anomalies in biophysical factors are accentuated by wide diversity in social, economic, and political factors. Some relevant technological innovations that have produced site-specific results include:

- No-till methods of seedbed preparation and crop residue management. No-till with crop residue mulch is a preferred technique for soils of the humid and sub-humid regions (Lal, 1982; 1989a), but not for the semi-arid and arid tropics (Charreau and Nicou, 1971; Charreau, 1972; Nicou, 1974; Lal, 1975),
- Mulch farming based on live mulch systems: This practice, based on the principle of mixed cropping of a legume cover at the same time with a cereal, may be applicable for humid regions of soil moisture surplus rather than for arid and semi-arid regions of moisture deficit (Lal et al., 1978; Akobundu, 1982),
- Mixed cropping: The practice is a risk-avoidance system for resource-poor farmers (Okigbo and Greenland, 1976; Okigbo, 1978).
- Alley cropping: It is based on growing woody perennials and food crop annuals, and is also applicable to fertile soils in humid and sub-humid regions but not in drought-prone soils of low fertility in arid regions (Lal, 1989b) or in acid Ultisols (Szott, 1987),
- Manual land clearing: Clearing manually or by shear blade followed by biomass burning, in-situ or in windrows, has been found useful in some cases but not in others (Lal and Ghuman, 1989),
- Conservation effectiveness of mechanical vs. biological measure: Usefulness of these methods of erosion control has been an open debate since the 1950s (Lal, 1976; Greenland and Lal, 1977).

The fact remains that technological options for enhancing agricultural production and managing soil and water resources are soil- and ecoregion-specific. Yet, data from well designed, properly equipped, and adequately managed experiments continued for 10 to 20 years do not exist. Several potentially useful experiments have been initiated in Sub-Saharan Africa, but were rarely conducted for a long enough time to establish yield trends with regard to dynamics of soil properties.

Keeping in view the importance of understanding soil's potential and constraints, a series of long-term soil management experiments were conducted at the research farm of the International Institute of Tropical Agriculture (IITA). These experiments were designed to study dynamics of soil properties under a wide range of soil and crop management systems. Important considerations in establishing these experiments was to identify critical soil factors that affect crop yields, and to establish critical levels of soil properties in relation to soil degradation and agricultural productivity.

The objective of this report is to assess technological options for sustainable management of soil and water resources in sub-humid regions of west Africa.

Sustainability was assessed by monitoring productivity and dynamics of changes in soil properties that influenced yield trends. In addition, long-term experiments were conducted to understand processes affecting soil erosion and degradation. Technological options for sustainable management were evaluated on the basis of a series of long-term (8-10 year) experiments conducted on Alfisols in western Nigeria.

## II. Materials and Methods with Regard to Long-Term Experiments

Long-term field experiments were conducted at the research farm of the International Institute of Tropical Agriculture (IITA), Ibadan, Nigeria (7° 30' N, 3° 54' E). The farm is located about 30 km south of the northern limit of the tropical rainforest. The region has a bimodal rainfall distribution with long-term mean average rainfall of about 1250 mm. The first and the longer growing season begins about the end of March and continues until mid-July. The shorter second season begins in late August and ends abruptly in early November.

Soils of the experimental site are classified as Oxic Paleustalf. These soils have low to medium inherent soil fertility and are characterized by a coarse-textured surface horizon and a clayey $B_{2t}$ sub-soil horizon. Soil profile is characterized with a gravelly horizon beginning 30 to 60 cm below the soil surface. The gravelly horizon varies from 20 to 50 cm in thickness, with gravel concentration ranging from 30 to 70% by weight.

Shifting cultivation, the predominant farming system of the region, is characterized by mixed cropping of root crops (e.g., *Manihoc esculenta* and *Dioscorea rotundata*) with maize (*Zea mays*) and cowpea (*Vigna unguiculata*). These crops are grown with little or no off-farm inputs in a partially cleared land where some economically useful trees are maintained and crops are grown in association with these trees. Common trees maintained include oil palm (*Elaies guineensis*), cocoa (*Theobroma cacao*), cola (*Cola nitida*), banana (*Musa* spp), and several other native trees.

When IITA's research farm was developed in early 1970s, secondary forest vegetation of about 20-year growth was cut manually with cutless, ax, chain saw, and diggers. Roots and stumps were excavated to about a 30-cm depth. Felled biomass was burnt in-situ, and unburnt material was removed to the field boundary. Graded channel terraces were constructed at 1 m contour interval, and terrace channels drained into specially constructed waterways grown to bahia grass (*Paspalum notatum*). The results of two long-term experiments are presented in this report.

## A. No-till Farming

These experiments, conducted from 1979 through 1987, were designed to evaluated the effects of residue retention and fertilizer application for two tillage methods on maize grain yield. Two tillage methods were no-till and plow-till. No-till system involved seeding maize directly into the residue of the previous crop but following applications of 2.5 L ha$^{-1}$ of paraquat (1-1-dimethyl 4.4 bipyridinium ion). The plow-till system of seedbed preparation involved disc plowing to about a 20-cm depth followed by harrowing. Tillage treatments were established on main    plots and 4 combinations of fertilizer and residue management as sub-plots. These combinations were (1) no fertilizer and residue removed ($F_oR_o$), (2) no fertilizer and residue retained ($F_oR_1$), (3) with recommended fertilizer and residue removed ($F_1R_o$), and (4) with recommended fertilizer and residue retained ($F_1R_1$). Two crops of maize were grown every year, and soil properties for 0- to 5-cm and 5- to 10-cm depth were measured after harvesting the second crop. Soil physical properties measured included infiltration rate, bulk density and penetration resistance measured *in situ*, and soil chemical properties measured in the laboratory on disturbed and sieved samples.

## B. Tropical Deforestation

This watershed management experiment was conducted from 1978 through 1987 (Lal, 1992). There were 14 watersheds of about 2 to 4 ha each. Each watershed was equipped with a rate measuring H-Flume, a water stage recorder, and a Coshocton Wheel Sampler. These watershed experiments were conducted in two phases:

Phase 1:   Principal objective of Phase 1 of the project, lasting three years from 1979 to 1981, was to study the effects of methods of land clearing and post-clearing land development-cum-tillage methods on runoff, erosion, and crop growth and yield. There were six land clearing and tillage methods.

Phase 2 :   Principal objective of Phase 2, lasting from 1982 through 1987, was to assess the effects of farming systems on soil properties, runoff and erosion and crop growth and yield. Detailed objectives of both phases were described by Lal and Couper (1990), and Lal (1992). This experiment was continued through 1988. Soil samples from 0- to 5-cm depth were obtained during 1990-91 and analyzed for chemical and physical properties to assess impact of management systems (Hulugalle, 1994).

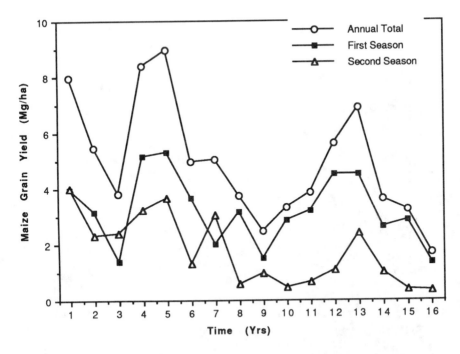

**Figure 1.** Temporal changes in maize grain yield with plow-till.

## III. Principal Results of Long-Term Experiments

### A. No-till Farming

The data of the effects of two tillage methods for $F_1R_1$ treatment combination on maize grain yield for 16 consecutive years are shown for the plow-till method in Figure 1 and for the no-till method in Figure 2. The wide variability in grain yield among seasons was attributed to variation in seasonal distribution and rainfall patterns. Low yield for the second season was mainly due to low total rainfall and erratic distribution. The data in Figure 1 show that maize grain yield in the first season exceeded that of the second season in 13 out of 16 years. The range of grain yield was 1.4 to 5.3 Mg ha$^{-1}$ for the first season and 0.35 to 3.15 Mg ha$^{-1}$ for the second. In addition to differences in yield among seasons, there was also a wide variation in annual total yield among years. The annual grain yield of maize ranged from 3.55 to 8.95 Mg ha$^{-1}$. Similar to the trends in seasonal yield, differences in annual total yield among years was also attributed to variations in rainfall patterns.

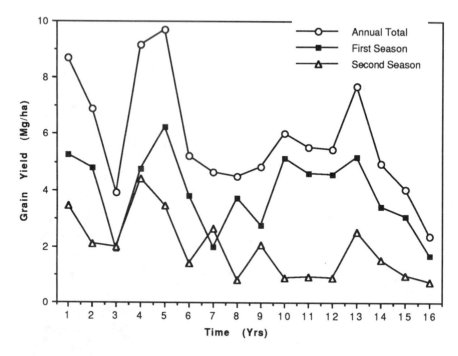

**Figure 2.** Temporal changes in maize grain yield with no-till.

The data in Figure 2 show variations in seasonal and annual total grain yield of maize for the no-till method. Similar to the yields in plow-till method, there were also wide variations in grain yield among seasons. Maize grain yield ranged from 1.85 Mg ha⁻¹ to 6.50 Mg ha⁻¹ in the first season, and from 0.70 Mg ha⁻¹ to 4.40 Mg ha⁻¹ in the second. Maize grain yield in the first season exceeded that of the second for 14 out of 16 seasons. Annual total grain yield in the no-till method ranged from 2.6 Mg ha⁻¹ to 9.7 Mg ha⁻¹.

The plot of seasonal data of maize yield do not clearly indicate yield trends over time. The data in Figure 3 is a plot of average yield for four consecutive seasons or years and show that grain yield declined with time. The rate of yield decline was more drastic for the second than for the first season, and more drastic for plow-till than for the no-till method of seedbed preparation (Figures 4 and 5). The data in Table 1 show regression equations showing temporal changes in maize grain yield for the plow-till system. The rate of yield decline was about 0.05 Mg ha⁻¹ yr⁻¹ in the first season compared with about 0.15 Mg ha⁻¹ yr⁻¹ for the second. In comparison, the data in Table 2 show that the rate of decline in maize grain yield with no-till method was about 0.06 Mg ha⁻¹ yr⁻¹ for the first season compared with 0.12 Mg ha⁻¹ yr⁻¹ for the second. Differences in the rate of decline in annual maize grain yield with regards to the tillage

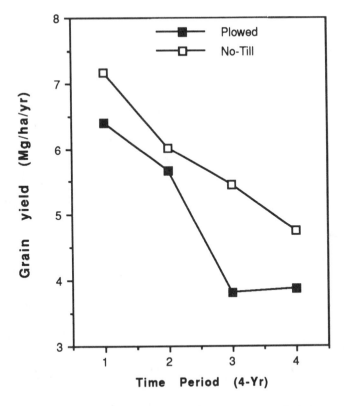

**Figure 3.** Changes in annual yield for four 4-year periods.

methods are shown by the data in Table 3. The rate of decline in annual grain yield was about 0.14 Mg ha$^{-1}$ yr$^{-1}$ for the no-till method compared with 0.17 Mg ha$^{-1}$ yr$^{-1}$ for the plow-till method of seedbed preparation.

It is apparent from the data in Figsures 1 to 5 and Tables 1 to 3 that the system based on continuous cultivation of monocrop maize for the 16 consecutive years was unsustainable even with fertilizer application at the recommended rate, and with no-till farming with crop residue mulch. No-till farming with residue mulch and fertilizer application, however, may have been sustainable by adoption of appropriate crop rotations or mixed cropping systems (Okigbo and Greenland, 1976; Okigbo, 1978).

There are several possible reasons for this decline in yield. Apparently, soil compaction, deterioration in soil structure, or soil erosion were not among the major reasons for decrease in the no-till system of seedbed preparation because there was no vehicular traffic and the crop residue mulch promoted activity of soil fauna especially that of the earthworms. However, crusting, compaction, deterioration in soil structure, and accelerated erosion were evident in plow-till systems probably due to decline in activity of soil fauna and exposure of the soil

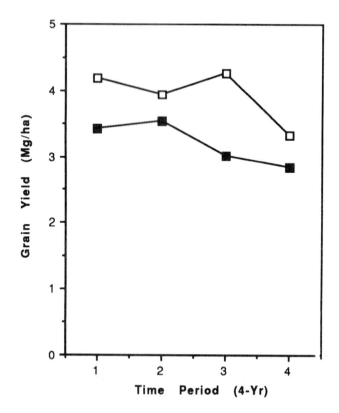

**Figure 4.** Temporal changes in grain yield of maize in the first growing season.

to climatic elements. Decline of grain yield in no-till plots may not have been due to changes in soil chemical properties either.

Changes in soil chemical properties were more drastic in plow-till compared with no-till systems of seedbed preparation (Table 4). For example, soil organic carbon content declined but little from 2.33% in 1970 to 2.15% in 1986 in no-till compared with a drastic decline of 2.33% in 1970 to 0.93% in 1986 in plow-till method of seedbed preparation. Similarly, decline in soil pH was also more in plow-till than no-till method of seedbed preparation. Decline in pH was from 6.8 in 1970 to 5.5 in no-till compared with 5.1 in plow-till in 1986. Tillage-related changes in cation exchange capacity (CEC) were in accord with those in pH. Decline in pH was from 7.30 Meq/100 g to 5.4 Meq/100 g in no-till compared with 3.0 Meq/100 g to 2.2 Meq/100 g in plow-till.

In addition to changes in soil properties, decline in maize grain yield may also have been due to rotation-related effects. Monoculture of maize based on 2 crops a year for 17 years can lead to build-up of pests including diseases and insects.

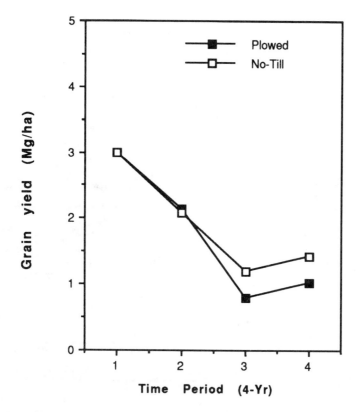

**Figure 5.** Temporal changes in grain yield of maize in the second growing season.

**Table 1.** Temporal changes in maize grain yield in the plowed system

| Season | Regression equation | $R^2$ |
|--------|---------------------|-------|
| First | $y = 3.57 - 0.05\ T$ | 0.20 |
| First | $y = 1.63 + 2.10\ T - 0.58\ T^2 + 0.054\ T^3 - 0.0016\ T^4$ | 0.57 |
| Second | $Y = 3.10 - 0.15\ T$ | 0.69 |
| Second | $Y = 1.44 + 1.25\ T - 0.31\ T^2 + 0.024\ T^3 - 0.0006\ T^4$ | 0.74 |

$y = $ Mg/ha/season; $T = $ years.

The data on changes in soil properties suggests that yield decline in no-till treatment may be primarily due to the lack of beneficial rotation effects, while that in the plow-till plots was due to a combination adverse changes in soil chemical and physical properties and to the lack of beneficial effects of crop rotation.

**Table 2.** Temporal changes in maize grain yield

| Season | Regression equation | $R^2$ |
|--------|---------------------|-------|
| First | $y = 4.22 - 0.06\ T$ | 0.21 |
| First | $y = 3.98 + 0.77T - 0.31\ T^2 + 0.035\ T^3 - 0.0011\ T^4$ | 0.58 |
| Second | $Y = 3.10 - 0.12\ T$ | 0.58 |
| Second | $Y = 1.77 + 0.76\ T - 0.16\ T^2 + 0.01\ T^3 - 0.0002\ T^4$ | 0.63 |

$y$ = Mg/ha/season; $T$ = years.

**Table 3.** Temporal changes in maize grain yield in the plowed system

| Season | Regression equation | $R^2$ |
|--------|---------------------|-------|
| No-till | $y = 7.07 - 0.136\ T$ | 0.36 |
| No-till | $y = 5.46 + 1.84\ T - 0.56\ T^2 + 0.053\ T^3 - 0.0016\ T^4$ | 0.47 |
| Plow-till | $Y = 6.46 - 0.17\ T$ | 0.41 |
| Plow-till | $Y = 2.85 + 3.85\ T - 0.95\ T^2 + 0.085\ T^3 - 0.0024\ T^4$ | 0.59 |

$y$ = Mg/ha/season; $T$ = years.

**Table 4.** Temporal changes in soil chemical properties

| Year | Organic carbon (%) NT | PT | pH (1:1 $H_2O$) NT | PT | Exchangeable Ca (meq/100 g) NT | PT |
|------|------|------|------|------|------|------|
| 1970 | 2.33 | 2.33 | 6.8 | 6.8 | - | - |
| 1981 | 2.12 | 2.12 | 6.5 | 5.9 | 7.3 | 3.0 |
| 1984 | 1.89 | 1.37 | 5.4 | 5.5 | 3.4 | 1.7 |
| 1986 | 2.15 | 0.93 | 5.5 | 5.1 | 5.4 | 2.2 |

## B. Tropical Deforestation

Detail description of the results for Phases I and II were presented by Lal (1992). Results of soil analyses reported by Hulugalle (1994) are summarized in Table 5. The data are presented in 3 sections dealing with the effects of land clearance methods, tillage systems, and cropping systems. Mechanized land clearance methods had significant effects on clay and sand content. In comparison with manual and traditional clearance methods, mechanized methods decreased sand content and increased clay content probably due to accelerated erosion and exposure of the clayey sub-soil. Infiltration rate was significantly more in traditional plots. There were significant effects of tillage methods on soil properties. Infiltration rate was significantly more in no-till and traditional treatments than in the plow-till method. Soil organic carbon content and CEC were also more in no-till compared with the plow-till method. In contrast, total acidity was much higher in plow-till than they were in the no-till system. Significant effects of cropping systems were observed only in pH and total acidity. Soil pH was the least in alley cropping and the highest in pastures. Accordingly, total acidity was the highest in alley cropping and least in pasture (Table 5).

Results of this long-term experiment presented here and elsewhere (Lal, 1992) show the adverse effects of mechanized land clearance methods, plow-till system of seedbed preparation, and continuous monoculture on soil properties and productivity. Despite the use of several innovative systems (e.g., no-till, alley cropping, mucuna fallowing), none of the systems tried was effective in reversing the soil degradative trends.

## IV. Identification of Soil-Related Constraints to Agricultural Productivity and Sustainability

The data of four long-term experiments presented in this report indicate that Alfisols of the sub-humid West African region are highly prone to soil degradation by physical, chemical, and biological processes. Principal among soil physical degradative processes is decline in soil structure. Soil structure is extremely susceptible to deforestation, especially by mechanical methods of forest removal, and mechanized tillage techniques e.g., plowing, discing, rotovation, etc. (Lal, 1984; Lal and Couper, 1990; Lal, 1992). Structural decline involves reduction in total aggregation and stability of aggregates, with attendant decreases in volume of macropores and biopores. Decline in soil structure, also caused by drastic reduction in activity and species diversity of soil fauna, sets in motion other degradative trends including crusting, compaction, reduction in infiltration rate, high runoff, and accelerated erosion.

Soil physical and biological degradation go hand-in-hand. Soil biological degradation involves decline in activity and species diversity of soil fauna, reduction in soil organic carbon content, and decrease in active or biomass carbon. Negative impact of soil biological processes on soil structure was particularly evident in the twin watershed experiment involving comparative assessment of two tillage methods (Lal, 1984; 1985). Soil application of furadan, eliminated soil biological activity and led to the collapse of soil structure (Lal, 1982; 1983). In addition to impact on soil physical properties, decline in soil biological properties also impacts soil chemical properties and processes.

Soil chemical degradation was set-in-motion by decrease in organic matter, loss of clay and other colloids, reduction in cation exchange capacity, loss of bases, and decrease in soil pH. Rapid decline in pH, even with modest or low rates of fertilizer application, is a strong indication of soil chemical degradation. Loss of bases is caused by crop uptake, leaching, or runoff and erosion.

Critical soil factors with strong impact on agronomic productivity and sustainability are outlined in Table 6 and Figure 6 (Lal, 1994). Soil structure is the most important critical factor in the sustainable use of soil and water resources in this ecoregion. However, it is difficult to determine the critical

**Table 5.** Effects of land clearing methods on soil properties

| Treatment | pH | Clay content (%) | Sand content (%) |
|---|---|---|---|
| A. Land clearance methods | | | |
| Manual (n = 9) | 5.6 ± 0.2 | 25.0 ± 1.4 | 63.2 ± 1.9 |
| Shearblade (n = 6) | 5.7 ± 0.2 | 30.5 ± 1.7 | 56.8 ± 2.3 |
| Treepusher/rootrake (n = 9) | 5.4 ± 0.2 | 36.3 ± 1.4 | 51.7 ± 1.9 |
| Traditional ( n = 6) | 5.6 ± 0.2 | 26.2 ± 1.7 | 61.0 ± 2.3 |
| AOV | NS | 0.001 | 0.01 |
| | | | |
| B. Tillage systems | | | |
| Plow-till (n = 9) | 5.3 ± 0.2 | 29.4 ± 2.0 | 58.1 ± 2.4 |
| No-till (n = 15) | 5.7 ± 0.1 | 31.3 ± 1.6 | 56.8 ± 1.9 |
| Traditional (n = 6) | 5.4 ± 0.2 | 26.2 ± 2.5 | 61.0 ± 2.9 |
| AOV | NS | NS | NS |
| | | | |
| C. Cropping systems | | | |
| Annual cropping (n = 9) | 5.4 ± 0.2 | 29.0 ± 2.1 | 59.2 ± 2.4 |
| Alley cropping (n = 6+ | 5.1 ± 0.1 | 32.0 ± 2.6 | 55.5 ± 3.0 |
| Pasture (n = 6) | 5.7 ± 0.1 | 29.2 ± 1.6 | 58.3 ± 1.9 |
| AOV | 0.05 | NS | NS |

AOV = one way analysis of variance expression as level of probability; NS = not significant.
(Adapted from Hulgalle, 1994.)

limits of this important factor. Critical limits refer to the range beyond which soil's functions are adversely affected. An important indicator of soil structure is infiltration capacity. Runoff and accelerated erosion become a serious problem in these soils when infiltration is below 50-75 mm ha$^{-1}$ (Table 7). Suggested range of critical limits of other agronomically important soil physical properties are bulk density of 1.35-1.45 Mg m$^{-3}$ and rooting depth of 15-20 cm. Because of edaphologically inferior sub-soil, high concentration of coarse fragments imbedded in the matrix of low-activity clay, low effective rooting depth becomes a critical factor by accelerated soil erosion. Therefore, tolerable range of soil loss may be barely 1 to 2 Mg ha$^{-1}$ yr$^{-1}$.

An important indicator of soil biological properties is the organic carbon content. The critical limit of soil organic carbon content for these Alfisols is 1.0 to 1.1%. However, more important than the total soil organic carbon content is the biomass or the active carbon content. Soil structure is easily degraded if the biomass carbon content falls below 25 to 33%. Reduction in biomass carbon is reflected in the drastic decrease in soil biodiversity especially that of macrofauna e.g., earthworms.

**Table 5.** continued--

| Infiltration rate (mm mm$^{-1}$) | Organic carbon (%) | Bray-1 P (mg kg$^{-1}$) | CEC (m mol kg$^{-1}$) | Total acidity (m mol kg$^{-1}$) |
|---|---|---|---|---|
| 0.7 ± 0.2 | 1.22 ± 0.12 | 33.4 | 51.6 ± 5.7 | 2.8 ± 0.6 |
| 0.5 ± 0.3 | 1.69 ± 0.14 | 20.3 | 69.5 ± 6.9 | 0.7 ± 0.8 |
| 0.7 ± 0.2 | 1.60 ± 0.12 | 20.5 | 62.9 ± 5.7 | 2.9 ± 0.6 |
| 2.3 ± 0.3 | 1.13 ± 0.14 | 23.6 | 52.0 ± 6.9 | 3.7 ± 0.8 |
| 0.001 | 0.05 | NS | 0.05 | NS |
| | | | | |
| 0.4 ± 0.2 | 1.29 ± 0.12 | 40.4 | 51.6 ± 5.7 | 2.8 ± 0.6 |
| 0.8 ± 0.2 | 1.59 ± 0.10 | 20.3 | 69.5 ± 6.9 | 0.7 ± 0.8 |
| 2.3 ± 0.3 | 1.13 ± 0.15 | 23.6 | 62.9 ± 5.7 | 2.9 ± 0.6 |
| 0.001 | 0.05 | 0.05 | 0.05 | 0.05 |
| | | | | |
| 0.8 ± 0.3 | 1.37 ± 0.14 | 20.5 | 61.9 ± 6.0 | 2.6 ± 0.6 |
| 0.5 ± 0.4 | 1.34 ± 0.17 | 48.4 | 51.5 ± 7.3 | 4.6 ± 0.8 |
| 1.3 ± 0.2 | 1.46 ± 0.11 | 22.8 | 59.5 ± 4.6 | 1.8 ± 0.5 |
| NS | NS | NS | NS | 0.05 |

Base saturation, pH, and CEC are important soil chemical properties. Critical limits of chemical properties for these soils include pH range of 5.5 to 5.7 (1:1in H$_2$O), CEC of 4 to 5 meq 100 g$^{-1}$, and base saturation of 40 to 50% (Table 6).

Soil's response to fertilizers and amendments may also decline when soil chemical properties fall below the ranges indicated in Table 6.

**Table 6.** Critical soil factors on Alfisols

| Soil structure | Infiltration capacity |
|---|---|
| | Macroporosity |
| | Crusting |
| | Bulk density |
| | |
| Soil biodiversity | Earthworm activity, and other soil fauna |
| | Soil organic matter content, biomass carbon |
| | |
| Soil chemical properties | Cation exchange capacity |
| | Base saturation |
| | pH |

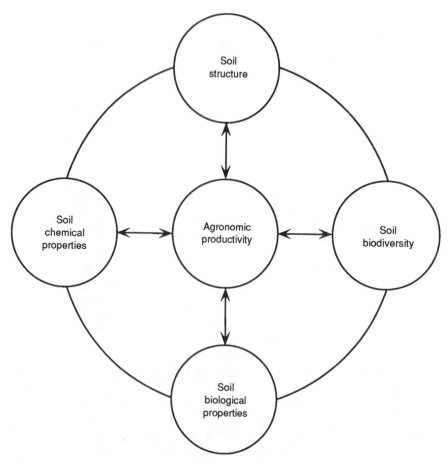

**Figure 6.** Critical soil factors affecting agronomic productivity of tropical Alfisols in southwestern Nigeria.

**Table 7.** Critical level of soil properties

| Property | Critical range |
|---|---|
| Organic carbon (%) | 1.0 - 1.1 |
| Biomass carbon | 25 - 33% of the total soil carbon |
| pH | 5.5 - 5.7 |
| Bulk density (Mg m$^{-3}$) | 1.35 - 1.45 |
| CEC (meq 100 g$^{-1}$) | 4 - 5 |
| Base saturation (%) | 40 - 50 |
| Infiltration capacity (mm hr$^{-1}$) | 50 - 75 |
| Soil loss tolerance (Mg ha$^{-1}$ yr$^{-1)}$ | 1 - 2 |
| Rooting depth (cm) | 15 - 20 |

# V. Limitations of Existing Long-Term Experiments

Soil management experiments described herein provided useful information on management effects on yield trends and changes in soil properties. Furthermore, analyses of the data obtained provided useful information on important soil degradative processes. However, these experiments had inherent limitations that makes generalization of the results to other regions questionable. Further, the cause-effect relationships were difficult to establish due to:

- short duration of the experiments. Preferred duration should be 25 years or more.
- non-independence of variables involved. Several properties and processes are inter-dependent.
- importance of exogenous factors (e.g., climate). The data from soil erosion experiments depended largely on rainfall intensity, frequency of rains, and antecedent moisture content.
- lack of inter-disciplinary support. Several key variables could not be studied because of the lack of support of some key disciplines e.g., soil biology, pesticide analyses, etc.
- change of treatments over time in some experiments e.g., tropical deforestation and watershed management project.

Nonetheless, these experiments provided some extremely useful data on soil-related constraints to intensive management. These experiments also highlighted the important of initiating new long-term soil management experiments that are specifically designed to establish the cause-effect relationship. These experiments should be designed in view of the constraints outlined above and focus on the following issues:

- defining critical limits of soil properties,
- standardizing methodology for quantitative assessment of soil degradation by different processes,
- assessing impact of different degrees of soil degradation on agronomic productivity in relation to land use and management systems,
- evaluating soil loss tolerance in relation to rate of weathering, and on-site and off-site effects of soil erosion, and
- quantifying the effects of land sue and management systems on greenhouse gas emissions from tropical ecosystems.

# VI. Conclusions

The data presented in this report from two long-term soil management experiments indicate the importance of soil physical properties and processes in sustainable use of Alfisols in the sub-humid region of western Africa. The data also highlight the significance of soil biodiversity and macrofauna on soil physical properties and several important soil processes. Although soil chemical properties and processes are also important, their relative significance to agronomic productivity and environmental quality is secondary to those of soil physical properties and processes. In addition, the data presented support the following conclusions:

- Mechanized methods of deforestation have severe detrimental impact on soil productivity and environmental quality,
- No-till farming based on residue mulching minimizes rate of soil degradation,
- Live mulches and agroforestry may retard on-set of soil degradative processes but also reduce yields through competition and alleleopathic effects,
- Soil erosion processes are not well understood especially with regards to climatic erosivity, slope length effect, and soil loss tolerance in relation to rate of weathering and impact on crop productivity,
- An important pre-requisite for development of economical and effective soil restorative measures is the understanding of critical limits of soil properties and processes in relation to agronomic productivity and environmental regulatory capacity.

## References

Akobundu, I.O. 1982. Live mulch for crop production in the tropics. *World Crops* 34:125-126, 144-145.

Blake, R.O., D.E. Bell, J.T. Mathews, R.S. McNamara, and M. Peter McPherson. 1994. Feeding 10 billion people in 2050: The key role of CGIAR's International Agricultural Research Centers, World Resources Institute, Washington, D.C.

Charreau, C. 1972. Probleme s poses par l'utilisation agricole des sols tropicaux par des cultures annueles. *Agron. Trop.* (Paris) 27:905-929.

Charreau, C. and R. Nicou. 1971. L'amerioration du profil cultural dens les sols sableux et sablo-argilleux de la zone tropicale seche Ouest Africaine et ses incidences agronomiques. *Agron. Trop.* (Paris) 26:1183-1247.

FAO. 1991. *Production yearbook. FAO*, Rome, Italy.

Greenland, D.J. and R. Lal (eds.), 1977. *Soil and Water Conservation in the Humid Tropics*. J. Wiley & Sons, Chichester, U.K., 278 pp.

Hulugalle, N.R. 1994. Long-term effects of land-clearing methods, tillage systems and cropping systems on surface soil properties of a tropical Alfisol in S.W. Nigeria. *Soil Use and Management* 10:25-30.

Lal, R. 1975. Role of mulching techniques in tropical soil and water management. *IITA Tech. Bull.* 1. 38 pp.

Lal, R. 1976. Soil Erosion Problems in Western Nigeria and Their Control. IITA Monograph 1. 208 pp.

Lal, R. 1982. No-till farming for soil and water conservation. IITA Monograph 3. Ibadan, Nigeria.

Lal, R. 1984. Mechanized tillage systems effects on soil erosion from an Alfisol in watersheds cropped to maize. *Soil and Tillage Res.* 4:349-360.

Lal, R. 1985. Mechanized tillage systems effects on properties of a tropical Alfisol in watersheds cropped to maize. *Soil and Tillage Res.* 6:149-162.

Lal, R. 1986. Effects of eight tillage treatments on soil properties and grain yield of maize. *J. Sci. Food and Agric.* 37:1073-1082.

Lal, R. 1989a. Conservation tillage for sustainable agriculture: Tropics vs. temperate environments. *Adv. Agron.* 42:85-197.

Lal, R. 1989b. Myths and scientific realities of agroforestry as a strategy for sustainable management for soils in the tropics. *Adv. Soil Sci.* 15:91-137.

Lal, R. 1992. Tropical agricultural hydrology and sustainability of agricultural systems. *The Ohio State University/IITA Tech. Bull.*, Columbus, OH. 303 pp.

Lal, R. 1994. Guidelines and methods for assessing sustainable use of soil and water resources in the tropics. SMSS-Ohio State University Report, 78 pp.

Lal, R. and D.C. Couper. 1990. A ten year watershed management study on agronomic productivity of different cropping systems in sub-humid region of Western Nigeria. *Topics in Applied Resource Management* 2:61-81.

Lal, R. and B.S. Ghuman. 1989. Biomass burning in windrows after clearing a tropical rainforest: Effects on soil properties, evaporation and crop yields. *Field Crops Res.* 22:247-255.

Lal, R., G.F. Wilson, and B.N. Okigbo. 1978. No-till farming after various grasses and leguminous cover crops. *Field Crops Res.* 1:71-84.

Nicou, R. 1974. Contribution on the study and improvement of the porosity of sand and sandy-clay soil in the dry tropical zone. *Agron. Trop.* (Paris) 29:110-127.

Okigbo, B.N. 1978. Cropping systems and related research in Africa. AAASA Occasional Publication Serios OT-1, 81 pp. Addis Ababa, Ethiopia.

Okigbo, B.N. and D.J. Greenland. 1976. Intercropping systems in tropical Africa. In R.I. Papendick, P.A. Sanchez, and G.B. Triplett, Jr. (eds.), *Multiple Cropping*. ASA Publication 27, American Society Agronomy, Madison, WI.

Ragland, J. and R. Lal (eds). 1993. Technologies for sustainable agriculture in the tropics. ASA Special Publication No. 56, American Society Agronomy, Madison, WI. 313 pp.

Szott, L.T. 1987. Improving the productivity of shifting cultivation in the Amazon Basin of Peru through the use of leguminous vegetation. Ph.D. Disst., North Carolina State Univ.

# Long-Term Fertilizer and Crop Residue Effects on Soil and Crop Yields in the Savanna Region of Côte d'Ivoire

Soumaïla Traoré and Peter J. Harris

ISBN 1-56670-076-0

## I. Introduction

Soils of the tropics have generally low inherent fertility, and the soils of the Côte d'Ivoire are no exception. Nitrogen and phosphorus are the major deficient nutrients. Soil fertility worsens with a combination of poor natural nutrient supply and mismanagement.

Traditional farming systems as resource management strategies in tropical Africa were based on extensive land use, high labor input, and indigenous tools (Lal, 1986). These systems were effective and stable in shifting cultivation, and were characterized by a few years of land use followed by a longer period of bush fallow (Ahn, 1979; Sanchez, 1976). The slash and burn system provided adequate nutrients from bush and tree ashes while the long fallow periods allowed the soil to replenish lost nutrients. However, the system requires the availability of large amounts of land while crop productivity remains low.

The decrease in available land and the need for higher crop productivity to provide food for a mounting population, are reducing the sustainability of the shifting cultivation system.

The obvious alternative, of shortening the fallow period, leaves no opportunity for land resource renewal; therefore food deficits, as a result of poorer land productivity, increase from year to year. In addition, there is the risk of rapid environmental degradation. In the absence of reliable information, generalizations, bordering on myths, have tended to develop in relation to the ability of tropical soils to sustain a permanent and productive agricultural system (Greenland et al., 1992; Mokwunye and Hammond, 1992; Sanchez and Logan, 1992).

Soil type and climate have a strong influence on agriculture systems in central Côte d'Ivoire where the vegetation is predominantly savanna (Le Buanec, 1979; Le Buanec and Didier de Saint Armand, 1975). Cotton is the major commercial crop in the savanna area although there is more intensive cotton cultivation in the northern part of the country. Central Côte d'Ivoire is traditionally the zone of arable crops, which are mainly yam followed by maize, peanut, and cassava. This is in contrast to the humid southern part of the Côte d'Ivoire which is largely dominated by "cash crops" such as cocoa, coffee, rubber, pineapple, coconut, and palm oil. The country is one of the world's leading producers of several of these cash crops. Consequently, farmers from the south are relatively better off as compared with their colleagues in the northern and the central areas.

Food production in a sustainable agriculture system in the savanna area and the increasing need to reduce the income gap between arable crop and cash crop farmers, is the prime mission of IDESSA. IDESSA, the Institut des Savannes, is the national research institute for agricultural and animal production in the savanna area. Its mission has been recently extended beyond the savanna area to include all research on food crops and animal production in the country.

To provide a basis for a sustained agricultural system, several long-term experiments were established in different ecosystems. Many of them are still running and deal with soil organic matter, soil nutrient depletion under continuous cropping, soil acidification in relation with fertilizer nitrogen sources, and the effect of different cropping systems on soil fertility under continuous soil use.

The long-term experiment to be described was established by French scientists in 1969.

## II. Description of the Experiment

### A. Objectives

In the savanna area, nitrogen deficiency is likely to appear during the first year when the land is cultivated, while deficiency may only occur after four to five years of land use after clearing from forest, as nutrients are progressively released from the litter and the soil organic matter (Sement, 1983). Poor crop responses to nitrogen fertilization in the West African sudan savanna have been reported by Christianson and Vlek (1991) despite the fact that, in lysimeter work, Chabalier (1980), reported little fertilizer nitrogen loss through leaching. It is therefore important to determine the reasons for poor crop response to nitrogen fertilizer.

In order to face the new challenges of sustainable agriculture in the savanna area, the long-term experiment was established to:

(1) evaluate crop yield performance under a continuous cropping system,

(2) monitor the effect of applied mineral nitrogen and crop residues and their interaction on crop yields and selected soil parameters,

(3) find means to improve management and nitrogen efficiency under local conditions,

(4) identify other factors that control crop productivity under continuous cropping systems, and

(5) propose suitable land management practices for sustainable agriculture for the area, with minimum inputs.

### B. Site Description

The experiment is situated at Bouaké, which is about 400 kilometres north from the Atlantic Ocean, the actual parameters are:

Latitude: $7^0$ 46' North

Longitude: $5^0$ 06' West

Altitude: 375 meters

The experimental field is part of the 150 ha of the Department of Arable Crops of the Central experimental station of IDESSA. It is 150 meters long and 50 meters wide giving a total area of 7500 $m^2$. The average monthly rainfall, the potential evapotranspiration, and the mean maximum and minimum air temperatures from 1969 to 1990 are presented in Table 1. The rainfall distribution in the

**Table 1.** Key climatic features averaged from 1969-1990 at IDESSA experiment station at Bouaké

|           | Rainfall | P.E.T. | Temperature °C | | |
|-----------|----------|--------|------|------|---------|
|           | mm       | mm     | Min  | Max  | Average |
| January   | 8.5      | 172.1  | 19.5 | 32.5 | 26.0    |
| February  | 43.1     | 170.8  | 20.5 | 33.5 | 27.1    |
| March     | 81.6     | 181.8  | 20.8 | 33.2 | 27.0    |
| April     | 121.3    | 166.0  | 20.8 | 31.9 | 26.3    |
| May       | 138.7    | 151.4  | 20.4 | 30.2 | 25.3    |
| June      | 130.9    | 124.3  | 19.9 | 28.7 | 24.3    |
| July      | 100.3    | 99.1   | 19.4 | 27.1 | 23.2    |
| August    | 158.2    | 90.9   | 19.3 | 27.0 | 23.2    |
| September | 163.9    | 105.8  | 19.5 | 27.9 | 23.7    |
| October   | 109.0    | 124.6  | 19.6 | 29.0 | 24.3    |
| November  | 24.5     | 120.6  | 19.9 | 30.3 | 25.1    |
| December  | 11.5     | 134.1  | 19.3 | 31.1 | 25.2    |

area is quite variable, and it has a bimodal distribution. The two broad peaks of the rainfall distribution curve each correspond to a crop-growing period. The first growing period starts with the establishment of the first rains, from March to April, and ends around June each year. The end of this cropping season coincides with the beginning of the small dry season, allowing safe harvesting of crops. The second part of the rainfall distribution curve corresponds to the most reliable cropping season because of higher and more evenly distributed rainfall (Gigou, 1973, Figure 1). The second cropping season, however, presents some disadvantages such as lower solar radiation and high insect and parasite pressure (Koné, 1991). During the first growing season, farmers face two main climatic constraints: the low total rainfall and the variability of the rainfall which can give rise to severe water stress during crop development. These periods of water stress may occur at any time after crop establishment.

The average annual rainfall measured from 1969 to 1990 on the experimental site is 1094.31 mm.

## C. Soil Type

Soils are typical desaturated ferrallitic soils in the French soil classification system (Chopart, 1989). Details of typical soil parameters are presented in Table 2, and are also described elsewhere (Chabalier, 1976; Chopart, 1984).

Long-Term Fertilizer and Crop Residue Effects in Côte d'Ivoire.

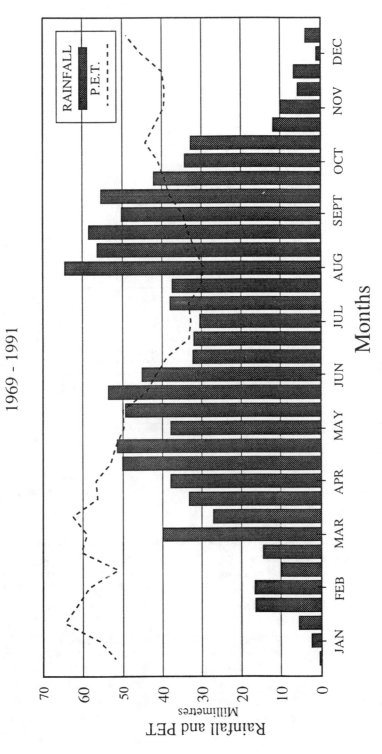

**Figure 1.** Average rainfall and potential evapotranspiration (PET).

**Table 2.** Physical and chemical soil parameters

| Parameters | 0 -20 cm | 20 - 40 cm |
|---|---|---|
| Clay (%) | 26.2 | 34.2 |
| Silt (%) | 9.6 | 5.5 |
| Fine sand (2mm-20µm) % | 18.6 | 18.7 |
| Coarse sand (>2mm) % | 45.6 | 41.6 |
| Carbon (%) | 1.40 | 0.84 |
| Total Nitrogen (%) | 0.12 | 0.07 |
| pH ($H_2O$) | 5.5 | 5.3 |
| Total Phosphorus (ppm) | 594 | 274 |
| Phosphorus Olsen (ppm) | 7 | - |
| Calcium meq/100g | 3.6 | 1.65 |
| Magnesium (meq/100g) | 1.3 | 0.5 |
| Potassium (meq/100g) | 0.6 | 0.09 |
| Sodium (meq/100g) | 0.05 | 0.02 |

**Table 3.** Treatments

| | Nitrogen rates | | |
|---|---|---|---|
| Crop residues rates | N0 | N1 | N2 |
| M0 | M0N0 | M0N1 | M0N2 |
| M1 | M1N0 | M1N1 | M1N2 |

Where    M0 = no crop residue, M1 = crop residue
N0 = no nitrogen, and N1 = first rate of nitrogen,
N2 = second rate of nitrogen

## D. Treatments

Three rates of nitrogen, as urea, were combined with and without incorporation of crop residue from the previous crop at the beginning of each cropping year. All treatments received equal amounts of lime, phosphorus, and potassium.

The experiment is a randomized complete block design with six treatments and six replicates. Each single plot measures 20 meters long and 7 meters wide. Treatments are presented in the Table 3.

Nitrogen application rate varied with the crop (Tables 4 and 5). Half of the nitrogen, in the form of urea, was applied shortly after crop emergence and the remainder at flowering. The nitrogen fertilizer was usually applied in a furrow 5 cm from the crop row and at about a 10-cm depth. Potassium and phosphorus were applied annually to all treatments at a rate of 90 kg $K_2O$ and 90 kg $P_2O_5$ per hectare respectively, broadcast before sowing. Potassium was applied as

Table 4. Summary of the principal features of maize management

| | 1970 | 1972 | 1974 | 1976 | 1978 | 1980 | 1982 | 1984 | 1986 | 1988 | 1990 |
|---|---|---|---|---|---|---|---|---|---|---|---|
| Variety | H507 | H507 | H507 | H507 | IRAT 83 | IRAT 83 | CJB | CJB | CJB | CJB | CJB |
| Date of sowing | April 02 | April 01 | April 12 | April 14 | April 05 | April 16 | March 29 | April 03 | March 01 | March 16 | April 24 |
| Date of harvest | Jul 30 | August 01 | August 02 | August 16 | August 03 | July 31 | July 19 | July 05 | July 02 | July 06 | August 08 |
| N0 | 0 | 0 | 0 | 0 | 0 | 0 | 0 | 0 | 0 | 0 | 0 |
| N1 (kgN ha$^{-1}$) | 100 | 100 | 100 | 100 | 100 | 80 | 80 | 80 | 80 | 80 | 80 |
| N2 | 200 | 200 | 200 | 200 | 200 | 160 | 160 | 160 | 160 | 160 | 160 |
| Date of 1st N application | NI | April 04 | April 09 | NI | April 17 | June 02 * | April 09 | NI | April 04 | March 16 | April 23 |
| Date of 2nd N application | NI | May 05 | June 06 | NI | May 26 | | May 11 | NI | April 15 | May 03 | June 11 |
| Rainfall (mm) SH | 599.4 | 420 | 381.7 | 362.1 | 516.5 | 617.5 | 390 | 331.3 | 447.1 | 558.3 | 332.6 |
| Rainfall (sow-mat) | 307 | 392.8 | 298.5 | 306 | 503 | 603 | 374 | 314 | 420 | 497 | 330 |
| Previous crop | rice | rice | rice | rice | rice | yam | yam | yam | yam | yam | fallow |

SH = sowing to harvest    NI = Non indicated    * = N was applied once    sow-mat = sowing to maturity

Table 5. Summary of the principal features of cotton management

| | 1970 | 1972 | 1974 | 1976 | 1978 | 1980 | 1982 | 1984 | 1986 | 1988 | 1990 |
|---|---|---|---|---|---|---|---|---|---|---|---|
| Sowing date | NI | August 30 | August 16 | July 05 | June 21 | July 22 | June 16 | August 21 | June 26 | June 20 | maize/ fallow |
| Variety | ALLEN | ALLEN | ALLEN | ALLEN | ALLEN | ISA205 | ISA205 | ISA205 | ISA205 | ISA205 | |
| Date of harvest | Dec 26 | Dec 26 | Dec 16 | Dec 20 | Dec 14 | Dec 28 | Nov 17 | Nov 19 | Dec 17 | Dec 13 | |
| kg N ha$^{-1}$ | | | | | | | | | | | |
| N0 | | 0 | 0 | 0 | 0 | 0 | 0 | 0 | 0 | 0 | |
| N1 | NI | 60 | 60 | 60 | 60 | 60 | 60 | 60 | 60 | 60 | |
| N2 | NI | 120 | 120 | 120 | 120 | 120 | 120 | 120 | 120 | 120 | |
| Date of 1st N | NI | August 28 | August 14 | October 08 | August 28 | August 23 | August 09 | August 27 | August 04 | July 29 | |
| date of 2nd N | NI | Oct 10 | Oct 17 | * | NI | Sept 17 | Sept 08 | Sept 18 | Sept 01 | August 18 | |
| Rainfall (mm) SH | | 207.9 | 445.9 | 469.2 | 386.8 | 869.6 | 460.5 | 270.4 | 532.9 | 816.2 | |
| Previous crop | maize | maize | maize | maize | maize | maize | maize | maize | maize | maize | fallow |

NI = Non indicated     * = N was applied once     SH = Sowing to harvest potassium chloride and phosphorus as single superphosphate, triple superphosphate, or calcium phosphate. Dolomite (30% CaO and 20% MgO) was applied at intervals to the plots at variable rates from 200 to 250 kg ha$^{-1}$. All nutrients, except nitrogen, were applied together before planting.

## E. Cropping System

From 1969 to 1978 the crop succession was as follows:
First year = Rice (which occupied the land during both the first and the second cropping periods);
Second year = Maize during the first cropping period and Cotton during the second period.
From 1979 to 1990 crop succession changed to:
First year = Yam (which occupied the land during both cropping periods); second year = Maize during the first cropping season and Cotton during the second.
Maize and cotton are cultivated in the year in a sequence. This succession ((rice and maize/cotton) or (yam and maize/cotton)) allows three crops to be cultivated in two years. These two cropping systems have been used during the experimental period. The first cropping system with rice will be referred to as the rice-based cropping system whereas the second with yam will be referred to as the yam-based cropping system.

## F. Cultivation Practices

The experimental field is ploughed every year to 25- to 30-cm depth. Maize and cotton are sown in rows of 20-meters length separated by 80-cm alleys. They were thinned to two plants per hole, with plant separation of 25 cm in the row. Each individual plot holds nine maize or cotton rows. Cotton is sown just before the maize is harvested, in the inter-rows. After the maize grain was harvested, the stalks were cut down on the inter-rows of the emergent cotton plants for the M1 crop residue treatment plots, for the M0 treatment the residues were removed. For rice the gap separating the rows is 40-cm wide. Yam is cultivated in ridge-tilled rows 1.35 m apart. Each plot contains nine rows for maize and cotton, 18 rows for rice, and five rows for yam. A summary of the principal features of crop management are presented in Tables 4 and 5. All crops were treated with pre-emergence herbicide whenever possible. Rice was treated with 4 l ha$^{-1}$ of *Ronstar* (250 g l$^{-1}$ of Oxadiazon; *Rhone-Poulenc*), and maize with 4 l ha-1 of *Primagram* (250 g l$^{-1}$ of Atrazine + 250 g l$^{-1}$ of Metolachlor; *Ciba-Geigy/Sochim*). Yam received 2 kg ha$^{-1}$ of *Sencor* (70 % of Metribuzine; *Bayer*). Cotton is treated with 4 l ha$^{-1}$ of *Cotodon* (160 g l$^{-1}$ of Metolachlor + 240 g l$^{-1}$ of Dipropetryne; *Ciba-Geigy/Sochim*) as pre-emergence herbicide. Cotton requires regular treatment with pesticides. *Decis* was used to treat cotton and other crops, when necessary, for insect attacks. Cotton was treated four to six times during the growth period.

### G. Harvesting

Maize was harvested as hard grain, allowing flexibility in the actual date of harvest; therefore, the length of the season fluctuated (Table 4). At harvest, only the three central rows were used for the calculation of yields; the last 50 cm from each end of the rows were also excluded. The total harvested area was 2.4 x 19 m = 45.6 m$^2$; grain and dry matter from this area was used for yield calculations. The harvest area is the same for cotton, although cotton is harvested two to three times as flowering is in two to three flushes. For rice, the six central rows were used excluding 1 m at each end of the rows. The harvest area used for yield calculations was 2.4 m x 18 m = 43.2 m$^2$. For yam, the three inner ridges were used for the calculations of yields, again the extreme 50 cm of each end of the ridge were excluded from the harvest area; the harvest area of yam was 1.35 m x 3 x 19 m = 76.95 m$^2$. This approach limited border effects from adjacent plots. The dry matter of each crop from the harvest area was weighed immediately after grain or boll harvest. A random sample of about one kilogram was collected, and oven dried for 48 hours at 80$^0$C for an estimate of moisture content. Yam was harvested as fresh tubers, lifted from the soil and weighed. The yields are expressed in tonnes fresh weight per hectare but tuber samples were taken also for the estimation of water content.

### H. Soil and Plant Analyses

Soil pH was measured using a standard glass electrode with a 1:1 soil water mixture. Soil and plant nitrogen contents were measured by Auto-analyzer after Keldjahl digestion (IDESSA, 1984). Soil organic carbon was determined using the dichromate method (Walkley and Black, 1934).

## III. Crop Yield Results

In this account, only the maize grain and cotton yields will be fully examined since they were grown throughout the experimental period.

### A. Maize Yields

Maize grain yield is presented in Table 6 and is characterized by its variability throughout the experimental period. Maize was cultivated every other year from

**Table 6.** Maize grain yield (t ha⁻¹)

| | 1972 | 1974 | 1976 | 1978 | 1980 | 1982 | 1984 | 1986 | 1988 | 1990 |
|---|---|---|---|---|---|---|---|---|---|---|
| M0N0 | 3.12 | 3.60 | 3.25 | 4.66 | 4.52 | 1.94 | 1.48 | 1.64 | 1.31 | 3.20 |
| M0N1 | 3.26 | 5.75 | 4.80 | 5.3 | 4.75 | 4.43 | 3.07 | 1.74 | 1.96 | 2.94 |
| M0N2 | 3.32 | 6.31 | 3.86 | 6.08 | 4.89 | 4.34 | 2.65 | 1.82 | 2.04 | 2.52 |
| M1N0 | 2.93 | 3.55 | 4.0 | 5.23 | 4.66 | 3.12 | 1.78 | 1.83 | 1.49 | 3.50 |
| M1N1 | 3.47 | 6.07 | 5.23 | 6.20 | 4.97 | 4.86 | 3.21 | 1.88 | 1.99 | 3.57 |
| M1N2 | 3.35 | 5.89 | 4.55 | 6.44 | 5.34 | 4.60 | 2.62 | 1.65 | 1.88 | 3.40 |
| SD | 0.28 | 0.69 | 0.54 | 0.80 | 0.37 | 0.68 | 0.28 | 0.36 | 0.31 | 0.40 |
| AV M0 | 3.19x | 5.22x | 3.97y | 5.35y | 4.72y | 3.57y | 2.40x | 1.73x | 1.70x | 2.89y |
| AV M1 | 3.20x | 5.17x | 4.59x | 5.96x | 4.99x | 4.19x | 2.54x | 1.78x | 1.79x | 3.49x |
| AV N0 | 3.03b | 3.58b | 3.63c | 4.95b | 4.59b | 2.53b | 1.63c | 1.73a | 1.40b | 3.35a |
| AV N1 | 3.37a | 5.91a | 5.02a | 5.75a | 4.86ab | 4.65a | 3.14a | 1.81a | 1.88a | 3.25a |
| AV N2 | 3.34a | 6.10a | 4.20b | 6.26a | 5.12b | 4.47a | 2.64b | 1.73a | 1.96a | 3.96a |
| CV % | 8.7 | 13.3 | 12.0 | 14.2 | 7.5 | 17.5 | 11.5 | 12.8 | 17.3 | 12.7 |
| SE | 0.13 | 0.20 | 0.17 | 0.21 | 0.14 | 0.19 | 0.12 | - | 0.13 | 0.15 |

M0N0 = no residue + no urea    M1N0 = crop residue + no urea
M0N1 = no residue + first rate urea    M1N1 = crop residue + first rate urea
M0N2 = no residue + second rate urea    M1N2 = crop residue + second rate urea
CV = coefficient of variance.    SE = standard error    SD = standard deviation
Values within the same column bearing same letter are not statistically different.

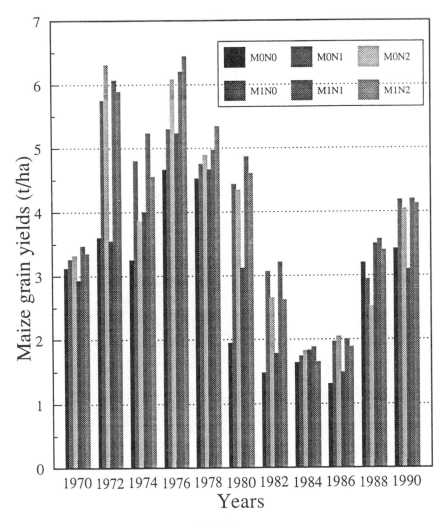

**Figure 2.** Maize grain yields, 1970-1990.

1970 to 1990. Maize grain yields range from 6 t ha$^{-1}$ at the best, to 1.3 t ha$^{-1}$
across all treatments.

Data presented in Figure 2 indicates that the general pattern of maize yields
may be subdivided into three parts. In the early stage of the experiment, from
1970 to 1980, grain yields were relatively good; they were drastically reduced
from 1982 to 1986, but then recovered in 1988 and 1990.

Yields consistently increase or decrease regardless of the treatments imposed.
Another important observation is the relatively good performance of the no
nitrogen (*N0*) treatment for most of the experimental period. In contrast to the
general concept of a rapid decline of crop yields under continuous cropping

systems in West Africa (Kang et al., 1977), the results from the present experiment indicate that yields were maintained at a relatively good level for more than ten years before they declined. This decline was then reversed, even without additional fertilizer, in 1988 (Table 6 and Figure 2).

The most obvious feature of the maize yields is the marked variability from year to year regardless of treatment. A general decrease in maize grain yield occurred after a change in crop rotation when yam was introduced in the system; although the upward trend in yield from 1988, when yam had been cultivated only a year before, probably indicates a more complex situation. Despite the high variability in yields, nitrogen and crop residue did, in some years, have a significant effect on maize yields.

## 1. Effect of Nitrogen Treatments on Maize Grain Yield

The influence of nitrogen on maize grain yields was apparent in 1970, the second year after the experiment started. Analysis of variance of maize grain yield showed significant yield differences due to urea treatment (P = 0.01) (Table 6). Until 1980, response of maize yields to applied mineral nitrogen, although not as high as expected, was consistent. In the early stage of the experiment, the nitrogen application rate was 100 kg N ha$^{-1}$ for N1 and 200 kg N ha$^{-1}$ for N2 (Table 4). The higher rate of nitrogen did not perform well, given the considerable amount of nitrogen applied to the soil. There was rarely evidence of a yield response to applications higher than the N1 level of 80 kg N ha$^{-1}$ adopted later. The response of maize grain yields to nitrogen applications may have been, to some extent, diminished by the relatively good performance of the no nitrogen (N0) treatments. The grain yield response to nitrogen fertilizer was influenced by overall yield fluctuations. During years of very low yields, such as in 1982, no effect of nitrogen was detected, probably because of insufficient plant growth.

In general, the response of maize grain yields to nitrogen is more evident in years of overall better yields and less significant when yields are low. Crop yields under the no nitrogen treatment are surprising and remained relatively good, even after years of continuous cultivation.

## 2. Effect of Crop Residue Return

Application of crop residue is an additional source of nitrogen, plus it gives other beneficial contributions to soil physical and chemical properties (Haggar et al., 1991, Tindall et al., 1991).

Application of crop residue increased the grain yield from the beginning of the experimental period until 1980, when grain yields were relatively good except for the first year of maize cultivation. The effect of applied crop residue was not significant in 1982, 1984, 1986, and 1990. In terms of absolute values, the

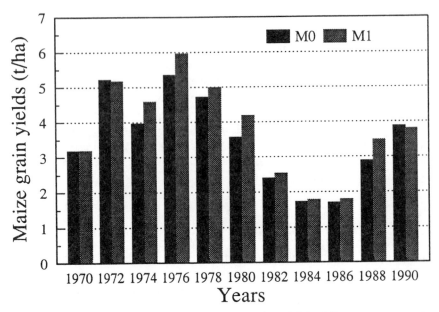

**Figure 3.** Effect of residues on maize grain yields (all urea treatments combined).

increase of yield due to the crop residue, at it best, reached 620 kg ha[-1] of grain compared to no crop residue treatments from the mean values in 1980 (Table 6). Except for 1974, better effects of crop residue on yields were obtained during years of better crop performance; no significant increase was promoted by crop residues in years of poor performance (Figure 2). Application of crop residue did not prevent the pronounced decrease of maize grain yields between 1982 and 1986. The factor that induced yield decreases during that period is not affected, at least not sufficiently enough, by the application of urea or by the restitution of residue from the previous crops to the soil. The effect of residue return for all nitrogen treatments combined is shown in Figure 3.

## B. Cotton Yields

Cotton yields were relatively low early in the experiment then steadily increased from 1976 onward (Figure 4, Table 7). The very low yield in 1972 was due to late sowing and the dry period which started when the plant was at the critical period of reproduction. In general, cotton performed better than maize and the yields were more regular. (Figure 4).

Cotton yields were relatively stable throughout the experimental period. The change in crop variety did not noticeably affect the overall yields. However, yields under the no nitrogen treatment have regularly produced under 0.5 T. ha[-1]

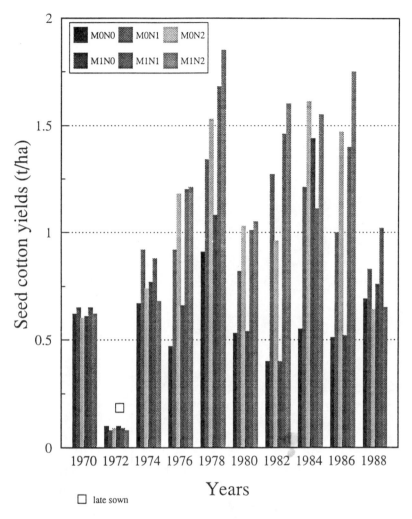

**Figure 4.** Cotton yields, 1970-1988.

from 1980 onward, except in 1987 (Table 7). This decrease of cotton yields under the control treatment resembles the decrease in yields of the maize crop at about the same period. Even so, the decrease of seed cotton yields was not pronounced under the no nitrogen treatment (Table 7). By and large, cotton under the no nitrogen treatment performed less well than maize under the same treatment when compared with their respective average yields. The maximum seed cotton yield never reached 2 t. ha$^{-1}$, whereas Créténet (1987) estimated that, with adequate fertilizer application, proper weed and pest control, cotton can produce over 2 t. ha$^{-1}$, in the most favorable areas of the Côte d'Ivoire.

**Table 7.** Seed cotton yield (t ha$^{-1}$)

|  | 1970 | 1972 | 1974 | 1976 | 1978 | 1980 | 1982 | 1984 | 1986 | 1988 |
|---|---|---|---|---|---|---|---|---|---|---|
| M0N0 | 0.62 | 0.10 | 0.67 | 0.47 | 0.91 | 0.53 | 0.40 | 0.55 | 0.51 | 0.69 |
| M0N1 | 0.65 | 0.08 | 0.92 | 0.92 | 1.34 | 0.82 | 1.27 | 1.21 | 1.00 | 0.83 |
| M0N2 | 0.60 | 0.09 | 0.74 | 1.18 | 1.53 | 1.03 | 0.96 | 1.61 | 1.47 | 0.64 |
| M1N0 | 0.61 | 0.10 | 0.77 | 0.66 | 1.08 | 0.54 | 0.40 | 1.44 | 0.52 | 0.76 |
| M1N1 | 0.65 | 0.09 | 0.88 | 1.20 | 1.68 | 1.01 | 1.46 | 1.11 | 1.40 | 1.02 |
| M1N2 | 0.62 | 0.08 | 0.68 | 1.21 | 1.85 | 1.05 | 1.60 | 1.55 | 1.75 | 0.65 |
| SD | 0.07 | 0.01 | 0.13 | 0.15 | 0.18 | 0.23 | 0.34 | 0.16 | 0.24 | 0.20 |
| AV M0 | 0.62d | 0.09d | 0.77d | 0.86e | 1.26e | 0.79d | 0.88e | 1.12d | 0.99e | 0.72d |
| AV M1 | 0.63d | 0.09d | 0.78d | 1.02d | 1.54d | 0.86d | 1.15d | 1.03d | 1.22d | 0.81d |
| N0 | 0.61a | 0.10a | 0.72b | 0.57c | 1.00c | 0.53b | 0.40b | 0.50c | 0.51c | 0.72b |
| N1 | 0.65a | 0.09b | 0.90a | 1.06b | 1.51b | 0.91a | 1.37a | 1.16b | 1.20b | 0.93a |
| N2 | 0.65a | 0.08b | 0.71b | 1.19a | 1.69a | 1.04a | 1.28a | 1.58a | 1.61a | 0.64b |
| CV % | 11.5 | 14.6 | 17.1 | 15.5 | 13.0 | 27.3 | 33.9 | 23.8 | 18.8 | 26.1 |
| SE | - | 0.024 | 0.09 | 0.09 | 0.10 | 0.11 | 0.14 | 0.12 | 0.11 | 0.11 |

M0N0 = no residue + no urea      M1N0 = crop residue + no urea

M0N1 = no residue + first rate urea      M1N1 = crop residue + first rate urea

M0N2 = no residue + second rate urea      M1N2 = crop residue + second rate urea

CV = coefficient of variance.      SE = standard error    SD = standard deviation

Values within the same column bearing same letter are not statistically different.

## 1. Effect of Nitrogen Fertilizer on Cotton Yield

Cotton responded positively to nitrogen fertilizer almost every year during the period of the experiment. Seed cotton was significantly increased by nitrogen in eight out of ten years (Table 7). In 1972 the yield of seed cotton was too low to be responsive to any fertilizer effect. In relation to yields with the first nitrogen rate ($N1$), the higher rate of nitrogen application ($N2$) produced greater effects with cotton than with maize. This result was consistent throughout the whole experimental period.

## 2. Effect of Crop Residue on Cotton Yield

The effect of crop residue return on seed cotton yield was not consistent. Crop residue tended to have a much greater effect on yields when they were good, except for 1984. In 1980, although the average yield was higher under treatments that received crop residue, this difference was not significant as a result of the higher coefficient of variance (Table 7). On the other hand, crop residue resulted in a lower yield in 1984 although again, it was not significant. This might suggest an interaction of residue treatments with another factor. The cotton crop benefited from the mulching effect of maize residue which was cut and put down between rows during cotton development. It could, therefore, have provided better moisture retention, possible lower soil temperature, and certainly an overall effect on nitrogen availability (Power et al., 1986). In the early part of the experiment, cotton yields were relatively low, which is probably the reason for the absence of an effect of crop residue on cotton yield during that period. The effect on cotton of residue return is shown in Figure 5.

## C. Discussion on Maize and Cotton Yields

During the experimental period, maize yields were not consistent with wide year-to-year variations. This variability affected all treatments (Table 6). The principal reason suspected for the pattern of variability is the climate and especially a lack of sufficient water during the periods of flowering and grain filling (Traoré, 1993). Rainfall is not regularly distributed throughout the first growing season. The effect of rainfall distribution on maize yield fluctuation was reported from experiments carried out in the same area by Gigou (1973), Chopart (1989), and Koné (1991). Although the literature suggests a strong effect of rainfall on yield variability, the influence of the introduction of yam into the cropping system cannot be ruled out entirely.

The second key observation is the lack of effect of the high nitrogen application rate on maize yield. The absence of a maize response to the high rate of nitrogen is probably due to a lack of sufficient water to the plant. Mughogho et al. (1986) reported that, in West African sudan savanna, maize response to

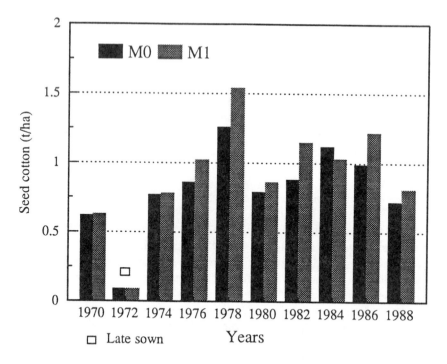

**Figure 5.** Effect of crop residue on cotton yields (all urea treatments combined).

nitrogen does not extend beyond 80 kg N ha[-1]. Although treatments with nitrogen performed better than those without, the latter gave surprisingly good maize yields despite so many years of cultivation. Maize, in order to be able to yield consistently, must be able to obtain nitrogen from somewhere. Two auxiliary nitrogen sources are possible. One is nitrogen washed into the soil from the air by the rain; although Bouaké is not an important industrial town, there may be a possibility of some nitrogen from smoke of bushes burnt during the dry season (Ahn, 1979; Wambeke, 1992). In the U.K., Jenkinson et al. (1986) reported an annual yield of wheat of from 1 to 2 tons per hectare on plots that had not received any applied nitrogen since 1843. In the same report, the authors explained this performance by the possible input of about 14 kg N ha-1 from the air, and other inputs from cyanobacteria. In the case of Bouaké, the second possible source of nitrogen for maize is from deep rooted weeds and wild legumes that develop during the dry season fallow. The exact quantity is yet to be assessed, but from observing the above-ground biomass, roots, and nodule size, there is evidence of nitrogen fixation and this nitrogen may be easily transferred to a subsequent crop.

Crop yield response to nitrogen and crop residue was better under cotton, 120 kg N ha[-1] significantly improved seed cotton yield, almost certainly because of better water availability (Entz and Fowler, 1989), and rainfall distribution during the cotton growing period. Better cotton response to a rate higher than 80 kg N ha[-1] of nitrogen (120 kg N ha[-1]) might suggest that the optimum rate for nitrogen in this area for cotton is not 80 kg but higher provided that climatic conditions are suitable. Mughogho et al. (1986) reported that as rainfall increases, the optimum rate of nitrogen for maize response increased to 120 kg N ha[-1] in the more humid zones of West Africa. Seed cotton response to urea was more consistent than maize grain yields and the application of maize stalk as mulch may also have contributed to the cotton performance.

The amount of mineralized nitrogen in the soil is less in the second growing period than in the first (Chabalier, 1976; Gigou, 1987). For that reason the *MONO* treatments performed less well under cotton despite the better water availability (Figure 3 and Table 7), therefore contributed to a higher perceived response of the crop to added nitrogen fertilizer.

The effects of crop residue on both cotton and maize were lower than expected.

The overall results indicated different crop sensitivities to soil and applied nitrogen and crop residues. They also suggest that the treatments that were applied were not the only factors that control to crop yields, particularly during the first growing period. The main factor that might have played a key role in determining the final yield is climate and especially rainfall and its distribution. The cumulative cotton and maize yield curves suggest little effect of crop residue on yields over time in relation to the amount of returned crop debris.

An important outcome from the analysis of crop yields is that a relatively good performance in terms of crop yields under continuous and intense land cultivation was obtained without any mineral nitrogen input. This finding is not unique, since Jenkinson et al., (1986), and Jenkinson, (1991) reported similar results under temperate conditions.

## 1. The Variability of Crop Yields and the Effect of Rainfall

The sensitivity of crops to factors other than crop residue and urea, can only be compared if all crops are cultivated during the same cropping season. There is, however, a need to investigate further the effect of rainfall on yields for the entire growing period, and also for portions of the growing period in order to detect periods sensitive to water stress. A detailed analysis of maize yield in relation to rainfall by Traoré (1993) indicated that maize grain yield is greatly affected by rainfall from 50 to 70 Days After Sowing (DAS). In the same report the author found that maize grain yield is good when the rainfall during the above period is higher than 100 mm, and low when the rainfall is lower than 80 mm.

The analysis of cotton yields did not reveal a strong effect of rainfall (Traoré, 1993), although previous experience in the area reported cotton to be sensitive to rainfall (Bigot, 1977; Lang and Bartsch, 1977; Créténet, 1987). Land management can affect cotton sensitivity to water stress. In the present experiment, maize stalks were applied as a mulch to cotton after the maize grain was harvested. It therefore might have improved soil moisture and possibly reduced soil temperature. The relationship between rainfall and cotton yield does not appear to be a straight line relationship. The response of crops to the return of crop debris was not consistent; it sometimes increased yields. Whenever adequate circumstances prevailed, crop residue application affected the overall performance, particularly by further improving the grain yield. In overall agronomic terms, the effect of crop residue on maize grain yield was not important, the increase in yield was small, even though the amount was enough to provide a statistical difference between the M0 and the M1 crop residue rates. For cotton the effect of crop residue was more likely to be due to the mulching effect of maize stalks than the nitrogen input from them.

# IV. Effects of Treatments on Soil Properties

## A. Soil pH

Soil samples were collected from the top 20 cm layer and analyzed for pH in 1970, 1980, 1984, and 1990. The results are presented in Table 8. The application of crop residue and of urea did not affect soil pH in 1970. But in 1980, 1984, and 1990, the return of crop residues to the plots significantly increased soil pH (P < 0.01). The soil pH was lower from 1970 to 1990 than in 1969, when the experiment started. This decrease was not consistent with time, for example the pH in 1984 was higher than the pH in 1980, and soil pH slightly increased in all treatments in 1990 (Table 8). By and large, the soil pH was lower after the land had been cultivated than before. The slight variations of the soil pH from year to year can probably usually be attributed to the time of sampling. When soil samples are collected in the dry season or during the rainy season, there could be a minor variation in pH due to nitrogen mineralization which is higher with the onset of rain (Gigou and Chabalier, 1987). The application of urea fertilizer to crops significantly decreased soil pH in 1980, 1984, and in 1990. The higher the urea application rate, the lower the soil pH, although all treatments received the same calcium and magnesium amendments.

## B. Soil Total Nitrogen

Soil total nitrogen content had decreased markedly by 1980 and further in 1990, regardless of the treatments. The application of fertilizer nitrogen or crop residue did not significantly affect the downward trend of soil nitrogen content. Total soil nitrogen decreased from 0.12 % in 1970 to 0.06 % (Table 8) in 1990 under the M1N2 treatment, which received both the higher rate of urea and crop residue. The pH was lower in 1984 (5.15 with M1N2). Soil pH remained relatively stable under the N0 nitrogen treatment (Table 8).

**Table 8.** Soil pH, total carbon and total nitrogen, as affected by treatments

| Treatment | pH 1970 | 1980 | 1984 | 1990 | %C 1970 | 1980 | 1990 | %N 1970 | 1980 | 1990 |
|---|---|---|---|---|---|---|---|---|---|---|
| M0N0 | 6.00 | 5.65 | 5.90 | 6.03 | 1.35 | 1.02 | 0.82 | 0.121 | 0.085 | 0.064 |
| M0N1 | 6.03 | 5.25 | 5.35 | 5.52 | 1.32 | 1.01 | 0.83 | 0.122 | 0.083 | 0.061 |
| M0N2 | 5.99 | 5.00 | 5.07 | 5.41 | 1.32 | 1.04 | 0.80 | 0.114 | 0.087 | 0.065 |
| M1N0 | 6.12 | 5.73 | 6.00 | 6.19 | 1.27 | 0.99 | 0.84 | 0.124 | 0.085 | 0.064 |
| M1N1 | 6.31 | 5.50 | 5.62 | 5.95 | 1.36 | 1.08 | 0.85 | 0.129 | 0.091 | 0.065 |
| M1N2 | 5.90 | 5.15 | 5.23 | 5.54 | 1.31 | 1.11 | 0.76 | 0.118 | 0.093 | 0.060 |
| STD | 0.26 | 0.11 | 0.15 | 0.19 | 0.13 | 0.12 | 0.05 | 0.13 | 0.09 | 0.09 |
| AV N0 | 6.06 a | 5.69 a | 5.95 a | 6.11 a | 1.31 a | 1.01 a | 0.83 a | 0.122 a | 0.085 a | 0.064 a |
| AV N1 | 6.17 a | 5.38 b | 5.48 b | 5.73 b | 1.34 a | 1.05 a | 0.84 a | 0.126 a | 0.087 a | 0.063 a |
| AV N2 | 5.95 a | 5.07 c | 5.15 c | 5.47 c | 1.32 a | 1.08 a | 0.78 b | 0.116 a | 0.090 a | 0.063 a |
| AV M0 | 6.01 x | 5.30 y | 5.44 y | 5.66 y | 1.33 x | 1.02 x | 0.82 x | 0.119 x | 0.085 x | 0.064 x |
| AV M1 | 6.11 x | 5.46 x | 5.62 x | 5.89 x | 1.32 x | 1.06 x | 0.82 x | 0.124 x | 0.090 x | 0.063 x |
| CV % | 4.3 | 2.0 | 2.8 | 3.4 | 7.2 | 8.2 | 6.1 | 10.5 | 10.4 | 13.5 |
| SDE | NS | 0.08 | 0.09 | 0.11 | NS | NS | 0.05 | NS | NS | NS |

SDE = Standard error, AV = Average, NS = Non significative; numbers with the same letters are not different within the same column at P = 0.05

M0N0 = no nitrogen + no residue    M0N2 = no nitrogen + second rate urea    M1N1 = crop residue + first rate urea
M0N1 = no nitrogen + first rate urea    M1N0 = crop residue + no urea    M1N2 = crop residue + second rate urea
STD = standard deviation

## C. Soil Carbon Content

Soil organic carbon content declined steadily throughout the experiment regardless of treatments. By 1980 soil carbon had decreased to 76.7 % of its original value with the *M0* treatment (no residue return) and to 80.3 % with the addition of crop residues. The final values were not statistically different, suggesting that neither urea nor crop residues had much effect. By 1990, the mean values of soil organic carbon had decreased to 61.6 % and 62.1% respectively, for the M0 rate of crop residues and the M1 rate of crop residues, from the 1970 value. Application of crop residues did not significantly influence soil organic carbon levels and fertilizer nitrogen did not affect soil organic carbon in the 1980 data (Table 8).

## D. Discussion on Soil pH, Nitrogen, and Carbon Content

### 1. Soil pH

The analysis of changes in soil pH, organic carbon, and nitrogen suggests that treatments and land use had a significant effect on the above soil parameters. As indicated earlier, nitrogen is reported to be one of the most limiting nutrients for sustained crop production in the humid and sub-humid tropics (Mughogho et al., 1986; Bouldin et al., 1979). For maize the suggested optimum nitrogen rate varies from 80 to 120 kg N ha$^{-1}$ (Christianson and Vlek, 1991; Stumpe and Vlek, 1992).

The source of nitrogen has a significant effect on soil pH. Miner et al., (1986), reported, from work carried out on the American Eastern Coastal plain, that soil pH decreased significantly when ammonium sulphate or diammonium phosphate were band applied as starter fertilizer, as compared to sodium nitrate. The decrease in pH was sufficient to allow some of the complexed manganese present in the original soil to be released. The latter authors also indicated that ammonium sulphate acidified the soil more than diammonium phosphate. In tropical Africa, ammonium sulphate decreased soil pH more rapidly than urea, while calcium ammonium nitrate (CAN) and urea showed similar rates of acidification under field conditions in northern Nigeria (Balasubramanian and Singh, 1982). In a network of trials across several West African countries, Mughogho et al., (1986) reported that the mean response curves of maize and millet revealed no significant difference between urea and CAN.

Urea is the most commonly used nitrogen fertilizer in West Africa, since its efficiency is similar to other nitrogen sources, provided management is adequate, and it has the added advantage of a high nitrogen analysis (46 % N). However, urea is quickly hydrolysed in the soil and converted to ammonium $NH_4^+$. When ammonium undergoes nitrification, hydrogen ions are released into the soil. As a consequence, continuous applications of urea in the experiment resulted in a significant decrease of soil pH, as compared to the *N0* urea rates. The decrease in soil pH was greatest with the *N2* urea rate. The regular application of dolomite almost certainly slowed this decrease. Without it, soil acidification would have been higher and aluminium toxicity might have resulted which would have been very detrimental to crop growth. Despite calcium and magnesium amendment, soil pH decreased to slightly above 5 with the *N2* rate of urea. This might have contributed to the poor response of maize at this nitrogen rate, since maize is reported to perform better if the soil pH remains above 5.5 (Sanchez, 1977; Gigou and Chabalier, 1987).

The results in Table 8 also indicate a significant effect of the return of crop residue on soil pH. Crops absorb nutrients such as Ca, K, and Mg from deeper soil layers during their growth; when these residues are returned to soil, the release of calcium and magnesium restores these cations to the top layer of the soil. As a consequence, the pH increases (Sanchez, 1976; Stumpe and Vlek, 1992). The relative increase of soil pH in 1990 is partly due to the fallow year in 1989. The weeds were returned to the *M1* crop residue plots and removed from the without-residue treatments, when land was returned to cultivation in 1990. The return of crop residue and the use of dolomite amendments contributed to a reduction of possible soil acidification. The results suggest that on intensively cultivated soils, regular application of lime and the return of crop residues are advisable to avoid a decline in productivity as result of rapid soil acidification.

## 2. Soil Nitrogen

The application of mineral nitrogen and the return of crop residues could not prevent soil nitrogen from falling with time (Table 8). It suggests that mineralization of nitrogen from organic sources was so rapid that the crops could not absorb all the released nitrogen. The loss was probably worsened because of a lack of synchrony with crop needs and because of poor crop growth due to adverse climatic factors. Since only a portion of the nitrogen was absorbed by the crop, less was available for eventual return to the soil and as the process continued with time the decrease of nitrogen in the soil system became more marked.

The application of mineral nitrogen did not improve the situation, since stress during the critical period of crop growth did not favor optimum crop productivity. In this particular situation, denitrification, and possibly volatilization, may be important as processes of nitrogen loss. There is a possibility that nitrate might accumulate deep in the profile, and become available to the crop in good years, according to the depth at which it is located. The ideal situation would be for the release of nitrogen into the soil to be in synchrony with the need and the demand of crops.

The introduction of yam to the system will have caused additional disturbance to the soil. As a result, nitrogen mineralization may have increased (Greenland et al., 1992), as well as opportunities for loss of N from the system.

3. Soil Organic Carbon

The carbon contents of the soil in 1970, 1980, and 1990 are presented in Table 8. The 1990 results show the same pattern for carbon as for total nitrogen. The 1980 carbon contents were derived from total N measurement and so are not independent observations. This fall in organic carbon is primarily due to the decreased annual returns of plant material under cropping, compared to returns under secondary woodland. This decrease of soil carbon will continue until a new steady state level is reached. When the soil organic matter is stabilized at a steady-state, the nitrogen released each year through decomposition of soil organic matter is exactly balanced by the nitrogen incorporated into new soil organic matter each year (Jenkinson and Rayner, 1977; Jenkinson and Ayanaba, 1977; Jenkinson, et al., 1986; Bouldin et al., 1979; Greenland et al., 1992).

# V. Modeling Soil Organic Carbon

Sustainable agriculture in the tropics, as elsewhere, will depend on the avoidance of soil degradation to a point that it will no longer support satisfactory crop growth. The importance of soil organic carbon is paramount as it will influence moisture retention, nutrient supply and retention, and susceptibility to erosion. The data available from the long-term experiment at IDESSA allowed the use of an established model with the aim of assessing how well it would predict the changes in soil carbon under tropical conditions.

## A. Description of the Model

The Rothamsted model for the turnover of organic matter in the soil (Jenkinson, 1990), is a descendant of earlier versions (Jenkinson and Rayner, 1977; Jenkinson et al., 1987). In this model (Figure 6) the authors simulated the behavior of soil organic matter by dividing it into compartments, independent of the data to be fitted. It is assumed that each compartment behaves as though it contains a single organic matter species undergoing first order biological decomposition processes.

Another assumption of the model is that the priming action is zero, *i.e* the addition of fresh plant residues does not alter the rate of decomposition of organic matter already in the soil, and amount of incoming organic matter that is decomposed after a given time is independent of the amount added.

The first two compartments are called Decomposable Plant Material (**DPM**) and Resistant Plant Material (**RPM**). They represent monthly additions of plant (**P**) carbon from crop residues at a set proportion of $P_{DPM}$ and $P_{RPM}$. The incoming organic carbon passes through these compartments once only, and all incoming carbon is assumed to belong to one or the other. Both **DPM** and **RPM** decompose to the same products, which are $CO_2$, lost from the system, microbial biomass (**BIO**), and humified organic matter (**HUM**). It is assumed that the ratio

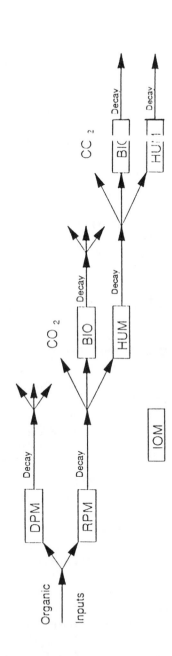

**Figure 6.** Flow of carbon through the Rothamsted organic matter turnover model. (DPM, decomposable plant material; RPM, resistant plant material; BIO, microbial biomass; HUM, humus; IOM, inert organic matter.) (From Jenkinson, 1990. With permission.)

of **BIO** to **HUM** is the same in all soils. **HUM** decomposes into $CO_2$ and more **BIO** and **HUM** are formed in the same proportion. It is also assumed that the soil contains a small compartment inert to biological attack called inert organic material (**IOM**).

The quantity of carbon in a specific soil layer is given by $ye^{-kt}$ at the end of a specific time (one month), where (**t**) is time, (**y**) the quantity of carbon present at beginning of the month, and (**k**) the rate constant for this compartment. The organic matter is moreover assumed to enter the soil monthly in a stepwise manner. The quantity of organic matter in the compartment at the beginning of the month is converted to that at the end of the month by a transition matrix in the model, so that it proceeds in a stepwise fashion rather than continuously.

## B. Key Parameters of the Model

The site specific parameters needed to run the model are soil temperature, soil moisture content, soil bulk density, soil clay content, monthly rainfall, and evaporation. Soil temperature and moisture content have an influence on organic matter decomposition by altering the rate constants so that decay during the month is given by $1 - e^{-abckt}$; where (**a**) and (**b**) are modifying factors for temperature and moisture respectively and (**c**) the plant retainment factor. The soil temperature is taken, for convenience, to be the same as the mean monthly air temperature.

The process of organic matter decomposition is assumed to be at the maximum rate when water tension reaches -100 kPa. The water deficit is calculated from the balance between the mean monthly rainfall and the mean monthly pan evapotranspiration.

The plant retainment modifying factor (**c**) is set in the model at 0.6, when the plants are actively growing, and at 1 when the soil is bare fallowed (Jenkinson, 1977).

The model needs to be adjusted for soil texture. The soil texture affects the rate constants of the different compartments indirectly. In practice, the rate constants are assumed not to alter with texture. However, the partition between the quantity of $CO_2$ evolved and the quantity BIO + HUM formed, is assumed to depend on the cation exchange capacity of the soil inorganic colloids. When the inorganic cation exchange capacity of the soil is not available, as in the present report, soil clay content is used as a substitute. The soil clay content in the present experiment is 26.2 %; but it is always preferable, whenever possible, to use cation exchange capacity because this takes account of clay mineralogy. Mean monthly air temperature, mean monthly rainfall and evapotranspiration are presented in Table 1, as are measurements of average plant above-ground biomass without grain (Table 9).

The model is programmed in FORTRAN 77 and is available from Rothamsted Experimental Station, Harpenden, United Kingdom.

**Table 9.** Above-ground crop dry matter (excl. grain) 1969 to 1990

| Year Crop | M0 N0 | M0 N1 | M0 N2 | M1 N0 | M1 N1 | M1 N2 | AV M0 | AV M1 | AV N0 | AV N1 | AV N2 |
|---|---|---|---|---|---|---|---|---|---|---|---|
| 69r* | 2.11 | 2.73 | 2.39 | 2.08 | 2.20 | 2.33 | 2.41 | 2.20 | 2.09 | 2.46 | 2.36 |
| 70m | 6.72 | 8.67 | 8.40 | 6.80 | 7.26 | 7.39 | 7.93 | 7.15 | 6.76 | 7.97 | 7.89 |
| 70c | 1.70 | 2.79 | 3.15 | 1.86 | 3.01 | 3.61 | 2.55 | 2.83 | 1.78 | 2.9 | 3.38 |
| 71r | 2.11 | 2.59 | 3.69 | 2.06 | 2.87 | 3.65 | 2.80 | 2.86 | 2.09 | 2.73 | 3.67 |
| 72m | 6.05 | 6.73 | 10.4 | 5.5 | 6.78 | 9.64 | 7.71 | 7.31 | 5.78 | 6.76 | 10.0 |
| 72c | 0.98 | 1.12 | 1.26 | 0.92 | 1.26 | 1.23 | 1.12 | 1.13 | 0.95 | 1.19 | 1.24 |
| 73r | 5.41 | 5.67 | 7.91 | 6.88 | 6.08 | 6.10 | 6.33 | 6.35 | 6.15 | 5.86 | 7.00 |
| 74m | 4.66 | 5.43 | 5.45 | 5.35 | 6.09 | 6.57 | 5.18 | 6.00 | 5.01 | 5.76 | 6.01 |
| 74c | 2.78 | 4.97 | 5.65 | 2.53 | 4.33 | 5.21 | 4.47 | 4.02 | 2.65 | 4.65 | 5.43 |
| 75r | 1.75 | 2.03 | 2.59 | 1.82 | 2.52 | 2.80 | 2.13 | 2.38 | 1.79 | 2.28 | 2.69 |
| 76m | 6.59 | 6.73 | 6.92 | 5.59 | 7.43 | 7.50 | 6.75 | 6.84 | 6.09 | 7.08 | 7.21 |
| 76c | 1.80 | 2.91 | 3.46 | 2.14 | 3.50 | 4.80 | 2.72 | 3.48 | 1.97 | 3.21 | 4.13 |
| 77r | 2.26 | 2.65 | 3.53 | 2.56 | 3.61 | 3.65 | 2.81 | 3.27 | 2.41 | 3.13 | 3.59 |
| 78m | 5.39 | 5.42 | 5.88 | 5.89 | 6.05 | 6.51 | 5.56 | 6.14 | 5.62 | 5.74 | 6.19 |
| 78c | 1.25 | 2.14 | 2.21 | 1.85 | 2.93 | 3.19 | 1.86 | 2.66 | 1.55 | 2.53 | 2.70 |
| 79y | 0.66 | 1.04 | 1.37 | 0.59 | 1.30 | 1.20 | 1.02 | 1.03 | 0.63 | 1.16 | 1.30 |
| 80m | 7.53 | 6.93 | 6.57 | 8.10 | 8.41 | 8.05 | 7.01 | 8.19 | 7.81 | 7.67 | 7.31 |
| 80c | 0.28 | 0.48 | 0.81 | 0.71 | 0.87 | 1.01 | 0.53 | 0.86 | 0.49 | 0.68 | 0.91 |
| 81y | 0.66 | 1.04 | 1.37 | 0.59 | 1.30 | 1.20 | 1.02 | 1.03 | 0.63 | 1.16 | 1.30 |
| 82m | 3.52 | 3.68 | 4.38 | 4.14 | 4.81 | 4.87 | 3.86 | 4.61 | 3.83 | 4.25 | 4.63 |
| 82c | 0.59 | 1.32 | 1.50 | 0.60 | 1.57 | 2.15 | 1.14 | 1.44 | 0.60 | 1.44 | 1.82 |
| 83y | 0.66 | 1.04 | 1.37 | 0.59 | 1.30 | 1.20 | 1.02 | 1.03 | 0.63 | 1.16 | 1.30 |
| 84m | 2.69 | 3.11 | 3.14 | 2.97 | 2.98 | 3.20 | 2.98 | 3.05 | 2.83 | 3.05 | 3.17 |
| 84c | 0.65 | 1.00 | 1.26 | 0.44 | 1.12 | 1.56 | 0.97 | 1.05 | 0.56 | 1.06 | 1.41 |
| 85y | 0.25 | 0.74 | 1.09 | 0.34 | 0.84 | 1.06 | 0.69 | 0.75 | 0.30 | 0.79 | 1.07 |
| 86m | 3.39 | 3.33 | 4.16 | 4.05 | 3.81 | 4.63 | 3.63 | 4.16 | 3.72 | 3.57 | 4.40 |
| 86c | 0.38 | 1.07 | 1.87 | 0.43 | 1.57 | 2.22 | 1.10 | 1.41 | 0.41 | 1.32 | 2.04 |
| 87y | 1.06 | 1.31 | 1.65 | 0.84 | 1.75 | 1.34 | 1.34 | 1.31 | 0.95 | 1.53 | 1.50 |
| 88m | 5.0 | 4.9 | 4.67 | 4.76 | 4.92 | 4.99 | 4.85 | 4.89 | 4.88 | 4.91 | 4.83 |
| 88c | 0.50 | 0.59 | 0.46 | 0.54 | 0.57 | 0.47 | 0.51 | 0.55 | 0.52 | 0.58 | 0.46 |
| 90m | 5.70 | 5.76 | 5.55 | 5.82 | 6.56 | 6.12 | 5.57 | 6.16 | 5.76 | 6.16 | 5.84 |

* r = rice;  m = maize;  c = cotton;  y = yam.

## C. Results from the Carbon Model

The primary aim was to fit the model to the soil carbon data and thus calculate the annual return of carbon to the soil in the different treatments. The organic carbon content of the soil in kg C ha$^{-1}$ was calculated at the beginning of the experiment in 1980, and in 1990, from the measured percent soil carbon content and the mass of soil in the top 25 cm, taking the soil mean bulk density to be 1.4 g cm$^3$, as measured in 1980. Soil organic carbon content calculated for the different treatments in 1970, 1980, and 1990 is presented in Table 8.

Carbon in plant material was calculated from dry matter production assuming 40 % of carbon in all crop residues. A summary of total dry matter production is presented in Table 9. Soil total carbon in the soil top 25 cm from 1970 to 1990 is presented in Table 10 for each cropping system.

It is apparent from Table 8 that the total soil carbon content in the top 25 cm decreased sharply during the experiment, whether crop residues were returned or not. The values in Table 8 were calculated across nitrogen treatments; the parameter of most interest in this situation being the effect of crop residue. Despite the quantity of carbon returned to the soil from the crop residues, the soil carbon content of plots receiving the *M1* crop residue rates did not significantly differ from those of plots not receiving residues (the *M0* crop residue treated plots), in the balance in 1990. Unless the soil is totally bare, there is always some carbon input with the *M0* crop residue rates, mainly from crop roots. The annual input of carbon in plots not receiving crop residues can be estimated by iterative adjustment of the annual input of the plant debris to the model, so as to get as close as possible to the measured carbon content of the soil. This value was finally estimated at 1.0 t C ha$^{-1}$yr$^{-1}$ under the rice-based cropping system and the same value with the yam-based cropping system. It therefore implies that a substantial quantity of carbon is returned to the soil even when the above-ground dry matter is removed from the plots, and also that the introduction of yam into the system did not influence the carbon input from below-ground dry matter and unharvested plant debris in the soil. This result also suggests that even when the above-ground residue is removed from the soil, there is a non-negligible source of mineralizable nitrogen for subsequent crops which is mainly from plant roots. The modeled carbon turnover with the *M0* crop residue rates is quite close to the measured values. In contrast, the modeled values of turnover for the *M1* crop residue rates treatments are higher than the observed values; the simulated and actual curves are presented in Figure 7. Most importantly, soil carbon in the soil decreased whether crop residue was returned to the soil or not. The amounts of carbon from the plant material returned to the soil are presented in Table 10. The modeled values were estimated on an annual basis. The annual carbon input is calculated as the total plant carbon from maize and cotton, plus plant carbon from rice, and the sum was divided by two. It gives the mean annual carbon input into the soil for the rice-based cropping system, since cotton and maize are cultivated in the same year. The same method was used for the yam-based cropping system. The results give the annual carbon

**Table 10.** Mean carbon returned to the soil by crop dry matter (t ha$^{-1}$ yr$^{-1}$)

| Crops | M1N0 | M1N1 | M1N2 | AVM0 | AVM1 |
|---|---|---|---|---|---|
| **Rice-Maize/Cotton Cropping system 1969 to 1978** | | | | | |
| Total above-ground dry matter (t ha$^{-1}$) [a] | 53.83 | 65.92 | 73.49 | 62.33 | 64.62 |
| Carbon from above-ground (t ha$^{-1}$ yr$^{-1}$) [b] | 2.15 | 2.64 | 2.97 | - | 2.58 |
| Carbon from below-ground (t ha$^{-1}$ yr$^{-1}$) [c] | 1.0 | 1.0 | 1.0 | 1.0 | 1.0 |
| Total C applied | 3.15 | 3.64 | 3.97 | 1.0 | 3.58 |
| **Yam-Maize/Cotton Cropping system 1979 to 1990** | | | | | |
| Total dry matter (t ha$^{-1}$) [a] | 35.51 | 43.68 | 45.27 | 37.24 | 41.52 |
| Carbon from above-ground (t ha$^{-1}$ yr$^{-1}$) [d] | 1.29 | 1.59 | 1.65 | - | 1.38 |
| Carbon from below-ground (t ha$^{-1}$ yr$^{-1}$) [d] | 1.0 | 1.0 | 1.0 | 1.0 | 1.0 |
| Total C applied | 2.29 | 2.59 | 2.65 | 1.0 | 2.38 |

[a] Calculated from Table 6.4.

[b] Calculated from the model taking soil bulk density = 1.4 g cm$^{-3}$ and soil depth = 25 cm.

[c] Total dry matter/number of cropping seasons, multiply by 40 % (% carbon assumed in the dry matter); number of cropping seasons = 10.

[d] Total dry matter/number of years, multiply by 40 % (% carbon assumed in the dry matter); number of years = 12 including 1989.

AVM0 = Average dry matter from M0 treatments; AVM1 = Average dry matter from M1 treatments.

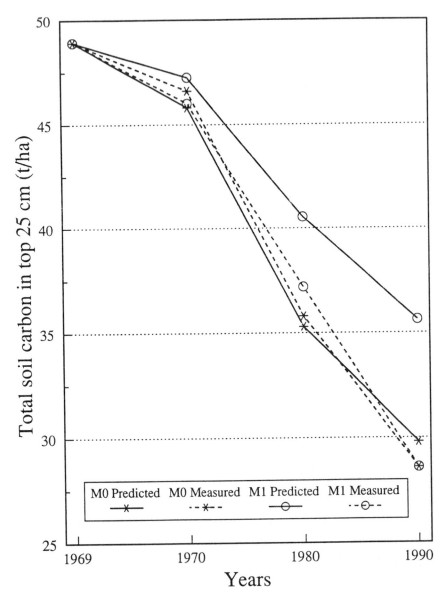

**Figure 7.** Actual and simulated soil carbon content.

input for each cropping system. To this was added input from the below-ground dry matter estimated at 1.0 t C ha$^{-1}$ yr$^{-1}$ for the rice-based cropping system and equally 1.0 t C ha$^{-1}$ yr$^{-1}$ for the yam-based cropping system.

**Table 11.** Soil total carbon as affected by crop residue and mineral nitrogen

| | 1969 | 1970 | | | 1980 | | | 1990 | | |
|---|---|---|---|---|---|---|---|---|---|---|
| | Total C * | % C | Predicted t C ha⁻¹ | Measured t C ha⁻¹ | % C | Predicted t C ha⁻¹ | Measured t C ha⁻¹ | % C | Predicted t C ha⁻¹ | Measured t C ha⁻¹ |
| N0 | 48.91 | 1.31 | 45.54 | 45.85 | 1.01 | 37.68 | 35.35 | 0.83 | 32.48 | 29.05 |
| N1 | 48.91 | 1.34 | 45.82 | 46.9 | 1.05 | 38.65 | 36.75 | 0.84 | 33.54 | 29.4 |
| N2 | 48.91 | 1.32 | 46.06 | 46.2 | 1.08 | 39.54 | 37.8 | 0.78 | 34.17 | 27.3 |
| M0 | 48.91 | 1.33 | 45.8 | 46.6 | 1.02 | 35.3 | 35.8 | 0.82 | 29.80 | 28.6 |
| M1 | 48.91 | 1.31 | 47.25 | 46.0 | 1.06 | 40.55 | 37.2 | 0.82 | 35.63 | 28.6 |

N0 = No nitrogen   N1 = First rate of nitrogen   N2 = Second rate of nitrogen
M0 = No crop residue return   M1 = Crop residue return
* = Total carbon predicted at equilibrium with IOM = 1.25 t C ha⁻¹ and Input = 6.54 t ha⁻¹ yr⁻¹
IOM = Biologically inert organic matter

There was a large decline in soil organic matter between the beginning of the experiment and the final year, 1990. After 23 years, the total carbon of the soil decreased from 46.6 t ha$^{-1}$ to 28.6 t ha$^{-1}$ with the no crop residue treatments (Table 11). Because of the lack of regular soil analyses, the decrease in soil total carbon was only detected in 1980 when an analysis was performed, although much of the decrease might have occurred earlier. An average of 3.58 t C ha$^{-1}$ yr$^{-1}$ was returned to the plots with the rice-based cropping system and also 2.38 t C ha$^{-1}$ yr$^{-1}$ was returned with the yam-based cropping system (Table 10). However, the soil carbon decreased as sharply with the *M1* rates of residues as it did with the *M0* rate of residue. The prediction of the model for the M0 crop residue rates was very close to the measured curve when the input was set to 1.0 t C ha$^{-1}$ yr$^{-1}$ of carbon from the roots and other non-fertilizer sources. The model predicted a slightly higher carbon content with the *M1* crop residue rates. The effects of the application of mineral nitrogen on soil carbon turnover was also assessed from the observed and predicted values. For the predicted values from the model, plant carbon was calculated using mean plant dry matter under each rate of urea, averaged over all treatments. In the same way as for the crop residue treatments, the estimated below-ground carbon was added to the total carbon input in the soil, 1.0 t yr$^{-1}$ for both the yam and the rice-based cropping systems.

## D. Comparisons with the Results from the Field Experiment

Field data from long-term experiments provide a very good test for simulations of soil carbon turnover and for investigations into how crop residue inputs influence the quantity of organic matter in a soil. In this investigation, interest was directed not only to the effects of crop residues on soil total carbon but also to the effects of applied urea. It is known that mineral nitrogen availability may affect soil organic matter, and thus indirectly, soil total carbon, by influencing the soil biota (Paustian et al., 1992). The rate of soil organic matter decomposition was far higher than carbon input to the soil, even with residue, so that no build-up of soil carbon was possible. Decreases of soil carbon in West African soils have been investigated by several authors and found to be quite rapid. Jenkinson and Ayanaba (1977) reported, from $^{14}$C work in Nigeria, that only 20% of applied $^{14}$C remained in the soil after one year. The rate of the decrease of carbon subsequently declined. The pattern of decrease of soil organic matter in the soil is the same in other environments except that the rates are different (Jenkinson, 1977; Greenland et al., 1992). The introduction of yam may have accelerated soil carbon depletion, because increased cultivation and tillage enhance the rate of soil organic matter decomposition (Seubert et al., 1977, Greenland et al., 1992).

From 1980 to 1990 the rate of nitrogen decrease was lower than from 1970 to 1980, suggesting that the soil organic matter is progressing toward a new equilibrium value. The Rothamsted model is known for its relative accuracy in predicting soil carbon turnover, provided that various input parameters are set correctly. These parameters are crucial for the accuracy of the predictions, as some of them have strong influence on the rate of mineralization. External factors such as temperature, rainfall, and clay content affect organic matter decomposition (Greenland et al., 1992; Parton et al., 1989). They must therefore be reliably measured. For soils of West Africa, Jones (1973) reported the following relation for carbon content:

$$\% \ C = 0.341 + 0.027 \ Y \quad \text{where } Y = \% \text{ clay content}$$

The best fitted values of IOM and soil annual carbon input at equilibrium under forest are 1.25 t C ha$^{-1}$ and 6.54 t C ha$^{-1}$ yr$^{-1}$, respectively. They gave an equilibrium carbon of 48.9 t C ha$^{-1}$ in the top 25 cm of the soil in 1969, when the experiment started. With 48.9 t C ha$^{-1}$ in the soil in 1969, the annual input of root material decreased to 1.0 t, IOM remaining the same. This value of 1.0 t C ha$^{-1}$ yr$^{-1}$ was obtained by iterative adjustment of the annual input until the output of the model gave the best possible fit to the changes in mean carbon contents of the combined *M0N0*, *M0N1*, and *M0N2* treatments over the 1969-1990 period as whole (Figure 7).

The annual input thus calculated is the carbon return from roots and unharvested plant debris to the plots not receiving residues. It would have been a little greater over the 1969-1978 period, when rice was grown every second year, and a little less during 1979-1990, when the rice was replaced by yam.

In the next stage of investigation, the annual return of carbon from roots was taken as 1.0 t C ha$^{-1}$ yr$^{-1}$ plus the mean carbon return in crop debris from 1969-1978 (taken as 2.58 t C ha$^{-1}$ yr$^{-1}$, Table 10) and 1.38 t C ha$^{-1}$ yr$^{-1}$ from 1979-1990. The carbon content of the soil, as calculated by the model for those inputs, are given in Table 11; They are far above the actual measured values for means of *M1N0*, *M1N1*, and *M1N2* treatments.

It is very difficult to explain why the addition of crop residues had no effect on the total carbon and nitrogen content of the soil. The model, as would be expected, indicates that the quantity of plant carbon added in crop residues was sufficient to cause an easily measurable increase in soil organic carbon, compared to the treatments without residue (Figure 7). Yet no such increase was observed.

Various explanations are possible. Thus erosion could have removed more top soil from the with-residues (*M1* rate of residues) plots than from the without-residues (*M0* rate of residue) plots, although the converse would seem far more likely. The hypothesis is also improbable since erosion was kept to a minimum on the experiment and soil phosphorus determinations indicate a negligible level of erosion on the whole experimental plot. Again, soil bulk density in the with-residues plots may have been less than in the without-residues plots so that the with-residue plots were sampled too shallowly. This is a more realistic possibility, although the differences in bulk density necessary to bring the modeled and measured curves in Figure 7 together would have been large.

Another explanation is that the return of root carbon, set at 1.0 t C ha$^{-1}$ yr$^{-1}$, is less with the crop residue than without, implying that residues depress root production. Welbank et al. (1973) reported that application of nitrogen decreased the root dry matter weight of cereals in the early stages of the growth. This would only have occurred in the present experiment if the crop debris had supplied significant additional nitrogen to the soil.

A further possible explanation is that most of the applied crop residues do not decompose in or on the plots, but are removed from the sampled soil layer by animal activity, *e.g.* by termites or earthworms. In the soils of the humid and sub-humid tropics, earthworm activities significantly affect soil organic matter dynamics by accelerating nutrient release (Lavelle et al., 1992). Termites feed, basically, on cellulosic materials; some species, the humivores, feed on organic matter within the soil. Termite mounds are enriched by organic carbon from the soil from which they are formed (Lavelle et al., 1992). Lepage (1972) reported that termites, in some parts of the Sahelian grassland of Sénégal, may consume up to 49 % of the annual herbage production, though the figure was much lower (5.4 %) for the entire area. In a wet savanna area of Lamto, Côte d'Ivoire, Lensi et al. (1992) reported a high potential denitrification, which required available organic matter, on fresh soils from the walls of fungus comb chambers of termites, and earthworm casts. The above findings suggest that soil fauna, termites and earthworms in particular, can have a significant impact on soil organic matter turnover. Still another possible explanation of the meagre effect of crop residues on the total carbon and nitrogen content of the soil, is that a substantial part of the return of carbon in the without-residues plots is from weeds, which are less prevalent on the with-residue plots. Marnotte (1981) reported similar results in the same area, in which higher fertilizer and land management had significantly affected weed type and population density. Téhia and Marnotte (1990) found that the application of urea and crop residues discriminates against some weed species. None of the above hypotheses are totally convincing individually. The reason why addition of crop residues had no effect on the soil organic carbon, while predicted by the model, may be a combination of several processes. It is more than likely that the factors of bulk density, soil fauna, and weeds have played a combined role. More attention to such diverse factors may be needed in future experiments.

# VI. General Conclusions on the Long-Term Experiment

One of the main objectives of the long-term experiment was to try to maintain soil fertility through the return of the above-ground crop residue. After more than twenty years of continuous return of crop residue to the soil, the results are disappointing. The level of soil organic carbon significantly decreased with both fertilizer nitrogen and crop residue. In an attempt to understand the process of organic matter turnover, the Rothamsted Station soil carbon model was used to predict soil carbon trends. Both the predicted and the measured soil carbon levels decreased with time. The model confirmed that insufficient carbon is being returned to the soil in relation to losses by decomposition. Soil organic matter is quickly decomposed following the cultivation of the land and the amount of carbon from the residue is not enough to balance the losses. The model indicates that the carbon input was sufficient enough to make a significant difference between the soil total carbon under the *M0* and the *M1* treatments. However, this prediction was not confirmed by the actual soil total carbon measurements; several hypotheses are put forward and are fully presented in previous sections.

The measurements of soil pH showed that continuous application of urea increased soil acidity significantly, most markedly with the addition of the highest rate of urea. This decrease, in spite of calcium and magnesium addition, suggests adopting caution in the use of ammonium nitrogen, for it can easily create an acidity problem. The application of urea hardly affected soil total nitrogen and had no effect on soil total carbon.

The use of a modern carbon model allowed the calculations of the annual return of carbon to soil from below-ground plant material. This quantity thus calculated, 1 t C ha$^{-1}$ yr$^{-1}$, would probably leave some 200-300 kg humified C in the soil at the end of each year (Jenkinson and Ayanaba, 1977). Such a quantity of humified carbon would contain 20-30 kg N ha$^{-1}$, taking the C/N ratio of humified carbon as 10:1, (Table 8). Under steady-state conditions (not yet reached in the present experiment), this quantity of N, at least, would be available to the crop each year. This goes some way to explaining the ability of the soil in the *N0* treatments to provide nitrogen for the level of yields observed throughout the experimental period. This mineral nitrogen will come from the original soil organic matter and from the below-ground plant material produced each year.

The effect of all treatments and climate on crop yields suggests that climate is the most important factor that controls crop yield in central Côte d'Ivoire, and that, for maize, fertilizer nitrogen application is not efficient beyond 80 kg N ha$^{-1}$. The return of crop debris is still to be recommended as it allows the recycling of nutrients such as calcium and magnesium. Calcium and magnesium play an important role in preventing soil pH reduction.

Yields obtained under the *M0N0* treatment indicate that, when soil erosion is kept at a minimum, the soil can provide adequate nitrogen for a longer period of time than usually reported in the literature, under tropical conditions. This result also suggests, however, that sources other than residue and fertilizer nitrogen are possible and should be investigated.

The analysis of maize and cotton yields suggests that the second growing period in the Bouaké area is, in agronomic terms, of more value for food production although some disadvantages exist.

The experimental site produced good crops, when rainfall permitted, even when no fertilizer or crop residue was returned. This has continued over 20 years but at the cost of a considerable decline in soil organic matter and total soil nitrogen.

There is no evidence that a new organic matter equilibrium has been reached in the soils investigated. The Rothamsted model suggests that the equilibrium level may not be reached until the organic matter level has fallen much further. It is not possible to predict accurately how far into the future this may occur, but it is possible that levels of soil organic matter and total nitrogen may be reached at which crop production begins to be seriously reduced. Low soil organic matter may be expected to cause problems through lower water retention, lower nutrient retention and poorer soil structure. The latter effect would enhance the possibility of soil erosion. The reduction in total soil nitrogen would be expected eventually to reduce yields on the plots to which no nitrogen or crop residues were returned although, under such conditions, a greater response to fertilizer nitrogen addition might be expected than was observed in the past twenty years.

A significant outcome of the present long-term experiment is to establish that, when rainfall is adequate, soil pH is controlled and potassium and phosphorus are supplied, the need for additional nitrogen, even after 20 years, is not as great as might have been expected. It suggests that the role of nitrogen in improving crop yields in West Africa, and possibly other tropical areas, may have been overemphasised in the past. There is little evidence from the present experiment to suggest that the continual addition of large amounts of fertilizer nitrogen can improve the long-term nitrogen status of the soil. The high rates of nitrogen supplied in the present experiment did not protect the soil from nitrogen depletion.

Crop yields in the long-term experiment have been sustained (if the effects of climatic factors on variability are excluded) for over twenty years. However, the soil organic matter has declined significantly and there has been a decrease in soil pH despite liming. It would be unwise to predict that cropping at the level observed could continue indefinitely even if the inputs were continued. The important over-riding question must still be — for how long could this soil continue providing reasonable crop yields? Is there a critical level of soil organic matter below which the soil suffers serious and economically irreversible physical degradation? The future of many agricultural economies in the tropics depends on an answer to these questions. Long-term field experiments will continue to be needed to provide the answers.

## VII. References

Ahn, P.M. 1979. The optimum length of planned fallows. In: H.O. Mongi, P.A. Huxley, and D. Spurgeon. (eds.), *Soil Research in Agroforestry.* Proceedings of an Expert Consultation. ICRAF, Nairobi, March 26-30, 1979.

Balasubramanian, V., and L. Singh. 1982. Efficiency of nitrogen fertilizer use under rainfed maize and irrigated wheat at Kadawa, Northern Nigeria. *Fertilizer Research* 3:315-324.

Bigot, Y. 1977. Fertilisation, labour et espèce cultivée en situation de pluviosité incertaine du Centre de Côte d'Ivoire. Synthèse des principaux résultats d'un test de differents systèmes culturaux de 1967 à 1974. *L'Agronomie Tropicale* XXXII:242-247.

Bouldin, D.R., W.S. Reid, and P.J. Stangel. 1979. Nitrogen as a constraint to non-legume food crop production. *Conference on Priorities for Alleviating Soil-Related Constraints to Food Production in the Tropics.* 4-8 June, 1979. IRRI, Los Baños, Laguna, Philippines.

Chabalier, P.F. 1976. Contribution à la connaissance du devenir de l'azote du sol et de l'azote-engrais dans un système sol-plante. Thèse $N^0 33$ Université Abidjan.

Chabalier, P.F. 1980. Utilisation de l'engrais par les cultures et pertes par Lixiviation dans deux agrosystemes de Côte d'Ivoire. p. 393-398 In: T. Rosswall (ed.), *Nitrogen Cycling in West African Ecosystems.*

Chopart, J.L. 1984. Soil erosion and control methods for upland rice cropping systems: some West African examples. p. 479-491. In: *An Overview of Upland Rice Research.* Proceedings of the 1982 Bouaké, Côte d'Ivoire, Upland Rice Workshop. IRRI, Los Baños, Lagüna, Philippines.

Chopart, J.L. 1989. Effect of Tillage on a Corn-Cotton Sequence in Côte d'Ivoire. In: *Proc. of an International Workshop on Soil, Crop and Water Management Systems for Rainfed Agriculture in the Sudano-Sahelian Zone.* ICRISAT Sahelian Center, Niamey, Niger. 7-11 January, 1987.

Christianson, C.B., and P.L.G. Vlek. 1991. Efficiency of Nitrogen Fertilizers Applied to Food Crops. p 45-58. In: U. Mokuwnye (ed.), *Alleviating Soil Fertility Constraints to Increased Crop Production in West Africa.* A. Kluwer Academic Publisher in Relation with the IFDC.

Crétenét, M. 1987. Aid for decision making on cotton fertilization in the Côte d'Ivoire. Coton et Fibres Tropicales. Vo.l. XLII, Fascicule 4, p. 245-254.

Entz, M.H., D.B. Fowler. 1989. Influence of crop water environment and dry matter accumulation on grain yield of no till winter wheat. *Canadian Journal of Plant Science* 69:369-375.

Gigou, J. 1973. Etude de la pluviosité en Côte d'Ivoire. Application à la Riziculture Pluviale. *L'Agronomie Tropicale* XXXII, $N^0 9$:858-875.

Gigou, J. 1987. L'adaptation des cultures dans le Centre de la Côte d'Ivoire. *L'Agronomie Tropicale* 42, $N^0 1$: 1-11

Gigou, J., and P.F. Chabalier. 1987. l'Utilisation de l'Engrais Azotee par les Cultures Annuelles en Côte d'Ivoire. *L'Agronomie Tropicale* 42,3: 1971-1979.

Greenland, D.J., A. Wild, and A. Adams. 1992. p. 17-33. In: *Organic Matter Dynamic in soil of the Tropics—From Mythe to Complex Reality.* Soil Science Society America Special Publication 29. Madison, WI.

Haggar, J.P., G.P. Warren, J.W. Beer, and D. Kass. 1991. Phosphorus Availability Under Alley Cropping and Mulched and Unmulched Sole Cropping System in Costa Rica. *Plant and Soil* 137:275-283.

Institut des Savannes (IDESSA). 1984. Méthodes d'analyses des Sols-Eaux-Végétaux. *Laboratoires d'Analyses Minérales* (Juillet, 1984).

Jenkinson, D.S. 1977. Studies on the Decomposition of Plant Material in Soil. V. The Effects of Plant Cover and Soil Type on the Loss of Carbon From [14]C Labelled Ryegrass Decomposing Under Field Conditions. *J. Soil Sci.* 28:424-434.

Jenkinson, D.S. 1990. The Turnover of Organic Carbon and Nitrogen in Soil. *Phil. Trans R. Soc. Lond.* B 1990 329:361-368.

Jenkinson, D.S. 1991. The Rothamsted Long-Term Experiments: Are They Still of Use ?. *Agron. J.* 83:2-10

Jenkinson, D.S., and A. Ayanaba. 1977. Decomposition of [14]C Labelled Plant Material Under Tropical Conditions. *Soil Sci. Soc. Am. J.* 41:912-915.

Jenkinson, D.S., and J.H. Rayner. 1977. The Turnover of Soil Organic Matter in Some of The Rothamsted Classical Experiments. *Soil Science* 123:298-305.

Jenkinson, D.S., P.B.S. Hart, J.H. Rayner, and L.C. Parry. 1987. Modelling the turnover of organic matter in long term experiment at Rothamsted. *INTECOL Bull.* 1987 15:1-8

Jenkinson, D.S., D.S. Powlson, and A.E. Johnson. 1986. The Nitrogen Cycle Under Continuous Winter Wheat. *Transaction of the XIIIth Cong. of the Int. Soc. of Soil Sci..* Vol. III:793-794.

Jones, M.J. 1973. The Organic Matter Content of Savanna Soils of West Africa. *J. Soil Sci.* 24:42-53.

Kang, B.T., F. Dondoh, and K. Moody. 1977. Soil Fertility Management Investigation on Benchmark Soils In The Humid Low Altitude Tropics of West Africa: Investigations on Egbeda Soil Series. *Agron. J.* 69:651-656.

Koné, D. 1991. Caracterisation du Risque Climatique della Culture du Mais en Zone Centre Côte d'Ivoire. In: Sivakumar, Wallace, Renard, and Giroux (eds.), *Soil Water Balance in The Soudano-Sahelian Zone. Proc. of Niamey Workshop.* IAHS Publication N[0] 199. Feb. 1991.

Lal, R. 1986. Impact of Farming Systems on Soil Erosion in The Tropics. *Transactions of XIII Cong. of the Int. Soc. of Soil Sci.* 1:97-111.

Lang, H. and R. Bartsch. 1977. Evaluation de l'Intérêt économique des Méthodes Culturales Améliorées en Conditions d'Incertitude Climatique Présentée à l'Exemple de la Région Centre en Côte d'Ivoire. *L'Agronomie Tropicale* Vol. XXXII, N[0]3: 248-256.

Lavelle, P., E. Blanchart, A. Martin, A.V. Spain, and S. Martin. 1992. Impact of Soil Fauna on the Properties of Soils in the Tropics. p. 157-185. In: Soil Sci. Soc. Am. Spec. Publ. 29,. Madison, WI.

Le Buanec, B. 1979. Intensification des Cultures Assolées en Côte d'Ivoire. Milieu Physique et Stabilité des Systèmes de Cultures Motorisées. *L'Agronomie Tropicale* XXXIV, 1:54-74

Le Buanec, B., and R. Didier de Saint Armand. 1975. Mise en Evidence d'une Carence en Phosphore sur les Sols Dérivés de Granites en Côte d'Ivoire et Contribution à la Mise au Point de Tests Permettant son Diagnostic. *Annales de l'Université d'Abidjan*, Serie C (Sciences) XI:103-122.

Lensi, R., A.M. Domenach, and L. Abbadie. 1992. Field Study of Nitrification and Denitrification in a Wet Savanna of West Africa (LAMTO, Côte d'Ivoire). *Plant and Soil* 147:107-113.

Lepage, M. 1972. Recherches Ecologiques sur une Savane Sahélienne du Ferlo Septentrional, Sénégal : Données Préliminaires sur l'Ecologie des Termites. *La Terre Vie* 26:383-409.

Marnotte, P. 1981. Modification des Populations de Mauvaises Herbes sous l'Effect des Facteurs d'Intensification des Cultures. p. 86-93. In: *Trans. of The First Int. Conf. of West African Weed Sci. Soc.*. Monrovia, Liberia. 3-5 August, 1981.

Miner, G.S., S. Traore, and M.R. Tucker. 1986. Corn Response to Starter Fertilizer Acidity and Manganese Materials Varying in Water Solubility. *Agron. J.* 78:291-295.

Mokwunye, A.U., and L.L. Hammond. 1992. Myths and Science of Fertilizer Use in the Tropics. p. 12-134. In: *Myths and Science of Soils of the Tropics.* Soil Sci. Soc. of Am. Spec. Publ. 29. Madison, WI.

Mughogho, S.K., A. Bationo, C.B. Christianson, and P.L.G. Vlek. 1986. Management of Nitrogen Fertilizers for Tropical African Soils. p. 117-172. In: A.U. Mokwunye and P.L.G. Vlek. *Management of Nitrogen and Phosphorus Fertilizers in Sub-Saharan Africa.* Martinus NIJHOFF, Publishers.

Parton, W.J., R.L. Sanford, P.A. Sanchez, J.W.B. Stewart. 1989. Modelling Soil Organic Matter Dynamics in Tropical Soils. p. 153-171. In: D.C. Coleman, J.M. Oades, and G. Uehera (eds.), *Dynamics of Soil Organic Matter in Tropical Ecosystems.* NifTAL Project Publishers.

Paustian, K., W.J. Parton, and J. Persson. 1992. Modelling Soil Organic Matter in Organic-Amended and Nitrogen-Fertilized Long-Term Plots. *Soil Sci. Soc. Am. J.* 56:476-488.

Power, J.F., L.N. Mielke, W.W. Wilhelm, J.W. Doran, and J.E. Gilley. 1986. Crop residue management and the soil environment. *Trans. of the XIII[th] Cong. of the Int. Soc. of Soil Sci.* Vol IV:1398-1399.

Sanchez, P.A. 1976. *Properties and Management of Soils in the Tropics.* Wiley, New York.

Sanchez, P.A. 1977. Advances in the Management of Oxisols and Ultisols in Tropical South America. p. 535-566. In: *Proc. of The International Seminar on Soil Environment and Fertility Management in the Intensive Agriculture.* Tokyo, Japan.

Sanchez, P.A., and T.J. Logan. 1992. Myths and Science about the Chemistry and Fertility of Soils in the Tropics. p. 35-46. In: *Myths and Science of Soils in the Tropics.* Soil Sci. Soc. Am. Spec. Publ. 29. Madison, WI.

Sement, G. 1983. La Fertilité des Systèmes Culturaux à Base de Cotonnier en Côte d'Ivoire. *Paris IRCT*, Documents Etude et Synthèses.

Seubert, C.E., P.A. Sanchez, and C. Valverde. 1977. Effects of Land Clearing Methods on Soil Properties of an Ultisol and Crop Performance in the Amazone Jungle of Peru. *Trop. Agri.* 54:307-321.

Stumpe, J.M., and P.L.G. Vlek. 1991. Acidification Induced by Different Nitrogen Sources in the Columns of Selected Tropical Soils. *Soil Sci. Soc. Am. J.* 55:145-151.

Téhia, K.E. and P. Marnotte. 1990. Observations des mauvaises herbes sur l'Essai Stokage d'Azote. *Note Technique Syst/27/90.* Filière Système IDESSA-DCV.

Tindall, J.A., R.B. Beverly, and D.E. Radcliffe. 1991. Mulch Effect on Soil Properties and Tomato Growth Using Micro-Irrigation. *Agron. J.* 83:1028-1034.

Traoré, S. 1993. Effect of Climate and Nitrogen Fertilization on Crop Yields in the Savanna Area of The Côte d'Ivoire. Ph.D. Thesis. Soil Science University, University of Reading, U.K.

Walkley, A. and C.A. Black. 1934. An examination of the Degtjareff Method for Determining Soil Organic Matter and a Proposed Modification of The Chromic Acid Titration Method. *Soil Sci.* 37:29-38.

Wambeke, V.A. 1992. *Soils of The Tropics: Properties and Appraisal.* McGraw-Hill, Inc. Publishers, New York.

Welbank, P.J., M.J. Gibb, P.J. Taylor, and E.D. Williams. 1973. Root Growth of Cereals Crops. *Rothamsted Report for 1973*, Part 2: 26-56.

# Yield Decline and the Nitrogen Economy of Long-Term Experiments on Continuous, Irrigated Rice Systems in the Tropics

K.G. Cassman, S.K. De Datta, D.C. Olk, J. Alcantara,
M. Samson, J. Descalsota, and M. Dizon

## I. Introduction

### A. Importance of Irrigated Lowland Rice Systems

Rice (*Oryza sativa* L.) has been grown as a food crop for more than 6000 years in China, India, and Southeast Asia (Harlan, 1992). Irrigated rice systems have supported sophisticated civilizations with high population density for more than 2000 years. Today, rice-based agricultural systems dominate in much of Asia where more than 90% of global rice supplies are produced and consumed, and rice contributes 30-75% of dietary calories (Table 1).

ISBN 1-56670-076-0
©1995 by CRC Press, Inc.

**Table 1.** Irrigated rice area and the contribution of rice to total caloric intake in human diets of South and Southeast Asia

| Country | Irrigated rice area[1] | | Calorie supply from rice[2] |
|---|---|---|---|
| | Area harvested | Proportion of total rice area | |
| | ha (x $10^6$) | (%) | % of total |
| Bangladesh | 1.96 | 19 | 71 |
| Cambodia | 0.25 | 16 | 76 |
| India | 18.08 | 44 | 30 |
| Indonesia | 8.01 | 81 | 56 |
| Laos | 0.04 | 6 | 70 |
| Malaysia | 0.35 | 54 | 32 |
| Myanmar | 0.83 | 18 | 74 |
| Nepal | 0.30 | 23 | 42 |
| Pakistan | 2.11 | 100 | 8 |
| Philippines | 2.02 | 58 | 41 |
| Sri Lanka | 0.59 | 77 | 44 |
| Thailand | 2.42 | 27 | 59 |
| Vietnam | 3.14 | 54 | 69 |

[1] Data based on most recent available data, 1987-1989.
[2] For the period 1986-1988.
(From IRRI, 1991.)

Irrigated rice systems account for 55% of total harvested rice area and 76% of global rice production (IRRI, 1993). About 40 million ha of irrigated rice are harvested each year in tropical and subtropical Asia (Table 1). Of this total, an annual double-crop rotation of irrigated rice and wheat accounts for approximately 12 million ha in northern India, Pakistan, Nepal, and Bangladesh. Most of the remaining 28 million ha of irrigated rice is produced in tropical environments where the climate permits two and sometimes three rice crops on the same land each year. Such systems dominate in the deltas and inland valleys of southern India, Indonesia, Vietnam, the Philippines, Sri Lanka, and Malaysia so that only 10 million ha are required for a total harvested area of more than 20 million ha with average yields of 4 to 6 t ha$^{-1}$. Indeed, intensive rice production on irrigated lowlands reduces the need for expansion of agriculture on more fragile uplands where rice yields are often less than 1 t ha$^{-1}$ and soils are prone to erosion.

## B. Intensification of Irrigated Lowland Rice Systems

Intensified rice systems as exist today evolved rapidly since the 1960s. This evolution depended on two factors: modern high-yielding varieties (HYVs) of short duration that were first released in the 1960s, and access to irrigation. When transplanted, HYVs reach maturity in 105 days or less compared to the traditional varieties of 150 days or more that were replaced. Furthermore, the new HYVs were more responsive to N fertilizer due to shorter stature and greater resistance to lodging than traditional landraces. Concurrent with the availability of HYVs, governments made large investments in irrigation infrastructure which promoted the expansion of irrigated rice in new areas. Together these developments led to a green revolution in rice production.

Present irrigated rice systems in the tropics are the most intensive cereal crop production systems in the world. Relatively high nutrient inputs are applied to each crop cycle, and most crop cycles receive at least one pesticide application. Further intensification of lowland rice systems will be needed to feed the rapidly growing urban populations. It is estimated that rice production must increase more than 50% between 1992 and 2020 (IRRI, 1993). Much of this increase must come from existing irrigated land that is already producing two rice crops per year because a net increase in irrigated area for lowland rice is not expected. At issue, therefore, is whether the biophysical resource base of intensive rice systems can sustain the increases in rice output that are required.

Although some lowland paddies have been in continuous rice production for hundreds and perhaps thousands of years, continuity and stability alone do not prove the capacity to sustain significantly higher yield levels and further increases in cropping intensity. Indeed, the structure and function of modern rice systems are vastly different from the traditional, single crop rice systems of the past which received little, if any, external inputs. And because modern systems have been in existence less than 30 years, it is only now becoming possible to obtain definitive evidence about future prospects. Anticipating the magnitude of change brought about by green revolution technologies, rice agronomists at the International Rice Research Institute (IRRI) in the Philippines had the vision to initiate several long-term field experiments on intensive rice systems in the 1960s. The purpose of this paper is to assess productivity trends in these long-term experiments, and where clear trends are evident, to consider potential causes of changes in productivity based on recent experimental evidence.

## II. Long-Term Experiments on Rice in the Philippines

### A. Design and Purpose of the Long-Term Rice Experiments

The first long-term experiment on modern rice systems in the tropics was established at the IRRI Research Farm in 1964 to quantify changes in soil fertility and rice yield in response to N, P, and K fertilizer inputs. Fertilizer treatments were applied to the same plots each crop cycle in a continuous rice double-crop system (Table 2). Treatments also included three varieties in factorial combination with the NPK fertilizer regimes, and the varieties used in a given season were the same at all sites. This experiment is called the Long-Term Fertility

**Table 2.** Description of the Long-Term Fertility Experiments (LTFE), the Long-Term N Response Experiments (LTNRE), and the Long-Term Continuous Cropping Experiment (LTCCE) in the Philippines

| Experiment | Sites | Duration | Rice cropping intensity | Experimental design | Fertilizer treatments $N^1$ DS/WS | P | K | Varietal entry treatments |
|---|---|---|---|---|---|---|---|---|
| LTFE | IRRI | 1964-present | Double-crop | Factorial RCB[2] | 0/0 | 0 | 0 | 3 |
| | | | | | 0/0 | 13 | 0 | |
| | | | | | 0/0 | 0 | 25 | |
| | | | | | 140/60 | 0 | 0 | |
| | | | | | 140/60 | 13 | 0 | |
| | | | | | 140/60 | 0 | 25 | |
| | PRRI | 1968-present | Double-crop | Factorial RCB | 0 | 0 | 0 | 3 |
| | BRIARC | 1968-present | | | 140/70 | 0 | 0 | |
| | | | | | 140/70 | 26 | 0 | |
| | | | | | 140/70 | 0 | 50 | |
| | | | | | 140/70 | 26 | 50 | |
| LTNRE | IRRI | 1966-1988 | Double-crop | Split-plot RCB | 0/0 | 13 | 17 | 18-28 |
| | PRRI | 1968-1988 | | N rates = MP[3] | 36/30 | 13 | 17 | |
| | BRIARC | 1968-1988 | | Varieties = SP[3] | 72/60 | 13 | 17 | |
| | | | | | 108/90 | 13 | 17 | |
| | | | | | 144/120 | 13 | 17 | |
| | | | | | 180/150 | 13 | 17 | |

**Table 2.** continued

| Experiment | Sites | Duration | Rice cropping intensity | Experimental design | Fertilizer treatments $N^1$ DS/WS | P | K | Varietal entry treatments |
|---|---|---|---|---|---|---|---|---|
| LTCCE | IRRI | 1968-present | Triple-crop | Split-plot RCB: | 0/0 | 13 | 25 | 6 (1968-91) |
| | | | | N rates = MP | 50/30 | 13 | 25 | 3 (1992-present) |
| | | | | Varieties = SP | 100/60 | 13 | 25 | |
| | | | | | 150/90 | 13 | 25 | |

[1] DS = dry season, WS = wet season. All N fertilizer is applied in two split applications, 2/3 incorporated before planting and 1/3 5 to 7 days before panicle initiation.

[2] RCB = randomized complete block design.

[3] MP = main plot treatments. SP = subplot treatments.

Experiment (LTFE). Similar experiments were established in 1968 at the Philippines Rice Research Institute (PRRI) in Munoz, Central Luzon, at the Bicol Regional Integrated Agricultural Research Center (BRIARC) in Pili, Camarines Sur, and at the Visayas Rice Experiment Station (VRES) on Panay Island. Irrigation supply became unreliable at VRES in the 1980s so that yield trend analysis at this site will not be presented. A brief description of soils and climate are given in Table 3.

A second set of long-term experiments was established to evaluate the response of elite varieties from the IRRI breeding program to fertilizer-N rates. These Long-Term N-Response Experiments (LTNRE) were initiated in 1966 at IRRI and 1968 at PRRI, BRIARC, and VRES. Main plots were N rates applied to the same plots in a continuous double-crop system. Subplots were 18 to 28 elite breeding lines and varieties (Table 2). These experiments were terminated in 1988. At each site, soil characteristics were similar to those in the LTFE (Table 3) which was located in a nearby field.

The most intensive rice system in a long-term experiment is a triple-crop system that was initiated in 1968 at IRRI. This experiment, called the Long-Term Continuous Cropping Experiment (LTCCE), recognized the feasibility of continuous, triple-crop rice systems made possible by the new short season HYVs and year-round access to irrigation. Today, triple-crop rice systems are common in south and central Vietnam and Malaysia, and to a lesser extent in the Philippines, Indonesia, and Southern India. Main plots are N-rates, subplots are six varieties and elite breeding lines, and broadcast or transplanted establishment methods are sub-subplots (Table 2). The soil at the LTCCE site is classified in the same subgroup as in the LTFE at IRRI (Table 3), but has a slightly greater clay content (63%) and less silt (28%). The 90th consecutive crop cycle in the LTCCE experiment was planted in the 1993 wet season.

In each of the long-term experiments, variety IR8 was included as one of the treatment entries since the initial crop cycle until the experiments were terminated (as for the LTNRE experiments) or until 1991 in the LTFE and LTCCE, which are still in progress. Since the 1991 wet season, however, IR8 has not been used because it became extremely susceptible to the rice tungro virus and provided a continuous source of inoculum for the vector of this disease. Other varieties included as treatments are changed regularly to include the highest yielding, most disease and insect resistant germplasm available at each point in time. Several of the breeding lines used in these experiments became IRRI varieties that were widely adopted by rice farmers throughout Asia. At all sites, permanent bunds of 20- to 30-cm height surround each fertilizer treatment plot to minimize soil, fertilizer, and floodwater movement between treatments. In each crop cycle, soil remains saturated without standing water from the time of puddling until 14 days after transplanting when a floodwater depth of 5-10 cm is maintained until the field is drained prior to harvest.

**Table 3.** Soil type and climatic conditions at three sites where the Long-Term Fertility Experiments (LFTE) are conducted

| Site | Soil | | | | Climatic conditions[1] | | | | | | | | | Annual |
| | Subgroup | Clay minerals | Clay | Silt | --Mean temp.--- | | | --Rel. humidity-- | | | -Solar radiation-- | | | Rainfall |
| | | | | | Jan | Apr | Sep | Jan | Apr | Sep | Jan | Apr | Sep | (mm) |
| | | | ----(%)---- | | -------(°C)------- | | | -------(%)------- | | | ---(MJ m$^{-2}$)------ | | | |
| IRRI | Andaaqueptic Haplaquoll | allophanes, amorphous | 61 | 30 | 25 | 28 | 27 | 90 | 85 | 90 | 13 | 21 | 16 | 2048 |
| PRRI | Vertic Tropaquept | montmorillonite vermiculite halloysite | 38 | 55 | 26 | 29 | 27 | 87 | 85 | 94 | 20 | 24 | 21 | 1656 |
| BRIARC[2] | Typic Pelludert | montmorillonite | 51 | 36 | 25 | 28 | 28 | 84 | 78 | 86 | na[2] | na | na | 2161 |

[1] Mean values for the period 1979-1990 at IRRI, 1986-1992 at PRRI, and 1975-1990 at the CSSAC weather station near BRIARC.

[2] Solar radiation data not available.

## B. Initial Conditions, Measurements, and Management

### 1. Previous History and Initial Soil Characterization

For long-term studies on rice-based systems, information on previous cropping history is of particular importance because soil properties change markedly when soils are puddled and cropped under submerged conditions. Records of previous cropping history at the LTFE and LTNRE sites are lacking. By contrast, the previous history at the LTCCE site is well documented with 13 rice crops in the 5-year period before the experiment was initiated in 1968 (IRRI, 1967).

Initial soil characterization was thorough in the LTFE sites. Soil measurements included pH, organic C by the Walkley Black Method, total N by the Kjeldahl method, particle size analysis, exchangeable bases extracted by neutral 1 M ammonium acetate, P by Bray-2 extraction, CEC, and clay mineral composition (De Datta et al., 1988). Less certain are the sampling procedures used to obtain the initial soil samples. Although sampling depth was specified at 0-20 cm, it is uncertain whether the initial sample was taken before or after puddling in preparation for planting the first crop, and samples were field composites so that confidence intervals for each soil property cannot be estimated. In the LTCCE study, soils were composited by block and standard deviations for initial soil organic C and total N can be estimated.

Physical properties other than particle size distribution were not measured initially. Some physical attributes such as aggregate stability and pore size are less meaningful for irrigated rice grown in puddled soil than for upland crops grown in aerated soil. Bulk density of the 0-20 cm sampled layer, however, is crucial for estimating the mass balance of soil organic C, total N, and other nutrients in the system. In 1991 and 1992, bulk density measurements were taken for the first time in the LTFE and LTCCE experiments.

### 2. Sampling, Processing, Measurements, Analysis, and Archives

Detailed records of sampling and processing procedures are extremely important in long-term experiments because interpretation of results will ultimately be made by scientists who were not involved with taking the samples. For soils, records should include the specific point in the cropping sequence when soil was sampled, the soil condition at time of sampling, depth sampled, number of cores taken per plot, and the instrument used to take the sample. Such records are available for soil samples taken since 1978 in the IRRI experiments. Precise depth of sampling is particularly important in puddled rice fields because bulk density, organic C and total N of the puddled layer can differ greatly from the unpuddled soil at lower depths. At the LTFE sites, for example, there is a marked decrease in the organic C and total N content of soil from 10-20 cm compared to the 20- to 30-cm layer (Figure 1), while bulk density increases by 15 to 25% in the 20- to 30-cm layer.

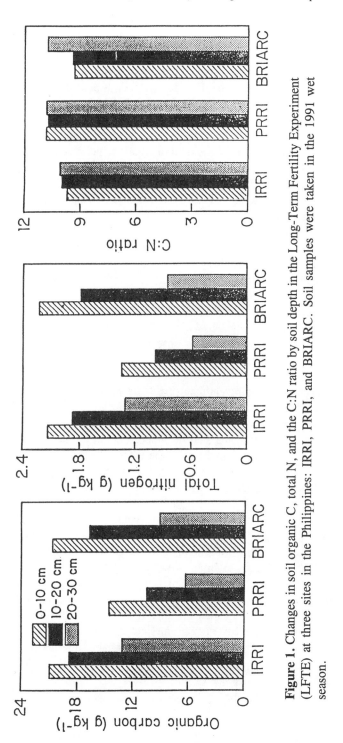

**Figure 1.** Changes in soil organic C, total N, and the C:N ratio by soil depth in the Long-Term Fertility Experiment (LFTE) at three sites in the Philippines: IRRI, PRRI, and BRIARC. Soil samples were taken in the 1991 wet season.

Surplus soil from each plot sample that is not needed for laboratory analyses is archived in air-tight plastic or glass containers. Unfortunately, all soils sampled before 1983 from the IRRI long-term experiments were discarded. Archived samples are crucial for making accurate comparisons of certain chemical properties in soil samples from different time periods because analyses can be performed by the same laboratory technician, using the same equipment and standards. These reanalyses are useful to quantify long-term trends when measurements involve titration methods with visual end points which are somewhat subjective, such as total N by Kjeldahl distillation and Wakley-Black organic C, or when the method of analysis has changed. Reanalysis of total soil N and organic C was recently performed on soil samples taken in 1983 and 1985 from the LTCCE, concommittant with analyses of soil samples taken in the 1991 wet season (see Section IVA).

Grain yields were determined in 5 $m^2$ or 10 $m^2$ harvest areas within each plot at maturity in each crop cycle. Components of yield were not routinely measured before 1991. Likewise, total nutrient content of grain and straw was not measured, nor was tissue nutrient status at critical stages of crop development. Since 1991, plant nutrient status at panicle initiation and flowering stages, nutrient accumulation in aboveground biomass at maturity, and components of yield are measured in each crop cycle. We anticipate, however, that these measurements can be taken on 3-year cycles once baseline values have been established in each long-term experiment.

Since 1991, estimates of nutrient uptake and components of yield are based on a 12-hill sample taken at physiological maturity. In experiments that include several varieties and fertilizer treatments, hills from each plot are sampled at physiological maturity when 90 to 95% of all grain have lost their green color. This avoids leaf loss due to wind, rain, and decomposition in humid tropical climates. Leaf losses due to field weathering can be large because differences in growth duration among varieties in the long-term experiments can be 14 days or more, and differences in treatments with and without fertilizer-N are from 5 to 7 days for the same variety.

Grain from the 12-hill samples is dried to constant weight at 70°C (oven-dry) before weighing. Each straw sample is weighed fresh, immediately subsampled (approximately 30% of the total), a subsample fresh weight taken, then oven-dried to constant weight for final weighing. Aboveground nutrient content is calculated from the nutrient concentration measured in grain and straw from the 12-hill samples, grain yield of the large 5 $m^2$ or 10 $m^2$ harvest area adjusted to oven-dry moisture content, and straw yield calculated from the oven-dry harvest index of the 12-hill sample and the oven-dry grain yield from the large harvest area. Because the harvest index remains relatively constant in modern HYVs, this method of estimating nutrient uptake avoids the greater variability of straw and grain yields based on small samples of only 12 hills.

**Table 4.** Changes in tillage operations and fertilizer incorporation in the Long-Term Continuous Cropping Experiment (LTCCE) and in the Long-Term Fertility Experiments (LTFE) conducted at three sites in the Philippines

| | Site | | Puddling operations | | Fertilizer |
|---|---|---|---|---|---|
| Expt. | (Plot size) | Period | Plowing | Harrowing | incorporation |
| LTCCE | IRRI | 1963-1985 | Carabao | Carabao | Carabao |
| | (25x25m) | 1986-1989 | Landmaster | Landmaster | Landmaster |
| | | 1989-present | Landmaster | Hydrotiller | Landmaster |
| LTFE | IRRI | 1964-1982 | Carabao | Carabao | Carabao |
| | (4x5m) | 1983-1990 | Carabao | Carabao | Power weeder |
| | | 1991-present | Landmaster | Hydrotiller | Power weeder |
| LTFE | PRRI | 1968-1983 | Carabao | Carabao | Carabao |
| | (3x7m) | 1983-1987 | Carabao | Carabao | Power weeder |
| | | 1988-present | Carabao | Hydrotiller | Power weeder |
| LTFE | BRIARC | 1968-1987 | Carabao | Carabao | Carabao |
| | (3x6m) | 1988-present | Carabao | Carabao | Rake |

## 3. Crop Logs and Non-Treatment Variables

Although researchers attempt to apply uniform management over time and space in long-term experiments, no two crops are managed the same due to variations in climate, pest pressure, and unexpected events. Crop management logs are needed to record the date and method of each operation from seedbed preparation to harvest and also in the transition period between crops. Such a log includes pest control measures, disease and insect damage, tillage operations, water management, and unusual climatic events that affect crop performance. Crop log records for the IRRI long-term experiments have been used to minimize the influence of year to year variation in climate, insect, and disease pressure on estimated yield trends (Flinn and De Datta, 1984).

Sometimes it is necessary to modify non-treatment management operations or the imposed treatments to improve the efficiency of managing a long-term experiment, to reduce management costs, or to make the experiment more relevant. An example of such a change in the IRRI experiments was the removal of IR8 as a varietal reference treatment because it was no longer a viable genotype and was a source of disease inoculum. Tillage operations, a non-treatment variable, were mechanized in the 1980s to mirror widespread adoption of small hand-held mechanized tractors and hydrotillers for plowing, harrowing, and fertilizer incorporation by Philippine rice farmers (Table 4).

Management guidelines for non-treatment variables that may confound treatment effects must also be documented for future reference. For example in the IRRI long-term experiments, control measures are taken as required to prevent yield loss from both insect and weed pests. Likewise, in the LTRNE and LTCCE which have fertilizer-N rates as treatments, both P and K inputs are applied to all plots in each crop cycle to avoid deficiencies. Sufficient S has been provided in sulfate-containing N and P fertilizers. Although Zn was not applied in the long-term experiments for many years, our guidelines for Zn and other non-treatment nutrients now specify that such nutrients should be applied as required to maintain soil or plant test values well above critical levels.

In contrast to insects and weed pests, no attempt was made to control diseases with fungicides or other chemicals in the long-term experiments. For the first time in 1992, however, fungicides were applied to prevent sheath blight (*Rhizoctonia solani*) because fertilizer-N rates were increased in the N-rate mainplot treatments of the LTCCE to test the hypothesis that N deficiency contributed to the observed yield declines in that experiment.

## III. Yield Trends in the Long-Term Experiments

### A. Occurrence of Declining Yield Trends

1. Yield Declines in the Long-Term N Response Experiments

The phenomenon of declining yields in continuously cropped, irrigated rice systems was first reported by Flinn et al. (1982) based on yield trends of the highest yielding entry (HYE) of the 28 varieties and elite lines grown in each crop cycle of the LTFNE at IRRI. In subsequent work, Flinn and De Datta (1984) expanded the analysis to evaluate yield trends of HYEs at all four LTFNE sites. In the latter analyses, the effects of yearly variations in disease incidence, solar radiation, and typhoon damage (wet seasons only) on yields were isolated from the overall yield trend so that the underlying trends under favorable growth conditions could be quantified. Based on these analyses, a significant negative yield trend was found at IRRI and VRES in the dry season, and in the wet season at IRRI and BRIARC. Where yield declines occurred, the rate of decrease was comparable for all fertilizer-N treatments, including control plots which did not receive N inputs. Rates of decline in grain yield ranged from 90 to 160 kg ha$^{-1}$ yr$^{-1}$.

2. Yield Declines in the Long-Term Fertility Experiments

Although the LTNREs were terminated in 1988, the LTFEs continued at the same four sites (Table 2). Yield trends at LTFE sites were evaluated by linear regression of yield versus year of continuous cropping. Whereas Flinn and De

Datta (1984) based their analysis of yield trends on the HYE regardless of whether the HYE yield was significantly better than one or more of the other 18-28 entries in the experiment, we evaluated trends in the LTFE based on the highest yielding cultivar in each crop cycle only when analysis of variance indicated a significant yield advantage compared to the other varieties in the experiment. When analysis of variance indicated no significant difference in yield between two or three varieties, then the mean yield of varieties with comparable yields in a given cropping cycle was used in the trend analysis.

Using this approach, a significant linear yield decline is evident at IRRI in the dry season in +NPK treatments (Figure 2). Wet season yield trends were also negative in the +NPK treatment at PRRI, and in the control plots without fertilizer-nutrient inputs at IRRI and PRRI in both seasons. The rate of decrease in yields ranged from 60 to 100 kg ha$^{-1}$-yr$^{-1}$ at IRRI from 1966 to 1991, and 50 to 70 kg ha$^{-1}$-yr$^{-1}$ at PRRI from 1968 to 1991. Although no significant yield trends are evident at BRIARC, initial yield levels at the beginning of the experiment were considerably lower than at IRRI. Average yield levels in +NPK and control plots over the past five years are similar at both sites.

## 3. Yield Declines in the Long-Term Continuous Cropping Experiment

The LTCCE included N-rate main plots, six varieties or elite lines as subplots, and transplanted versus broadcast establishment methods as sub-subplots. Analysis of variance by year and season indicated that the establishment methods explained only a small portion of the total variation in yield. In the dry season, for example, establishment methods and their interactions with N-rate and variety accounted for less than 4% of the total variation in grain yield when averaged over all years from 1968-1991. By contrast, N-rate treatments accounted for 52%, varieties 21%, and the interaction of NxV 4% of total variation in dry season grain yields. Thus, to simplify the analysis of yield trends, only transplanted yields are considered and the experiment is analyzed as a split-plot design with N-rate main plots and genotypes as subplots (Table 2).

In most years, there was no significant difference in yield of two or three of the six varieties or elite lines included in the study. Yield trends were evaluated by regression of yield of the three highest-yielding varieties on year of continuous cropping. Based on these analyses, significant yield declines are apparent in each of the three cropping seasons, and in all N-rate treatments (Figure 3). Greater deviation from linear trends in the late wet season reflects the influence of poorer growth conditions and the occurrence of typhoons in some years. In treatments without fertilizer-N inputs, a linear-plateau regression model provided a best-fit to the data in the dry and in the late wet seasons. In general, however, the magnitude of the yield decline was comparable in treatments with or without applied N so that the increase in yield from fertilizer-N remained relatively constant (Figure 4).

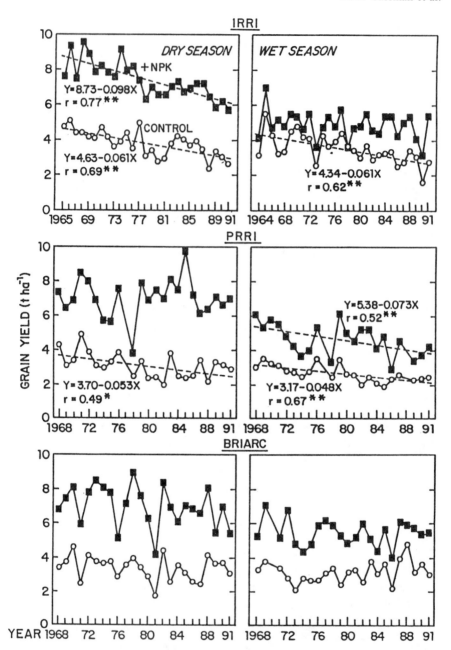

**Figure 2.** Rice yield trends for the dry and wet seasons in the Long-Term Fertility Experiments conducted at three sites in the Philippines: IRRI, PRRI, and BRIARC. Trends are based on the highest yielding varieties in each cropping cycle for each season as discussed in Section III.A.2.

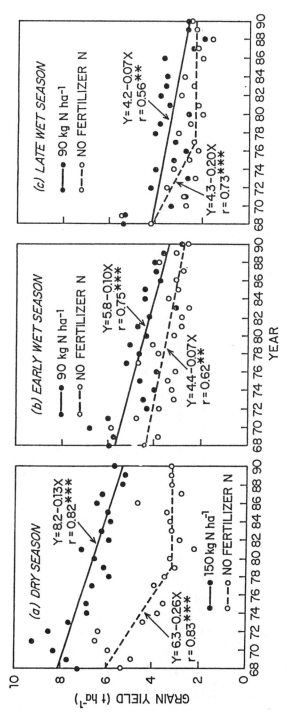

**Figure 3.** Rice yield trends in the triple-crop system of the Long-Term Continuous Cropping Experiment (LTCCE) conducted at the IRRI Research Farm. Trends are based on the three (of six) highest yielding varieties in each year for each crop cycle as discussed in Section III.A.3.

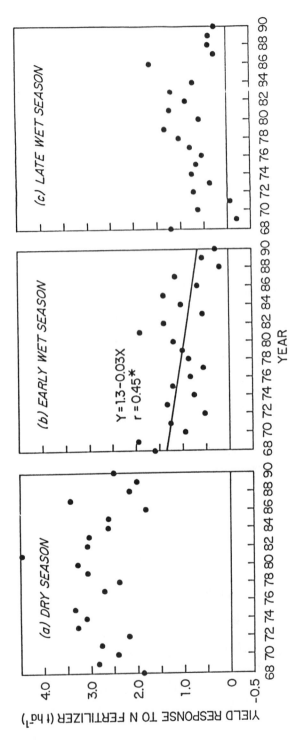

**Figure 4.** Magnitude of the grain yield response to the highest rate of applied fertilizer-N in the Long-Term Continuous Cropping Experiment (LTCCE) conducted at the IRRI Research Farm. The yield response is calculated as the difference between the yield of the highest N-rate treatment and the control treatment without fertilizer-N addition, based on yields of the three highest-yielding varieties in each year for each crop cycle.

The only variety that was included in all years from 1968-1990 was IR8. Comparison of yield trends for IR8 and the mean of the three highest-yielding varieties in any given season (Figure 5) illustrates the impact of insect and disease resistance that was incorporated into more recent IRRI varieties (Khush and Coffman, 1977). In the early years of the LTCCE, IR8 was often among the three varieties with the highest yields; in the mid-1970s and thereafter, the performance of IR8 was erratic due to viral diseases such as grassy stunt, ragged stunt, tungro, and bacterial leaf blight (Flinn et al, 1982). Indeed, the highly variable nature of disease pressure is evident in the yield pattern of IR8, which became susceptible to the insect and disease populations at the experimental site. Incorporation of disease resistance into more recent varieties reduced the deviation about the yield trend line, and decreased the rate of decline.

## 4. Declining Yields in Replicated Varietal Trials

Since the early 1960s, plant breeders have conducted replicated field trials at the IRRI Research Farm that included standard varieties as checks and a large number of elite breeding lines which reflect the most promising germplasm at any point in time. These trials are not conducted in the same field and crop management has changed as improved technologies became available. Despite changes in field location and crop management, yield trends in these replicated trials are relevant to the issue of declining yields because they represent the "state-of-the-art" in agronomic practices, including germplasm, over time.

At IRRI, yield of the HYE in replicated yield trials which include 300-400 entries each year have decreased from 8.8 t ha$^{-1}$ in 1966-1972 to 7.2 t ha$^{-1}$ from 1986-1992 (pers. comm., G.S. Khush). At issue, therefore, is whether there has been a decrease in the yield potential of more recent IRRI varieties, and whether such putative changes in yield potential have contributed to declining yield trends documented in long-term experiments. Recent evidence, however, clearly demonstrates that there has been no decrease in yield potential (see section III.C.1. below).

## B. Causes of the Yield Decline: Initial Hypotheses

### 1. Soil-Related Problems

#### a. Zinc Deficiency and Boron Toxicity at IRRI

Reasons for declining yields at the LTNRE sites were never determined, and the experiments were terminated in 1988. Flinn and De Datta (1984) speculated that boron toxicity, the result of irrigation with alkaline water pumped from an aquifer high in boron, and zinc deficiency contributed to the yield declines in the

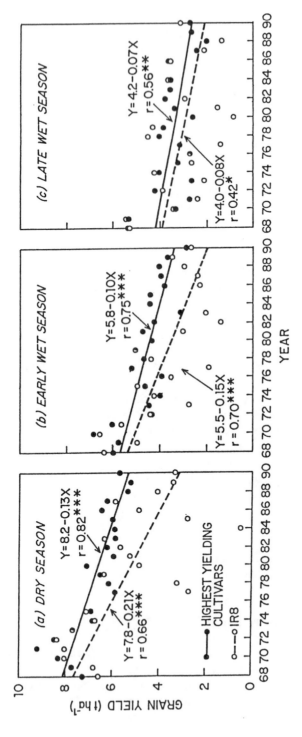

**Figure 5.** Comparison of trends in yields of the three highest-yielding varieties and IR8 in each year for each crop cycle of the Long-Term Continuous Cropping Experiment (LTCCE) at IRRI. Trends are based on yields in the highest N-rate treatment in each cropping season.

LTNRE and LTFE experiments at IRRI. Subsequent research, however, failed to demonstrate a consistent relationship between plant or soil boron status and yield loss under field conditions (Cayton, 1985). Soil pH of the 0- to 20-cm layer in LTFE at IRRI has remained constant in control plots without fertilizer inputs, and decreased from 6.0 in 1968 to 5.7 in 1991 in +NPK treatments. Slight acidification is also occurring in the LTCCE where mean soil pH in the highest N input treatment decreased from 6.5 in 1979 to 6.2 in 1991.

Recent evidence suggests that zinc deficiency is not likely in the long-term experiments at IRRI for two reasons. First, zinc concentration measured in plant tissues sampled in the 1991 dry season tests well above critical thresholds reported for rice (Tanaka and Yoshida, 1970; Rashid and Fox, 1992). Second, Zn inputs to the long-term experiments have not reversed declining yield trends. In response to the suggestion of a Zn deficiency in the LTCCE, seedling beds were treated with a 2% zinc oxide suspension before transplanting from 1981 to 1989. Previous research demonstrated that addition of Zn to the seedling bed and ZnSO4 application to the field were equally effective in correcting Zn deficiency on Philippine soils where yield reductions from Zn deficiency were documented (IRRI, 1978; 1979). In the LTFE at IRRI, ZnSO4 has been applied at regular intervals since 1986, and, more recently, it has been applied at the other LTFE sites and in the LTCCE.

### b. Phosphorus and Potassium Availability

At BRIARC and VRES, Flinn and De Datta (1984) speculated that changes in nutrient availability due to intensive cropping were potential contributors to negative yield trends in the LTNRE at these sites. Soil analyses were not performed to test this hypothesis. At IRRI, declining yields in the control treatment without fertilizer-nutrient inputs in the LTFE (Figure 2) cannot be attributed to P or K deficiency because a comparable linear yield decline occurs in treatments that only receive fertilizer-P or K addition (data not shown), and there has not been a yield response to K or P additions in treatments with fertilizer-N addition since the inception of this experiment (De Datta et al., 1988). Moreover, inputs of P and K are made in each crop cycle in the LTCCE, and soil test values and leaf tissue analysis indicate P and K well above sufficiency levels.

In contrast to IRRI, there is a large yield response to P and K addition at PRRI based on comparisons of the +N, +NP, and +NPK treatments. The wet season yield decline in +NPK treatments, however, does not result from P deficiency: both soil test P values and plant tissues measured in the 1991 wet season and the 1992 dry season indicate adequate P supply (data not shown). The soil at PRRI contains vermiculite. Extractable K in the +NPK treatments at this site indicate a decrease from 0.50 cmol K kg$^{-1}$ in 1968 to 0.11 cmol K kg$^{-1}$ in 1991. This depletion was confirmed by measurement of K removal with harvested grain and straw, which exceeded the rate of K addition in the 1991 dry season. While K inputs have failed to maintain soil K status, K deficiency does not account for declining wet season yields in +NPK treatments at PRRI because tissue K concentration at panicle initiation and flowering stage were well above sufficiency levels, and because there is no negative yield trend in the dry season at higher yield levels when crop K demand is much greater (Figure 2).

Yield declines also occurred in the control plots without fertilizer-nutrient additions at PRRI. Potassium deficiency is not likely the limiting factor because extractable K in control treatments was 0.12 cmol K $kg^{-1}$ in 1991, which is comparable to levels in the +NPK treatments where no yield decline occurred in dry season. Extractable soil P levels in the control treatment remained relatively constant from 1977 to 1988, then it dropped to very low values in the 1991 sample. This time-trend for soil P availability, however, differs markedly from the steady decline in yields of the control plots during this period (Figure 2). Analysis of vegetative tissues at panicle initiation and flowering in 1991 and 1992 indicated adequate K (17 to 21 g $kg^{-1}$), marginally sufficient P (1.3 to 1.9 g $kg^{-1}$), and deficient N with only 13 to 14 g N $kg^{-1}$ at flowering in both years.

### c. Comparison of Soil Properties at Sites with Different Yield Trends

Although yields did not decline in the LTFE at BRIARC (Figure 2), there was a significant yield decline in the LTNRE at this site (see section III.A.1.). It is possible that previous cropping history and initial soil conditions differed in the LTNRE and LTFE fields at the BRIARC site but records of previous history are not complete, and soil analyses from the LTNRE are not available.

Comparison of soil nutrient status, pH, or bulk density in the IRRI and PRRI LTFE where yields decline with soil properties at BRIARC where no decline is apparent (Figure 2) does not suggest reasons for the different yield trends at these sites. Soil organic C and total N at both IRRI and BRIARC, for example, are comparable and have remained relatively stable since 1977 (Table 5). At PRRI, both soil organic C and total N have increased since the initial crop cycle. Although there is evidence of a plowpan at PRRI where bulk density of the 20- to 30-cm zone is 1.17 versus 0.95 g $cm^{-3}$ in 10- to 20-cm interval at the bottom of the puddled layer, there is little compaction in the profile at IRRI where bulk density of the 20- to 30-cm layer is only 0.75 g $cm^{-3}$.

### 2. Increased Disease and Insect Pressure

While it is clear that continuous rice monoculture in tropical environments fosters conditions that are conducive for disease and insect outbreaks, the contribution of pest pressure to the yield decline phenomenon is uncertain. Flinn and De Datta (1984) summarized the issue as follows: "Other factors such as diseases, insects, and lodging have been observed. Their effects on yield decline need confirmation."

**Table 5.** Stability of soil organic C and total N in the 0-20 cm topsoil of treatments with (+NPK) and without (0-0-0) fertilizer inputs to each crop cycle in the Long-Term Fertility Experiments with continuous double-crop irrigated rice systems at three sites

| Site | | ---Organic C--- | | ----Total N---- | | ---C/N ratio--- | |
|------|------|------|------|------|------|------|------|
| (Municipality) | Year | 0-0-0 | +NPK | 0-0-0 | +NPK | 0-0-0 | +NPK |
| | | ----------------g kg$^{-1}$---------------- | | | | | |
| IRRI | 1964[+] | 20.0 | 20.0 | 1.78 | 1.78 | 11.2 | 11.2 |
| (Los Baños) | 1982 | 18.9 | 20.4* | 1.99 | 2.09* | 9.5 | 9.8 |
| | 1984 | 21.3 | 24.0* | 1.88 | 1.98* | 11.4 | 11.1 |
| | 1988 | 19.3 | 21.2* | 1.86 | 2.01* | 10.4 | 10.5 |
| | 1991 | 18.2 | 19.9* | 1.87 | 2.01* | 9.8 | 9.9 |
| | | | | | | | |
| PRRI | 1968[+] | 8.7 | 8.7 | 0.80 | 0.80 | 10.9 | 10.9 |
| (Muñoz) | 1977 | 10.6 | 12.0* | 0.97 | 1.20* | 10.9 | 10.0 |
| | 1982 | 11.2 | 12.6 | 0.84 | 1.00* | 13.3 | 13.1 |
| | 1984 | 10.4 | 12.4* | 0.83 | 1.01* | 12.5 | 12.2 |
| | 1988 | 10.8 | 12.5* | 0.92 | 1.03 | 11.8 | 12.1 |
| | 1991 | 10.4 | 12.6* | 0.96 | 1.17* | 10.7 | 10.8 |
| | | | | | | | |
| BRIARC | 1968[+] | 25.5 | 25.5 | 2.22 | 2.22 | 11.6 | 11.6 |
| (Pili) | 1977 | 17.4 | 19.0 | 1.70 | 1.80 | 10.2 | 10.5 |
| | 1982 | 17.9 | 19.6 | 1.92 | 2.14* | 9.3 | 9.2 |
| | 1984 | 21.0 | 22.2 | 1.83 | 2.03 | 11.5 | 11.0 |
| | 1988 | 19.7 | 20.5 | 1.98 | 2.09 | 10.0 | 9.8 |
| | 1991 | 18.3 | 20.1* | 1.71 | 1.83* | 9.7 | 9.3 |

[+] A single composite soil sample was taken in the first cropping cycle. In other years, treatment plots were sampled and analyzed individually.
* For the year sampled, indicates a significant difference (P<0.05) between the control treatment without fertilizer inputs and the +NPK treatment.

Disease and insect pressure are extremely variable in time and space. Temporal and spatial variation in disease pressure reflects year to year and site to site variation in the quantity of inoculum, vector population, climatic factors, stage of crop development when infection occurs, and interactions among these factors (Ou, 1985). The erratic yield trend of IR8 is a case in point (Figure 5). For example, a yield of IR8 was less than 1 t ha$^{-1}$ in 1984 dry season when pest pressure was high, but in the 1983 and 1987 dry seasons, a yield of IR8 was comparable to the highest yielding varieties when pest pressure was low. Even in years with low pest pressure, however, the yield of IR8 fell on the yield trend line of the highest yielding varieties and did not attain the yield potential achieved in earlier years.

Damage from stemborers (*Scirphophaga incertulas* and *Chilo suppressalis*) and brown plant hopper (*Nilaparvata lugens*), bacterial leaf blight, leaf streak (*Xanthomonas campestris*), and the occurrence of viral diseases were recorded in the long-term experiments at IRRI. Unlike IR8, yield reductions from pest damage were not found to have a significant influence on yield trends of the HYEs due to their greater resistance (Flinn and De Datta, 1984). This finding is consistent with the uniform plant growth that is observed within treatment plots of the long-term experiments where yield declines have occurred; there is no patchiness in the HYE plots that is typical of disease or insect problems.

More recently, high numbers of the root nematodes (*Hirschmanniella* spp.) have been found in rice fields at the IRRI farm and elsewhere in the Philippines (Prot et al., 1992; Hendro et al., 1992). Likewise, the incidence of sheath blight and stem rot diseases were not monitored in the long-term experiments, and both diseases are visible toward the end of grain filling. Research is in progress to determine the potential contribution of these pests to the yield decline phenomenon (see section III.C.1. below).

## C. Recent Evidence on the Role of Nitrogen Supply

1. Reversing the Yield Decline with Increased N Inputs

Rates of fertilizer-nutrients applied as treatments in the long-term experiments were initially selected to achieve maximum yield. For N, a large number of field studies were conducted in the 1960s to identify the appropriate rate and timing. Based on these studies, a maximum N rate of 150 kg N ha$^{-1}$ was selected for the dry season in the LTCCE, and 140 kg N ha$^{-1}$ was used in the LTFE (Table 2). These rates were applied in two splits, with two-thirds broadcast-incorporated before planting and one-third topdressed at panicle initiation.

Observation of leaf color made in 1991 dry season, however, indicated the possibility of suboptimal leaf N status before panicle initiation and at flowering stage, even in treatments that received the highest N rates of the LTCCE. Flag leaf samples taken at flowering stage indicated an N concentration that was only marginally sufficient, and differences in flag leaf N concentration among N-rate treatments that ranged from 0 to 150 kg N ha$^{-1}$ were not significant. Measurement of N accumulation in aboveground biomass from flowering to physiological maturity indicated extremely low N uptake during this period regardless of the applied N rate (Table 6). Soil samples taken at flowering and immediately extracted for inorganic N indicated no difference in $NH_4$-N content among the N-rate treatments in both the 1991 dry and wet seasons. Lack of detectable residual N from fertilizer-N applied at panicle initiation, 25 days before flowering, is consistent with the lack of increased crop N accumulation after flowering in treatments with high rates of applied N.

These findings suggested an inadequate N supply, particularly in the late season after booting stage when the crop was dependent on the soil N supply.

**Table 6.** Nitrogen accumulation ($\Delta$N) during the grain-filling period from flowering to physiological maturity in the Long-Term Continuous Cropping Experiment at IRRI

| N Input | | -------------1991 DS[1]------------ | | | -------------1991 WS[1]---------- | | |
|---|---|---|---|---|---|---|---|
| --Treatment-- | | Total N | | $\Delta$N | Total N | | $\Delta$N |
| DS | WS | Flowering | Physiol. Mat. | | Flowering | Physiol. Mat. | |
| | | | | ----kg N ha$^{-1}$---- | | | |
| 0 | 0 | 49d[2] | 55d | 6a | 53c | 62c | 9a |
| 50 | 30 | 69c | 76c | 7a | 67b | 76b | 9a |
| 100 | 60 | 88b | 93b | 5a | 80a | 83a | 3b |
| 150 | 90 | 107a | 112a | 5a | 84a | 85a | 1b |

[1] Values represent the mean of four cultivars at each N rate in 1991 Dry Season (DS), and the mean of three cultivars in 1991 Wet Season (WS).
[2] Means within columns followed by a different letter differ significantly by DMRT, $P < 0.05$.

This conclusion is consistent with recent evidence of a significant yield response to N applied at flowering stage in field experiments conducted at the IRRI Research Farm (Cassman et al., 1994: Kropff et al., 1994). Beginning in the 1991 WS, broadcast-seeded and transplanted sub-subplots were eliminated so that all plots were transplanted, and three varieties instead of six were used as subplots. Selected varieties continue to represent the best available genotypes, but only varieties with a similar maturity are used. From 1970-1972, for example, mean days to maturity for the varieties used in those years was 124 days (+/- 4 SD) versus a mean maturity of 112 days (+/- 7 SD) in the 1989-1991 period. This compares with a mean maturity of 115 days (+/- 1 SD) in 1992-1993. More uniform maturity and elimination of broadcast-seeded sub-subplots facilitate greater precision in both water management and timing of N application in relation to the stage of crop development, both of which contribute to increased fertilizer-N use efficiency. These changes have been maintained in subsequent seasons.

In the 1992 and 1993 dry seasons, fertilizer-N rates were also increased. Rates of 0, 63, 127, and 190 kg ha$^{-1}$ were applied with the standard timing in 1992. In 1993, N rates of 0, 72, 144, and 216 kg N ha$^{-1}$ were applied in four split applications at transplanting, mid-tillering, panicle initiation, and booting stage. Additional N was applied at booting stage in 1993 because estimates of leaf N content based on measurements with a chlorophyll meter (Peng et al., 1993) indicated leaf N deficiency in treatments which had already received a total of 180 kg N ha$^{-1}$ in equal splits at transplanting, mid-tillering and panicle initiation. Prophylactic applications of fungicide were made in both 1992 and 1993 to protect against sheath blight, a disease which is more damaging to rice crops with N-rich leaf canopies (Mew, 1991).

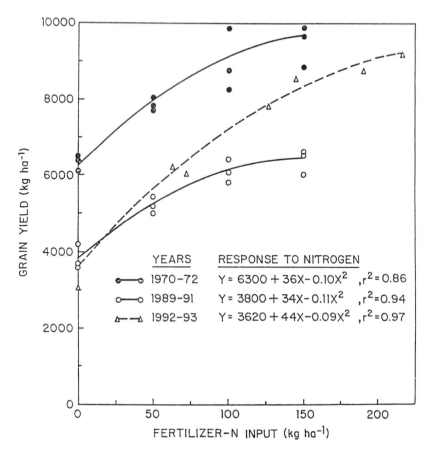

**Figure 6.** Yield response to fertilizer-N rates in three periods of the Long-Term Continuous Cropping Experiment (LTCCE) at IRRI. Responses are based on the highest-yielding variety in each season as shown in Table 7.

The mean grain yield attained with the highest N rate in the 1992 and 1993 dry seasons was 40% greater than the highest yields achieved in the previous three-year period from 1989-1991 (Figure 6). In both periods this comparison is based on the yield of IR72, and only a small fraction of the differences in yield in 1991 and 1992 can be attributed to differences in solar radiation or temperature regime based on estimates of yield potential predicted by the ORYZA1 simulation model (Kropff et al., 1993b). The highest yield levels of IR72 in 1992 and 1993 were comparable to the mean yield levels achieved in the early years of the LTCCE from 1970-1972 (Table 7). In those early years, however, the highest yielding varieties required 127 days from sowing to maturity compared to 115 days for IR72. Yields of 9.5-10.3 t ha$^{-1}$ were also achieved with IR72 in several other replicated experiments at other sites on the IRRI farm

**Table 7.** Yield response and agronomic fertilizer-N use efficiency in three periods of the Long-Term Continuous Cropping Experiment at IRRI

| Period (Cultivar) | Days to maturity (d) | N fertilizer rate (kg ha$^{-1}$) | Grain yield (kg ha$^{-1}$) | Yield increase ($\Delta$kg ha$^{-1}$) | Agronomic N fertilizer use efficiency ($\Delta$kg grain kg$^{-1}$ N) |
|---|---|---|---|---|---|
| 1970-72 | 127 | 0 | 6300c[+] | ---- | ---- |
| (IR8, IR24) | | 50 | 7870b | 1570 | 31.4 |
| | | 100 | 8870a | 2570 | 25.7 |
| | | 150 | 9430a | 3130 | 20.9 |
| | | | | | |
| 1989-91 | 115 | 0 | 3800c | ---- | ---- |
| (IR72) | | 50 | 5210b | 1410 | 28.2 |
| | | 100 | 6100a | 2300 | 23.0 |
| | | 150 | 6380a | 2580 | 17.2 |
| | | | | | |
| 1992-93 | 115 | 0 | 3640d | ---- | ---- |
| (IR72) | | 68 | 6090c | 2450 | 36.2 |
| | | 135 | 8170b | 4530 | 33.5 |
| | | 203 | 8930a | 5290 | 26.1 |

[+] For each time period, means followed by different letters indicate a significant mean separation by DRMT, $P < 0.05$.

in the 1992 and 1993 dry seasons. Thus, based on the highest mean yields of the LTCCE in 1970-72 and 1992-1993, yield potential on a growth duration basis is actually 5% greater for IR72 than the early IRRI varieties (Table 7), and it is clear that a decrease in the yield potential of more recent IRRI varieties has not occurred.

## 2. Yield Potential, N Uptake, and N Use Efficiency

For irrigated rice, yield potential is determined by genotype, temperature regime, solar radiation, and N supply when the availability of other nutrients is adequate and pest damage does not limit crop growth. The relationship between grain yield and N accumulation in aboveground biomass at maturity is tightly conserved in modern rice varieties grown in similar environments. For IR72, there is a close association between grain yield and N uptake based on measurements from the LTFE, LTCCE, and other experiments at IRRI in 1991 and 1992 (Figure 7). Almost identical relationships were found for the 1992 dry season in on-farm study in Central Luzon which included 44 farmers' fields and 9 varieties (Cassman et al., 1993), and in the wet season in the tropics of Brazil (Cassman and Plant, 1992). The consistency of this relationship reflects a

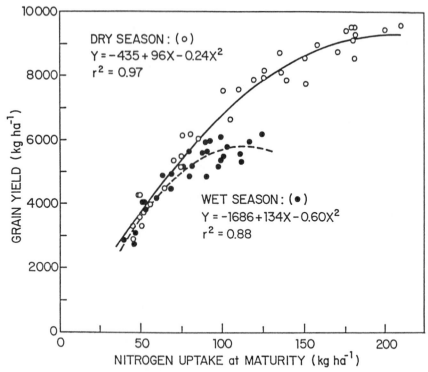

**Figure 7.** The relationship between grain yield and total N accumulation in aboveground biomass of IR72 at physiological maturity. Data were obtained from the LTCCE, the LTFE, and other field experiments conducted at IRRI in 1991 and 1992.

relatively stable harvest index, and small differences in grain protein content among modern rice varieties with similar yield potential grown in comparable environments.

Greater N uptake in 1992 than in 1989-1991 resulted from the increased rates of N application and greater apparent uptake efficiency from applied fertilizer although N uptake in plots without N addition was similar in 1991 and 1992 (Figure 8). Fertilizer-N was applied with the same timing in 1992 as in 1989-1991. We speculate that the increase in fertilizer-N uptake efficiency in 1992 reflected better control of floodwater depth in combination with greater precision in the timing of N application at panicle initiation made possible by the simplified treatment regimes (see Section IIIC1), and prevention of sheath blight in treatments with high rates of N input. Even with fungicide applications made at panicle initiation stage and before heading, 9% of the hills from the harvested areas had severe symptoms of sheath blight in treatments with the highest N rate. Total yield loss from sheath blight was estimated at 3% in the high-N

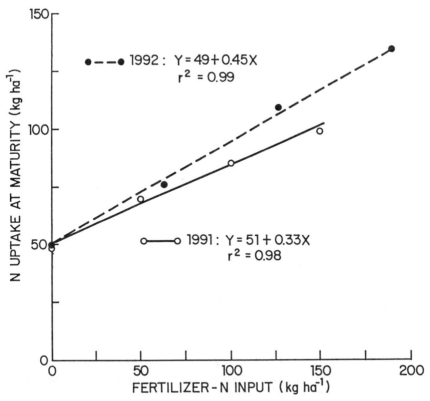

**Figure 8.** The relationship between total N accumulation in aboveground biomass of IR72 at physiological maturity and the rate fertilizer-N applied in the Long-Term Continuous Cropping Experiment (LTCCE) in the dry season of 1991 and 1992 at IRRI.

treatment based on separate grain yield measurements from diseased and healthy hills.

Although grain yields with high N inputs in 1992 and 1993 were comparable to the initial years of the LTCCE (Figure 6), grain yields in plots without fertilizer-N inputs were similar to those attained in 1989-1991, but 2600 kg ha$^{-1}$ less than in 1970-1972 (Table 7). This reduction represents a decrease in the "effective" N supply in treatments without added N which is equivalent to a decrease of 40 kg N ha$^{-1}$ in actual N uptake (Figure 7). Because grain yield is closely associated with N uptake, the reduction in grain yield from plots without fertilizer-N indicates that the N supply from indigenous soil resources has decreased, or the capacity of the rice root system to acquire soil N has decreased. Agronomic efficiency from applied N in 1989-1991 was 18% less than in 1970-1972 (Table 7), which accounts for a yield difference of only 500 kg ha$^{-1}$ — far less than the yield decline that occurred between 1970 and 1991.

By contrast, greater agronomic efficiency from applied N was achieved in 1992-1993 than in 1970-1972 at comparable N input rates and also at comparable yield levels. Greater efficiency in 1992-1993 than in 1970-1972 suggests that the decrease in N uptake in plots without inputs of fertilizer-N does not result from a reduction in the capacity of the root system to acquire N from soil. It is therefore, unlikely that root nematodes, or a decrease in root activity for any reason, contributed to the yield decline in the LTCCE.

In summary, it appears that the greater fertilizer-N input requirements to achieve maximum yield levels in 1992-1993 versus 1970-1972 result from a decrease in soil N supplying capacity because the yield response to applied N has remained relatively constant (Figure 4), agronomic efficiency from applied N has not decreased (Table 7), and the soil N supply estimated by crop uptake in the late season was comparable or lower in treatments which receive fertilizer-N inputs (Table 6).

## 3. Soil N Supplying Capacity, N Reserves and Organic C

Although yield declines are evident in both the LTFE and LTCCE at IRRI, and the LTFE at PRRI (Figures 2 and 3), soil organic C and total N content remain relatively constant or increase at each of these sites. At PRRI where 30 cm of standing rice stubble is recycled to soil after harvest, organic C and total N have increased by 20% in control plots without fertilizer-nutrient inputs, and by more than 40% in the +NPK treatments since 1968 (Table 5). Based on linear regression of soil bulk density on total N of the 1991 soil samples, only 4% of the increase in total N since 1968 was in unfertilized treatments, and 5% of the difference between control and +NPK treatments in 1991 can be attributed to a decrease in bulk density that accompanies organic matter enrichment. No trends are evident in the C/N ratio.

At the LTCCE site, measurements on composite samples from each block in 1963 indicate organic C content of 18.3 g kg$^{-1}$ (+/-1.0 SD) and total N of 1.94 g kg$^{-1}$ (+/-0.16 SD). From 1963 to the end of 1967, 13 consecutive rice crops were grown with a 2-month break for a green manure crop in April and May of 1964. The triple-crop rice system was initiated in 1968, and all aboveground plant material has been cut at the soil surface and removed at harvest. Measurements of total soil N and organic C were not made during the first ten-year period. Measurements made in 1978, and at regular intervals thereafter, indicate that organic C and total N concentration have increased since 1963 in treatments receiving high rates of fertilizer-N addition, and have remained relatively constant in treatments without N addition (Figure 9). Differences between years in the plots without fertilizer-N inputs are not significant and probably result from small differences in the depth of sampling and the abrupt decrease in soil C and N at the lower boundary of the puddled layer (Figure 1).

Soil bulk density was not measured in the early years of the LTCCE so that it is not possible to estimate a mass balance for soil C or N since the beginning

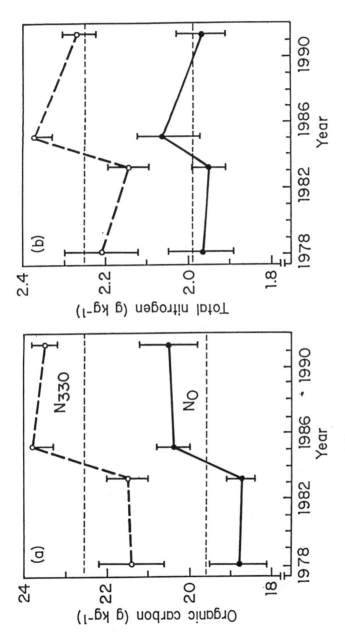

**Figure 9.** Organic C and total N concentration in the 0-20 cm topsoil from 1978 to 1991 in the treatments without fertilizer-N addition (No) and with the highest rate of N addition (N330) which received annual fertilizer-N inputs of 150, 90, and 90 kg N ha⁻¹ applied to dry, early wet and late wet seasons, respectively, in the Long-Term Continuous Cropping Experiment (LTCCE) at IRRI. Bar intervals indicate one standard error, and the dashed line represents the mean of the values measured in 1978, 1983, 1985, and 1991.

of the experiment. For irrigated rice, soil puddling is practiced to facilitate transplanting and to reduce water losses from percolation. The puddling operation creates a slurried soil by physical destruction of macropores and aggregates, and it results in a lower bulk density than for plowed or dry soil under upland conditions. Transition from a single rice crop system with one puddling operation to a double- or triple-crop system with two or three puddling operations each year leads to a decrease in bulk density of the puddled soil. It is likely, however, that the effects of such a transition on bulk density would occur within the first few years and stabilize thereafter. Thirty consecutive rice crops were grown in the LTCCE from 1968-1978. Assuming constant bulk density during 1978-1991 period, then total soil N and organic C in the 0-20 cm puddled topsoil have remained constant in plots that did not receive fertilizer-N (Figure 9) on a mass basis despite complete removal of all straw at harvest in this intensive, triple-crop system.

## IV. Homeostasis of Total Soil C and N, and the N Supplying Capacity of Rice Soils

### A. Sources, Balance, and Mineralization of C and N

Conservation of soil C and N is unique to lowland paddy soils of the tropics where high temperatures, intensive cropping, and little recycling of crop residues would otherwise deplete reserves of organic matter under upland soil conditions. The stability of soil N and C in lowland rice soil is thought to result from a slower rate of humus decomposition than in aerated soils (Dei and Yamasaki, 1979), and relatively large inputs of C and N from biological activity in the soil-floodwater system.

1. N Inputs to Flooded Rice Soils

Estimates of N inputs from biological $N_2$ fixation (BNF) from blue-green algae in floodwater and heterotrophic bacteria in soil and the rice rhizosphere range from 15-50 kg N ha[-1] per crop cycle (Koyama and App, 1979). Equilibrium in the total soil N content of the control plots without fertilizer-N inputs from 1978-1991 in the LTCCE (Figure 9) allows estimation of N inputs from BNF to this continuous, triple-crop rice system.

Assuming constant bulk density and total soil N during this 13-year period and accounting for the N contribution from rainfall and irrigation water, N inputs from BNF must equal the removal and losses of N (Table 8). The quantity of N in harvested biomass can be estimated from the relationship between grain yield and aboveground plant N at maturity (Figure 7) because all aboveground biomass was removed in the LTCCE. Nitrogen inputs from irrigation water are

**Table 8.** Estimates of N inputs from biological nitrogen fixation (BNF) in plots without fertilizer-N in the Long-Term Continuous Cropping Experiment at IRRI

| Cropping season | Mean grain yield (1978-1991) | Estimated N removal (grain + straw) | Estimated external N inputs[1] | Estimated BNF-N inputs |
|---|---|---|---|---|
| | ---------------------------kg ha[-1]--------------------------- | | | |
| Dry season | 3200 | 42 | 10 | 32 |
| Early wet season | 3200 | 46 | 5 | 41 |
| Late wet season | 2400 | 37 | 5 | 32 |
| Total | 8800 | 125 | 20 | 105 |

[1] Based on measurements of N inputs from rainfall of App et al. (1984), N inputs from irrigation assuming applied water depth of 125, 50, and 40 cm in the dry, early and late wet seasons, respectively, and a mean N concentration of 0.08 mg N L[-1]. Estimates also assume no losses of N from leaching, runoff, volatilization or denitrification.

calculated from recent measurements of the $NH_4$- and $NO_3$-N in canal water entering the LTCCE field, and estimates of consumptive water use based on seasonal patterns of rainfall and potential evaporation. Atmospheric inputs at the site were measured by App et al. (1984). Based on these estimates, BNF inputs of 32-41 kg N ha[-1] per crop cycle are required to maintain soil N content, with total BNF inputs of 105 kg N ha[-1]-yr[-1]. This estimate is conservative in that losses from leaching, runoff, volatilization, or denitrification are not considered. Such losses, however, are likely to be small in treatments without fertilizer-N inputs.

## 2. C inputs to Flooded Rice Soils

Primary productivity of the photosynthetic biomass in rice field floodwater is estimated to range from 500-1000 kg C ha[-1] per crop cycle (Yamagishi et al, 1980; Vaquer, 1984). At the IRRI farm, C input of 500-600 kg C ha[-1] from the floodwater community was estimated in a 90-day cropping cycle from transplanting to harvest (Saito and Watanabe, 1978). To construct a C budget for the LTCCE plots without fertilizer-N addition, an input of 700 kg C ha[-1] from aquatic biomass in each crop cycle was used to account for additional C inputs during the period between crop cycles when soil is puddled before transplanting, and for a longer crop growth duration of 115 days (Table 9).

Only roots and the basal stump of the stem cut at the soil level are recycled in the LTCCE. Root dry matter excluding the basal portion of the stem represents from 17-29% of the aboveground biomass at flowering stage when vegetative dry matter reaches a maximum (Yoshida, 1981; Yamaguchi and

**Table 9.** Estimated annual C balance and N mineralization in treatments without fertilizer-N addition in the Long-Term Continuous Cropping Experiment (LTCCE) at IRRI

| Source of C | Annual C input | C/N ratio | Annual turnover | Annual mineralization C | Annual mineralization N | Net C input |
|---|---|---|---|---|---|---|
| | (kg ha⁻¹) | | (%) | -------kg ha⁻¹------ | | kg ha⁻¹ |
| Aquatic photosynthetic biomass | 2100 | 12 | 80 | 1680 | 140 | 420 |
| Rice root biomass | 1000 | 40 | 60 | 600 | 15 | 400 |
| Root exudates and turnover | 700 | 30 | 70 | 490 | 16 | 210 |
| Soil organic C[1] | ---- | 10 | 4 | 1030 | 103 | (-1030) |
| Totals | 3800 | | | 3800 | 274 | 0 |

[1] Total soil organic C and N are assumed to remain constant (Figure 9) so that estimated net C inputs equal C mineralization from soil organic matter. Total organic C in the 0-20 cm topsoil is 25,500 kg C ha⁻¹ based on a mean organic C content of 19.6 g C kg⁻¹ and a bulk density of 0.65 g cm⁻³.

Tanaka, 1990). In those studies, measurements of root biomass were made on plants with vegetative dry matter yields comparable to the plants in the LTCCE treatments without fertilizer-N. Recycled roots and stem bases are therefore estimated to be 22% of the aboveground biomass at flowering stage in the LTCCE, and C inputs are calculated assuming 40% C content in these tissues (Table 9). The C input from turnover of fine roots and root exudates during the growing season is more difficult to estimate. Sauerbeck and Johnen (1977) found rhizodeposition of C from root turnover and root exudates equivalent to more than 20% of total C assimilation by wheat (*Triticum aestivum* L.), while Shamoot et al. (1968) measured rhizodeposition values of 5-20% of aboveground biomass in a study with 11 plant species. The C balance for the LTCCE assumes that C input from root turnover and root exudates represents 15% of maximum aboveground vegetative dry matter, contains 40% C, and has a C/N ratio of 30. Total annual C inputs to the LTCCE plots without added N are estimated to be 3800 kg ha⁻¹.

## 3. Nitrogen Mineralization Rates in Submerged Soils

Japanese soil scientists have reported that soil N supplying capacity decreases with continuous rice cropping. Dei and Yamasaki (1979) summarized this early work and concluded that organic matter accumulates in paddy soils because the rate of organic matter decomposition is slower than in aerated soil. Furthermore, these authors suggested that the quantity of total soil N decreases during aerated crop cycles but the amount of readily decomposable organic matter increases due to improvement in the quality of the soil humus. More recent reports indicate decreases in size of the mineralizable soil N pool after each of three consecutive rice crops (Saito, 1991). Although these reports are based on single rice-crop systems in a temperate climate, the findings are consistent with the apparent decline in the soil N supply observed in the long-term rice experiments in the Philippines (see section III.C.2.).

Despite these reports, definitive evidence is lacking to support the hypothesis that the rate of soil N mineralization is slower in anaerobic versus aerated soil. Direct comparisons are difficult because of differences in the methods used to measure N mineralization. In aerated soil, $NO_3$-N is monitored over time to estimate the rate of N mineralization, and $NO_3$-N is not readily transformed when soil remains aerobic. In submerged soils, $NH_4$-N must be monitored to estimate N mineralization, and $NH_4$-N can be lost during anaerobic incubation via denitrification or volatilization (Tusneem and Patrick, 1971). Likewise the build-up of $NH_4$-N in incubated soil without plant uptake can result in $NH_4$-N fixation in minerals such as illite and vermiculite that fix $NH_4$-N at interlayer sites between clay sheets. Thus, while it is relatively easy to measure net N mineralization in aerated soil, the methods used to make similar measurements in submerged soil are less certain (Keeney and Sahrawat, 1986).

Microbial biomass plays a pivotal role in governing the rate of N mineralization in both aerated and anaerobic soils. Inubushi and Watanabe (1986) measured the microbial biomass N pool in flooded soils collected from rice paddies in the Philippines using the chloroform-incubation method. In these experiments, soils were placed in large concrete holding tanks or pots and cropped with rice. They found a smaller microbial N pool size than values reported for temperate upland soils, but the rate of N turnover through the microbial pool was rapid in the submerged paddy soils. Uptake of N by rice in these experiments was closely correlated with the size of the microbial N pool. A relatively small mass of microbial biomass N in puddled soil partly reflects lower bulk density and a smaller soil mass in the topsoil volume of lowland rice fields than in soil that is plowed under aerated conditions with moisture content well below field capacity. More recent field experiments at IRRI also indicate a relatively small microbial biomass N pool in the puddled soil layer that ranged from 40 kg N ha$^{-1}$ during the early vegetative growth period to 20 kg N ha$^{-1}$ at flowering stage in treatments without N fertilizer (personal comm., J. Gaunt).

4. Nitrogen and C Balance in the Long-Term Continuous Cropping
Experiment

Annual N mineralization can be calculated from estimates of C inputs, the C/N
ratio of these inputs, the C/N ratio of soil organic matter, and turnover rates of
these C pools. Such an estimate was constructed for treatments without
fertilizer-N addition in the LTCCE (Table 9) based on the assumption that soil
organic C concentration and bulk density have not changed significantly from
1978-1991 (see section III.C.3). The greatest uncertainty associated with the
amount of C input to this system is the C input from root turnover and exudates.
Annual turnover rates for each source of C input are also uncertain although
previous studies on rice straw decomposition (Neue and Scharpenseel, 1987) and
estimates of the C/N ratio of different C inputs provide some guidance for
estimating turnover values. The aquatic photosynthetic biomass is assumed to
turnover rapidly because blue-green algae dominate floodwater communities in
field plots without fertilizer-N inputs (Simpson et al., 1993) and the,se
organisms have a relatively low C/N ratio (Roger and Watanabe, 1986). A
turnover rate of 60% is assigned to recycled root biomass based on evidence
from recent field studies on decompositon of buried rice roots in submerged soil
(personal comm., I. Simpson). At issue, then is how much of the stable soil
organic matter is degraded each year because it has a C/N ratio of 10 (Figure
9) and contributes relatively large quantities of mineralized N per unit of
decomposed substrate.

If total soil organic C remains constant, then C mineralized from the
decomposition of soil organic matter must equal the the net inputs of C from
floodwater photosynthetic biomass, recycled roots, and root turnover and
exudation . Approximately 4% of the soil organic C pool, or 1030 kg ha$^{-1}$, is
estimated to turnover each year to balance net C inputs in the LTCCE (Table 9).
This C balance assumes no losses due to leaching of soluble organic substrates
with percolation, and it is subject to the uncertainties in each of the estimated
parameters. The C budget is useful, however, to examine the sensitivity of N
mineralization to changes in the estimated parameters and the underpinning
assumptions.

The quantity of mineralized N predicted by the C balance is 274 kg N ha$^{-1}$
yr$^{-1}$, equivalent to an average of 91 kg ha$^{-1}$ for each of three crop cycles in the
LTCCE. In each crop cycle, a total of 88 kg N ha$^{-1}$ can be accounted for by a
mean accummulation of 42 kg N ha$^{-1}$ in aboveground biomass (Table 8), mean
microbial biomass N during the growing season of 30 kg N ha$^{-1}$ (see section
IV.A.3.), and a total of 16 kg N ha$^{-1}$ N per crop cycle in root biomass, in-season
root turnover, and root exudates (Table 9). This estimate of N mineralization,
however, is most sensitive to the C inputs and turnover rate of the photosynthetic community in floodwater and the turnover rate for soil organic matter. For
example, if C input from the aquatic photosynthetic biomass is 800 kg C ha$^{-1}$
per crop cycle instead of 700 kg C ha$^{-1}$, as used in the C budget of the LTCCE
(Table 9), an additional 25 kg N ha$^{-1}$-yr$^{-1}$ of mineralized N would be produced

from the greater total turnover of the aquatic biomass and increased turnover of soil organic C to offset the additional net C input. Likewise, percolation of floodwater through the profile would influence the C balance through leaching of soluble organic substrates which may exist for short periods at mM concentrations in the soil solution (Tsutsuki, 1984). Although puddled soils typically have low percolation rates, even a daily rate of 2 mm as found in some fields at the IRRI farm (Wopereis et al., 1992) would result in a total leaching volume of $7 \times 10^6$ L per year in the continuous flooded rice culture of the LTCCE. Leaching losses of organic solutes would have to be offset by a comparable decrease in the turnover rate of soil organic C or an increase in the net C inputs to the system.

Whether the rate of soil organic matter decomposition is slower in anaerobic soil than in aerobic soil still remains a key issue. An estimated value of 4% annual turnover of the soil organic C pool with a C/N ratio of 10 (Table 9) does not seem high for a fertile soil in an environment with a mean annual air temperature of 27 °C, that is thoroughly churned by puddling three times a year and never dries to moisture levels that inhibit microbial activity. By contrast, annual turnover of soil organic C can exceed 5% in a tropical forest soil with an udic moisture regime as estimated by the data of Nye and Greenland (1960), and also in an irrigated cotton field in California as estimated by the data of Cassman et al.(1992). A higher annual turnover rate in the submerged soil of the LTCCE, however, would mean that net C inputs must be greater than estimated or that losses of C from leaching are significant.

## B. Formation and Quality of Humus in Submerged Soils

Submergence causes marked changes in the chemical, physical, and biological attributes of the soil ecosystem. While there is a considerable body of literature on the formation and decomposition of soil humus, most of this literature concerns aerated soils. Where comparisons have been made between aerated and anaerobic conditions, much of this work targets the anoxic conditions in lake sediments or natural wetlands rather than the lowland rice systems in which soils are waterlogged for most of the year although oxygen availability fluctuates during fallow, puddling, and submergence cycles of the cropping season. Direct comparisons of the same soil cropped for long periods of time under aerated versus waterlogged conditions are lacking. Comparative studies of different soils from aerated or flooded agroecosystems suggest differences. In one such comparison, the organic C contained in the humin fraction of soil organic matter represented a relatively large proportion of total organic C in paddy soils compared to grassland or forest soils (Watanabe and Kuwatsuka, 1991). Humin is thought to be tightly bound to clay particles and more recalcitrant to degradation than other fractions of soil humus.

While the rate of decomposition of fresh organic matter like rice straw or manure is initially similar in both waterlogged and aerobic soil conditions

(Tusneem and Patrick, 1971; Neue and Scharpenseel, 1987), degradation of the more recalcitrant components of fresh organic matter may differ depending on the availability of oxygen. Lignin, for example, is an aromatic compound that is slow to degrade even in aerated soils. It comprises about 12% of rice roots (pers. comm., M. Becker). Unlike polysacharries and other more readily degraded organic molecules, a large proportion of the C in lignin that remains in soil gets incorporated into the humic acid fraction (Martin et al., 1980; Stott et al., 1983). Oxygenase enzymes facilitate the initial steps in the decomposition of oxygen-poor aromatic compounds and lipids, and molecular oxygen is the preferred hydroxylating agent in the first steps of degradation for such compounds (Hayaishi and Nozaki, 1969). Rates of lignin decomposition are extremely low under anaerobic conditions (Benner et al., 1984), and it is argued that degradation of lignin does not occur at all in an anoxic environment (Kirk and Farrell, 1987).

In the low oxygen environment of flooded muck soils, Tate (1979) found the rate of catabolism of several carbon substrates decreased markedly compared to rates in the same soil in an aerobic environment. The rate of catabolism for amino acids, glucose, and acetate, however, was less sensitive to flooding than aromatic compounds such as salicylate. Moreover, high $Fe^{++}$ activity in the solution-phase of flooded soils (Ponnamperuma, 1972) may chemically stabilize the precursor substrates of humus, and even the humic substances themselves. In one study, stability constants for complexes between $Fe^{++}$ and fulvic acids were found to be unusually high (Schnitzer and Skinner, 1966).

That there are major differences in the processes that govern C turnover in aerated and anaerobic soils is clear. We do not yet know whether these differences in microbial ecology and biochemical processes of organic matter formation and decomposition affect the chemical nature of soil humus fractions and the mineralization of N from them.

## C. Abiotic Immobilization of N by Soil Organic Matter

The most obvious difference between N mineralization in aerated versus anaerobic soil is the end-product of the mineralization process. Compared to $NO_3$-N, $NH_4$-N is more readily incorporated through abiotic immobilization into organic substrates which are similar in structure to humic acids (Ozbek, 1977). Abiotic incorporation of $NH_4$-N occurs in soil (Burge and Broadbent, 1961; Schimel and Firestone, 1989), and in laboratory studies under controlled conditions with raw humus (Nommik, 1970). These reactions are facilitated when pH is neutral or alkaline (Broadbent et al., 1960) so that the increase in pH that often occurs as soil becomes reduced after flooding would favor abiotic incorporation.

Chemical immobilization of N in humic substances may also occur through direct incorporation of amino acids and amino sugars that are released in the turnover of microbial biomass. In such reactions, the incorporated N becomes more recalcitrant to subsequent degradation when the organic substrates are rich in phenolic and aromatic compounds (Bondietti et al., 1972; Verma et al., 1975). As mentioned in section IV.B., decomposition of lignin and other aromatic compounds is reduced in low oxygen environments, and this may increase the potential for chemical immobilization of $NH_4$-N and amino compounds in submerged rice soils. Likewise, evidence of rapid turnover in the microbial biomass N pool of submerged soil (Inubushi and Watanabe, 1986) suggests greater exposure of cellular constituents such as amino acids and amino sugars to abiotic immobilization. With time, as the quantity of soil organic C increases, as it did in the LTFE at PRRI (Table 5), the potential for abiotic immobilization of $NH_4$-N would also increase. The result would be manifested by a decrease in the apparent soil N supplying capacity without a decrease in total soil N. Even when soil organic matter content remains constant, the chemical structure of humic acids and other humus fractions may assume a more aromatic core, which may increase the potential for chemical immobilization of $NH_4$-N and amino compounds (Flaig et al., 1975).

# V. Summary and Conclusions

A yield decline occurs in most long-term experiments where continuous, irrigated rice is grown in the Philippines. At the IRRI Research Farm, this decline is associated with a decrease in the effective N supply from soil although total soil N remains constant. The yield decline can be reversed by increasing the quantity of N applied to the system. Nitrogen use efficiency does not decrease so that a reduction in root activity due to nematodes or disease is not likely, and there is no evidence of deficiencies for nutrients other than N, or toxicities. At PRRI, the direct role of N supply is less certain, but a yield decline also occurs in treatments where other nutrient deficiencies are not apparent and despite a large increase in total soil N and organic C.

Identifying the causes of the yield decline phenomenon at IRRI will require a better understanding of the processes and mechanisms that govern N-supplying capacity of soil that is intensively cropped under submerged conditions. Key issues involve the rate of organic matter turnover in relation to the formation and chemical properties of humic substances, the quantities and chemical composition of C inputs and outputs from the system, the regulation of N cycling processes by microbial biomass, flora and fauna in the soil-floodwater continuum, and the role of chemical immobilization of mineralized N. Understanding the effects of an aerated crop cycle or drying period on the processes that control soil N and C transformations is also pivotal for developing mitigation strategies that increase the effective soil N supply of intensive lowland rice systems over the long term.

In our analysis of the yield decline phenomenon, we have given less attention to factors other than soil N balance and N-supplying capacity which may also contribute to sustaining yields in continuous irrigated rice systems. Potential changes in subsoil or rooting volume that could influence total nutrient supply from the soil profile are two obvious concerns. Likewise, strategies to minimize crop losses from non-specific, endogenous pathogens such as sheath blight will be crucial to maintain rice yields in systems where large inputs of fertilizer-N are needed to achieve high yield levels.

The importance of rice to the human food supply dictates that sustainable increases in rice yield must be achieved on irrigated land in the tropics. Recent evidence from on-farm monitoring studies in the Philippines suggest that the partial factor productivity from N fertilizer applied by rice farmers has declined significantly in the last decade (Cassman and Pingali, 1994). Productivity trends in long-term experiments also indicate the difficulty in sustaining yield gains and a need for greater N inputs to maintain yields. At best, we have been able to keep rice yields from decreasing despite considerable investment in breeding efforts and agronomic research to improve crop management. There remains an exciting challenge ahead for agricultural scientists concerned with rice production and global food supplies.

# References

App, A., T. Santiago, C. Daez, C. Menguito, W. Ventura, A. Tirol, J. Po, I. Watanabe, S.K. De Datta, and P. Roger. 1984. Estimation of the nitrogen balance for irrigated rice and the contribution of phototropic nitrogen fixation. *Field Crops Research* 9:17-27.

Benner, R., A.E. MacCubbin, R.E. Holson. 1984. Anaerobic biodegradation of the lignin and polysaccharide of lignocellulose and synthetic lignin by sediment microflora. *Appl. Environ. Microbiol.* 47:998-1004.

Bondietti, E., J.P. Martin, and K. Haider. 1972. Stabilization of amino sugar units in humic-type polymers. *Soil Sci. Soc. Am. Proc.* 36:597-602.

Broadbent, F.E, W.D. Burge, and I. Nakashima. 1960. Factors influencing the reaction between ammonia and soil organic matter. *Trans. 7th Int. Congr. Soil Sci..* Madison, WI. 2:509-516.

Burge, W.D. and F.E. Broadbent. 1961. Fixation of ammonia by organic soils. *Soil Sci. Soc. Am. Proc.* 25:199-204.

Cassman, K.G., M.J. Kropff, J. Gaunt, and S. Peng. 1993. Nitrogen use efficiency of rice reconsidered: what are the key constraints? p. 471-474. In: N.J. Barrow (ed.) *Developments in Plant and Soil Science Vol. 54, Plant Nutrition: from Genetic Engineering to Field Practice.* Kluwer Academic Publishers, Dordrecht.

Cassman, K.G., M.J. Kropff, and Yan Zhende. 1994. A conceptual framework for nitrogen management of irrigated rice in high-yield environments. In: *Proc. of the 1992 International Rice Research Conference.* International Rice Research Institute, Los Banos, Philippines. (in press)

Cassman, K.G., and P.L. Pingali. 1994. Extrapolating trends from long-term experiments to farmers fields: the case of irrigated rice systems in Asia. In: V. Barnett, R. Payne, and R. Steiner (eds.) *Agricultural Sustainability in Economic, Environmental, and Statistical Terms*. John Wiley & Sons, Ltd., London, U.K. (in press)

Cassman, K.G. and R.E. Plant. 1992. A model to predict crop response to applied fertilizer nutrients in heterogeneous fields. *Fertilizer Res.* 31:151-163.

Cassman, K.G., B.A. Roberts, and D.C. Bryant. 1992. Cotton response to residual fertilizer potassium as influenced by organic matter and sodium in a vermiculitic soil. *Soil Sci. Soc. Am. J.* 56:823-830.

Cayton, M.T.C. 1985. Boron toxicity in rice. *IRRI Research Paper Series*, number 113. International Rice Research Institute, Los Baños, Philippines.

De Datta, S.K, R.J Buresh, M.I. Samson, and Wang Kai-Rong. 1988. Nitrogen use-efficiency and nitrogen-15 balances in broadcast-seeded flooded and transplanted rice. *Soil Sci. Soc. Am. J.* 52:849-855.

De Datta, S.K., K.A. Gomez, and J.P. Descalsota. 1988. Changes in yield response to major nutrients and in soil fertility under intensive rice cropping. *Soil Sci.* 146:350-358.

Dei, Y., and S. Yamasaki. 1979. Effect of water and crop management on the nitrogen-supplying capacity of paddy soils. p. 451-463. In: *Nitrogen and Rice*. International Rice Research Institute, Los Baños, Laguna, Philippines.

Flaig, W., H. Beutelspacher, and E. Rietz. 1975. Chemical composition and physical properties of humic substances. p. 1-212. In: J.E. Gieseking (ed.), *Soil Components Vol. 1: Organic Components*. Springer-Verlag, New York.

Flinn, J.C., and S.K. De Datta. 1984. Trends in irrigated-rice yields under intensive cropping at Philippine research stations. *Field Crops Res.* 9:1-15.

Flinn, J.C., S.K. De Datta, E. Labadan. 1982. An analysis of long-term rice yields in a wetland soil. *Field Crops Res.* 5:201-216.

Harlan, J.R. 1992. *Crops and Man*. Second Edition., Amer. Soc. Agron., Madison, WI.

Hayaishi, O., and M. Nozaki. 1969. Nature and mechanisms of oxygenases. *Science* 164:389-396.

Hendro, M.E., J.C. Prot, and C.P. Madamba. 1992. Population dynamics of *Hirschmanniella mucronata* and *H. oryzae* on *Sesbania rostrata, Aeschynomene afraspera* and rice cv. IR58. *Fundam. Appl. Nematol.* 15:167-172.

Inubushi, K. and I. Watanabe. 1986. Dynamics of available nitrogen in paddy soils. II. Mineralized N of chloroform-fumigated soil as a nutrient source for rice. *Soil Sci. Plant Nutr.* 32:561-577.

IRRI. 1967. *Annual Report*, p. 150. International Rice Research Institute, Los Baños, Philippines.

IRRI. 1978. *Annual Report*, p. 259-262. International Rice Research Institute, Los Baños, Philippines.

IRRI. 1979. *Annual Report*, p. 319-320. International Rice Research Institute, Los Baños, Philippines.

IRRI. 1991. *World rice statistics 1990.* International Rice Research Institute, Los Baños, Philippines.

IRRI. 1993. *Rice Research in a time of change: IRRI's Medium-Term Plan for 1994-1998.* International Rice Research Institute, Los Baños, Philippines

Kaneta, Y., T. Kodama, and H. Naganoma. 1989. Characteristics of the rice plants's nitrogen uptake patterns in rotational paddy fields: effect of paddy-upland rotation on the productivity of rice in Hachirogata reclaimed fields (Part 1). *Japanese J. Soil Sci. and Plant Nutr.* 60:127-133 (in Japanese with English summary).

Keeney, D.R., and K.L. Sahrawat. 1986. Nitrogen transformations in flooded rice soils. *Fertilizer Research* 9:15-38.

Khush, G.S., and W.R. Coffman. 1977. Genetic Evaluation and Utilization (GEU), the rice improvement program at the International Rice Research Institute. *Theor. Appl. Genet.* 51:97-110.

Kirk, T.K., and R.L. Farrell. 1987. Enzymatic combustion: The microbial degradation of lignin. *Ann. Rev. Microbiol.* 41:465-505.

Koyama. T., and A. App. 1979. Nitrogen balance in flooded rice soils. *Nitrogen and rice*, p. 95-103. International Rice Research Institute, Los Banos, Laguna, Philippines.

Kropff, M.J., K.G. Cassman, and H.H. van Laar. 1994. Quantitative understanding of the irrigated rice ecosystem for increased yield potential. In: *Proc. of the 1992 International Rice Research Conference*, International Rice Research Institute, Los Baños, Philippines. (in press).

Kropff, M.J., H.H. van Laar, and H.F.M. ten Berg. 1993. *ORYZA1: a basic model for lowland rice production.* International Rice Research Institute, Los Baños, Philippines.

Martin, J.P., K. Haider, and G. Kassim. 1980. Biodegradation and stabilization after 2 years of specific crop, lignin and polysaccharide carbons in soils. *Soil Sci. Soc. Amer. J.* 44:1250-1255.

Mew, T. 1991. Disease management in rice. p. 279-299. In: *D. Pimentel (ed) CRC Handbook of Pest Management in Agriculture*, Second Edition, Vol. III. CRC Press, Inc., Boca Raton, FL..

Neue, H.U., and H.W. Scharpenseel. 1987. Decomposition pattern of $^{14}$C-labeled rice straw in aerobic and submerged rice soils of the Philippines. *Sci. Total Environ.* 62:431-434.

Nommik, H. 1970. Non-exchangeable binding of ammonium and amino nitrogen by Norway spruce raw humus. *Plant Soil* 33:581-595.

Nye, P.H., and D.J. Greenland. 1960. *The Soil Under Shifting Cultivation.* Technical Communication No. 51. Commonwealth Bureau of Soils, Harpenden, England. 156 pp.

Ozbek, H. 1977. Effect of nitrogen on the formation of pyrocatechin-humic acid and the nitrogen linkage characteristic of this acid. *International Symposium on Soil Organic Matter Studies*, International Atomic Energy Agency, Vienna. Vol. 2:59-66.

Ou, S.H. 1985. *Rice Diseases.* Second Edition. CAB International, U.K.

Peng, S., F. Garcia, R. Laza, and K.G. Cassman. 1993. Adjustment for specific leaf weight improves chlorophyll meter's estimate of rice leaf nitrogen concentration. *Agron. J.* (In press)

Ponnamperuma, F.N. 1972. The chemistry of submerged soils. *Adv. Agron.* 24:29-96.

Prot, J.C., I.R.S. Soriano, D.M. Matias, and S. Savary. 1992. Use of green manure crops in control of *Hirschmanniella mucronata* and *H. oryzae* in irrigated rice. *J. Nematology* 24:127-132.

Rashid, A., and R.L. Fox. 1992. Evaluating internal zinc requirements of grain crops by seed analysis. *Agron. J.* 84:469-474.

Roger, P.A., and I. Watanabe. 1986. Technologies for utilizing biological nitrogen fixation in wetland rice: potentialities, current usage, and limiting factors. *Fertilizer Research* 9:39-77.

Saito, M. 1991. Soil management for the conservation of soil nitrogen. *Extension Bull. No. 341.* Food and Fertilizer Technology Center, Taipei, Taiwan.

Saito, M., and I. Watanabe. 1978. Organic matter production in rice field floodwater. *Soil Sci. Plant Nutrition* 24:427-440.

Sauerbeck, D.R., and B.G. Johnen. 1977. Root formation and decomposition during plant growth. p. 141-148. In: *International Symposium on Soil Organic Matter Studies*, FAO/IAEA/Agrochimica. (Braunschweig, 1976), Germany.

Schimel, J.P. and M.K. Firestone. 1989. Inorganic N incorporation by coniferous forest floor material. *Soil Biol. Biochem.* 21:41-46.

Schnitzer, M., and S.I.M. Skinner. 1966. Organo-metallic interaction in soils: 5. Stability constants of $Cu^{++}$-, $Fe^{++}$-, and Zn-Fulvic Acid complexes. *Soil Sci.* 102:361-365.

Shamoot, S., L. McDonald, and W.V. Bartholomew. 1968. Rhizo-deposition of organic debris in soil. *Soil Sci. Soc. Amer. Proc.* 32:817-820.

Simpson, I.C., P.A. Roger, R. Oficial, and I.F. Grant. 1993. A study of the effects of nitrogen fertilizer and pesticide management on floodwater ecology in a wetland ricefield: I. Experimental design and dynamics of the photosynthetic aquatic biomass. *Biol. Fertil. Soils*: (In press).

Stott, D., G. Kassim, W.M. Jarrel, J.P. Martin, and K. Haider. 1983. Stabilization and incorporation into biomass of specific plant carbons during biodegradation in soil. *Plant and Soil* 70:15-26.

Tanaka, A. and S. Yoshida. 1970. Nutritional disorders of the rice plant in Asia. IRRI Tech. Bull. No.10. International Rice Research Institute, Los Baños, Philippines.

Tate, R.L. 1979. Effect of flooding on microbial activities in organic soils: carbon metabolism. *Soil Sci.* 128:267-273.

Tsutsuki, K. 1984. Volatile products and low-molecular-weight phenolic products of the anaerobic decomposition of organic matter. p. 329-343. In: *Organic Matter and Rice*, International Rice Research Institute, P.O. Box 933, 1099 Manila, Philippines.

Tusneem, M.E., and W.H. Patrick. 1971. Nitrogen transformations in waterlogged soil. *Louisiana State Univ. Agric. Experiment Station Bull. No.657.* Baton Rouge, Louisiana.

Vaquer, A. 1984. La production algae dans les rizieres de Carargue pendant la periode de submersion. *Ver. Internat. Verein. Limnol.* 22:1651-1654.

Verma, L., J.P. Martin, and K. Haider. 1975. Decomposition of carbon-14-labelled proteins, peptides, and amino acids-, free and complexed with humic polymers. *Soil Sci. Soc. Am. Proc.* 39:279-294.

Watanabe, A., and S. Kuwatsuka. 1991. Triangular diagram for humus composition in various types of soils. *Soil Sci. Plant Nutri.* 37:167-170.

Wopereis, M.C.S., J.H.M. Wosten, J. Bouma, and T. Woodhead. 1992. Hydraulic resistance in puddled rice soils: measurement and effects on water movement. *Soil & Tillage Research* 24:199-209.

Yamagichi, T., K. Okada, and Y. Murata. 1980. Cycling of carbon in a paddy field. *Jpn. J. Crop Sci.* 49:135-145.

Yamaguchi, J., and A. Tanaka. 1990. Quantitative observation on the root system of various crops growing in the field. *Soil Sci. Plant Nutr.* 36:483-493.

Yoshida, S. 1981. *Fundamentals of Rice Crop Science.* International Rice Research Institute, P.O. Box 933, 1099 Manila, Philippines. 269 pp.

# B. Sub-Humid and Semiarid Tropics

# Long-Term Soil Management Experiments in Semiarid Francophone Africa

Christian Pieri

## I. Introduction

Long-term agronomic experiments raise a renewed interest among the agricultural community. Indeed, soil scientists, agriculturists, and/or rural development planners may find in these long-term experiments a practical approach to address the difficult issues associated with quantitative assessment of sustainability in agriculture (Herdt and Steiner, 1992).

However, such experiments are costly to maintain over time, as they require constant and valid management as well as accuracy and adequacy in data collection and management. This financial burden may be considered excessive particularly by national, regional, or even international research organizations working in developing countries which prefer to address the most pressing agronomic issues with short-term trials. Then the question is, "What do we specifically learn from long-term agronomic experiments that is important enough to warrant long-term funding commitment?"

ISBN 1-56670-076-0

This paper attempts to answer this question by focusing on the results from Long-Term Soil Management (LTSM) experiments initiated in the mid 50's in Semiarid Francophone Africa (SAF).

A comprehensive literature survey conducted recently (Pieri, 1989, 1992) showed that these LTSM experiments provide a unique referential framework to assess the major soil fertility processes which control soil productivity in SAF. Data from those LTSM experiments provide guidelines on the prospects for improved soil management practices adapted to the semiarid savannah south of the Sahara.

Consequently, the best existing LTSM experiments, despite their inherent limitations, should be unconditionally maintained. Furthermore, a set of new LTSM experiments might be usefully proposed to address new pressing issues of soil management in SAF and to evaluate the potential and impact of new promising practices.

The scope of this review paper is limited to SAF for two reasons. First up to now most of the related information is not readily available outside the francophone area in contrast with works done in anglophone countries, e.g., in Nigeria at Samaru Research Station or more recently in the Sahelian Research Center at Sadore in Niger where ICRISAT and IFDC are presently conducting experiments on soil fertility. This information gap is partially explained by the language barrier in spite of notable exceptions (Fournier, 1967; Charreau, 1975; Tourte, 1984). The second reason is the lack of comprehensive interpretation of the research data provided by many LTSM experiments in SAF.

In the first section, the status of the analyzed LTSM experiments in SAF is briefly presented, showing how and why 24 LTSM experiments were finally selected, and how have the data been interpreted.

The second section focuses on the main scientific results provided by these LTSM: i) to measure changes over time in crop productivity and in input efficiency related to change in soil fertility; and ii) to assess soil quality changes and the related agronomic causes which have induced soil fertility change.

The last section draws lessons from these scientific results to identify new soil management components which could promote the desired sustainable increase in productivity of African savannah soils. In conclusion, some of the conditions required to promote a fruitful generation of new LTSM experiments adapted to Semiarid Africa are briefly discussed in the light of the experience acquired and of the current situation of agronomic research in Sub-Saharan Africa.

## II. The Status of Long-Term Experiments (LTSM) on Soil Management in Semiarid Francophone Africa (SAF)

Long-term experiments in sub-saharan countries are difficult to interpret, particularly those from countries which came through fundamental change over the last decade in the status of agricultural research.

To draw general conclusions about soil management and soil fertility change it is necessary to (i) lay down criteria for the acceptance of LTSM experiments for inclusion in the survey and ii) define a framework for the interpretation of the selected LTSM experiments.

## A. Selection of Experiments

The selection of the interpreted LTSM experiments has been done on the basis of four sets of criteria.

## 1. Site

LTSM experiments are located in a zone roughly comprising Sudano-Sahelian areas with climax vegetation of grass savannah and deciduous woody plants, with an annual rainfall from 500 to 1200 mm distributed throughout one short rainy season. LTSM experiments located in areas either outside rainfall regimes or with bimodal distribution( e.g., Bouake, Central Cote d'Ivoire) were not incorporated in the survey. It was anticipated that the pattern of rainfall distribution has an important impact on the nature and intensity of soil processes such as organic matter oxidation, and consequently LTSM belonging to areas with two different water regime would not be compatible.

## 2. Soil Types

LTSM experiments have been established on Ferruginous Tropical Soils (FTS), according to the French Soil Classification. These soils are most characteristic of the area. The FTS may be best described in the U.S. Taxonomy as Ustropepts, Haplustalfs, Plinthustalfs, and Paleustalfs. They are derived from gritty or sandy continental deposits often found in the northern parts of the region, and from Precambrian granite and gneiss. Over time the parent rocks were modified by alternation of wet and semiarid climates. The final result is the wide-spread occurrence of soil characterized by a C-horizon (alterite) formed from pockets of kaolinitic clay and iron concretions (Plinthite), and a B-horizon with an hardening of the plinthite into concretions or laterite. The surface horizon of most of the upland soils is dominated by quartz and sands. The more clayey alluvial and vertic soils, found in the inland valleys and depressions, are not considered in this survey.

FTS have a moderate or marginal agricultural value. The amount of organic matter in the top layer is low (less than 3% under permanent natural cover) and decreases sharply when cultivated (0.5%, Siband 1974). The mineral colloids are exclusively nonexpanding kaolinites, iron and aluminum hydroxides and silica. The cation exchange capacity is very low (usually less than 2meq/100g).

Deficiency of phosphorous is very common. The superposition of horizons of differing porosity leads to irregular moisture profiles. Paradoxically, while water deficit is a major constraint, these soils often suffer from excess moisture because the upper horizons may be quickly saturated during the rains, while the restoration of water content at depth is achieved only with difficulty. The soil is subject to capping and risk of runoff and erosion is increased.

## 3. Relevance of Experimental Treatments

The treatments tested involve tillage methods, crop residue management, manuring, fertilizer application, crop rotation, and fallow management. The objective and the design of the LTSM experiment were targeted since the very beginning of the work toward an assessment of the sustainability of different soil management systems.

## 4. Quality Criteria

These criteria were critical in the selection process:

- adequate knowledge of the actual experimental conditions, such as site description, design and treatments description, non-treatment or extraneous factor management, and knowledge of any change during the course of the experiment;

- amount and quality of the information recorded on crops (yield and yield components, crop health) and, on soil surface characteristics (texture and basic chemical data, at least for the 0- to 20-cm layer).

Twenty four LTSM experiments were considered suitable according to those criteria. Location of these LTSM experiments is shown in Figure 1.

## B. Data Interpretation

To get a common understanding on the change in soil productivity as affected by differing cultural management in a semiarid environment, it is necessary to (i) bring together the data and information collected for each experiment under a common format; (ii) select a few analytical criteria which can be significantly related to change in soil fertility; and (iii) identify the effect of annual rainfall on year-to-year variation in yield and the change in the soil productivity that induce change in yields over the long term.

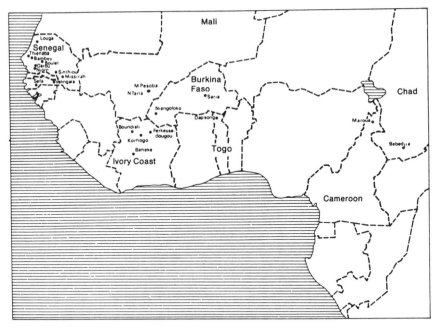

**Figure 1.** Sites of long-term experiments in francophone West Africa.

## 1. Data Format

Among 24 LTSM experiments, some yielded a wealth of information, others produced little data because of the lack of frequent assessment of changes in soil properties over time or because of major changes in soil management (e.g. subsoiling of the experiment "Fertilizer and Manure," N'Tarla, Mali, 1980).

For a proper common interpretation of the data each LTSM was considered individually and the information compiled under a unique format comprising the following:

- Site, Name, Duration, References,
- Treatments and Management,
- Observations and Changes in the Experiment,
- Yields and Interpretation
- Soil Properties and Interpretation
- Conclusion

Subsequently, the results jointly were interpreted after grouping experiments in three categories based on similarity among cropping and soil management systems tested:

a)  Manual cultivation and Low Intensity Systems: little fertilizer, seed treated with fungicide, seedbed preparation and hoeing, and traditional cereal and groundnut crops succession with inclusion of grass fallow period. Three experiments belong to this group.

b)  Manual Cultivation with Intensification: higher rate of fertilizer applied on continuous annual cropping (no fallow period), mostly cereal groundnut crops and with a progressive inclusion of a cotton crop. This group consisted of two experiments located in North Togo and one multisite (7) experiment with standard design located in Senegal.

c)  Intensive Systems: based on animal or motorized (tractor) traction for plowing and seedbed preparation; application of both manure and mineral fertilizer; continuous cropping, including improved varieties of maize, sorghum, cotton, groundnut; and careful control of weeds and pests (on cotton principally). A total of 12 experiments belong to this last group.

## 2. Analytical Criteria of Soil Fertility

The soil science literature is replete with soil attributes that affect soil fertility and crop growth. However, two soil parameters, a) soil organic matter (SOM) content and b) soil acidity measurements, are the most significant attributes for assessing soil fertility change in the zone under consideration.

a)  In the sudano-sahelian zone, particularly for the areas dominated by sandy FTS, soil organic matter is a key soil attribute for the maintenance of soil fertility (Charreau, 1971; Greenland and Lal, 1977). Because soil organic matter changes rapidly during cultivation cycle under tropical conditions (Sanchez, 1976), changes in soil organic matter over years is also a sensitive indicator.

In this survey two parameters were used to characterize soil organic matter:

•   The annual rate of soil organic matter (SOM) loss (k%) calculated according to the formula proposed by Jenny (1980).

$$k\% = [1 - e^{\frac{\log C_o - \log C_n}{n}}]\ 100 \qquad\qquad Eq.\ 1$$

where $C_o$ is the initial carbon content and $C_n$ the content after n years.

It should be noted that k% is a measure of the annual rate of SOM loss arising from both mineralization and other causes such as runoff and soil erosion.

- $S_t\%$, structural index, (Pieri, 1989) is an index to measure the susceptibility of soil to structural stability or disintegration:

$$St\% = SOM\% \ / \ (clay+silt)\% \quad 100 \qquad\qquad Eq.\ 2$$

The choice of the $S_t\%$ index is justified as follows on the basis of the climatic characteristics of SAF. The climate of the area with severe desiccation and torrential rains does not favor the maintenance of stable aggregates and crumbs in the surface soil. Neither do the properties of soil mineral colloids dominated by Kaolinite and Oxides and Hydrous-Oxides of Iron and Aluminum. Organic colloids act as a cement in the soil aggregates, preventing the mineral skeleton from breaking down. For a cultivated soil, this means that there must be sufficient SOM in relation to the surface area of the mineral colloids where it is adsorbed. As others have remarked (Emerson, 1967, Jones, 1976), it is logical to think of a critical ratio between organic colloids and mineral colloids to act as cement and to form soil microaggregate. The results of LTSM experiment supported this hypothesis and led to identifying a critical threshold value below which there is a risk of structural decline (cf III.C.2.).

b)  The results of LTSM experiments in SAF show that the change in soil acidity is a very significant indicator of soil productivity change. Soil pH is a practical indicator of acidity, but it can be both imprecise and insufficiently informative, particularly on poorly buffered sandy soils from sub-saharan Africa. The pH measured in concentrated medium (normal KCl) gave better estimates of acidity than that measured in water. When exchangeable aluminum in soil has been measured, which was not common procedure 15 years ago in SAF, percent aluminum saturation in the soil was preferred. This index is better than pH measurements.

## 3. Rainfall Distribution and Yield Variability

Because of yield variability due to vagaries of rainfall, it is difficult to identify the yield trend. Therefore, it is desirable that the effects of rainfall variability be eliminated so that a series of yield records reflect a trend in soil fertility.

The relationship between crop yield and daily rainfall has been studied by ORSTOM, CIRAD and its partners. Forest and Lidon (1984) used data from the Saria (Burkina Faso) LTSM experiment, to develop a relationship between plant—available water and crop yield. They observed a very close relationship

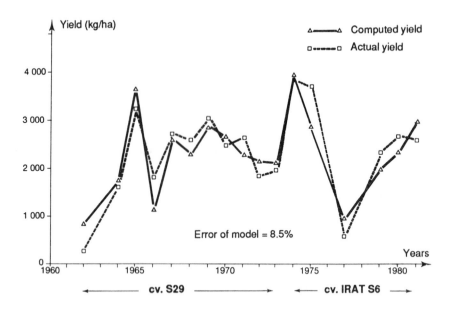

**Figure 2.** Yields calculated from a water-balance model compared with actual recorded yields. Sorghum monoculture at Saria, Burkina Faso. (From Forest and Lidon, 1984.)

between sorghum yield and the ratio of actual to potential evapotranspiration (ETR/ETM) (Figure 2).

Chopart and Nicou (1987) came to an even more precise conclusion in Bambey (Senegal). Since 1982, two methods of tillage applied to pearl millet and groundnut rotation were compared. Annual groundnut yields could only be predicted by including the cropping duration along with water balance factors (ETR, ETM, and DR, mm of drained water during the crop cycle), which strongly suggest that the inherent soil fertility changed over time.

By taking no account of cropping duration, the following equation accounted for 48% of the groundnut yield variation in treatments:

$$Yield \ (kg/ha) = 3670 \ ETR/ETM - 3.22 \ DR - 418 \qquad Eq. \ 3$$
$$(R^2 = 0.48)$$

By taking into account the number of years (n) under crop 64% of the variance was accounted for:

$$Yield \ (kg/ha) = 4621 \ ETR/ETM - 44n - 1.44DR - 1075 \qquad Eq. \ 4$$
$$(R^2 = 0.64)$$

This experiment, conducted at Bambey (Senegal) provides another interesting result. It shows that the annual rate of decline in yield was 44 Kg/ha with animal cultivation compared with only 11 Kg/ha with hand cultivation.

From these empirical models of the effect of rainfall distribution on cereal and groundnut crops, it is apparent that inter-year variation in the yield of rain-fed crops is largely explained by fluctuation in annual water supply to crops, the change in yields over the long term is related to change in soil productivity due to different management systems.

The daily rainfall data required for applying such simple models are rarely available for the LTSM surveyed. Consequently, both yield and rainfall data are presented throughout as moving averages over 3 or 5 years to overcome, partially at least, the problem of year-to-year variation caused by variation in annual rainfall distribution.

## III. Assessment of Agricultural Sustainability in LTSM Experiments

LTSM experiments contribute to the assessment of agriculture sustainability in so far as they provide quantitative information with regard to change in *soil productivity* as it relates to change in *soil qualities.* In addition, LTSM play an important role: in the evaluation of the nature and the intensity of soil degradation and in the understanding of the *major processes* involved.

### A. Assessing Soil Productivity Changes over Time

The change in soil productivity can be assessed by analyzing the trend in annual series of crop yields and by evaluating change in efficiency of inputs used.

### 1. Yield Trend

Clearly yield trend over a sufficient period (10 years or more) is the major indicator of change in soil productivity particularly on such LTSM experiments where technological packages are maintained unmodified over time.

Figure 3 shows the result for a low-intensity cropping system with hand cultivation at Darou, Senegal. The Darou (Sine Saloum, Senegal) experiment "Optimum length of fallow" was laid down on fallow land in 1952 and was continued without change in treatment by hand cultivation for 22 years.

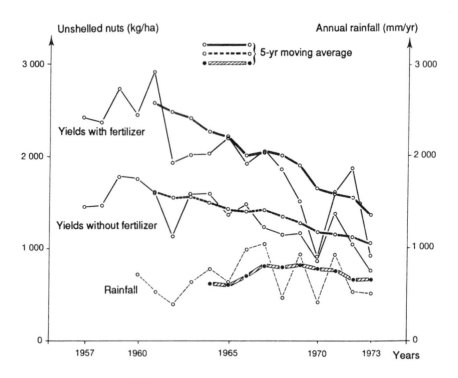

**Figure 3.** Groundnuts grown in rotation with millet or sorghum followed by 2-yr fallow cut and left as mulch over the dry season. Yields over 16 years at Darou, Senegal. (IRHO Annual reports on experiments, 1956-1974.)

There were three lengths of fallow (2, 3, and 6 years) under different treatments (burned just before cultivation, burned every year, cleared and composted every year, cleared and left in the field as mulch every year) followed by a rotation of 2 years groundnut with 1 year of millet or sorghum.

The compost was applied each year before cropping or to "dry" fallow on the plot from which it originated. Fertilizer was applied to half of each plot The other half was maintained as a control. Fertilizer was applied as follows:

- groundnut (1st year):
  8 kg ha$^{-1}$ N and 8 kg ha$^{-1}$ P$_2$O$_5$ as sulfate of ammonia and dicalcium phosphate;
- to cereal and groundnut (3rd year):
  8 kg ha$^{-1}$ N, 15 kg ha$^{-1}$ P$_2$O$_5$, and 12 kg ha$^{-1}$ K$_2$O as muriate of potash.

Nonintensive systems clearly could not maintain soil productivity, regardless of the length of fallow (up to 6 years) and rainfall distribution. Similar results

**Table 1.** Groundnut yield (y, kg ha$^{-1}$) in relation to number of years (n) under crop following a 6-year fallow in Darou, Senegal

| Treatment | Regression equation | Annual yield reduction (%) |
|---|---|---|
| Without fertilizer | y = 1820.1 - 46.9 n with r = 0.57 | 3.0 |
| With fertilizer | y = 3002.7 - 111.8 n with r = 078 | 4.8 |

have been obtained with other LTSM experiments under low-intensity systems of traditional crops in Senegal and Burkina Faso (Pieri, 1989).

Although the general yield level was higher with fertilizer than it was without it, the annual rate of yield decrease was higher on the no-fertilizer treatment (Table 1) suggesting that temporary higher soil productivity was gained at the expense of higher rate of soil fertility depletion.

The results are quite different when kraal earth and/or cattle manure was used. A similar experiment held at Niangoloko, Burkina Faso indicates (Figure 4) that not only the quantity but also the form of organic matter restitution play an important role in the maintenance of soil fertility. The decline in soil productivity in the Niangoloko experiment was much less than in the preceding experiment. Applying farmyard manure slowed down the rate of productivity decline.

The second group of LTSM experiments, i.e. manual cultivation systems partially intensified with a higher rate of fertilizer applied only to cotton, shows a more fuzzy picture. The cotton yields ranging from 1,700 to 2,300 kg/ha (cotton seeds) have been stable from the mid 1960s to 1980 when a balanced fertilization was applied equivalent to 53 kg ha$^{-1}$ N, 58 kg ha$^{-1}$ P$_2$O$_5$, 60 kg ha$^{-1}$ K$_2$O, 10 kg ha$^{-1}$ S, and 2.2 kg ha$^{-1}$ B$_2$O$_3$. In contrast sorghum and groundnut, which were not receiving any fertilizer, had low, yet gradually declining, yields.

The third group of LTSM experiments (with mechanical tillage, fertilizer, and organic matter restitution by incorporation of crop residues or cattle manure) provides complementary information.

Results summarized in Table 2 lead to three main conclusions:

a) Continuous cropping, without any fallow period, is technically achievable using research information available while increasing substantially the yields (Figure 5);

b) However, soil productivity cannot be maintained at a satisfactory level (more than 1t/ha of grain) without annual application of mineral fertilizers and with periodic liming (Saria, Korhogo Figure 6);

c) The combination of mineral fertilizer, liming, and periodic application of cattle manure brings several benefits, increasing both yield level and year-to-year yield stability. The yield stability is particularly increased

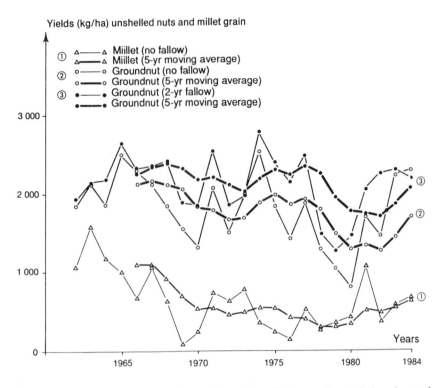

**Figure 4.** Groundnuts grown in rotation with millet or without fallow burned before sowing with fertilizer and 5 t ha⁻¹ kraal earth to the groundnuts. Yields over 23 years at Niangoloko, Burkina Faso. (IRHO, 1969; and Annual Reports on 1969-1984 experiments.)

when water harvesting techniques are used in combination with fertilizer applications (Table 3).

In addition, it is observed that plowing under of cereal straw in these sandy FTS (cf. Saria, Bambey) may increase yield by 10% on a short-term basis. However, contrary to cattle manure application, plowing under some straw doesn't prevent the declining yield trend over time.

In soils with relatively high SOM, such as the experimental site at Bebedja in Chad for which the St ratio equals 16% in the top soil layer, sustainable high yields can be achieved by fertilizer application alone without any need of organic matter application or liming. The cause of this distinct behavior is analyzed in the section on soil quality changes.

**Table 2.** Impact of soil management treatments on crop yield in LTSM experiments

| Site | Experiment | (Clay + Silt) % 0-20 cm | C % 0-20 cm | Mineral fertilizers | Crop residues | Cattle manure | Liming | Yield trends |
|---|---|---|---|---|---|---|---|---|
| SARIA (Burkina Faso) | Maintenance of fertility | 12.0 | 0.29 | + |  |  |  | ---[1] |
|  |  |  |  | + | + |  |  | -- |
|  |  |  |  | + |  | + |  | -[2] |
|  |  |  |  | + |  | + | + | S[3] |
|  |  |  |  | + |  | + | + | S |
| BAMBEY (Senegal) | Potash and straw | 3.1 | 0.21 | + | + | + | + | - |
| THILMAKHA (Senegal) | Cultural techniques | 2.4 | 0.14 | + |  |  |  | - |
|  |  |  |  | + |  |  | + | S |
| BEBEDA (Chad) | Experiment X | 11.0 | 1.02 | + |  |  |  | S |
| KORHOGO (Cote d' Ivoire) | Rate of fertilizer | 16.8 | 0.84 | + |  |  |  | - |
|  |  |  |  | + |  |  | + | S |

[1] (---) = Sharp decline; [2] (-) = Measurable decline; [3] (S) = Stable yields.

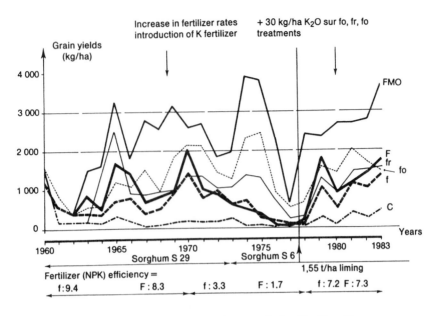

**Figure 5.** Yields of monocropped sorghum at Saria, Burkina Faso.

## 2. Fertilizer Use Efficiency

Fertilizer use efficiency declines appreciably over time as a consequence of decline in crop growth and/or steady increase in fertilizer rates. The Saria experiment shows clearly the change in fertilizer efficiency which is calculated by using Equation 5.

$$\frac{(Yield\ with\ fertilizer\ -\ Yield\ without\ fertilizer)\ kg/ha}{kg/ha\ fertilizer\ applied\ (N+P_2O_5+K_2O}\qquad Eq.\ 5$$

The initial efficiency equaled 8.3 during the period 1963-1970, then dropped down to 1.7 during the "acidification" period (1971-1978), and rose again to 7.3 after liming. Similar data have been obtained under partially intensified systems such as the cotton-based systems in the SAF (Sement, 1983, Gakou, 1986, cited in Pieri, 1989). Under low intensity systems, such as the LTSM study at Darou, a 6-year grass fallow was not able to restore fertilizer efficiency (Table 4).

The common belief in the virtue of fallow to regenerate soil fertility is true for humid tropical zones but not so for the semiarid and arid Sub-Saharan savannahs. The primary production of degraded FTS (Haplustalfs, Plinthustalfs) can be as low as 2 to 5 t ha⁻¹ yr⁻¹ of dry matter (Charreau and Nicou, 1971). In contrast, under a perhumid environment, such as the Guinean savannah of Grimari (Central African Republic), Morel and Quantin (1972) showed that a

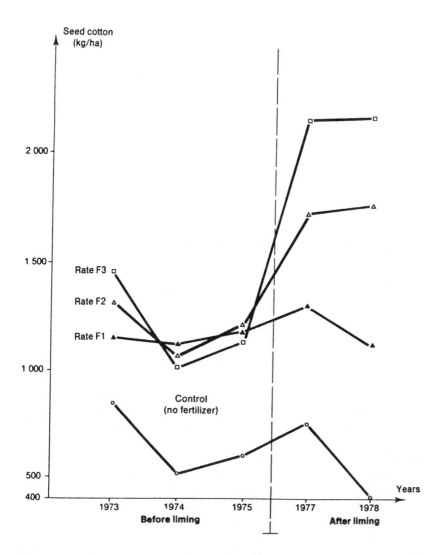

**Figure 6.** Cotton yields as affected by fertilizer treatment and liming at Korhogo, Côte d'Ivoire.

vigorous and deeply rooted grass cover, producing more than 12t ha$^{-1}$ yr$^{-1}$ of dry matter, could restore the qualities of a desaturated ferrallitic soil (Haplorthox) in a period of 2 to 6 years for a full recovery of its physical, chemical and biological properties.

**Table 3.** Average yields of *Pennisetum typhoides* (Var. Souma III) as affected by water harvesting and balanced fertilizer application, at Bambey (Senegal), during 6 years (1973-1978)

| Millet | Potash and straw experiment (1973-1978) | | | Farmers' field | Rainfall (mm) |
|---|---|---|---|---|---|
| | Control | NP | $NPK_{90}$ | | |
| Yield (kg ha$^{-1}$) | 1278 | 2246 | 2515 | 450 | $\bar{x} = 452$ |
| SE | 265 | 261 | 67 | - | SE = 84 |
| SE % | 20.7 | 11.9 | 2.7 | - | SE % = 18.5 |

**Table 4.** Fertilizer (N + $P_2O_5$) efficiency on groundnut at Darou in the groundnut-millet-groundnut-fallow rotation

| Period | Fertilizer efficiency 2-year fallow | 6-year fallow |
|---|---|---|
| 1957-1962 | 25.7 | - |
| 1959-1964 | - | 26.7 |
| 1969-1973 | 8.5 | 9.5 |

## 3. Relevance

LTSM results lead to the conclusion that there is a high risk of soil productivity decline over time in SAF. How relevant such a statement is remains a matter of contention, particularly for those who take into consideration the steady increase in total agricultural production experienced by Sub-Saharan countries during the last two decades, according to FAO Production Yearbooks. It is beyond the scope of this paper to fully address the complex issue of the relevance of experiments in comparison with local and national farming conditions. However, the following remarks are put forward to partially address this issue.

Whatever reservations there are as to the reliability of national statistics, evidence from yield trends, assessed at national scale (FAO data base), gives some ground for optimism about the state of soil productivity. Yields of sorghum, pearl millet, and groundnut follow a rather stable pattern. Cotton yields have significantly increased during the past decades, though rainfall records do indicate that interannual variation in yield may be explained by erratic rainfall.

However, a more detailed examination of yield trend assessed at village level and watershed level, which is rarely done, reveals that the prediction from LTSM is reasonably relevant for areas where the density of rural population is well above the 20 habitants km$^2$ current average. It is apparent that there are no direct relationships between rural population increase and soil degradation (cf. Figure 7).

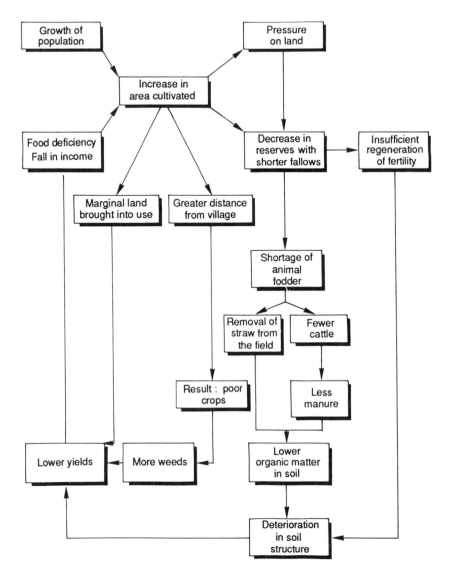

**Figure 7.** Degradation in a traditional farming system when pressure on land approaches saturation.

Yet rural population densities are close to or higher than 100 habitants km$^2$ in areas such as in South Mali, in North Togo and in Central Senegal. Here suitable land is no longer available for increasing production through land expansion and shifting agriculture (World Bank, 1989).

For instance, the development of cotton production together with seed planter adapted to animal traction have greatly increased the area of farming land

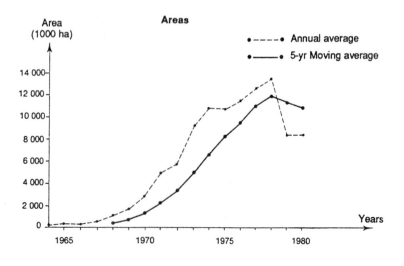

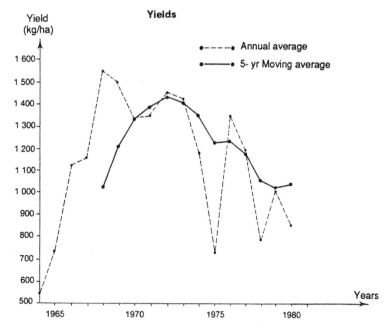

**Figure 8.** Area planted and yield of cotton (seed) in Upper Casamance, 1964-1980. (From CFDT, 1964-1973; Sodefitex, 1964-1980.)

exploited by a fast growing population (2.8% to 3.2% annual growth rate). Figure 8 illustrates this cotton expansion in Upper Casamance, a region in Western Senegal, with the resulting decrease in yields. A survey made, by the organization in charge of cotton production in Senegal (SODEFITEX) to

**Table 5.** Course of yields with fertilizer treatment and fertilizer efficiency at Young Farmer Centres in Burkina Faso.

| Rotation no. (course no.) | Yield (kg ha$^{-1}$) | Fertilizer rate (kg ha$^{-1}$) | | | Efficiency[1] |
|---|---|---|---|---|---|
| | | N | P$_2$O$_5$ | K$_2$O | |
| 1 (1st) | 2060 | 20 | 31 | 30 | 25.4 |
| 2 (4th) | 1546 | 20 | 31 | 30 | 19.1 |
| 3 (7th) | 1437 | 69 | 82 | 66 | 6.6 |
| 4 (10th) | 1514 | 107 | 116 | 96 | 4.7 |
| 5 (13th) | 1445 | 107 | 116 | 96 | 4.5 |

[1] Yield / (N + P$_2$O$_5$ + K$_2$O)
(From Hien et al., 1984.)

determine the cause of this decline led to the following conclusion. "The difficulty of the Cotton Project since 1977 cannot be ascribed to vagaries of the climate nor to the introduction of new insecticides. The main cause of the yield decline is most probably a decline of soil fertility," (Ange, 1984).

Similarly decline in fertilizer efficiency is often observed at farmers' field level. It may be hidden by apparent yield stability. Yields of cotton decreased by 25% in Young Farmers Centres in Burkina Faso although the rates of fertilizer application had been quadrupled by the 13th year (Table 5).

The limited role played by natural grass fallow to regenerate degraded soil fertility is also consistent with farmers' field observations. Charreau and Nicou, 1971, measured no improvement of physical properties of Senegalese soils under fallow, (e.g. water permeability, in aggregation, and organic matter content) unless land is fallowed for a period of 10 to 15 years. Peltre-Wurtz and Seck, 1979, observe that a fallow period of thirty to forty years follows 3 to 4 years of cultivation in the traditional cropping systems used by Senoufo people in Northern Côte d'Ivoire.

Finally, many factors are presently acting to assert that most of the agricultural land in Semiarid Africa is under the threat of severe soil degradation. The most critical situation arises in areas, such as South-West Burkina, where a rapid growth of population is associated with an influx of migrants who have an insatiable appetite for land and have no respect for the ancestral customs concerning conservation of natural resources (Belem, 1985).

Thus, the degradation of the Sudano-Sahelian environment should no longer be a matter of contention. It should be recognized as a blunt fact, yet assessed by land and vegetation degradation occurring all over the Sub-Saharan savannah zone (World Resource Institute, 1992).

## B. Assessing Soil Quality Changes over Time

Soil analysis shows that decline in soil productivity related to experimental soil management systems in LTSM experiments can be ascribed to two major processes: the decline of SOM and soil acidification.

Analyses of LTSM experiment results lead to a first approximation of the annual rate of SOM loss. These results indicate that some common beliefs, such as "soil acidification is not a major constraint in the semiarid African environment" or "soil incorporation of cereal crop residues is recommended to maintain or increase SOM content" are simply not substantiated and lead to adoption of unsustainable soil management practices.

### 1. Change in SOM Content

#### a. Data

Values of the annual loss of SOM, k%, are given in Table 6. Without exception, only the methods with manure application prevent a decline in SOM content. On very sandy soils, such as in the Bambey site (Senegal), the combination of annual plowing with fertilizers leads to a loss of SOM at an annual rate of 5% or more. On better textured soils, clay+silt above 10%, like those of Saria (Burkina Faso) and Bebedjia (Chad), k is often about 2% but management practices have considerable effect on this average value.

In comparing the development of SOM content on plots with and without fertilizer application, it is found that there are two quite different situations:

- In sorghum monocropping (Saria), the higher the N fertilizer used, the higher the k%: 1.5% without fertilizer, 1.9% with moderate rates, 2.6% at high rates of fertilizer. However, in this case, fertilizer is mainly nitrogen which apparently accelerates the loss of SOM.

- The opposite happens in the groundnut-millet rotation at Bambey, and in various cotton-based systems tested in Cote d'Ivoire, Cameroon and Chad. Here, the application of balanced NPKS fertilizer reduced the rate of loss of SOM.

Thus, the relation between fertilizer management and SOM is not so simple. Nitrogen fertilizer may have a detrimental impact when applied alone. However, the assertion that inorganic NPK fertilizer "burns up" SOM is not only oversimplistic but also wrong.

The results of this survey suggest the following possibilities for the maintenance of SOM:

1) Depending on soil texture and the rate used, application of farmyard manure with a C/N ratio of about 10 is the most effective method of

SOM maintenance. This practice may at least slow down the rate of loss in very sandy soils and often maintain or increase SOM in heavy-textured soils with unrealistic rates of manure applications (Saria, 20t ha$^{-1}$ yr$^{-1}$).

2) Substituting crop rotation for monocropping also limits SOM loss.

3) Good quality fallow may contribute to SOM conservation if the conditions favor the biomass production, e.g., when the soil is fertile enough for the establishment of a vigorous natural cover (Bebedjia, rotations B, C, and D), and when "improved" grass fallows are introduced by fertilizer application and legume fallows are developed with appropriate species (Sement, 1983; cited in Pieri, 1989).

In addition, LTSM experiment results show that some currently used cultivation techniques, such as plowing and crop residues incorporation, are not good enough to halt the loss of SOM and might well aggravate SOM depletion:

1) The impact of plowing on SOM content, specifically of inverting the top layer of soil with disks or moldboard plow, is often a controversial matter. Data from "Ameliorations Foncieres" LTSM experiment in Senegal do not prove that plowing induces, on average, an increase in the loss of SOM (Table 7). However, there are differences, especially relative to increase in crop yield due to plowing: the more the increase in yield, the more the favorable impact on SOM (Table 8.a.), and vice versa (Table 8.b.).

Bebedjia LTSM experiment data suggest that the annual rate of SOM loss increases with increasing frequency of plowing:

- plowed every year, k% = 2.2 (rotation B),
- plowed 3 years in 4, k% = 1.1 (rotation C),
- plowed 3 years in 6, k% = 0.5 (rotation D),

This comparison is partially biased because the organic residues returned to the soil differ between three rotations (D > C > B).

2) Surprisingly enough, LTSM experiment results raise doubts as to the usefulness of restoring crop residues to the soil. At Bambey, incorporating millet straw raised k by 1.3% from 4.3 to 5.6%. Similarly, at Saria, plowing under sorghum straw increased k from 1.9 to 2.2% (on plots receiving the moderate fertilizer dressing).

Then, further experimentation and detailed data analyses have been realized to strengthen the interpretation of these results and draw definite conclusions and make practical recommendations,

**Table 6.** Annual rate of organic matter loss measured in the field in the SAF region

| Place and source | Dominant rotation | Caly + silt (%) (2-20 cm) | Annual rate of loss k (%) | Annual rate of loss no. years | Notes |
|---|---|---|---|---|---|
| BURKINA FASO | | | | | |
| Saria INERA-IRAT | Sorghum monocropping | 12 | 1.5 | 10 | Plowed No fertilizer |
| | Sorghum monocropping | 12 | 1.9 | 10 | Low rate manure |
| | Sorghum monocropping | 12 | 2.6 | 10 | High rate manure |
| | Sorghum monocropping | 12 | 2.2 | 10 | Manure + crop residues |
| CFJA INERA-IRCT | Cotton-cereal | 19 | 6.3 | 15 | Much erosion |
| CAMEROUN | | | | | |
| Maroua, IRCT-IRA | Cotton-cereal | 17 | 3.2 | 5 | No fertilizer |
| (Richard, 1983) | Cotton cereal | 17 | 2.9 | 5 | Fertilizer |
| Not cited in text | Cotton-cereal | 17 | 2.5 | 5 | Fertilizer + kraal earth |
| COTE D'IVOIRE | | | | | |
| Boundiali, | Cotton-cerea | - | 2.6 | 5 | Low rate manure |
| IDESSA-IRCT | | | | | |
| Beheke | Cotton-cereal | - | 2.3 | 3 | Low rate manure |
| Ferkessedougou | Cotton-cereal | - | 0.4 | 3 | Improved fallow |

**Table 6.** continued -

| Place and source | Dominant rotation | Caly + silt (%) (2-20 cm) | Annual rate of loss k (%) | no. years | Notes |
|---|---|---|---|---|---|
| **SENEGAL** | | | | | |
| Bambey, ISRA-IRAT | Millet-groundnut | 3 | 7.0 | 5 | Plowed |
| (k x straw) | Millet-groundnut | 3 | 4.3 | 5 | No fertilizer |
| | Millet-groundnut | 3 | 6.0 | 4.5 | Fertilizer |
| Bambey, ISRA-IRAT | Millet monocropping | 4 | 4.6 | 3 | Fertilizer + straw |
| (Role of organic matter) | Cereal-legume | 11 | 3.8 | 17 | PK fertilizer |
| Nioro-du-Rip | Cereal-legume | 11 | 5.2 | 17 | F0T0 |
| IRAT-ISRA | Cereal-legume | 11 | 3.2 | 17 | F0T0 |
| | Cereal-legume | 11 | 3.9 | 17 | F2T2 |
| | Cereal-legume | 11 | 4.7 | 17 | F1T1 |
| **CHAD** | | | | | |
| Bebedjia, IRCT-IRA | Cotton-cerea | 11 | 2.8 | 20 | Plowed, very fertile soil |
| (Expt. X) | | | 2.4 | 20 | Cotton monoculture |
| | | | 1.2 | 20 | Cotton-cereal |
| | | | 0.5 | 20 | + 2-year fallow |
| | | | | | + 4-year fallow |
| **TOGO** | | | | | |
| Fosse aux lions, IRCT-DRA | Cotton-cereal | 10 | 2.4 | 20 | Low rate fertilizer (D) |
| | Cotton-cereal | 10 | 1.1 | 20 | High rate fertilizer (1.5D) |

**Table 7.** Organic carbon content (%) of surface soil under manual (T0), light mechanized (T1), and heavy (T2) mechanical cultivation, mean of all manuring treatments. "Améliorations foncières" experiment, Senegal.

| Site | T0 0-8 cm | T0 8-30 cm | T1 0-8 cm | T1 8-30 cm | T2 0-8 cm | T2 8-30 cm |
|---|---|---|---|---|---|---|
| Thiénaba | 1.59 | 1.56 | 1.62 | 1.56 | 1.75 | 1.64 |
| Nioro | 3.59 | 2.95 | 2.79 | 2.76 | 2.84 | 2.89 |
| Boulel | 2.58 | 2.30 | 2.33 | 2.02 | 2.02 | 1.95 |
| Sinthiou | 3.02 | 3.18 | 3.89 | 3.22 | 3.09 | 3.36 |
| Vélingara | 4.90 | 3.65 | 4.43 | 3.89 | 3.88 | 3.44 |
| Missirah | 3.95 | 3.43 | 3.21 | 3.40 | 3.14 | 3.14 |
| Séfa | 4.73 | 4.30 | 4.83 | 4.60 | 5.50 | 4.75 |
| Mean | 3.2 | 3.1 | 3.3 | 3.1 | 3.2 | 3.0 |

(From Rabot, 1984.)

**Table 8a.** Relation between annual organic matter loss (k), cultural technique, and production index at Nioro-du-Rip, Senegal 1967-1983

| Cultivation | Fertilizer | Code | C(%) 0-8cm | C(%) 8-30cm | k(%) 0-8cm | k(%) 8-30cm | Crop index[1] |
|---|---|---|---|---|---|---|---|
| Manual | 0 | F0T0 | 3.6 | 2.9 | 3.1 | 4.4 | 1 |
| Deep plowing | 0 | F0T2 | 2.5 | 2.6 | 5.3 | 5.0 | 1.39 |
| Manual | High | F2T0 | 3.9 | 3.2 | 2.6 | 3.8 | 2.39 |
| Deep plowing | High | F2T2 | 3.0 | 3.3 | 4.2 | 3.6 | 2.78 |
| Superficial | Moderate | F2T2 | 2.7 | 2.8 | 4.8 | 4.6 | 2.31 |

[1] Crop production index (1 = 48.2 t/ha total yield of 4 crops over period 1967-1983)
(From Sarr, 1981; Rabot, 1984.)

**Table 8b.** Relation between carbon in soil, cultural method, and yield level at Sefa (Senegal) 1967-1982

| Cultivation | Fertilizer | Code | C (%) 0-8 cm | C (%) 8-30 cm | Crop index[1] |
|---|---|---|---|---|---|
| Manual | 0 | F0T0 | 3.8 | 3.8 | 1 |
| Deep plowing | 0 | F0T2 | 6.4 | 5.1 | 3.6 |
| Deep plowing | Moderate | F1T2 | 5.4 | 5.0 | 14.3 |
| Deep plowing | High | F2T2 | 4.7 | 4.2 | 13.8 |

[1] Crop production index (mean maize yield over 1967-1982; index 1 = 198.3 kg ha[-1] grain).
(From Rabot, 1984.)

**b. Interpretation**

Several pot and short-term field experiments were conducted to evaluate the impact of applying organic materials of different C/N ratios on the organic balance in cultivated soils at Bambey and Saria sites (Gueye and Ganry, 1978; Feller et al., 1981, 1983; Sedogo, 1981).

At both sites the incorporation of cereal straw led to a decline in soil organic carbon content (Table 9.a.) over one or two years, because it greatly stimulates microbial activity and mineralization. At the onset of the rainy season, the more carbon is available, the more are the heterotrophic microorganisms decomposing not only carbon from the straw but also soil organic carbon (9.b.). This is especially true when soil carbon is not well protected by abundant mineral colloids, which is the case of most FTS in Sub-Saharan countries.

Less lignified residues rich in decomposable carbon (low NDF/CC), regardless of the C/N ratio, favor the overall immobilization of carbon. This temporarily may decrease the carbon photosynthesized by the crop because of the lack of nitrogen in soil absorbed by the microfauna to metabolize their own "microbial protein".

Consequently, in the long run, incorporation of straw in the sandy ferruginous soil of Semiarid Africa may result in a negative organic matter balance and decrease of the humic carbon fraction adsorbed on soil mineral colloids. This humic fraction is most important for ensuring structural stability.

The impact of other cultural practices on the SOM balance can be evaluated by considering two effects:

- The direct effect of the cultural practices on biological activity in soils and on the rate of SOM mineralization: plowing, for instance, accentuates the rate of decomposition of organic residues and SOM by improving temporarily, at the onset of the rainy season, soil properties which stimulate microbiologial heterotrophic activity. On the contrary microfauna activity which indicates the process of humification is constrained by textural discontinuity (hardpan) resulting from tillage (Lal, 1987, Chopart cited by Pieri, 1989).

- The additive or subtractive effects of cultural practices on the yields and eventually on the amount of organic inputs to the soil. Application of inorganic and organic fertilizers, crop management and crop rotation, and plowing improve biomass production and vice versa the amount of residue returned to the soil.

**Table 9a.** Effects of manure and urea on sorghum yield (var. E 351) and carbon content of surface soil at Saria1, Burkina Faso

| Organic Materials | Kg/ha N as urea | Yield index[2] | C(%) (0-20 cm) |
|---|---|---|---|
| 0 | 0 | 1.00 | 0.41 |
| 0 | 60 | 1.57 | 0.40 |
| Straw (C/N = 93) | 0 | 0.86 | 0.38 |
| Straw (C/N = 93) | 60 | 1.65 | 0.38 |
| Straw (C/N = 15) | 0 | 1.34 | 0.40 |
| Straw (C/N = 15) | 60 | 1.64 | 0.40 |

[1] Short-term experiment, one-year duration (1980)
[2] Index 1 = 5,400 kg/ha straw + 3,150 panicles
(From Sedogo, 1981.)

**Table 9b.** Effects of various organic materials on carbon content of surface soil (0-20 cm) at Bambey[1], Senegal

| Treatment | Properties of C/N % | Organic materials MDF/CC%[2] | Soil organic C (%) Total C | Humic C |
|---|---|---|---|---|
| Control | 0 | 0 | 1.86 | 1.38 |
| Sorghum straw | 53 | 0.8 | 1.59 | 1.22 |
| Composted groundnut husk | 52 | 28.1 | 2.72 | 2.00 |
| Groundnut husk | 48 | 3.6 | 2.38 | 1.44 |

[1] Short-term experiment, one-year duration (1978)
[2] NDF: Neutral Detergent Fiber content, CC = Cellular Content
(From Gueye and Ganry, 1978.)

## 2. Soil Acidification

### a. Data

The LTSM experiments are unanimous in showing that when cultivated soils receive inorganic fertilizer and no manure, soil acidification is a general consequence. The major results of the survey are as follows.

The more sandy the soils and the higher the rates of fertilizer, the more marked is the soil acidification. In Bambey (Sénégal) the Al saturation of the top soil layer, initially nil (pH 5.8), reaches 40% after 5 years of cultivation. It is common to obtain 30% Al saturation in most soils in the cotton belt after 3 years of cropping (Sement, 1983). Nitrogenous fertilizer, recommended for cereals, plays an important role in enhancing soil acidification, as was clearly shown by the Saria (Burkina Faso) experiment under continuous sorghum cropping system. In this experiment when low and high rates of N fertilizer were applied the Al saturation reached respectively 28.7% and 35.6% after 18 years of cropping. In comparison, Al saturation was 9.3% on the control plots (no fertilizer) and no more than 5.4% and 1.0% on manured plots (5 and 40t/ha/yr of cattle manure during the period 1960-1978).

However soil acidification may not be a matter of concern if cultivated soils are rich in soil organic matter content. Such soils are very rare in the SAF region. They are found only on specific sites, such as i) valley bottom soils, basins or hollow sites (e.g. Bebedjia site, Chad) where the top soil layer has been enriched for years by organic colloids washed down from the upper parts of the toposequence and ii) below the canopy of legume trees e.g. *Acacia albida* (Charreau & Vidal, 1965).

Data from these LTSM confirm the following:

- The combined use of manures and inorganic fertilizers is more beneficial in maintaining soil pH than that of fertilizers alone. Because acidification can easily be controlled by liming and it would be wrong, therefore, to prefer one method over another (fertilizer and amendment or fertilizer plus manures), as long as SOM content is maintained above a satisfactory level;

- Incorporation of crop residue (straw) reduced Al saturation from 50 to 35% for millet (*Pennisetum sp.*) at Bambey and from 29 to 23% for sorghum at Saria (Pichot et al., 1981).

### b. Interpretation

Soil acidification results from two major interrelated processes i.e., soil nutrient depletion induced by fertility-mining cropping systems and leaching of Calcium and Magnesium as a consequence  downward fluxes of nitrate in soil profiles.

LTSM in SAF provide a fruitful framework to assess accurate nutrient balances in soils under a wide range of cropping systems (Pieri, 1983). It is possible to compare results of nutrient balance measured in cropped soil (Figure 9) with those of analytical measurements of the changes occurred over time in the nutrient content of the cultivated soil layers (Pieri, 1989).

Readers are referred to the literature cited above regarding details on the findings. These data have been further broadened to encompass all Sub-Saharan Africa (FAO, Winand Staring Center, 1990, Smaling, 1993).

Analyses presented here confirm that nutrient balances are generally negative, especially when one realizes that the average fertilizer usage is no more than 10 kg/ha of cropped land/yr of nutrients ($N + P_2O_5 + K_2O$). Nitrogen balance appears to be one of the most negative ones, even when a leguminous crop such as groundnut is cultivated. Careful complementary investigations have been carried

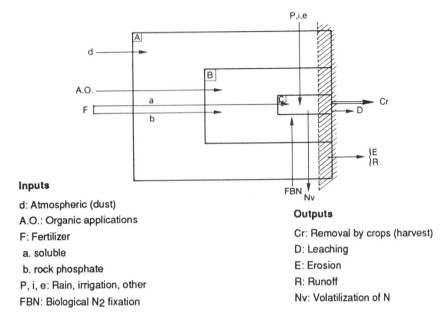

**Inputs**

d: Atmospheric (dust)
A.O.: Organic applications
F: Fertilizer
  a. soluble
  b. rock phosphate
P, i, e: Rain, irrigation, other
FBN: Biological $N_2$ fixation

**Outputs**

Cr: Removal by crops (harvest)
D: Leaching
E: Erosion
R: Runoff
Nv: Volatilization of N

**Figure 9.** Representation of nutrient balance in a cropped soil.

**Table 10.** Nitrogen balance under millet-groundnut rotation in semi-arid conditions.

| Gains (kg ha$^{-1}$) in (2 years) | | Losses (kg ha$^{-1}$) in (2 years) | |
|---|---|---|---|
| Atmosphere | Tr | Net removal by millet | -33 |
| Fertilizer to millet | +80 | Net removal by groundnut | -109 |
| Fertilizer to groundnut | +15 | Denitrification and | |
| Symbiotic Fixation | +82 |   volatilization: | |
| | |   Soil N (2 years) | -50 |
| | |   Organic N residues (2 years) | -8 |
| | |   N fertilizer (2 years) | -28 |
| | |   Leaching (2 years) | -20 |
| Net balance: -71 (2 years) | | | |

(From Wetselaar and Ganry, 1982.)

out at Bambey (Senegal) using $^{15}$N (Wetselaar and Ganry, 1982, Ganry, 1990). The major conclusion is that a groundnut crop actually causes nitrogen impoverishment of the soil (Table 10) under the current condition of farming (i.e. no return of groundnut husks to the field, apart from 10% which falls naturally).

Finally, LTSM experiment results show that, while assessing N balance in cropped soils, N input from biological fixation is often overestimated especially under the conditions of water deficit as is the case in the SAF region. Calcium and magnesium leaching from the soil upper layers is highly correlated with nitrate leaching. Measurement by ceramic cup lysimeters made at different depths on the LTSM experiment established at Bambey (Senegal) showed that each $NO_3^-$ ion in the soil solution is accompanied by a little more than its equivalent in cations (Pieri, 1983):

$$[Ca^{++} + Mg^{++}] \; meq/1 = 0.24 + 0.86 \; [NO_3^-] \; meq/l, \; (r = 0.9) \qquad Eq. \; 6$$

Work by Poss and Saragoni (1992) on "terre de barre", Eutrustox, in Southern Togo, lead to a similar conclusion. They found that Ca and Mg were mobilized by nitrate, and to some extent by chloride when KCl fertilizer was applied. They developed the following relationship on the basis of 186 measurements of soil solution composition:

$$[Ca^{++} + Mg^{++}] \; meq/l = 0.71 \; [NO_3^-] \; meq/l + 0.07 \; [Cl-] \; meq/l, \; (r = 0.9)$$

$$Eq. \; 7$$

Nitrate leaching, and subsequently soil acidification, are highly affected by rainfall distribution, particularly by rainfall occurring early in the growing season. Early rains accentuate SOM oxidation (Blondel, 1971) at the time when N uptake by the plants is low.

Root system development, its vigor, and depth are also important in controlling leaching loss. In this respect, the traditional cereal crops like millet and sorghum have an advantage over newly introduced crops with shallow and less prolific root systems (Table 11).

The forms of organic inputs to cultivated soils, in relation to their composition and resistance to biodegradation, are also important. The data in Figure 10 shows that root residues, besides organic manure, are key inputs in implementing sustainable cropping systems in the SAF region dominated by sandy FTS.

## C. Developing a General Framework for Sustainable Use of Ferruginous Tropical Soils in Semiarid Africa

In an attempt synthesize all the data presented, a general framework of soil productivity changes under cultivation in Semiarid Africa is proposed here with an objective to identify some key indicators of soil fertility decline related to SOM.

**Table 11.** Nutrient losses by leaching under crops in Senegal, Cameroun, and Ivory Coast

| Site | Annual rainfall, mm (year) | Crop (variety) | Drainage, mm | Losses (kg ha$^{-1}$ yr$^{-1}$) N | CaO | MgO | K$_2$O | P$_2$O$_5$ |
|---|---|---|---|---|---|---|---|---|
| Bambey (Senegal) | 507 (1981) | Millet (Souna III) | 9.5 | 0.3 | 0.8 | 0.4 | 0.3 | Tr |
| | | Groundnut (55-437) | 100.6 | 25.1 | 54.1 | 13.6 | 5.2 | Tr |
| Maroua Cameroun | 705 (1975) | Sorghum (IRAT 55) | 2 | Tr | 0.1 | 0.1 | Tr | Tr |
| | 683 (1977) | Cotton (BJA) | 83 | 2.1 | 43.7 | 12.3 | 1.7 | Tr |
| Bouaké Ivory Coast | 633 (1981) | Maize (CJB) | 210 | 6.1 | 36.4 | 26.2 | 2.4 | Tr |
| | 532 (1981) | Cotton (BJA) | 260 | 7.1 | 18.0 | 6.6 | 2.0 | Tr |

(From Chabalier, 1978a: Gigou, 1982; and Pieri, 1983.)

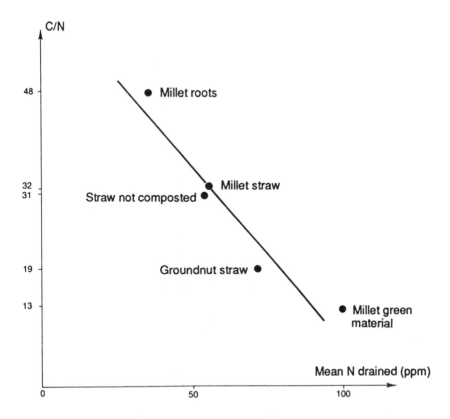

**Figure 10.** Mineralization of various farm residues in lysimeter, Bambey, Senegal. (From Ganry, 1977.)

## 1. Soil Productivity Deterioration

Figure 11 presents a schematic showing, at the farm scale, the cause-effect relationship to govern changes in soil fertility, their effects at soil and at crop levels, and the impact on above and below ground biomass production.

Based upon LTSM experiment results and complementary works, the schema indicates two fundamental causes of decline in soil productivity of rainfed agriculture:

a) Accelerated soil erosion causes the loss of soil material, particularly the more active components, i.e., clay and organic fractions;

b) Under current farming practices, there is a deficit organic balance, because the rate of native SOM mineralization exceeds the rate of accumulation of "humus" or stable soil organic fractions.

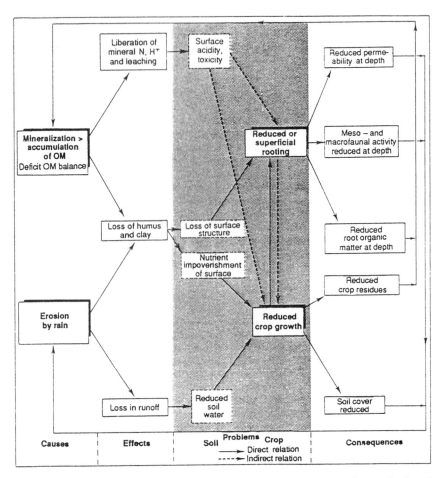

**Figure 11.** The causes of problems and their consequences for agricultural productivity in rain-fed farming in the savannah.

These two causes are triggered by climatic characteristics, topographical and soil features, and biological factors namely the "explosion" of heterotrophic microbiological activity at the onset of the rainy season.

This decline in soil productivity can be ascribed to four major problems: water deficit, nutrient deficit, physical barrier resulting from the breakdown of the structure of soil surface and chemical barrier to root development due to nutrient imbalance in the sub-soil. These four problems reduce crop productivityroot biomass. The savannah ferruginous soils are then locked into a declining vicious cycle of low productivity and poor soil quality.

A complementary schematic on a broader scale, (soil toposequence, watershed, and landscape units) is necessary for a better understanding of edaphological causes of land degradation. These land degradative processes are beyond the scope of most of LTSM experiments, because other socio-economic factors and land tenure systems are not considered.

However, this schematic clearly shows the agronomic causes of soil degradation at the farm scale. It is only the elimination of the causes which can solve the problems of soil degradation. Technologies designed to address most obvious soil and crop management problems, such as nutrient deficiency, may improve soil productivity on a short-term basis, but do not lead to sustainable soil use if they do not eliminate factors of soil degradation.

Finally, this schematic may help in identifying the best indicators to assess the inherent fertility status of cultivated soils, e.g. SOM which plays a seminal role in the maintenance of soil structure and porosity, soil buffering capacity, cation exchange capacity, water infiltration rate, and energy source for microbiological activity.

## 2. Critical Value of SOM

The rate of loss of SOM $(k\%)$ and its temporal variation, in spite of its inherent limitations and flaws, is an integrated index of soil degradation processes. The following calculation of a value $k_s\%$, which applies to soil managed by different technologies, has several limitations. However, if we are to progress from purely descriptive agronomy to a more quantitative assessment of the impact of specific recommendations on the SOM balance, it is necessary to establish such critical values.

With this objective, we can relate the value for $k_o\%$, the constant for a cultivated soil managed nonintensively (manual), to $k_s\%$ values for management systems described in the Table 12. Each coefficient applies to a specific component of soil intensification technology, such as fertilizer application, plowing etc., (cf. III.B.1.b.).

Additional research is needed to completely model SOM balance (see Pieri, 1989). This is an empirical approach devised to bring to the attention of project managers and policy makers the long-term impact of technical recommendations on soil productivity.

Then the question of "critical values" arises because they are indispensable in practical agriculture. The international soil science literature is replete with statements of critical levels of soil nutrients required for good crop growth, but information on critical levels of SOM is scanty (Sanchez and Miller, 1986).

Some agronomists are convinced that if SOM is important in soil quality, then the higher the content, the better it is. This view might apply if farming standards were high and the farmer were in full control. It certainly does not apply in semiarid Africa where there are so many technical and economic constraints to crop performance. Under these circumstances, it is fruitless to aim for an SOM content above a certain critical value. When the SOM content is below this critical value, soil structural stability is threatened and crop yields are reduced drastically.

**Table 12.** Values of multipliers used for calculating $k_s$ from $k_0$

| | Basic rate of loss $k_0$ | Effect of plowing | Effect of straw incorp. | Efect of rotation | Effect of complete fertilizer |
|---|---|---|---|---|---|
| Soil type | (%) | $\alpha$ | $\beta$ | $\gamma$ | $\delta$ |
| Very sandy | 4 | 1 to 1.6 | 1.4 | 0.9 | 0.8 |
| Loamy sand | 2 | 1 to 1.2 | 1.2 | 0.9 | 0.8 |

$$(k_s = k_0 \cdot \alpha \cdot \beta \cdot \gamma \cdot \delta)$$

It is also important to develop an empirical relation between the amount of SOM in a soil and the surface area of the mineral adsorbent, beyond which maintenance of soil structure is difficult to achieve. For the SAF region, similarities in soil properties and mineralogical composition have enabled computation of $S_t\%$ as shown in Part II.B.2. of this paper.

Values of $S_t\%$ were calculated for LTSM experiment sites and other sites for which a good description of the structural status was available, including poorly structured/eroded sites (Yatenga in Burkina Faso, Sine Saloum in Senegal, N'Tarla in Mali), sites in satisfactory physical conditions (South-East Benoue valley in North Cameroon), or very favorable sites [Bebedjia in Chad, compounds areas in Burkina Faso].

The data in Figure 12 justifies the following conclusions:

- if $S_t < 5\%$, the soil structure is degraded and the susceptibility to water erosion is high;

- if $5\% < S_t > 7\%$, the soil is structurally unstable and there exists a high risk of soil surface erosion;

- if $S_t > 9\%$, the soil structure is relatively stable.

This ratio appears to have a good predictive value, and may be a useful tool for agronomists and project developers involved in diagnosing and monitoring soil productivity. However, more work has to be done on $S_t\%$, particularly on non-African semiarid environments and soils in other ecoregions.

## IV. Toward Improved Soil Management Systems

Improved soil management systems can be usefully derived from LTSM results. There is a wide array of technological options available for enhancing the sustainability of agriculture in semiarid environments. Three major production systems can be considered, i.e., crop-based systems, livestock-based systems, and tree crop-based systems (Srivastava et al., 1993). Technological options

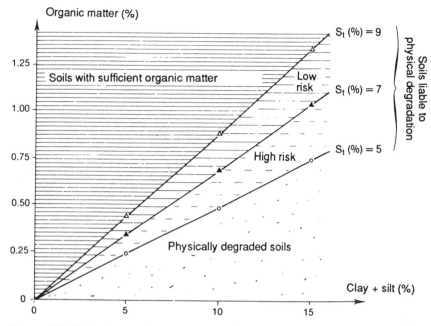

**Figure 12.** Critical organic matter levels for maintenance of physical stability.

should address the issues of water control and water harvesting, erosion control, and enhancement and maintenance of soil fertility. In this report, the contribution of improved soil management systems derived from LTSM results in SAF is primarily oriented to i) management of SOM in cultivated soils and ii) control of nutrient balances and soil acidification in crop-based systems.

## A. Management of Soil Organic Matter

Improvement and maintenance of SOM is one of the most challenging aims to be reached by newly proposed cropping systems adapted to the semiarid tropical environment. Conversion to agricultural landuse usually increases the rate of loss of SOM. The rate of loss is particularly high in the semiarid Tropics.

Improving and maintaining high SOM content can be achieved by:

- increasing biomass production per unit cropped area;
- increasing biomass restitution to cropped area; and
- decreasing SOM loss per unit cropped area.

## 1. Increasing Biomass Production

There is no alternative to sustainable agriculture in semiarid Africa but to increasing production of vegetative biomass by unit area. Beyond this option there is no solution to the compulsive need of increasing production of biomass for food, fiber, fodder, energy, and returning organic residue to soil.

Under the current circumstances of shortage of prime agricultural land in the SAF zone, this first objective can be principally achieved by technologies which can enhance water and nutrient availability to crops. Such technologies entail water-harvesting techniques and use of inorganic fertilizers.

Results from the LTSM experiment at Bambey (Senegal) shown in the Table 3 indicate that application of these technologies can increase both yield and yield stability.

The plowing at the end of rainy season creates a rough soil surface which protects the soil from wind erosion during the dry season and allows storage of residual water in the soil profile (10 to 40 mm according the annual rainfall distribution, unpublished data of Dancette). This complementary water supply allows the crop to withstand the frequent dry spells experienced at the onset of the rains. Adoption of this technology is, however, constrained by the uncontrolled grazing during the dry season.

## 2. Increasing Biomass Return to the Land

It has been demonstrated that some form of organic matter residues are better than others for improving SOM. The materials used should favor both organic matter accumulation and humification. Under local conditions, only specific crop (manure or compost) and root residues possess the desirable qualities (correct C/N ratio and the proportion of lignified fiber and soluble cell content) to contribute to improve and stabilize SOM content.

While applying cattle manure may be a solution in local-specific cases (Berger, 1991), many socio-economic factors (e.g. ethnic separation between herders and farmers, as well as the availability of equipment for carrying and incorporating bulky material to fields) limit the scope of manure application most of the SAF region.

There is a need to screen germplasm of improved plant species and crop association which favor the quick establishment of a dense and deep root system. Indigenous species adapted to the SAF region, although not highly productive, possess these qualities (Figure 13).

## 3. Decreasing the Loss of SOM

Results from LTSM experiments and other complementary research show that the rate of loss of SOM is directly influenced by selective process of soil

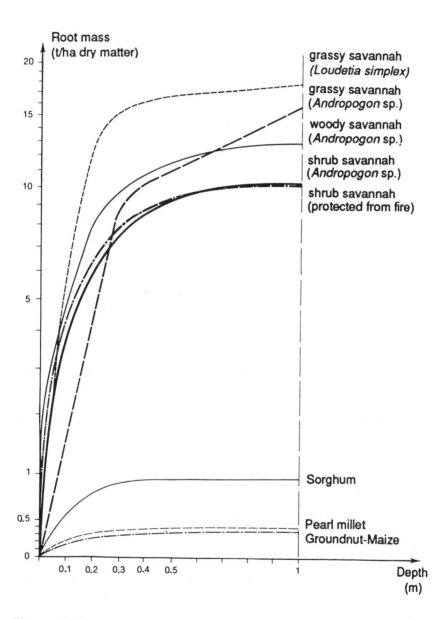

**Figure 13.** Root mass distribution under natural cover compared to crops. (Adapted from Chopard, 1980; Lamotte and Bourliere, 1978.)

erosion by water which carries away organic colloids and by the activity of heterotrophic microorganisms which oxidize soil organic components. Little applied research has been done to test adapted technologies at farmers' field level which can control the rate of SOM loss. LTSM results from semiarid Africa show a clear distinction between sites characterized by sandy surface soils with limited soil moisture regime and heavy-textured soils with a favorable moisture regime.

In the first case, the most significant results in terms of net SOM accumulation have been observed with the *Acacia albida*-annual crops-livestock system. This system has been studied and partially improved by IRAT, CTFT and, ICRISAT. The microbiological activity under the tree canopy has been extensively studied by Dommergues (1970, 1982), showing that organic C accumulation is favored by local improvement in soil moisture and thermal regimes.

In the second case, which is more favorable for agricultural production, recent experimental results obtained by CIRAD and IDESSA (Charpentier et al., 1991) at Tcholelevogo (North Cote d'Ivoire) show the good agronomic potential of different no-tillage systems. Use of direct drilling systems through crop residue and living mulch cover (e.g. *Macroptyllium atropurpureum, Tephrosia pedicellata*), results in an efficient control of soil erosion and in distinct accumulation of SOM in the soil surface.

LTSM results and associated works support the urgent need for more advanced and coordinated research on application of conservation tillage systems to FTS in semiarid Africa. Quantitative assessment of the impact of direct drilling on the decrease in rate of SOM loss, the accumulation of organic C, the fauna and microbiological activities, and their consequences on soil qualities and soil productivity should be made and compared with data already available for humid tropical Africa (Lal, 1987; 1989).

## B. Controlling Nutrient Balances

There is no doubt that sustainable growth in agricultural production is urgently needed in semiarid Africa, where the population is presently doubling every 20 years. The mining of the soil nutrients resource is no longer a practical solution to the challenge of increasing agricultural output. If the target is a more productive agriculture, the enhancement and the maintenance of soil productivity require that the soil nutrient balances should be tuned to the higher nutrient flux generated by higher yields produced for local consumption and for income generation. These objectives are achievable from technological innovations that:

- increase nutrient inputs to the soil,
- increase nutrient use efficiency, and
- decrease soil nutrients loss.

Results obtained from LTSM experiments in SAF and related research offer technological options to achieve these.

## 1. Increasing Nutrient Inputs

LTSM results clearly identified the need for more intensive use of fertilizers, inorganic and organic, to remedy the major soil nutrient deficiency, particularly in phosphorus, and to supply the major elements N, P, K, and S to the crops. There is also a need to supply trace elements like boron for cotton and molybdenum for groundnut. Liming may also be quite necessary under continuous cropping.

The increase in inorganic fertilizer use is constrained by economic limitations. Concentrated compound fertilizers offer an economic advantage over straight materials as the transport cost over large distances can be an important factor in the price offered at the farm (Schultz & Parish, 1989). However, local deposits of rock phosphates (Sénégal, Mali, Niger, Burkina Faso, Togo) and liming products (e.g. calcium carbonate in Sénégal and dolomite at Tiara in Burkina Faso, may be alternatives which are agronomically sound (Truong, 1984, Vlek & Mokwunye, 1989) and worth economic feasibility study.

In areas with mixed farming systems, cattle manure is certainly an excellent source of plant nutrients. Among the limitations to increasing use of manures, besides the ones mentioned above, it should emphasized that water availability is often a major drawback to producing good quality manure and/or compost at farm level in most SAF regions.

## 2. Decreasing Nutrients Loss

Controlling soil erosion is extremely important in reducing the loss of nutrients adsorbed on fine soil particles. The recycling of crop residues is another relevant strategy. However, there is also a high risk of soil structure degradation associated with plowing under cereal straw. Returning crop residues to the soil may be best achieved in the form of manure in the SAF and as mulch cover in the humid regions.

Soil management systems based upon direct drilling through living mulch cover, which favors the maintenance of SOM, are also recommended to decrease leaching losses (Moraes, 1993) and increase nutrient uptake by permanent deep rooted plants (Derpsch et al., 1991, Crovetto, 1992). Similarly the permanent vegetative cover which decreases soil surface temperature and improves soil moisture regime may alleviate some of the physical factors which limit the efficiency of biological nitrogen fixation by leguminous crops such as groundnut and cowpea.

# V. Conclusions

The survey and interpretation of data from 24 Long-Term experiments on Soil Management (LTSM) conducted in semiarid Africa (initiated in 1960), have produced unique information on two fundamental soil processes, soil organic matter decline and soil acidification. These processes control the sustainability of crop productivity under continuous cropping systems in the Sudano-Sahelian countries. Results obtained from short-term experiments cannot provide the appropriate scientific information, and often produce misleading technological recommendations, such as the plowing under of crop residues to improve soil organic matter status in sandy Ferruginous Tropical Soils. Such comprehensive information generated by LTSM is of particular importance now since traditional farming systems based upon shifting agriculture are no longer applicable and there is a compelling need for increasing agricultural productivity.

In addition, LTSM experiments provide guidelines on the prospects for improved soil management practices based upon the enhancement of the role of soil fauna and microbiological activity which can increase water and nutrient use efficiency. Permanent vegetative soil cover and better crop root development may be key components of new soil conservation tillage systems that should be tested throughout the region. These new experiments should complement three or four existing LTSM experiments which must be maintained.

Results from these LTSM experiments in SAF show that to get the most of new LTSM experiments, an initial planning phase should be carefully conducted on an eco-regional basis. Basic scientific hypotheses to be tested should be clearly spelled out and an agreement on congruous experimental designs, formats, and monitoring should be reached to allow fruitful regional linkages, common data interpretation, and identification of needs for training and analytical facilities. Long-term funding is required and likely to be obtained from several diverse sources. This would imply a decentralized fund management for each experimental site and a periodic scientific and administrative evaluation by external panels.

# References

Belem, P.C. 1985. Coton et systèmes de production dans l'ouest du Burkina Faso. Thèse 3e cycle, géographic de l'amé nagement de l'espace rural. Univ Paul Valéry, Montpellier III. 322 pp.

Berger, M. 1991.La gestion des résidus organiques à la ferme. p.293-316. In: Pieri, C. (ed.), *Savanes d'Afrique, Terres Fertiles?* CIRAD, Ministere de la Cooperation et du Developpement, Paris.

Blondel, D. 1971. Contribution à la connaissance de la dynamique de l'azote minéeral en sol sableux (Dior) au Sénégal. *Agron. Trop.* 26(12):1303-1333.

Charpentier, H., F. Coulibaly, et L. Kafao. 1991. Expérimentation sur le terroir de Tcholelevogo. Note Technique 40/90 IDESSA-DCV.

Charreau, C. and P. Vidal. 1965. Influence de l' *Acacia albida Del.* sur le sol, nutrition minérale et rendements des mils *Pennisetum* au Sénégal. *Agron. Trop.* 20(6-7):600-625.

Charreau, C. 1975. Organic matter and biochemical properties of soil in the dry tropical zone of West Africa. FAO Soils Bulletin 27. Rome: FAO, 313-335.

Charreau, C. and R. Nicou. 1971. *L'amélioration du profil cultural dans les sols sableux et sablo-argileux de la zone tropicale séche ouest-africaine et ses incidences agronomiques.* Irat Paris Bull. Agron. 23. 254 pp.

Chopart, J.L. and R. Nicou. 1987. *Vingt années de culture continue avec et sans labour en sec dans un sol sableux dunaire au Sénégal.* Irat, Montpellier. 36 pp.

Crovetto, C.L.. 1992. *Rastrojos Sobre el Suelo: una Introduccion a la Zero Labranza.* Universidad Autonoma, Santiago de Chile, Chile. 340 pp.

Derpsch, R., C.H. Roth, N. Sdiras and U. Köpke. 1991. *Controle da erosao no Parana, Brasil: Sistemas de cobertura do solo, plantio direto e preparo conservacionista do solo.* Sonderpublikation der GTZ, No.245.

Dommergues, Y. and F. Mangenot. 1970. *Ecologie microbienne du sol.* Masson, Paris. 795 pp.

Dommergues, Y. and H.G. Diem. 1982. *Microbiology of tropical soil and plant productivity.* Developments in Plant and Soil Sciences, vol 5. Nijhoff: Junk, The Hague. 328 pp.

Emerson, W.W. 1967. A classification of soil aggregates based on their coherence in water. *Aust. J. Soil Res.* 5(1).

Forest, F. and B. Lidon. 1984. Simulation du bilan hydrique pour l'explication du rendement et l'appui aux producteurs. p 55-65. In: *La sécheresse en zone intertropicale. Pour une lutte intégrée. Colloque Résistance à la sécheresse en milieu intertropical: quelles recherches pour le moyen terme?,* Dakar, Senegal, 24-47 Sept. CIRAD-CILF, Paris.

Fournier, F.G.A. 1967. Research on soil erosion and soil conservation in Africa. *African Soils.* 12: 53-96.

Ganry, F. 1990. Application de la méthode isotopique à l'étude des bilans azotés en zone tropicale sèche. Thèse doctorat d'Etat. Sciences naturelles, Univ. Nancy-1, France. 335 pp.

Greenland, D.J. and R. Lal. 1977. *Soil Conservation and Management in the Humid Tropics.* Chichester, John Wiley & Sons, England.

Herdt, R.W. and R.A. Steiner. 1992. Two approaches to measuring agricultural sustainbility using long-term agronomic trials. Unpublished paper. The Rockefeller Foundation.

IBSRAM. 1991. *Evaluation for Sustainable Management in the Developing World.* IBSRAM Proc. No. 12, Volumes I-III.

Jones, M.J. 1976. The significance of crops residues to the maintenance of fertility under continuous cultivation at Samaru, Nigeria. *J. Agric. Sci.* 77(3):473-482.

Lal, R.1987. *Tropical Ecology and Physical Edaphology*. John Wiley & Sons, Chichester, England. 732 pp.

Lal, R. 1989. Conservation Tillage for Sustainable Agriculture: Tropics versus Temperate Environments. *Advances in Agron*. 42. London/New York. Academic Press.

Moraes Sa, J.C. 1993. *Manejo da Fertilidade do Solo no Plantio Direto*. Fundacao SBC, Castro, Parana, Brazil.

Morel, R. and P. Quantin. 1972. Observations sur l'évolution à long terme de la fertilité des sols cultivés à Grimari (République centrafricaine). Résultats d'essais de culture mécanisée semi-intensive, sur des sols rouges ferrallitiques moyennement désaturés en climat soudano-guinéen d'Afrique centrale. *Agron. Trop*. 27(6-7):667-739.

Pichot, J., M.P. Sedogo, J.F. Poulain, and J. Arrivets. 1981. Evolution de la fertilité d'un sol ferrugineux tropical sous l'influence de fumures minérales et organiques. *Agron. Trop*. 36(2):122-133.

Pieri, C. 1983. Nutrient balances in rainfed farming systems in arid and semi-arid regions. p. 181-209. In: 17th Coll. Intl. Potash Inst. Rabat, Marrakech, Maroc. 2-6 May, 1989. IPI, Berne.

Pieri, C. 1989. *Fertilite des Terres de Savanes: Bilan de trente de recherche et de developpemnt agricole au Sud du Sahara*. CIRAD, Ministere francais de cooperation et du developpement. Montpellier, Paris, France.

Pieri, C. 1992. *Fertility of Soils: A Future for Farming in the West African Savannah*. Springer Verlag, Berlin. 348 pp.

Sanchez, P.A., 1976. *Properties and Management of Soils in the Tropics*. New York. John Wiley & Sons. 618 pp.

Sanchez, P.A. and R.H. Miller. 1986. Organic matter and soil fertility management in acid soils of the tropics. In: 13th Intl. Cong. Soil Sci. Hamburg, Germany.

Srivastava, J., P. Tamboli, J. English, R. Lal. and B. Stewart. 1993. *Best Practices for Soil Moisture and Fertility Conservation*: Technologies for natural resource management, focusing on the soils of the warm seasonally dry tropics (WSDT). World Bank Technical Paper, Washington D.C.

Stoorvogel, J.J. and E.M.A. Smaling, 1990. *Assessment of soil nutrient depletion in Sub-Saharan Africa: 1983-2000*. The Winand Staring Center, Report 28, Vol. I, Wageningen. 137 pp.

Tourte, R. 1984. Introduction. In: P.J. Matlon (ed.), *Coming Full Circle: Farmers' Participation in the Development of Technology*. IDRC Publication No.189. Ottawa, Canada.

Wetselaar, R., Ganry, F., 1982. Nitrogen balance in tropical agrosystems. p. 1-36. In: Y.R. Dommerques and H.G. Diem. (eds.), *Developments in Plant and Soil Sciences*, Vol. 5: *Microbiology of tropical soils and plant productivity*. Nijhoff: Junk, The Hague.

World Resources Institute. 1992-93. *A Guide to the Global Environment*. New York, Oxford University Press. 385 pp.

# Long-Term Experiments on Alfisols and Vertisols in the Semiarid Tropics

K. B. Laryea, M. M. Anders, and P. Pathak

## I. Introduction

It is often argued that long-term field experiments (i.e., experiments whose original treatments are repeated every year and whose lifespan extend beyond 10 years) are essential to our understanding of the complex environmental, management and biotic interactions which affect sustainable agricultural production (Brown, 1991; Frye and Thomas, 1991; Pieri, 1992). However, depending on interpretation, agricultural research with a sustainability perspective may or may not imply long-term experimentation. For example, if sustainable agriculture is interpreted as the ability of an agricultural system to "maintain its productivity when subject to stress or perturbation", i.e., the resilience of the agricultural system (Conway, 1991), then experiments involving systems aimed at achieving sustainability perspectives in agriculture can be less than 5 years. In some cases, different types of stress depending on objectives (e.g., drought, disease, and/or pests) can be imposed in short-term experiments (i.e., experiments continued for less than 5 years duration) to assess the system's

ISBN 1-56670-076-0

resilience. On the other hand, if sustainable agricultural production is interpreted either in terms of intergenerational equity especially with emphasis on stewardship or proper care and protection of the soil resource (Barker and Chapman, 1988) or in terms of the need to have continued growth in agricultural productivity while maintaining the quality of soil resource (Harrington, 1992), then agricultural research with sustainability perspectives necessarily requires long-term field experimentation to produce realistic results. Long-term experiments give trends in production and if measured, trends in the state of the resource base. These two trends in measurement are central to the assessment of sustainable agricultural systems.

A number of long-term experiments have and continue to be conducted in the semiarid tropics (SAT) which are defined as the regions of the world where rainfall exceeds potential evaporation for only two to four and half months (dry semiarid) or four and half to seven months (wet semiarid) in any particular year (Troll, 1965). Its extent is estimated to be about 19.6 million km$^2$ (Ryan et al., 1974) and is comprised of all or part of 48 countries located in Africa, Asia, Australia, and Latin America. It is inhabited by over 800 million people who rely mostly on rainfall that has high variability in onset and distribution, for their livelihood. Most of the soils in the SAT are infertile, fragile and structurally unstable when wet. Using the Soil Survey Staff (1975) soil classification, the soils occupying about 80% of the arable land area in the SAT are Oxisols, Alfisols (15.6%), Entisols (12.6%), Ultisols (10.6%), and Vertisols (7.3%) (Swindale, 1982).

During their existence, most of the old long-term experiments in SAT have experienced various changes (e.g., replacement of old cultivars with new improved types, changing fertilizer rates, and/or pest management) as a result of advances in technical and analytical methods relating to their objectives. For example, regression analysis, a technique which assumes that interacting effects of environmental factors can be described using empirical additive or multiplicative models, was the principal method of analyzing yield and other crop parameters ten or more years ago. This type of analysis did not adequately explain mechanisms involved in interactions between factors. Presently, in considering the interactions of weather, diseases, pests, and soil factors in crop yield analysis, mechanistic crop models which simulate many of the processes occurring in the soil, crop, and atmosphere and requiring a number of input parameters are needed. These input parameters might not have been measured in the initial stages of long-term experiments. Long-term experiments must be designed to accomodate future expansion and permit incorporation of changes as technology and objectives change.

In this paper, we review long-term experiments conducted on Alfisols and Vertisols in the SAT, focusing mainly on (a) their initial objectives and changes if any, (b) management problems associated with some of them, (c) the role modeling could play to reduce the length of these long-term experiments and (d) considerations that need to be included in the designs of such experiments to enable future changing scenarios to be incorporated.

## II. How Long Is Long-Term?

Different disciplines have different perceptions of time scales for long-term events. For example, in the field of finance, long-term capital gain will be "generated by assets held for longer than six months" while long-term bonds will constitute a "financial operation or obligation based on a considerable term and especially one of more than 10 years" (Webster's Ninth New Collegiate Dictionary, 1986). In agriculture, the classical Rothamsted manurial trials started by Lawes and Gilbert in 1843 are the world's oldest long-term experiments. In view of that, the question has often been asked as to the time scales for long-term experiments. A number of researchers have arbitrarily set periods beyond either 10 or 20 years for experiments to be classified as long-term (e.g., Frye and Thomas, 1991; Harmsen and Kelly, 1993). Monteith (1990) used the equation for standard deviation S(f) of estimated trends in an index of production X, (e.g., yield, organic carbon content) given by regression theory as:

$$S(f) = \{S(X)/X(0)\} / \{\Sigma(n^2) - (\Sigma n)^2/t\}^{0.5} \tag{1}$$

in an attempt to define time scale required for long-term experiments to assess sustainability of agricultural systems. In Equation 1, f is the change of X in one year expressed as a fraction of its initial value X(0), t is the total number of measurements, S(X) is the standard deviation in the measurement of X due to changes in the environment (assumed to be normally distributed), and n is the sequential number of measurements (i.e., 1, 2, 3, etc.) such that its values define the series 1, 2, . . . . . t. Therefore the difference in the two terms in the denominator is equal to $\{(t^3 - t)/12\}^{0.5}$ which is within 4% of $t^{1.5}/(12)^{0.5}$ when t > 3. Consequently equation 1 was rewritten as (Monteith, 1990),

$$S(f) = 3.46 \{ (S(X)/X(0)\} / t^{1.5} \tag{2}$$

Equation 2 can be manipulated to obtain an estimate of the number of measurements that will be required to obtain S(f) as a specified fraction of the trend f when the coefficient of variation (CV = S(X)/X(0)) is stipulated. Thus, to obtain a 68% chance for the true value of f to lie between 0.67 and 1.33 of the estimate, S(f) should not exceed f/3. Therefore replacing S(f) in Equation 2 by f/3 and solving for t, Monteith (1990) obtained the expression,

$$t = (10.4 \, CV/f)^{0.67} \tag{3}$$

Extending the analysis to the more general case of measurements repeated at intervals of m years where m is not necessarily an integer, Monteith (1990) showed that since the number of measurements made in y years will be equal to (y/m + 1) and that the trend measured as a change in an index of productivity over m years will be equal to mf, substitution into Equation 3 yields the expression,

$$y = m^{0.33} (10.4 \text{ CV}/f)^{0.67} - m \tag{4}$$

In order to estimate the number of years y(m) required to obtain a stipulated CV, Monteith (1990) suggests evaluating y(m)/y(1) since for values of m that are of practical interest, i.e., from 2 to 5 years, y(m)/y(1) is approximately equal to 1 + 0.15m. If for example m = 5 years, the required number of years will be 1.75 y(1) corrected to the nearest multiple of 5. Therefore suppose f = 0.05 per year and the accuracy one requires is a CV of 0.1, substitution into Equation 1 gives y(1) = 7 years so that using the relationship y(m)/y(1) = 1 + 0.15m, we obtain y(5) equals 12.25 rounded up to 15 years. Using this approach, the number of years required for a specified level of accuracy can be estimated to facilitate the design of long-term field experiments. A major assumption in Monteith's analysis is that trends from long-term experiments will be constant over the period of measurement. This may not be the case particularly in accelerating trends. The only remedy then would be to examine the mechanisms responsible for the changes of the trend with time.

## III. Long-Term Experiments on Alfisols

### A. Extent and Main Characteristics of Alfisols in SAT

There are two conflicting estimates of the extent of Alfisols in the SAT. Kampen and Burford (1980) using the soil map of the humid tropics by Aubert and Tavernier (1972) estimated that Alfisols and Oxisols are about 33.0 and 9.0% of the area of the SAT respectively. These estimates have been cited by El-Swaify et al. (1985), Stewart et al., (1991) and Ryan (1992). On the other hand, using the FAO-UNESCO (1974 to 1978) soil map of the world with a scale of 1:5,000,000, Swindale (1982) estimated that Luvisols which are equivalent to Alfisols in the Soil Taxonomy (FAO-UNESCO, 1974, Volume 1, Legend; Soil Survey Staff, 1975; Sanchez, 1976, pp. 64-69) are about 15.6% while Ferralsols (equivalent to Oxisols in the Soil Taxonomy) constitute 33.5% of the area of the SAT. The estimates by Kampen and Burford (1980) were the first approximations of the extent of the various soil orders of the Soil Taxonomy in the SAT and it is possible that their estimates are inaccurate (J.R. Burford, personal communication). Also since some of the soil orders in the FAO-UNESCO classification do not correlate very well with those in Soil Taxonomy as seen for example in Larson (1986) where Acrisols, Nitosols, and Planosols are all classified as Alfisols, it is possible that Kampen and Burford (1980) may have included some of these and some Oxisols in their estimation of the area under Alfisols in the SAT. Generally, it is accepted that Alfisols are extensive in the SAT.

Characteristically, they have red, reddish-brown or yellowish-brown colors and have clay content which increases with depth. Thus Alfisols possess an argillic horizon whose compactness may restrict movement of fluids, roots, micro- and macro-organisms. Because this soil order is prone to erosion, surface layers may have accumulation of coarse fraction resulting from selective removal of fine particles by overland flow of water. Most Alfisols under cultivation in the SAT have unstable soil structure when wet because of (a) prevailing farming practice whereby all above-ground crop residue is either removed and utilized by the farmer or grazed by animals roaming freely in the fields after harvest and (b) the inactivity of the dominant kaolinitic clay mineral.

## B. Long-Term Experiments on Alfisols in the SAT of Asia

Long-term experiments involving soil physical management effects on crop production in the SAT are very scarce. On the other hand, there are several long-term experiments in soil fertility management using organic and inorganic fertilizers in the SAT. A few of the long-term experiments in soil fertility management cover superficially some aspects of soil physical properties particularly in discussions on the role of farmyard manure on soil aggregation. A comprehensive overview of long-term fertilizer experiments in India, covering parts of the SAT and dating back to 1885 has been given by Nambiar and Abrol (1989). Limited information on the effects of soil physical properties and processes can be obtained from some of these long-term fertility experiments. Two such long-term plant nutrition experiments still in existence are the permanent plot trials on Alfisols started in Coimbatore (Tamil Nadu), India in 1909 in the case of the "Old Manurial Experiment" and 1925 for the "New Manurial Experiment". Crops were irrigated in the former experiment until 1937 after which they were grown under rainfed conditions because of scarcity of irrigation water. Since the inception of the New Manurial Experiment, crops have been grown under irrigated conditions. The main objective of these experiments has been to investigate the effects of continuous use of inorganic and organic (farmyard manure) fertilizers on yield (biomass and grain) and soil properties. Treatments under examination are, control (i.e., no manure), N, NK, NP, NPK, KP, K, P, cattle manure, and cattle manure residue. The N, P, K, and combinations thereof are applied every cropping season at 25.2 kg N ha$^{-1}$, 29.6 kg P ha$^{-1}$, and 50.3 kg K ha$^{-1}$. Cattle manure is applied in the old manurial experiments at 12.6 t ha$^{-1}$. One block of the new manurial experiment has been receiving a basal dressing of cattle manure at 2.2 t ha$^{-1}$ because of its lower fertility compared to the other block (Krishnamoorthy and Ravikumar, 1973). The crops that have been cultivated in these experiments are sorghum (*Sorghum bicolor* (L.) Moench.), pearl millet (*Pennisetum glaucum* (L.) R.Br.), foxtail millet (*Setaria italica,* Beavu.), proso millet (*Panicum milliaceum* (L.)), finger millet (*Eleusine coracana,* Gaertn.), cotton (*Gossypium hirsutum* (L.)), wheat (*Triticum aestivum* (L.)), and maize (*Zea mays* Linn.). The main drawback of these experiments is that it was not replicated, and there appears to have been no discernible crop rotations followed. Therefore, one cannot statistically compare treatments in any single year or between years. Here we note that earlier work by Cochran (1939) and cited by Brown (1991) points out that "The standards of amount of replication are of course not necessarily the same as in single annual experiments....there is a certain amount of replication provided by the results in different years." Brandt (1945) also makes a similar point to the effect that, "It is better to reduce the number of replicates so that each crop of every rotation can appear each year than not to have every crop and treatment. In fact, in some cases it might be permissible to use but one plot of each crop and treatment." In these experiments, the crop rotation followed is neither clear nor is it clear if each crop of every rotation appears each year. However, Krishnamoorthy and Ravikumar (1973) used the years in which the crops were grown as replicates to statistically analyze information on grain and straw yields.

Analysis of soil samples from the old manurial experiments after 44 years (i.e., 1916 to 1959), (Krishnamoorthy and Ravikumar, 1973), indicated that (a) there was an appreciable increase in N content in the control, NK, NPK, KP, and K treatments, and a general increase in the total P content of plots that had received P, NPK, KP, and cattle manure. It is surprising that there was an increase in N and P contents of control plots which had not been fertilized; (b) differences in soil organic carbon content between plots that received cattle manure and those which received inorganic fertilizers was not significant. Cattle manure was inferior to inorganic P fertilizers in increasing P status of Alfisol; (c) there was no appreciable difference in the various soil physical properties (e.g., bulk density, particle size distribution) at the old site during the 44-year period except that plots receiving cattle manure or NPK showed better aggregation with more than 10% of the mass of soil having aggregate size greater than 4.0 mm. The dominant aggregate sizes in all treatments were in the 2.0 to 1.0 mm and 1.0 to 0.5 mm range; and (d) the population of bacteria, fungi, actinomycetes, and azotobacter in various soil depths down to 30.5 cm was larger in all plots than the control. Maximum microbial population was observed in the 0 to 7.6 cm depth in plots which received cattle manure or phosphate fertilizers, while phosphate and potash fertilizers enhanced microbial population at 7.6- to 15-cm depth.

Another source of information on the influence of long-term fertility management (in progress since 1971) in intensive cropping on soil properties and plant nutrient balance is the All India Coordinated Research Project on Long-Term Fertilizer Experiments of the Indian Council of Agricultural Research (ICAR). Experiments are conducted at 11 centers located on different soils and agroclimatic zones with the major objectives of (a) studying the effects of continuous application of organic and inorganic forms of plant nutrients (either singly or in combination) including secondary and micronutrients (if and as required) on crop yields, and uptake in multiple cropping systems, and (b) monitoring changes in physical, chemical, and microbiological properties of soils as a result of continuous use of chemical fertilizers and manures. The experimental design included optimal NPK levels as recommended according to initial soil test values (=100% NPK), half that amount (= 50% NPK) which was designated as suboptimal, one and a half times that amount (=150% NPK as calculated using the initial soil test values) designated as super-optimal, 100% NP, 100% N, treatments with optimal NPK plus zinc, lime, farmyard manure (FYM) and a control treatment without fertilizer (ICAR, 1989). Six experimental fields were Inceptisols, three were Alfisols, and one each was located in Vertisols and Mollisols. The experiments at Bangalore (Karnataka) and Ranchi (Bihar) are on Alfisols and that at Jabalpur is on a Vertisol in the semiarid tropical regions of India. Average crop yields from plots receiving 100% NPK were about 2 to 4 times those from plots which have not been fertilized during the 16 years. There is no information on changes in soil physical characteristics from these experiments. Even though some inferences can be made on soil structure changes from the total soluble salts and the organic carbon content changes, by and large no soil physical properties and processes are considered in these experiments.

On an Alfisol at ICRISAT Center, a long-term experiment currently in its sixth year, compares the effects of tillage (i.e., no-till, 10-cm deep, and 20-cm deep tillage), amendment (i.e., bare soil, rice straw mulch applied at 5 t ha$^{-1}$ yr-1, and farmyard manure at 15 t ha$^{-1}$ yr$^{-1}$), and perennial species (e.g., perennial pigeonpea, *Cenchrus ciliaris,* and *Stylosanthes hamata* alone or in combination). Results so far indicate that straw mulch and farmyard manure consistently reduced runoff compared with bare plots. For example, averaged over 1991 season, reductions in runoff from straw- and farmyard manure-amended plots compared to the bare plots were 70% and 50% respectively (Resource Management Program Quarterly Report, January – March, 1992). Tillage produced variable responses in that for a short time after tillage (approximately 6 weeks) there were reductions in runoff from plots which had been tilled (10- and 20-cm deep tillage) compared to the untilled plots but thereafter tilled plots had more runoff than no-tilled plots during the remainder of the cropping season (Smith et al., 1992).

## C. Long-Term Experiments on Alfisols in Sub-Saharan SAT

A number of long-term studies conducted by French scientists in francophone West Africa are reported by Pieri (1992). As indicated in Appendix III.2. of Pieri (1992, page 181) the main objectives of most of these experiments were related to soil fertility. A common cropping system prevalent in West Africa and some parts of Asia is "shifting cultivation" which involves a cultivation phase followed by a fallow period. The latter commences when perceptible yield decline during the cultivation phase is observed by the farmer. As long as the fallow period is long enough to permit soil fertility recovery, this system is quite sustainable. However, population growth in these regions has resulted in shortening of the fallow period and intensification of farming. Experiments were therefore conducted by the Institut de Recherche pour les Huiles et Oléagineux (IRHO) on Alfisols in Darou (Senegal) from 1956 to 1984 and in Niangoloko (Burkina Faso) from 1961 to 1970 to obtain long-term yield trends in shifting

cultivation systems in order to estimate the optimum length of the fallow period
(Peiri, 1992). In Darou, three fallow periods (2, 3, and 6 years) proceeding
circumstances wherein the vegetation on the land was (a) burned just before
cultivation, (b) burned every year, (c) cleared and composted every year, and
(d) cleared and left in the field as mulch each year, were studied using a rotation
of two years groundnut (*Arachis hypogeae* (L.)) and one year millet or sorghum.
In this experiment, the compost was applied every year either before cropping
or to fallow plot from which it originated. Field plots were split and half served
as control while the other half received fertilizer at the rate of 8 kg N ha$^{-1}$ as
sulphate of ammonia together with 30 kg P$_2$O$_5$ ha$^{-1}$ as dicalcium phosphate in
the case of the first year groundnut and 8 kg N ha$^{-1}$, 15 kg P$_2$O$_5$ ha$^{-1}$, and 12 kg
K$_2$O ha$^{-1}$ to the cereal and groundnut in the second and third years. In
Niangoloko the fallow periods were similar to those at Darou, but the
experiment was continued for 10 years (1961 - 1970) after an initial cropping
phase to stabilize the site. The 2- and 4- year fallow periods were followed by
millet and groundnut mixture planted on ridges without and with fertilizer at 15
kg P$_2$O$_5$ ha$^{-1}$ and 25 kg N ha$^{-1}$ in two split applications to the millet. A 5-year
moving average of yields obtained over 16 years at Darou showed a decrease
in yield with similar rate of yield decline for both the 2- and the 6- year fallow
periods indicating that the latter was not superior to the former fallow as one
would expect. In the 6-year fallow treatments, the percent annual groundnut
yield reduction for the control plots (i.e., without fertilizer) was 3.0, while that
for the fertilized plots was 4.8. The moving average for rainfall during the
period under consideration showed that rainfall did not decrease. Yields at
Niangoloko were stable for the first 5 years and then declined. As expected,
yields in fertilized plots were consistently higher than in control plots (Pieri,
1992).

Measurements of the effect of different crop canopies and management on
runoff and erosion on Alfisols in Sefa (Senegal) averaged over thirty-two years
(1954 to 1986), indicate that under different crop canopies and management,
between 18 to 34% of rainfall was lost as runoff causing erosion of about 8 t ha$^{-}$
$^1$ yr$^{-1}$. Assuming a bulk density of 1.5 Mg m$^{-3}$ for these soils, it will take about
100 years for this rate of erosion to remove about 5 cm of topsoil. However, the
impact of erosion in terms of yield reduction of crops because of removal of
plant nutrients from the infertile soils can be remarkable. Though it can be
argued that the loss of plant nutrient can be replaced through fertilizer
application, resource-poor farmers in the region cannot afford it. Also, in a
region where rainfall is scarce and erratic, a loss of 18 to 34% of it as runoff
has tremendous consequences on the agriculture in that region. Pieri (1992)
indicated that plowing Alfisols in Senegal increased millet, sorghum, groundnut,
and cotton yields by about 20%. Yield increase in maize was as high as 50%
and that of upland rice (*Oryza sativa* (L.)) over 100%. Yield increases due to
plowing were ascribed to its resultant improvement in rooting of crops and
increased profile water storage.

# IV. Long-Term Experiments on Vertisols

## A. Extent and Main Characteristics of Vertisols in the SAT

Vertisols are deep black soils which, by definition, contain more than 30% clay. Their primary diagnostic features are swelling upon wetting and development of deep, wide cracks when dry. This swelling and shrinking process results in profile inversion over time. The wide cracks are also responsible for the high initial infiltration of about 100 mm $h^{-1}$, the rates in saturated soils are extremely low (about 0.2 mm $h^{-1}$) (El-Swaify et al., 1985). They have a large water holding capacity. For example, Russell (1980) estimated the average field capacity of a 1.85-m profile at ICRISAT Center to be 810 mm, the lower limit of plant available water to be 590 mm, so that the profile holds 220 mm of plant available water. Its smectite content imparts a considerable activity to these soils. Thus the cation exchange capacity (CEC) of Vertisols in India ranges between 47 and 65 cmol $kg^{-1}$ (Ray and Barde, 1962). Calcium is usually the dominant cation on the exchange complex and may constitute 52 to 85% of the total exchangeable ions. Vertisols occupy about 7.3% of the SAT and occur in large areas of central India, northern Australia, Ethiopia, Sudan, and in scattered areas throughout eastern, central, and sub-Saharan Africa, particularly in Chad.

## B. Long-Term Experiments on Vertisols in SAT of Asia

Due to unreliability of rainfall and the risk aversion of farmers in most rainfall zones of the SAT, Vertisols in those zones are traditionally fallowed in the rainy season. During the fallowing period the land is occasionally harrowed to control weeds (Krantz and Russell, 1971; Binswanger et al., 1980). Crops are grown in post-rainy season on the stored moisture in the soil profile. In high rainfall zones (average annual rainfall greater than 1200 mm) rainy season fallowing may be practised because cropping is risky from the standpoint of waterlogging, and occurrence of crop pests and diseases. Furthermore, difficulties encountered in preparing the hard clay soil prior to the commencement of rains or sticky wet soil after its onset are some of the reasons for rainy season fallow in high rainfall zones (Michaels, 1982). In these traditional crop production systems, about 28% of total rainfall has been found to be lost as runoff, 24% as evaporation from the bare soil during the fallow period, and 9% as deep percolation, with only 39% utilized by a post-rainy season crop as evapo-transpiration (Pathak et al., 1987).

In an attempt to develop an improved crop production system that enables crops to be grown in both the rainy and the postrainy seasons, and improves rainwater-use efficiency, a long-term experiment was initiated on the Vertisols at the International Crops Research Institute for the Semi-Arid Tropics (ICRIS-AT) in 1976. The initial objectives of the experiment were to (a) study alternative cropping systems on a realistic area of land (> 3.5 ha) in order to evaluate the production effects and water utilization patterns of the different crop and land management systems; and (b) obtain quantitative information on all components of the water balance and other issues related to resource development and management. Subsequently, these laudable objectives, even though broad, have been modified to reflect and take into account new knowledge emanating from the world scientific community.

The experiment involved a comparison between an "improved" package of soil and crop management practices and the traditional fallow system. The improved package of soil and crop management included (a) land preparation immediately after harvesting the previous season's crop when there is some moisture in the soil to facilitate tillage; (b) tillage operations with improved animal drawn wheel tool carrier; (c) dry seeding a few days before the onset of the rainy season; (d) a land configuration called the broadbed and furrow (BBF) system involving graded wide beds separated by furrows which drain into grassed waterways; (e) high yielding varieties (HYVs); (f) improved plant protection; (g) basal application of 12 kg P ha$^{-1}$ as diammonium phosphate and 75 kg ha$^{-1}$ of either sulphate of ammonia (18% N) or urea (46% N) in addition to 67 kg N ha$^{-1}$ top dressed 3 weeks after planting pearl millet or 87 kg N ha$^{-1}$ top dressed between sorghum rows; and (h) growing crops in both the rainy and post-rainy seasons either as intercrops, (e.g., sorghum/pigeonpea or maize/pigeonpea) or as a sequential crop (e.g., sorghum + chickpea (*Cicer arietinum* (L.)) or maize + chickpea) where one of the components is harvested earlier so that the other (i.e., pigeonpea in the case of the intercrop or chickpea in the sequential cropping) continues in the post-rainy season to utilize the remainder of the stored water in the soil. For optimum performance in improving surface drainage while minimizing at the same time runoff and soil loss, the BBF system with dimensions of 100-cm wide beds and 50-cm wide furrows is usually laid on a slope of 0.4 to 0.8%.

The traditional system included (a) fallowing in the rainy season and growing a crop on flat seedbed during the postrainy period; (b) applying farmyard manure in alternate years at about 10 t ha$^{-1}$, i.e., 50 cartloads of dry farmyard manure ha$^{-1}$; (c) land preparation with locally manufactured wooden or iron plow (desi plow) consisting of a stick with a hardened point after the rainfall softens the soil; and (d) one weeding done manually. Insect control was similar on both the improved package and the traditional production systems.

Total productivity from the improved package for any year was substantial (Tables 1 and 2), averaged over the period a particular crop was grown, the yields were: 3.23 t ha$^{-1}$ for maize and 1.16 t ha$^{-1}$ for chickpea (averaged over 10 years) in a maize+chickpea sequential cropping system; 3.34 t ha$^{-1}$ for maize and 1.29 t ha$^{-1}$ for safflower (*Carthamus tinctorius* (L.)) (averaged over 5 years) for maize+safflower sequential system; 2.60 t ha$^{-1}$ for maize and 1.10 t ha$^{-1}$ for pigeonpea (averaged over 9 years) in maize/pigeonpea intercropping system; and 3.27 t ha$^{-1}$ for sorghum and 0.70 t ha$^{-1}$ for pigeonpea (averaged over 7 years) in a sorghum/pigeonpea intercropping system. Corresponding yields for the traditional system were 0.72 t ha$^{-1}$ for sorghum (averaged over 12 years),

**Table 1.** Grain yields (t ha⁻¹) under improved crop production system on Vertisol at ICRISAT (1976-1988)

| Year | Sequential cropping | | | | Intercropping | | | |
|---|---|---|---|---|---|---|---|---|
| | Maize + Chickpea | | Maize + Sunflower | | Maize / Pigeonpea | | Sorghum / Pigeonpea | |
| 1976-77 | 3.12 | 0.65 | n.c.ᵇ | n.c. | 3.29 | 0.78 | n.c. | n.c. |
| 1977-78 | 3.34 | 1.13 | n.c. | n.c. | 2.81 | 1.32 | n.c. | n.c. |
| 1978-79 | 2.15 | 1.34 | n.c. | n.c. | 2.14 | 1.17 | n.c. | n.c. |
| 1979-80 | 3.03 | 0.59 | n.c. | n.c. | 1.95 | 0.89 | n.c. | n.c. |
| 1980-81 | 4.19 | 0.79 | n.c. | n.c. | 2.92 | 0.97 | n.c. | n.c. |
| 1981-82 | 3.27 | 1.35 | 3.24 | 0.98 | 2.70 | 0.98 | 4.18 | 0.68 |
| 1982-83 | 3.43 | 1.43 | 3.38 | 1.31 | 2.76 | 1.12 | 3.61 | 0.66 |
| 1983-84 | 3.22 | 2.17 | 3.19 | 1.55 | 2.24 | 1.81 | 1.06 | 1.92 |
| 1984-85 | 2.32 | 1.38 | 2.64 | 1.50 | 2.58 | 0.82 | 3.94 | 0.78 |
| 1985-86 | n.a.ᵃ | n.a. | 1.48 | 0.49 | n.a. | n.a. | 1.72 | 0.12 |
| 1986-87 | n.a. | n.a. | 3.27 | 0.00 | n.a. | n.a. | 4.45 | 0.38 |
| 1987-88 | 4.26 | 0.80 | 4.26 | 1.10 | n.c. | n.c. | 3.92 | 0.35 |
| 1988-89 | 4.71 | 0.65 | 4.71 | 0.55 | n.c. | n.c. | 3.23 | 1.20 |

ᵃn.a. = not available; ᵇn.c. = not cropped.

**Table 2.** Grain yields (t ha$^{-1}$) under traditional crop production system on Vertisols at ICRISAT (1976-1988)

| Year | Sorghum | Chickpea | Safflower |
|---|---|---|---|
| 1976-77 | 0.66 | 0.52 | n.c.[a] |
| 1977-78 | 0.37 | 0.76 | n.c. |
| 1978-79 | 0.50 | 0.45 | n.c. |
| 1979-80 | 0.76 | 0.66 | n.c. |
| 1980-81 | 0.60 | 0.56 | n.c. |
| 1981-82 | 1.55 | 1.23 | 1.10 |
| 1982-83 | 1.22 | 1.38 | 1.29 |
| 1983-84 | 0.97 | 0.60 | 0.48 |
| 1984-85 | 2.60 | 1.34 | 1.59 |
| 1985-86 | 0.76 | 0.84 | 0.60 |
| 1986-87 | 0.37 | 1.27 | 1.06 |
| 1987-88 | 0.80 | 0.92 | 0.20 |
| 1988-89 | 0.65 | 1.15 | 0.98 |

[a]n.c. = not cropped.

0.85 t ha$^{-1}$ for chickpea (averaged over 12 years), and 0.75 t ha$^{-1}$ for safflower (averaged over 7 years).

Runoff and soil loss were monitored in this experiment for 11 years for the whole watershed containing a number of the improved systems and for the watershed having the traditional system. Even though ideally more information would have been obtained if runoff and soil loss had been monitored from the individual plots having specific cropping system within one watershed, Table 3 (ICRISAT, 1988) clearly indicates that together, the improved systems in the watershed reduced runoff and soil loss.

However, confounding factors in the design of the experiment, changes in management practices, changes in genotypes, and the absence of a discernible crop rotation do not allow for quantitative evaluation of the improved package vis-a-vis the traditional system in terms of single or combined components in order to ascertain their contribution to yield improvement and/or interactions. For example, in the quest to obtain a management system superior to that used by farmers in the region, fertilizer rates on the maize+chickpea sequential system was changed from 125 kg diammonium phosphate (DAP) ha$^{-1}$ plus 375 kg ammonium sulphate ($(NH_4)_2SO_4$) ha$^{-1}$ in 1975, to 100 kg DAP ha$^{-1}$ plus 400 kg $(NH_4)_2SO_4$ ha$^{-1}$ in 1976, 75 kg DAP ha$^{-1}$ plus 330 kg $(NH_4)_2SO_4$ ha$^{-1}$ in 1977, etc. The fertilizer rates were changed every year until 1981. Between 1981 and 1987, the same rate of fertilizer was applied, but in 1988 and thereafter a rate of 100 kg 28:28:0 (N, $P_2O_5$, $K_2O$) ha$^{-1}$ plus 70 kg urea ha$^{-1}$ were applied each year. With the exception of the traditional system where the management was constant, fertilizer rates and cultivars in the improved system changed every year until 1989 when the experiment was restructured. Similarly,

**Table 3.** Runoff and soil loss under improved and traditional systems on Vertisols in 11 successive years at ICRISAT (1976-1987)

| Year | Cropping period rainfall[a] (mm) | Improved system (double cropping) | | Traditional system (single crop) | |
|---|---|---|---|---|---|
| | | Runoff (mm) | Soil loss (t ha$^{-1}$) | Runoff (mm) | Soil loss (t ha$^{-1}$) |
| 1976-77 | 708 | 73 | 0.98 | 238 | 9.20 |
| 1977-78 | 616 | 1 | 0.07 | 53 | 1.68 |
| 1978-79 | 1089 | 273 | 2.93 | 410 | 9.69 |
| 1979-80 | 715 | 73 | 0.70 | 202 | 9.47 |
| 1980-81 | 751 | 116 | 0.97 | 166 | 4.58 |
| 1981-82 | 1073 | 332 | 5.04 | 435 | 11.01 |
| 1982-83 | 667 | 10 | 0.20 | 20 | 0.70 |
| 1983-84 | 1045 | 154 | 0.80 | 289 | 4.70 |
| 1984-85 | 546 | 11 | N[b] | 75 | N |
| 1985-86 | 477 | 4 | N | 18 | N |
| 1986-87 | 538 | 35 | N | 114 | N |
| Mean | 748 | 98 | 1.46 | 184 | 6.38 |

[a] Average rainfall for Hyderabad (29 km from ICRISAT Center) based on 1901-1984 data is 784 mm with a CV of 27%.
[b] Measurements were not taken.
(From ICRISAT, 1988.)

**Table 4.** Summary of revised treatments and management for the improved system[a]

| Treatement | Rotation | Crops | Management |
|---|---|---|---|
| 1 | 1 | fallow + sorghum[b] | All on |
| 2 | 1 | fallow + chickpea[b] | broadbeds and |
| 3 | 2 | fallow + sorghum[c] | furrows, |
| 4 | 2 | fallow + chickpea[c] | 50% fertilizer |
| 5 | 3 | maize + chickpea[c] | 50% farmyard |
| 6 | 3 | maize + safflower | manure |
| 7 | 4 | sorghum / pigeonpea | |
| 8 | 4 | G. gram + sunflower | |
| 9 | 5 | soybean / pigeonpea | |
| 10 | 5 | millet + safflower | |
| 11 | 6 | millet + chickpea[c] | |
| 12 | 6 | G. gram + safflower | |

[a]Design is split-split plot; [b]traditional cultivar; [c]improved cultivar.

**Table 5.** Summary of revised treatments and management for the traditional system[a]

| Treatement | Rotation | Crops | Management |
|---|---|---|---|
| 1 | 1 | fallow + sorghum[b] | All on flat |
| 2 | 1 | fallow + chickpea[b] | seedbed, |
| 3 | 2 | fallow + sorghum[c] | 50% fertilizer, |
| 4 | 2 | fallow + chickpea[c] | 50% farmyard |
| 7 | 4 | sorghum / pigeonpea | manure |
| 8 | 4 | G. gram + sunflower | |
| 11 | 6 | millet + chickpea[c] | |
| 12 | 6 | G. gram + safflower | |

[a]Design is split-split plot; [b]traditional cultivar; [c]improved cultivar.

the maize genotype in this sequential system changed from var. DH101 in 1975-1976 and 1979-1983 to var. SB23 in 1978, and var. DH103 in 1984-1988. Furthermore, the gradient of 0.4 to 0.8% of the furrows in the BBF may be high in low-rainfall environments where lower gradients (i.e., 0.2 to 0.4%) may be suitable. Hence the objectives that were initially set were partially achieved.

In 1989, this experiment was redesigned (Tables 4 and 5) in order to permit statistical analysis of the individual components of the package and to ensure that each cropping phase of a rotation appears every year. The main treatments in the modified experiment are flat seedbed and BBF with cropping systems rotations and all cultural operations similar in the two landforms. Each landform has both organic (farmyard manure at 10 t ha$^{-1}$ dry weight every two years) and basal inorganic (diammonium phosphate at 100 kg ha$^{-1}$) fertilizer applied to all cereals and legumes in the subplot in addition to a topdressing of 46 kg N as urea ha$^{-1}$ for the cereals. The treatments are replicated three times in a split-split plot design, and there is emphasis on monitoring soil biological, chemical, and physical changes as well as above-ground plant parameters like crop canopy and crop growth. In this revised experiment, another set of the improved system under supplemental irrigation has been included in order to compare its performance with that under rainfed conditions. This way we hope to obtain in a reasonable length of time a comparison of BBF versus flat seedbed, high input (i.e., NPK) versus low input (FYM), irrigated versus rainfed on the BBF system, and improved versus traditional cultivars and management.

Analysis of 3 years (1989 - 1991) measurements (Figures 1 and 2) from this revised experiment shows some interesting results. Comparisons of the high and low fertility (i.e., farmyard manure versus NPK), irrigated versus rainfed, broadbed and furrow (BBF) versus flat seedbed, and improved versus traditional sorghum genotypes for that period indicate (Report of M.M. Anders, Resource Management Program, 1991), that (a) sorghum grain yields of both traditional (M 35-1) and improved (CSH 9) genotypes grown on all plots (except flat seedbed in 1989) were significantly ($P < 0.05$) higher than those receiving farmyard manure; (b) of the 18 comparisons between improved and traditional genotypes, the traditional sorghum grain yield was equal to or greater than the improved sorghum grain yield in 10 cases; and (c) the improved sorghum grain yield was higher than the yield for the traditional genotype only on the flat seedbed. On BBF, the traditional genotype has so far performed better than the improved sorghum genotype. An interesting observation from Figures 1 and 2 is that because of the fertility management history of the BBF plots (i.e., the area had had significant fertilizer applications for the preceding 15 years) sorghum grain yield which was significantly higher on BBF than the flat seedbed in 1989, had declined by 1991 resulting in very little or insignificant difference in traditional genotype grain yields between the BBF and the flat seedbed. With the improved sorghum genotype the difference in grain yield between BBF and flat seedbed in 1989 had reversed — the grain yield being significantly higher for flat seedbed than for the BBF under both rainfed and irrigated conditions in 1991. Similar results were obtained for postrainy season traditional (Annigeri) and improved (ICCC 37) chickpea genotypes (Resource Management Program Quaterly Technical Report, July – September, 1992, p. 6).

Khiani and More (1984) report on a tillage and manurial experiment that has been in progress on a Vertisol since 1932 at the Agricultural College farm at Pune (Maharashtra). The objective of the experiment was to examine the long-term effects of deep and shallow tillage with and without farmyard manure on soil properties and on yields of cotton and sorghum grown in rotation continuously for the preceding 45 years under rainfed conditions. The experiment has four treatment combinations, viz., (a) shallow tillage up to 10-cm depth and manuring with 6.2 t ha$^{-1}$ farmyard manure, (b) deep tillage up to 20-cm depth and application of 6.2 t ha$^{-1}$ farmyard manure, (c) shallow tillage, (d) deep

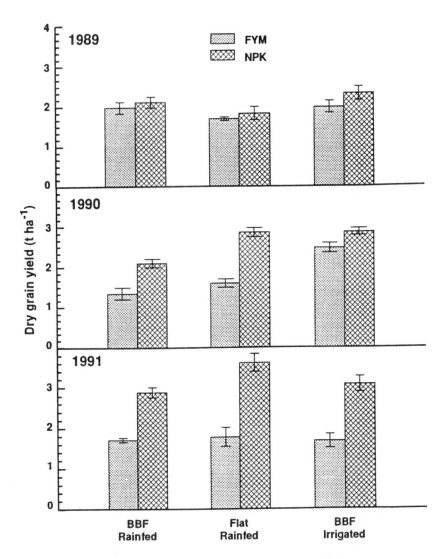

**Figure 1.** Dry grain yield for an improved genotype (CSH 9) grown in two fertility levels (FYM, NPK), two land forms (BBF, flat) with and without irrigation.

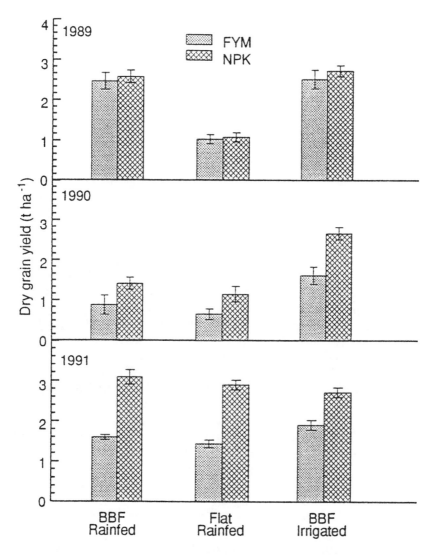

**Figure 2.** Dry grain yield for a traditional genotype (M 35-1) grown in two fertility levels (FYM, NPK), two land forms (BBF, flat) with and without irrigation.

tillage with five replications laid out in a randomised block design. Their soil physical and chemical determinations made after 45 years of experimentation indicated that (a) the water holding capacity and water content at 0.33 bar suction were significantly increased by the continual addition of farmyard manure, (b) deep tillage helped to improve the level of available P and K. It also increased the grain and fodder yield of postrainy season sorghum. However, it did not have any effect on seed cotton, possibly due to the tap root system of cotton which facilitated extraction of nutrients from deeper soil layers in all the treatments.

Soil nitrogen and phosphorus are the major plant nutrients usually lacking in Vertisols in the SAT. Because resource-poor farmers cannot afford the cost of inorganic fertilizers, a long-term experiment was initiated in 1983 on Vertisols at ICRISAT to examine the effectiveness of different types and proportions of grain legumes in different cropping system rotations that will maintain crop productivity and soil nitrogen levels under rainfed conditions. In its tenth year now, a grain legume intercropping system (cowpea/pigeonpea) has consistently benefited the subsequent sorghum crop to the level equivalent to 40 kg fertilizer N $ha^{-1}$ . The benefits from a cereal/grain legume intercrop system (sorghum/pigeonpea) or cereal+grain legume sequential system (sorghum + chickpea) grown in the preceding year was equivalent to 25 and 10 kg fertilizer N $ha^{-1}$ respectively. Pigeonpea had substantial effects on the surface soil (0- to 15-cm depth) resulting in a continuous sorghum/pigeonpea system increasing soil N content by 140 $\mu$g $g^{-1}$ while the soil nitrogen in a continuous sorghum+chickpea system increased by only 25 $\mu$g $g^{-1}$ in 8 years (Table 6). In contrast a continuous nonlegume system (sorghum+safflower) resulted in a loss of 25 $\mu$g N $g^{-1}$ of soil, indicating unambiguously the long-term benefits to both soil and nonlegumes from inclusion of grain legumes, e.g., pigeonpea, cowpea (*Vigna unguindata* (L.), chickpea, in cropping systems rotation (Rego and Burford, 1992).

## C. Long-Term Experiments on Vertisols in the Australian SAT

In the semiarid tropical region of Queensland, Australia, a long-term experiment was initiated at the Hermitage Research Station (28° 12'S and 152° 06'E) on a fine textured Vertisol (650 g clay $kg^{-1}$ soil) in 1969, to investigate the effects of tillage and stubble management practices on wheat yields, N fertilizer requirements and soil properties (Marley and Littler, 1989). The 12 treatments applied every year in four randomised blocks are factorial combinations of either conventional or no-tillage, stubble burnt or retained, and 0, 23, and 46 kg N $ha^{-1}$ $yr^{-1}$ (69 kg N $ha^{-1}$ $yr^{-1}$ since 1977). In examining the effects of these treatments on organic carbon content, total N, mineralizable N, pH, electrical conductivity, chloride, exchangeable sodium percentage (ESP), and aggregation index (i.e., undispersed fraction <20 micrometers of silt and clay) from measurements obtained after 13 years of the experiment, Dalal (1989) and also Thompson

**Table 6.** Total N content ($\mu$g g$^{-1}$) of the 0-15 cm depth of soil under different cropping systems

| Rotation | 1983 | 1991 | Change |
|---|---|---|---|
| Legumes every year | | | |
| S/PP  S/PP | 546 | 686 | 140 |
| S/PP  S + CP | 566 | 635 | 69 |
| S + CP  S + CP | 536 | 561 | 25 |
| | | | |
| Legumes in alternate years | | | |
| C/PP  S + SF | 543 | 640 | 97 |
| S/PP  S + SF | 559 | 610 | 51 |
| S + CP  S + SF | 540 | 548 | 8 |
| | | | |
| No legume | | | |
| S + SF  S + SF | 537 | 513 | -24 |
| | | | |
| Standard error $\pm$ | | 19.9 | 18.0 |

S = sorghum; PP = pigeonpea; CP = chickpea; SF = safflower; + = sequential; / = intercrop.
(From T.J. Rego (personal communication) and Resource Management Program, 1992.)

(1992) found that (a) plots having a combination of no-tillage, crop residue retained, and fertilizer N applied had highest concentrations of organic carbon and total nitrogen in the surface soil (0 to 10 cm); (b) soil pH from soil surface to 30-cm depth and the electrical conductivity to 120-cm depth were significantly lower under no-tillage than under conventional tillage; (c) the aggregation index of the surface soil (0 to 10 cm) was higher under no-tillage than conventional tillage; (d) the exchangeable sodium percentage (ESP) of the surface soil (0 to 4.0 cm) was lowest in plots having no-tillage with crop residue retained; (e) no-tillage plots contained less NaCl-equivalent salts (0.8 Mg ha$^{-1}$) compared with conventionally-tilled plots in which crop residues were burned (7.3 Mg ha$^{-1}$); (f) available water and nitrate were lower in the surface soil and greater at depths (60 to 120 cm) under no-tillage than in tilled soil; (g) numbers of root-lesion nematodes (Pratylenchus thornei Sher and Allen) and stunt nematodes (Merlinius brevidens Siddiqi) in the soil after a wheat crop were substantially greater in no-tilled than in the conventionally-tilled plots. Earthworm numbers were increased by stubble retention when combined with no-tillage. Thus tillage and crop residue management affect organic matter and, therefore, microbial activity in the surface layers. Tillage and crop residue management also affect water relations and salt movement in the profile of Vertisols to at least 120-cm depth.

After 22 years of the same experiment, further analysis of soil samples indicate (Dalal, 1992) that (a) total soil nitrogen in 0 to 10 cm depth declined under all treatments at an overall rate of $25 \pm 2$ kg N ha$^{-1}$ yr$^{-1}$ although the soil under zero-tillage, stubble retention, and in the 69 kg ha$^{-1}$ yr$^{-1}$ still contained higher total soil nitrogen than the soil under the conventional tillage, burned stubble or the 23 kg N ha$^{-1}$ yr$^{-1}$ treatments; (b) apparent fertilizer N recovery in the soil-plant system was poor (34 to 64%) under the conventional tillage-burned stubble, conventional tillage-stubble retention, zero-tillage-burned stubble, and zero-tillage-stubble retention treatment combinations, because N removed by the wheat crop was equivalent to less than 20% of fertilizer N in the first 12 years of management practices as a result of crop diseases; (c) under current cultural practices, total soil N in the 0- to 10-cm depth may decline further to a steady state of about 1000 kg ha$^{-1}$. Disease control and optimum utilization of soil water to minimize $NO_3$ - N losses through leaching will enhance the apparent nitrogen recovery. The effect of treatments on soil physical properties and processes have not yet been reported.

## V. The Role of Modelling in Long-Term Experiments

Crop performance and ultimately crop yield are normally affected by soil management because of its overriding influence on soil processes and properties. Even though long-term experiments integrate weather variations, fluctuations of output caused by weather are common and the results are at best, location specific and may apply only to the management and the specific agroecology in which measurements were made. Therefore, generalization and extrapolation of such results to other locations may be tenuous (Hadas et al., 1988; Laryea et al., 1990).Though long-term field research has merits in studies involving management (of both crop and soil), effects on soil processes, diseases and pests and therefore on crop production, they are expensive and time consuming, since a period of a decade or more may be required for definite and reliable conclusions to be made (Monteith, 1990). Furthermore such field research involves the cooperation of different disciplines in soil and crop sciences and usually a large number of variables and measurements. Consequently, field designs, layout and execution on a large scale to accomodate all measurements by different disciplines become very difficult and sometimes impractical. In such situations, computer simulations of management effects on soil-crop systems can be used to test and select the different management combinations, provide guidelines on parameters to measure and site characteristics that are important. Thereafter, long-term effects of the important management combinations can be predicted using the models together with the necessary stochastic data bases.

A number of models can be used for preliminary screening of treatments in management studies and for long-term predictions of management effects on soil properties and processes and crop production. Notable among simulation models are the Decision Support System for Agrotechnology Transfer (DSSAT) (IBSNAT, 1989), the Erosion Productivity Impact Calculator (EPIC) (Williams et al., 1984) and the Productivity, Erosion, Runoff Functions to Evaluate

Conservation Techniques (PERFECT) (Littleboy et al., 1989). These and others will simulate the effect of different soil and crop managements on yield and soil processes. They, however, may need to be validated with information from some of the long-term experiments already in existence.

# VI. Problems Associated with Long-Term Experiments in the SAT

As indicated by Monteith (1990), long-term trials are usually designed to (a) track changes of yield from year to year in response to weather superimposed on changes related to fertility, damage by pests, etc., and (b) measure of factors likely to be responsible for these changes, e.g., soil pH, infiltration rates, or nematode populations. For a field study to be considered as long-term, primary treatments must be maintained for some arbitrary length of time (at least longer than 10 years) that will be accepted as long-term by most agricultural researchers (Monteith, 1990; Frye and Thomas, 1991). Most experiments on Alfisols and Vertisols in the SAT that we examined do not meet this important criterion. For most, the objectives, primary treatments, and management changed as and when the experiment fell under the administrative or management control of new scientists and this is frequent in most countries located in the SAT. Even in the International Agricultural Research Institutes, this occurrence is frequent because of the high turnover of research scientists and also the fact that most donors wish to see their funds translated into a tangible technology that is increasing the productivity of resource-poor farmers in about five years. Because of frequent change of primary objectives, treatments, and management, all that has occurred is a series of short-term experiments conducted on the same site for a long period of time. As indicated by Frye and Thomas (1991) the three most important considerations that will enable the primary objectives (i.e., good measurements that can be statistically analyzed, interpreted, and used to test hypothesis or infer scientific principles) of long-term experiments to be realised are (a) the statistical design and analysis to be used, (b) selection of the site, and (c) the probability of future expansion of the experiment. These considerations are unfortunately lacking in most of the experiments we examined. It is important that the design for a long-term experiment should lend itself to modification of some of the secondary objectives without invalidating its long-term goals. This may be possible if the plot size is so large that a split plot design can be used to add more treatments in case this is required. Major restrictions to large plot size are the operational costs and spatial variability in soil properties. The latter can be examined and mapped using kriging methods available in geostatistics, in order to facilitate the site selection. Lastly, it is necessary that management practices such as tillage system and cropping system, which may produce drastic effects if changed, remain unaltered throughout the duration of the experiment.

## VII. Conclusions and Recommendations

We note that in the SAT, there is a paucity of long-term experiments on changes in soil properties and processes as well as crop productivity as a result of imposition of treatments involving soil and crop management options. Most of the few long-term experiments on Alfisols and Vertisols in the SAT have experienced changes in the primary objectives, design, and management during their existence. Those that have not, were not designed to facilitate statistical analysis and interpretation of within and between-year variations because they were not replicated. Long-term implies that primary objectives, treatments, and management are not changed during the period under consideration. Otherwise, it becomes a series of many short-term experiments conducted on the same site for long periods of time. The lack of long-term experiments may be attributed to the cost of running these studies over long periods, and lack of a commitment on the part of both scientists and donors because they want to have new technologies transferred to farmers as quickly as possible. With the advent of powerful computers and sophisticated models, it should be possible to simulate long-term effects of soil and crop management systems using results from short-term experiments and initial soil properties. The output from such models can then be validated using the results from existing long-term experiments in which appropriate measurements have been made, after which extrapolations to other soils and agroecological zones can be done with suitable parameters from those regions. In areas where models cannot be validated or calibrated with results from the existing long-term experiments because of the deficiencies in statistical design and appropriate measurements, computer simulation may be used to facilitate selection of treatments and measurements so that new long-term studies can be initiated.

## References

Aubert, G. and R. Tavernier. 1972. Soil Survey. p. 17-44. In: *Soils of the humid tropics*. U.S. National Academy of Sciences, Washington, D.C.

Barker, R. and D. Chapman. 1988. The economics of sustainable agricultural systems in developing countries. *Agricultural Economics Working Paper* 88-13, Cornell University, Ithaca, NY.

Binswanger, H.P., S.M. Virmani, and J. Kampen. 1980. Farming systems components for selected areas in India: Evidence from ICRISAT. *Research Bulletin no. 2*, Patancheru, A.P. 502 324, India: International Crops Research Institute for the Semi-Arid Tropics, 40 pp.

Brandt, A.E. 1945. Principles of experimental design applied to long-term rotations. *Soil Sci. Soc. Am. Proc.* 10:306-315.

Brown, J.R. 1991. Summary: Long-term field experiments symposium. *Agron. J.* 83:85.

Cochran, W.G. 1939. Long-term agricultural experiments. *J. Roy. Stat. Soc.* Supplement 6:104-140.

Conway, G.R. 1991. Sustainability in agricultural development: Tradeoff with productivity, stability, and equitability. p. 1-22. In: *Proceedings 11th Annual AFSRE/E Symposium*, 5-10 October, Michigan.

Dalal, R.C. 1989. Long-term effects of no-tillage, crop residue, and nitrogen application on properties of a Vertisol. *Soil Sci. Soc. Am. J.* 53(5):1511-1515.

Dalal, R.C. 1992. Long-term trends in total nitrogen of a Vertisol subjected to zero-tillage, nitrogen application and stubble retention. *Aust. J. Soil Res.* 30(2):223-231.

El-Swaify, S.A., P. Pathak, T.J. Rego, and S. Singh. 1985. Soil management for optimized productivity under rainfed conditions in the semi-arid tropics. *Adv. Soil Sci.* 1:1-64.

FAO-UNESCO (Food and Agricultural Organization/United Nations Educational, Scientific, and Cultural Organization). 1974-1978. Soil maps of the world. UNESCO, Paris.

Frye, W.W. and G.W. Thomas. 1991. Management of long-term field experiments. *Agron. J.* 83:38-44.

Hadas, A., W.E. Larson, and R.R. Allmaras. 1988. Advances in modeling machine-soil-plant interactions. *Soil and Tillage Res.* 11:349-372.

Harrington, L.W. 1992. Measuring sustainability: issues and alternatives. *Journal of Farming Systems Research-Extension* 3(1):1-20.

Harmsen, K. and T.G. Kelly. 1993. Natural resource management research for sustainable production. Chapter 4. In: *Report of the Joint TAC/CDC Working Group on Ecoregional Approaches to International Agricultural Research* (under preparation).

IBSNAT (International Benchmark Sites for Agrotechnology Transfer). 1989. Decision Support System for Agrotechnology Transfer (DSSAT), *DSSAT Users Guide*, version 2.1, IBSNAT Project, Department of Agronomy and Soil Science, University of Hawaii, Hawaii. 56 pp.

ICAR (Indian Council of Agricultural Research). 1989. All India Coordinated Research Project on Long-term Fertilizer Experiments. *Annual Report 1985-86 & 1986-87)*. Indian Agricultural Research Institute, New Delhi 110012, India, 152 pp.

ICRISAT (International Crops Research Institute for the Semi-Arid Tropics). 1988. *Annual Report 1987*. Patancheru, A.P. 502 324, India: ICRISAT, 390 pp.

Kampen, J. and J.R. Burford. 1980. Production systems soil-related constraints and potentials in the semi-arid tropics with special reference to India. p. 141-145. In: *Proc. of the symposium on priorities for alleviating soil-related constraints to food production in the tropics*. IRRI (International Rice Research Institute), Los Banos, Laguna, Philippines.

Khiani, K.N. and D.A. More. 1984. Long-term effect of tillage operations and farmyard manure application on soil properties and crop yield in a Vertisol. *J. Ind. Soc. Soil Sci.* 32:392-393.

Krantz, B.A. and M.B. Russell. 1971. Avenues for increased wheat production under Barani condition. Presented at Wheat Symposium, U.P. Agricultural University, Pantnagar, Uttar Pradesh, India.

Krishnamoorthy, K.K. and T.V. Ravikumar. 1973. Permanent manurial experiments conducted at Coimbatore, Tamil Nadu Agricultural University Offset and Printing Press, Coimbatore - 641003, India, 58 pp.

Larson, W.E. 1986. The adequacy of world soil resources. *Agron. J.* 78:221-225.

Laryea, K.B., J.L. Monteith, and G.D. Smith. 1990. Modeling soil physical processes and crop growth in the semi-arid tropics. p. 399-421. In: M. Eshan Akhtar and M.I. Nizami (eds.) *Proc. International Symposium on Applied Soil Physics in Stressed Environments*, 22 - 26 January 1989, Islamabad, Pakistan, Barani Agricultural Research and Development Project, Pakistan Agricultural Research Council, Islamabad, Pakistan.

Littleboy, M., D.M. Silburn, D.M. Freebairn, D.R. Woodruff, and G.L. Hammer. 1989. PERFECT. A computer simulation model of Productivity Erosion Runoff Functions to Evaluate Conservation Techniques. *Queensland Department of Primary Industries Bulletin* QB 89005, Brisbane, Australia, 119 pp.

Marley, J.M. and J.W. Littler. 1989. Winter cereal production on the Darling Downs — An 11 year study of fallowing practices. *Aust. J. Exptal. Agric.* 29:807-827.

Michaels, G.H. 1982. The determinants of kharif fallowing on the Vertisol in semi-arid tropical India. Ph.D. Thesis, University of Minnesota, MN. 191 pp.

Monteith, J.L. 1990. Can sustainability be quantified? p. 88-114. In: *The Proceedings of the first International Symposium on Natural Resources Management for a Sustainable Agriculture*, Vol. I, 6 - 10 February, 1990, New Delhi, India, Indian Society of Agronomy, New Delhi, India.

Nambiar, K.K.M. and I.P. Abrol. 1989. Long term fertilizer experiments in India: An overview. *Fert. News*, 34:11-20.

Pathak, P., S. Singh, and R. Sudi. 1987. Soil and water management alternatives for increased productivity on SAT Alfisols. p. 533-550. In: *Soil Conservation and Productivity. Proceedings of the IV International Conference on Soil Conservation*, 3 - 9 November 1985, Maracay, Venezuela, Soil Conservation Society of Venezuela.

Pieri, C.J.M.G. 1992. Fertility of Soils: A future for farming in the West African Savannah. Springer-Verlag, NY. 348 pp.

Ray, B.B. and N.K. Barde. 1962. Some characteristics of the black soils of India. *Soil Sci.* 93:142-147.

Rego, T.J. and J.R. Burford. 1992. Sustaining crop productivity on rainfed Vertisols through grain legumes. p. 289. In: *Agron. Abstr.* Annual meetings of ASA, CSSA, SSSA, CMS, 1 - 6 November 1992, Minneapolis, MN.

Resource Management Program. 1992. *Quart. Tech. Rep.* January – March. ICRISAT (International Crops Research Institute for the Semi-Arid Tropics), Patancheru, A.P. 502 324, India, 17 pp.

Russell, M.B. 1980. Profile moisture dynamics of soil in Vertisols and Alfisols. p. 75-78. In: *Agroclimatology Research Needs of the SAT: Proceedings of an International Workshop.* 22 – 24 November 1978, ICRISAT Center, Patancheru, A.P. 502 324, India.

Ryan, J.G., M. von Oppen, K.V. Subrahmanyam, and M. Asokan. 1974. Socio-economic aspects of agricultural development in the semi-arid tropics. p. 389-431. In: *Proceedings of the International Workshop on Farming Systems,* 18-21 November 1974. ICRISAT (International Crops Research Institute for the Semi-Arid Tropics), Patancheru, A.P. 502 324, India.

Ryan, J.G. 1992. Agricultural productivity and sustainability in the semi-arid tropics: Is there such a thing as a free environmental lunch? Presented on 28 August 1992 at the Panel Discussion on Sustainability and the Environment in the Semi-Arid Tropics: Food Production in the Future. On the occasion of ICRISAT's 20th Anniversary, ICRISAT Center, Patancheru, India, 18 pp.

Sanchez, P.A. 1976. *Properties and Management of Soils in the Tropics.* J. Wiley & Sons, NY. 618 pp.

Smith, G.D., K.J. Coughlan, D.F. Yule, K.B. Laryea, K.L. Srivastava, N.P. Thomas, and A.L. Cogle. 1992. Soil management options to reduce runoff and erosion on a hardsetting Alfisol in the semi-arid tropics. *Soil Tillage Res.* 25:195-215.

Soil Survey Staff. 1975. Soil Taxonomy: A basic system of soil classification for making and interpreting soil surveys. *United States Department of Agriculture (USDA) Handbook* 436, 754 pp.

Stewart, B.A., R. Lal, and S.A. El-Swaify. 1991. Sustaining the resource base of an expanding world agriculture. p. 125-144. In: R. Lal and F.J. Pierce (eds.), *Soil Management for Sustainability.* Soil and Water Conservation Society. Ankeny, IA.

Swindale, L.D. 1982. Distribution and use of arable soils in the semi-arid tropics. In: Managing Soil Resources. Plenary Session Papers. p. 67-100. *Trans. 12th Int. Cong. Soil Sci.* Indian Agricultural Research Institute. 8-16 February 1982. New Delhi 110 012, India.

Thompson, J.P. 1992. Soil biotic and biochemical factors in a long-term tillage and stubble management experiment on a Vertisol. 2. Nitrogen deficiency with zero-tillage and stubble retention. *Soil and Tillage Res.* 22:339-361.

Troll, C. 1965. Seasonal Climates of the Earth. In: E. Rodenwaldt, and H. Jusatz (eds.) *World Maps of Climatology.* Springer-Verlag, Berlin.

Webster's Ninth New Collegiate Dictionary. 1986. Merriam-Webster Inc., Springfield, Massachusetts. 1563 pp.

Williams, J.R., C.A. Jones, and P.T. Dyke. 1984. A modeling approach to determining the relationship between erosion and soil productivity. *Trans. Am. Soc. Agric. Eng.* (ASAE) 27:129-144.

# Improved Crop-Livestock Production Strategies for Sustainable Soil Management in Tropical Africa

I. Haque, J.M. Powell, and S.K. Ehui

## I. Introduction

The human population of sub-Saharan Africa is expected to reach 676 million by the turn of this century and almost double by the year 2025. Widespread economic stagnation and poverty, political unrest, and declining per capita agricultural production complicate the picture. Population growth and economic changes in the region are increasing the demand for food of animal and plant origin and exerting pressure on land and water resources (McNamara, 1992; Winrock, 1992).

As a result, current patterns of resource-base use and production practices are becoming unsustainable. Soil erosion, nutrient depletion of soils, deforestation,

ISBN 1-56670-076-0
©1995 by CRC Press, Inc.

desertification of over-cultivated and overgrazed lands, and overuse and pollution of surface and groundwater resources all contribute to the degradation of the environment (ECA/FAO, 1991; Lal, 1987; Lal, 1990). The relationship between population pressure and soil productivity is shown in Figure 1.

Most livestock (75%) are kept by agropastoralists and smallholder crop-livestock farmers. Crop-livestock systems are constrained by wide fluctuations in the quality and quantity of feed resources during the year and by the low availability of nutrients, especially nitrogen and phosphorus, in most soils which threatens the sustainability of primary production (FAO, 1986; Haque and Tothill, 1988; ILCA, 1993a; 1993b).

In the process of land use intensification, an important role for livestock arises from the income they generate; food crops are consumed largely within the family household, while the animal products are sold. In practice, the commercialization of livestock production, particularly milk sales by smallholders, usually provides the most effective and low cost means available for small farmers to generate an increased cash flow. The funds produced by the sale of milk, meat and fibre can then be used to provide the investments needed to stimulate crop production and to increase the capacity of a unit of land to support an increasing population (Brumby, 1991).

Soil, water, and nutrients are the primary resources essential for sustained agricultural production. To maintain or enhance agricultural productivity it is necessary to ensure that these resources are not degraded. As human and animal populations increase, soil and water resources are placed under ever increasing pressure. Ecologically-sustainable production systems that minimize soil erosion, replenish nutrients and conserve water must also be economically viable if they are to appeal to producers (Haque and Tothill, 1988).

In many parts of sub-Saharan Africa agricultural productivity is declining. Traditional land management systems that allowed soils to rejuvenate naturally (eg. fallowing, transhumant livestock movement) are being abandoned as producers exploit more land to increase production. Land-use systems that degrade the resource base must be replaced by sustainable systems. Improved soil, water and nutrient management technologies are required to satisfy the greater food demands. Soil, water, and nutrient management research in mixed crop-livestock systems must be holistic, multilocational, process oriented, relevant, and sensitive to socio-economic, policy, and environmental issues (Haque, 1993).

## II. The Role of Legumes in Sustainable Soil Management and Crop-Livestock Production

Forage legumes (herbaceous, tree, shrub) can play a vital role in sustaining the productivity of crops and livestock in many farming systems of SSA. Legumes

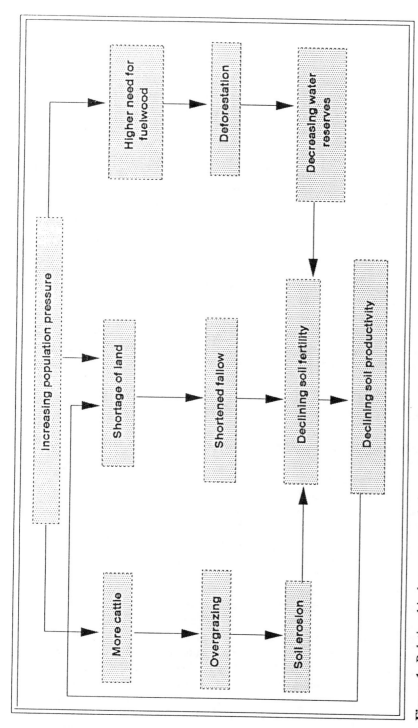

**Figure 1.** Relationship between population pressure and soil productivity. (From Prinz, 1986.)

can increase soil productivity through nitrogen (N) fixation and by reducing wind and water erosion; they also provide high quality feed for livestock.

A legume-based production soil management strategy for sustainable crop-livestock production is emerging as a result of multidisciplinary teamwork of International Livestock Centre for Africa (ILCA) and national agricultural research systems (NARS) scientists in various ecological zones of sub-Saharan Africa. Various components of the strategy are outlined in the following pages.

## A. Strategic Use of Legumes

Inadequate nutrition in the dry season is the main constraint to increased livestock production in various production systems of sub-Saharan Africa (ILCA, 1988a). During the dry season, the crude protein content of natural pasture, crop residues and stubble is below 7%. As a result, livestock lose body weight, milk production is low, calf mortality is high and conception is low (Otchere, 1986).

Mixed crop-livestock systems can be advantageously linked through the use of improved forages, particularly legumes which fix N (Haque and Jutzi, 1984; Haque et al, 1986; Nnadi and Haque, 1988a). This linkage can enhance both the level and rate of nutrient cycling in the system, leading to increased soil fertility and improved animal nutrition. Soil-moisture use efficiency is also increased through improvements in soil structure resulting from higher biological activity in the soil (Haque and Tothill, 1988).

### 1. Legume Ley Farming

One of the ways by which crop and livestock production can be improved is the use of herbaceous legumes which improve dry season feed for livestock and produce N for crops. The beneficial residual effects of legumes on the yield of subsequent crops have been demonstrated in many studies. These residual effects were noted when legumes were incorporated as green manure (Heichel, 1987), grazed by animals (Watson, 1963; Mohamed-Saleem and Otsyina, 1986; Tarawali, 1991) harvested for hay (Haque, 1990; Papastylianou and Samios, 1987), for grain (Blumenthal et al, 1988), or sod seeding (Lal et al., 1978; Lal et al., 1979; Wilson et al., 1982).

The concept of fodder banks (small fenced pastures of legumes used to supplement grazing animal diet during the dry season) was tested at four locations (Makurdi, Abet, Kontagora and Ganawuri) in the sub-humid zone of northern Nigeria. The residual effects of the legume *Stylosanthes* increased subsequent maize yields more than under natural vegetation. This may be attributed to an increase in total N fixed by the legume and improved physical properties of the soil. The maize planted outside fodder banks required 45 kg

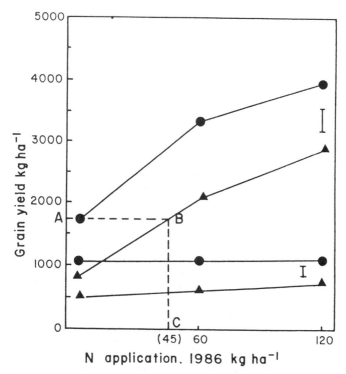

**Figure 2.** Direct and residual effects of nitrogen on grain yield of maize inside and outside fodder banks (averaged over sites).

                              • Direct inside ▲ Direct outside

                              • Residual inside  ▲ Residual outside

The bars represent the LSD for P = 0.05.

(From Tarawali, 1991.)

N ha⁻¹ to produce the same yield as that of maize on unfertilized plots within the legume pastures (Figure 2).

There was no significant response to residual fertilizer N for maize planted inside or outside the fodder banks. Yields were lower in the second year than in the first year. Although the overall yields were lower in the second year, the proportional increase in yield due to the previous legume fodder bank over no legume was similar to that in the first year. This suggests that there was still some residual N from legumes but it was insufficient for optimum growth of maize (Tarawali, 1991).

The symbiotically fixed N of various forage legumes was also evaluated on an Alfisol in the Ethiopian highlands. The proportion of total N fixed by symbiosis for *Vicia dasycarpa, Trifolium steudneri,* and *Medicago polymorpha,* was 78%, 78%, and 73%, respectively. This represented 77 to 131 kg N ha⁻¹. The dry matter and grain yield of wheat following *V. dasycarpa, T. steudneri,* and *M. polymorpha* was significantly higher than that of wheat following oats. Wheat following legumes or fallow accumulated between 12 and 31 kg more N than wheat following oats (Haque, 1990).

It is disputed whether the beneficial effect of legumes to subsequent cereals is due to N contribution or is a net rotation (Heichel, 1987) effect combining N availability, disease control, improvement of soil physical properties (Lal et al., 1979; Wislon et al., 1982; Papastylianou and Puckridge, 1983; Mohammed-Saleem et al., 1986; Adeoye, 1987; Hulugalle, 1989) or increased earthworm activity (Lal et al., 1978 and Wilson et al., 1982). Danso and Papastylianou (1992) reported that most of the increased assimilation of N in barley following vetch (compared to barley after oats) was attributable to a greater availability of soil N arising from the lower soil N uptake by vetch than oats.


## 2. Intercropping Forage Legumes with Cereals

The integration of forage legumes in crop-livestock systems can lead to an increase in production especially where land is in short supply (Gryseels and Anderson, 1983). The inter- and relay-cropping of forages with cereals has been reviewed by Nnadi and Haque (1986; 1988a) and only the recent results are reported here.

Forage legume/cereal intercropping often increased the quantity and feeding value of crop residues but decreased the yield of companion cereal crop (Mohamed-Saleem, 1985; Kouamé, 1991; Kamara and Haque; ILCA, 1992). Millet/stylo intercrops gave biomass yields similar to those of sole cropped millet however, crude protein yields were four times as much as millet on its own. Thus, the quality of feed available for livestock increased dramatically. Added to the boost in feed supplies was the residual effect of stylo on subsequent millet yields. Sole millet grown on plots that had previously been under stylo for one or two years yielded up to half a tonne more grain than millet grown on millet only plots. This is equivalent to the increase that would be obtained from applying up to 30 kg fertilizer N ha$^{-1}$ and compensated for the loss of grain yield in the intercrops. Water use efficiency calculated on a total dry matter basis indicated that millet/stylo intercrop produced significantly more per unit of water uptake than the pure crop of each intercrop component (ILCA, 1991b; and Kouamé, 1991).

In the Ethiopian highlands, intercropping of maize with *Vicia dasycarpa* (vetch), *Lablab purpureus*, *Trifolium steudneri* (clover), and *Medicago polymorpha* (medic) produced crude protein yields of 619, 585, 505, and 324 kg ha$^{-1}$, respectively, compared with only 269 kg ha$^{-1}$ for pure maize. The N contents of roots of legume intercrops were 2.32, 1.69, 1.20 and 1.09% for vetch, clover, lablab, and medic, respectively. These values were less than those of the pure legumes. Rapid root decomposition in soil requires a minimum N concentration of about 2%. Vetch with N concentration of more than 2% in its roots in both pure and intercropped systems, could decompose rapidly and make its N available to cereal crops (Nnadi and Haque, 1988b; 1990).

Mixed cropping clover and medic in small grain cereal crops is another option to increase the feed value of cereal crop residues and to sustain soil fertility. Field studies were conducted at various sites on Vertisols of Ethiopia to assess the effects of mixtures of clover on grain, straw and total biomass production of several wheat varieties. The presence of the clover mixture under wheat did not reduce grain yield significantly across locations and seasons. Mixed cropping of wheat with clover significantly increased the amount of crop residue produced per unit area (Abate Tedla et al., 1992; Abate Tedla et al., 1993).

Biological nitrogen fixation was greatest by clover harvested as hay (135 kg N ha$^{-1}$). Clover harvested as seed fixed 72 kg N ha$^{-1}$ while clover mixed crop in wheat fixed 54 kg N ha$^{-1}$. Harvesting pure clover at 50% podding fixed significantly higher N than when harvested at maturity and mixed crop in wheat. Generally, dry matter and grain yields of wheat were higher following clover harvested at 50% podding and grazed relative to other treatments. This might have been due to higher N contribution in pure clover and the compound effect of legume N, manure, and urine on grazed plots (ILCA, 1991a).

In another trial in the Ethiopian highlands, mixed cropping medic in wheat produced higher combined (wheat and medic) dry matter yields than pure wheat. Weed dry matter produced was lower in wheat and medic mixed cropping relative to pure medic and pure wheat. No significant effect on wheat grain yield was observed between mixed cropping and pure wheat (ILCA, 1992).


## 3. Sequential Cropping

Sequential cropping is defined as growing more than one crop on the same piece of land a different times of the year. Sequential cropping is possible where the rainy season is fairly long, supplemental irrigation is available, or where the soil has a high moisture-retention capacity, as is the case with Vertisols. Deep rooted legumes and cereals capable of utilizing sub-soil moisture are preferred as sequential crops. Crops used for sequential cropping should also be capable of utilizing nutrients efficiently, especially phosphorus (P).

Sequential cropping on Vertisol is facilitated by the broadbed and furrow (BBF) technology. The results of a study showed that the land-equivalent ratio (LER) of maize intercropped with medic and sequentially cropped with chickpeas was higher than that of other cropping systems (Nnadi and Haque, 1988b).

Sequential cropping provides feed early in the wet season, while feed from intercropping becomes available later. Using a combination of the systems would thus boost feed production at different times, smoothing the pattern of feed availability and broadening the livestock production and land management options available to farmers. For example, while fattening sheep for a particular market period requires feed over only a relatively short period, a farmer keeping dairy cattle will need to produce adequate supplies of feed throughout the year (ILCA, 1991b).

The early season forage crops/oats in pure stand or mixed with vetch (*Vicia dasycarpa*) yielded an average of nearly four tonnes of dry matter per hectare. Crude protein yields averaged about 280 kg per hectare for oats alone and over 520 kg per hectare for the oats/vetch mixtures. Yields of the four traditional crops grown after the forage crops/chickpea, India pea (*Lathyrus sativus*), and two local durum wheat (*Triticum durum*) cultivars were similar to those of crops that followed the fallow period. Including wheat/vetch forage crop in the cropping system provides enough high quality feed to support a crossbred dairy cow producing an average of 4 kg of milk a day for up to 15 months (ILCA, 1991b).

Studies have shown that highland Vertisols remain moist or wet throughout the year below a depth of 50 cm. The influence of cracks (up to 45 cm deep) on soil moisture depletion during the fallow period is negligible below 50 cm. Thus, these soils should be able to support a second crop if the top 50 cm can be wetted to maximum recharge (98 mm) with supplemental irrigation (Kamara and Haque, 1988). If supplementary irrigation is possible most of the highland crops can be grown (Abiye Astatke et al., 1989; 1991).

## 4. Alley Farming

Alley cropping, an agroforestry system developed as an alternative to shifting cultivation (Wilson and Kang, 1980; Kang et al., 1981a; 1981b), involves the cultivation in rows, usually 4-5 m apart, of fast-growing legume trees within cropping fields. When livestock production is integrated into alley cropping the package of crops, trees, and livestock is referred to as alley farming (Sumberg, 1985; Atta-Krah et al., 1986).

Alley farming can provide (i) mulch for companion crops, (ii) weed suppression through application of mulch and shade during fallows, (iii) favourable conditions for macro- and microorganisms, (iv) better soil conservation, (v) pruning of browse for livestock, (vi) staking material and firewood, and (vii) improvement of some soil properties (Mulongoy, 1986; Mulongoy and Kang, 1986; Wilson et al., 1986; Yamoach et al., 1986; Tothill, 1987; Sumberg et al., 1987; Atta-Krah and Sumberg, 1988; Sumberg and Atta-Krah, 1988; Sanginga et al., 1989; Atta-Krah, 1990; Hulugalle and Kang, 1990; Kang et al., 1990; Lal, 1991; Kang and Mulongoy, 1992).

In an alley farming grazing trial the combined effects of fallow and grazing sheep had been studied in southern Nigeria. Fallowing and grazing had significant effects on soil fertility and subsequent crop yield. In the first year of cultivation after a two-year fallow, crop yields were 56% higher than those of the continuously farmed alley plots. By the third year the yield advantage declined to 22% (ILCA, 1989a).

**Table 1.** Maize grain yields in alley farming as affected by level of mulch application, Ibadan, Nigeria

| Treatments | Maize grain yield (t/ha)[1] | |
|---|---|---|
| | Leucaena | Gliricidia |
| Mulch levels[1] | | |
| (% of tree productivity) | | |
| 0 | 1.53 | 1.58 |
| 25 | 1.95 | 1.79 |
| 50 | 2.29 | 2.15 |
| 75 | 2.15 | 2.24 |
| 100 | 2.76 | 2.52 |
| LSD (0.05) | 0.36 | 0.43 |

[1] Mean foliage yields for Leucaena and Gliricidia over the period were 6.2 and 3.5 t ha$^{-1}$ y$^{-1}$, respectively. Foliage not used for mulch was fed to sheep and goats.
(After ILCA, 1987.)

**Table 2.** Effect of mulching with *Leucaena* foliage from different prunings on maize yield, Ibadan, Nigeria

| Prunings applied as mulch | Maize yield (t ha$^{-1}$)[1] | |
|---|---|---|
| | Unfertilized | Fertilized[2] |
| None | 3.10 | 4.86 |
| First (preplanting) | 4.35 | 4.94 |
| First two | 4.68 | 5.32 |
| All three | 4.84 | 5.31 |

[1] Total grain yield for first and second cropping seasons; [2] 45 kg ha$^{-1}$ of 15:15:15 compound, 15 kg at planting and 30 kg 6 weeks later.
(After ILCA, 1988c.)

In alley farming tree foliage can be used either for soil fertility maintenance or as a feed for livestock; there is trade-off between the two uses. The effect of removing various proportions of tree foliage for mulch increased from zero to 100% of total tree foliage production (Table 1). The results indicated that using part of the tree foliage as feed results in some loss of maize yields (ILCA, 1987).

The effect of mulching with *Leucaena* foliage from different prunings on maize was studied in fertilized and unfertilized plots. In unfertilized plots application of foliage from first prunings significantly increased maize yield relative to no mulching. The effect of applying foliage from subsequent prunings as mulch were not significant (Table 2). The results suggest that, where no fertilizer is used, mulch applied before planting has the greatest positive effect on crop yields. Fodder for livestock should be taken from later prunings to reduce conflict with crop needs for mulch (ILCA, 1988c).

An economic analysis showed that, at low crop yield levels, feeding small ruminants with part of the foliage was profitable. Using 50% of the first season foliage as mulch and feeding the rest gave the highest total return from crop and animal. At higher yield levels feeding animals was either marginally profitable or not profitable at all (Jabbar et al., 1992).

Another trial examined the effect of the timing of mulching with *Sesbania* prunings on the yield of sorghum on an Alfisol in the Ethiopian highlands. Dried prunings (5 t ha$^{-1}$) were applied on each treatment plot on one of five occasions (Figure 3). Applying *Sesbania* prunings before planting the sorghum significantly increased dry matter yield relative to the unmulched control. Applying prunings at planting or after planting did not significantly increase dry matter yield. The response of grain yield was similar except that prunings applied after planting increased yield significantly.

## 5. Pastures

The introduction of legumes into native pastures has proved to be difficult in much of sub-Saharan Africa because of management problems associated with communal use of grazing lands (Tothill, 1986). Management of livestock is crucial to the maintenance of legumes in such systems, but in dry pastoral lands it can also affect the persistence of grasses, because legumes can be a vehicle to increase grazing pressures on the grasses. Since most African cattle are herded daily, the development of specialized feed resources for strategic feeding of livestock is a manageable option. Wildfire is a further hazard to the maintenance of legumes in many pastoral systems.

Jutzi et al. (1985) outlined the difficulties of achieving real benefits from various improvements to highland native pastures in Ethiopia. These included fertilization, the introduction of legumes and soil aeration. Research on the inclusion of legumes into native pastures or in planted grass/legume pastures dates back to the 1940s in many parts of Africa, (Kategile, 1985; Haque et al., 1986). Planted grass/legume pastures have found a place in commercial farms in a number of countries, particularly for dairy production. However, legume oversowing of native pastures used largely in communal grazing systems has been of very limited application. Where grazing management control can be carried out, such as the recently implemented stock exclusion zones on degraded pastoral areas in Ethiopia (Jutzi et al., 1987), considerable success may be achieved.

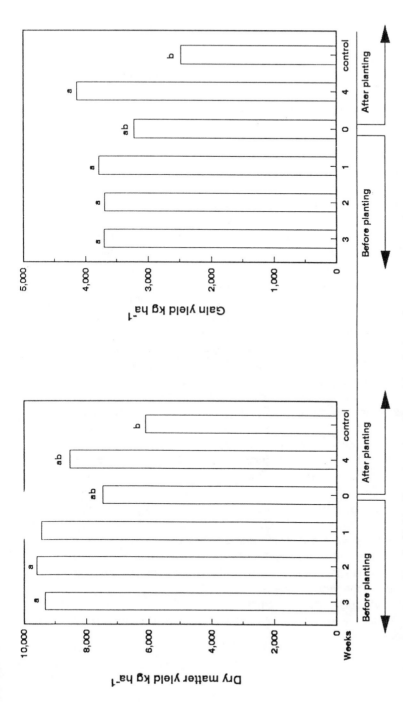

**Figure 3.** Dry matter and grain yield of sorghum as affected by timing of pruning applications on an Alfisol, Debre Zeit, Ethiopia. (From ILCA, 1988b.)

## B. Selection of Forage Legumes to Sub-Optimal Soil Conditions

The basic principle of this component is to use plants adapted to the soil and climatic limitations rather than change the soil to meet the nutritional requirements of the plants (Spain et al., 1975). The use of such plants reduces the rates of fertilizer needed, but does not necessarily eliminate the need to fertilize (Sanchez, 1982). Several forage and tree legume species and accessions that are tolerant to *aluminium* (Al) toxicity and low available soil P, have been identified.

### 1. Aluminum Tolerance

Soil acidity is a major constraint to plant growth on more than 400 million hectares of acid soils in sub-Saharan Africa (Sanchez and Nicholaides, 1982). Aluminum toxicity is a production problem in these soils. Correcting for low pH and excessive Al is not always economically viable with conventional liming practices. However, plant species and genotypes within species differ widely in tolerance to excess Al and some of these differences are genetically controlled (Foy, 1988). The main component of managing soil acidity for plant growth is the selection of productive varieties that are tolerant to Al toxicity (Sanchez and Salinas, 1981). Tolerances of *Medicago sativa* L., *Lablab purpureus* L., and *Sesbania* to Al and low P in soils and nutrient solutions have been recently reported (Mugwira and Haque, 1993a; 1993b; 1993c; 1993d).

Twenty accessions of *Sesbania* have been evaluated in two Ultisols deficient in P and high in exchangeable Al and manganese (Mn) (Table 3). These soils were treated with two rates of lime (4000 and 5800 mg kg$^{-1}$ soil) and three rates of P (37.5 and 25 and 37.5 mg kg$^{-1}$ soil). The sesbanias were grouped according to their performance on these two soils to facilitate selections for different conditions of soil acidity and P status. For the limited resource producer, commonly found in most African countries, ideal sesbanias may be those which yield reasonably well on soils that have not been amended by liming and fertilization (Group I). Second choices may be those accessions which produce fairly well in unlimed soils with P fertilizer (Group 2) or those which produce fairly well in limed soils without P fertilizer (Group 3). Alternatively, choices may be based on high responses to lime without P or high responses to P without lime (Group 4). For producers who have access to resources both lime and P fertilizer may be used for maximum yield (Group 5).

Evaluation of forage and tree legumes on soils with low levels of available P and those that have a low pH and Al and Mn toxicities is being carried out by ILCA (Haque, 1993).

**Table 3.** Sesbanias selected   for different conditions of soil acidity and phosphorus status

| Group | Soddo soil (Ultisol)[1] | Chencha soil (Ultisol)[2] |
|---|---|---|
| **Limited Resources** | -------------Sesbania varieties[3]--------------- | |
| 1. Perform relatively well without lime  or P | 15025-P, 15021-U 15023-H,  15036-U 15020-K,  15019-C | 15036-U |
| 2. Perform relatively well with P applied to unlimed soil | 15020-U, 15022-R 1203-T, 13144-U, 15019-C, 15021-U, 15025-P, 15036-U 15037-E | 15021-U, 1203-T 1265-T, 13144-K, 15022-R, 15025-P 15036-U |
| 3. Perform relatively well with limed soil without P fertilizer | NS | 1203-T, 15025-P |
| 4. High response to lime without P | 1177-T, 1291-T, 10865-El, 13144-K, | 9265-M, 10375-ET 15021-U, 15023-H |
| or | *S.goetzii* 14957 | |
| High response to P without lime | 1291-T, 9265-M, 10865-ET | 13144-K, 15021-U |
| **Adequate Resources** | | |
| 5. Perform well when both lime and P are applied | 15022-R, 15021-U, 1203-T, 15019-C 15025-P, 15036-U 15037-E | 15021-U, 1203-T, 13144-K, 15023-H, 15025-P |

[1] pH 4.8 (1:1 soil water ratio); Bray II extractable P 0.84 ppm; [2] pH 4.1 (1:1 soil water ratio); Bray II extractable P 0.67 ppm; [3] Numbers refer to ILCA accessions and letters refert to country of origin: C = Congo E = Egypt ET = Ethiopia, K = Kenya, H = Hawaii, M =  Mali, P = Pakistan, U = Ugands, R = Rwanda, T = Tanzania
(Adapted from Mugwira and Haque, 1993c.)

## 2. Low Levels of Available Soil Phosphorus

Phosphorus is grossly deficient in more than 60% of African soils (FAO, 1986), the most important nutrient needed for the successful establishment of forage and tree legumes. To maximize efficiency of the low amounts of fertilizer P available to producers, it is possible to select plants that have a lower requirement of P for maximum growth than those commonly used (Sanchez, 1982). A better understanding of differences in P plant nutrition may help in adapting forage and tree legume species and cultivars to marginal areas where fertilizers are not readily available.

Phosphorus management with special reference to forage legumes in sub-Saharan Africa has been reviewed by Haque et al. (1986). The effect of different fertilizer P levels on the growth and N and P nutrition of 20 accessions of 11 Ethiopian highland clover species was investigated on a Vertisol. Shoot and root dry matter yields plus N and P concentration were subjected to numerical classification which separated the accessions into four main groups (Figure 4). These groups ranged from accessions which had low yield and low response to P to accessions with high yields and high response to P. The most responsive group also had high uptake and utilization of N and P, but relatively low N and P concentrations in plant tops and shoot/root ratios. Accessions in the two highest yielding groups responded to each increment of additional P rate while accessions in the least responsive group did not respond to fertilizer applications above 25 mg P kg$^{-1}$ soil. The fourth group had intermediate response between extremes. Critical P levels of 0.016 to 0.017 ug P ml$^{-1}$ have been established for the native Ethiopian clovers, i.e. *Trifolium quartinianum*, *T. tembense*, and *T. steudneri* (Nnadi and Haque, 1985). The very low P requirements of these clovers indicate that they can obtain maximum yield with little P fertilization. These clovers tolerant to low P are likely to have low P concentration in their tissues. Their nutritive value may thus be lower than that of other cultivars/species and direct supplementation of ruminant livestock diet in the form of salt licks to offset P deficiency may be needed.

Sanginga et al. (1991) examined 23 provenances of *Gliricida sepium* and eleven isolines of *Leucaena leucocephala* at a low and at a high phosphorous level (20 and 80 mg P kg$^{-1}$ soil) for growth and P uptake and use efficiency. *L. Leucocephala* isolines and *G. sepium* provenances had large differences in their growth response to P. Mean shoot dry weight at low P varied from 1.30 to 3.01 g per plant for *L. leucocephala* and from 1.44 to 3.07 g per plant for *G. sepium* (Figure 5a and b).

Nine highland *Leucaena* accessions were screened in potted soil (Vertisol) in the greenhouse at 0, 10, 20, 60, and 80 ug P kg$^{-1}$. The idea was to classify them into tolerant, moderately tolerant and least tolerant to P-deficiency groups. The application of P up to 60 ppm resulted in higher top growth and increased total leaf chlorophyll and carotene content. Classifying the nine accessions into P-responsive groups based on the dry-matter yield showed that *L. leucocephala 71*, *L. leucocephala 14198*, *L. diversifolia 11676*, and *L. leucocephala 14200* were tolerant to low P and responded least to P fertilization. *L. pallida 14189*, *L. revoluta 14201* and *L. pallida 14203* were least tolerant to low P and responded positively to increasing rates of P fertilization. *Leucaena shannonii 14194* and *L. pulverulenta 14197* were only moderately tolerant to growth in the low P Vertisol (Aduayi and Haque, 1992).

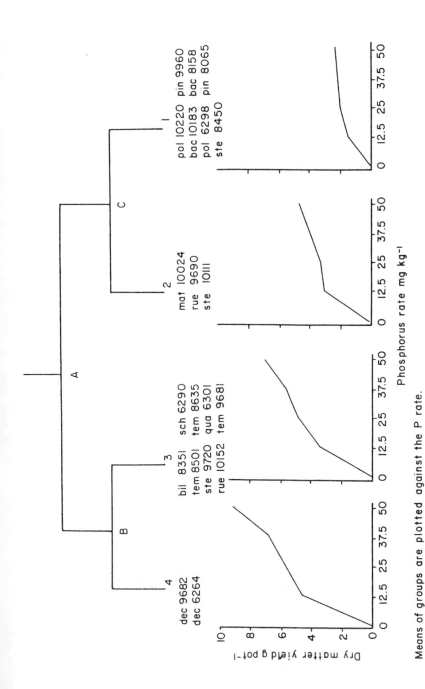

**Figure 4.** Hierarchy for the classification of clover dry matter yield of plant tops. (From Mugwira and Haque, 1991.)

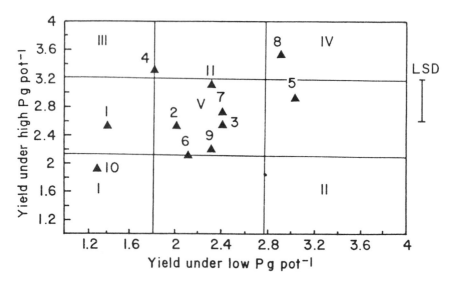

**Figure 5a.** Genetic variation in shoot weight of 11 provenances of *Leucaena leucocephala* grown under low (20 mg P kg$^{-1}$) P conditions. (From Sanginga et al., 1991.)

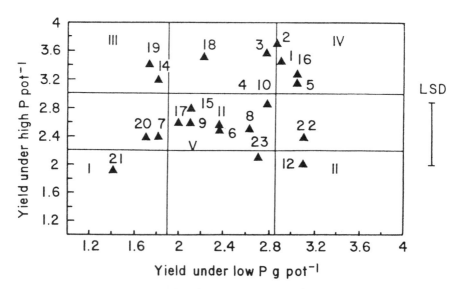

**Figure 5b.** Genetic variation in shoot weight of 23 provenances of *Glicridicia sepium* grown under low (20 mg P kg$^{-1}$) and high (80 mg P kg$^{-1}$) P conditions. (From Sanginga et al., 1991.)

## 3. Soil Salinity

There are approximately 79.5 million hectares of salt and sodium affected soils in sub-Saharan Africa (FAO, 1986). Salinity and sodicity affect soil chemical and physical properties, making it unfit for conventional crop production. A complete solution of the problem of waterlogging and salinity requires comprehensive drainage to remove the excess groundwater. This involves the construction of a very large network of drainage canals and is a major task made more difficult by the small gradient to the sea (Malik et al., 1986).

A possible approach could be to tailor plants to suit this environment. Studies indicate that growing forages on salt-affected soils to remove excess salt is feasible and economical (Malik et al., 1986; Kumar and Abrol, 1986; Szabolcs, 1992; Malcolm, 1992; Davidson and Galloway, 1993). Given the recent advances in biotechnology for the selection of salt-tolerant plants, it is likely that this major obstacle will be overcome by the use of these techniques.

## 4. Water Stress/Excess

Although both water stress and excess water can have adverse effects on root nodulation and N fixation by legumes they are among the most neglected areas of study on legume-rhizobium association. A desirable characteristic of drought resistant forages is their ability to acquire water under moisture-limiting conditions. Another is their ability to efficiently use absorbed water. Water-use efficiency (WUE) is usually expressed as the ratio of cumulative dry matter or grain produced over the amount of water used. A strong negative correlation has also been observed between $C^{13}$ isotope discrimination ($\Delta$) and total dry matter yield and N- fixation in legumes (Wright et al., 1988; Kumarasinghe et al., 1992). Based on these findings it has been proposed that $\Delta$ may be a useful tool in the selection of genotypes with high water-use efficiency, high dry matter yield and high N-fixation. The measurement of $\Delta$ in leaves collected during crop growth or at maturity is used for screening genotypes for water-use efficiency (Hubick et al., 1986).

Flooding/excess water tolerance of forage legumes has been reported (Marshall and Millington, 1967; Francis and Devitt, 1969; Francis and Poole, 1973; McIvor, 1976; Couto, 1983; Whitman et al., 1984; Wondimagegne Shiferaw et al, 1992). Forage and tree legume tolerance to flooding is being evaluated at ILCA (Haque, 1993).

There is great potential to select legumes that can withstand soil-water stress or excess found in various production systems of sub-Saharan Africa. Genetic resistance or tolerance to environmental stresses can be especially important in resource poor environments where low cost and low-input technologies are often seen as the most efficient approach to improving forage production.

## C. Use of Appropriate Rhizobium

There has been some controversy over the need to inoculate tropical legumes to improve N-fixation. Many tropical legumes will nodulate without inoculation (Norris, 1966) and there is evidence to support this statement (Bumpus, 1957; Adegbola, 1964; Boultwood, 1964; Horrell and Court, 1965; Oke, 1967; Anon, 1969; de Souza, 1969; Wendt, 1971; Thomas, 1973; Keya and van Eijnathan, 1975). However, wide variations exist in the effectiveness of rhizobia isolated from different sites and some of the indigenous strains may be of limited value to the host plant. Rhizobium strains also differ in their rates of N fixation and just because a legume has formed nodules does not mean that it will not respond to inoculation with a more effective N-fixation strain.

In field trials, inoculation with elite strains gave forage yields similar to those obtained with fertilizer applied at 150 kg N ha$^{-1}$, suggesting that nitrogen fixation was sufficient to provide the plants with their necessary requirements. The use of appropriate rhizobial strains has been found to improve the nitrogen-fixation and subsequent growth of *Leucaena* in soils deficient in N and *Leucaena* rhizobia (Sanginga et al., 1988a) .

A pot study was conducted to assess the residual effect of rhizobia-fixed nitrogen and fertilizer-N on *Leucaena* yield. Soil from inoculated field plots contained the most rhizobia and promoted increased nodulation and shoot dry matter production (Table 4). Nodule typing of the introduced rhizobium strain IRC 1050 indicated that only 17 and 7% of the nodules were due to the native rhizobia in the uninoculated and inoculated plots, respectively. The inoculated plots containing the rhizobium strain IRC 1045 formed 89% of the nodules and only 2% were due to indigenous rhizobia. Nine per cent of the nodules were due to the IRC 1050. It can be concluded that if adequate strains of rhizobia are introduced in the soil, the population will survive and eventually multiply over the years under the continuous *L. leucocephala* cropping without additional inoculation (Sanginga et al., 1988b).

The effect of prunings from *Leucaena* inoculated with the rhizobium strain IRC 1045 on the increase of maize grain yield was equal to that of fertilization with 80 kg N ha$^{-1}$. The maize grain yield obtained after using prunings from uninoculated *Leucaena* was low but comparable to that produced by N fertiliser at 40 kg N ha$^{-1}$. Generally, the yield of maize was higher in the plot inoculated with rhizobium IRC 1045 than in the uninoculated plots and those inoculated with rhizobium IRS 1050 (Sanginga et al., 1988c).

Cobbina et al. (1988) revealed that *Leucaena* inoculation with the two strains of rhizobium which improved the shoot dry matter and N uptake, was not always as effective as using fertilizer N. Therefore, it is necessary to identify more effective strains of rhizobium for the acid soils of humid Nigeria.

Sanginga et al. (1989a) further reported that estimates using the $^{15}$N dilution method gave a nitrogen fixation of 134 kg ha$^{-1}$ in six months when *Leucaena* was inoculated with the rhizobium strain IRC 1045 and 98 kg ha$^{-1}$ for *Leucaena* inoculated with the rhizobium strain IRC 1050.

**Table 4.** Effect of previous inoculation treatments on nodulation and growth of Leucaena 8 WAP

| Previous inoculation treatment | Number of nodules/plant | Nodule fresh weight mg/plant | Shoot dry weight g/plant | Height cm |
|---|---|---|---|---|
| Uninoculated | 18 | 105.00 | 1.67 | 55.50 |
| Rhizobium IRc 1050 | 28 | 200.00 | 2.27 | 65.53 |
| Rhizobium IRc 1045 | 28 | 232.50 | 2.79 | 74.75 |
| LSD (5%) | 5 | 62.50 | 0.27 | 4.46 |

(After  Sanginga et al., 1988b.)

Rhizobium strains isolated from *Leucaena leucocephala, Sesbania rostrata, S. grandiflora, S. punctata, Tephrosia vogelii, Acacia albida and Vigna unguiculata* in Nigerian soils were characterised and tested for their ability to nodulate and fix atmospheric nitrogen with *L. leucocephala. Rhizobium* isolates IRC 1045 and IRC 1050 obtained from *L. leucocephala* grown at Fashola and the International Institute of Tropical Agriculture (IITA) were found to be the most effective rhizobia on this host. In addition to their effectiveness and competitiveness, they survived well in the field a year after their establishment (Sanginga et al., 1989b). If adequate strains of rhizobia are introduced, the populations will survive and eventually multiply over the years under continuous *L. leucocephala* cropping without additional inoculation.

Various strains of rhizobium (USDA 3786, 3110, 3781, 3782, and 3117) were compared on *Sesbania sesban* grown on highland Vertisol in the greenhouse. Strain 3117 significantly increased the number of nodules and shoot dry weight compared to other strains. The various strains had no significant effect on the shoot dry weight as compared with the control (ILCA, 1989b).

## D. Response of Legumes to Phosphorus and Evaluation of Rock-Phosphates

Phosphorus is the most important nutrient in the successful establishment of legumes. Phosphorus often increases nodulation and crude protein content. Phosphorus application may also increase the digestibility of forages. The response of forage legumes to applied has been reviewed by Haque and Jutzi (1984) and Haque et al. (1986).

In acid soils that fix large qualities of P, applying rockphosphate (RP) is often more effective and more economical than applying superphosphate. Rock-phosphates are reactive in acid soils and costs per unit of P may be as low as one third to one fifth those of superphosphate. The effectiveness of RP depends on its solubility, fineness, time of reaction, and on its soil pH (Sanchez and Uehara, 1980). Some of the rockphosphate deposits in sub-Saharan Africa are

**Table 5.** Some phosphate deposits located in countries of sub-Saharan Africa.

| Country | Location | Total reserves and resources (t x $10^6$) | Type |
|---------|----------|-------------------------------------------|------|
| Angola | Cabinda | 120 | Sedimentary |
| Burkina Faso | Kodjari | 1500 | Sedimentary |
| Liberia | Bomi Hills | 1.5 | Sedimentary |
| Mali | Tilemsi Valley | 20 | Sedimentary |
| Mauritania | | 5 | Sedimentary |
| Niger | Parc W | 100 | Sedimentary |
| | Tahoua | - | Sedimentary |
| Senegal | Taiba | 1100 | Sedimentary |
| | Thies | 2090 | Sedimentary |
| Tanzania | Minjingu | 10 | Sedimentary |
| Togo | Hahotoe, Kpogame | 300 | Sedimentary |
| Uganda | Sukulu | 200 | Igneous |
| Zaire | Matadi | 83 | Sedimentary |
| Zambia | Kaluwe, Nkomba | 400 | Igneous |

(After Hammond et al., 1986.)

shown in Table 5. These deposits exhibit a wide range of chemical composition, solubility and potential for agronomic use. The evaluation of rockphosphates in sub-Saharan Africa has been reviewed by Hammond et al. (1986) while crop responses to rockphosphates in Eastern Africa and West Africa have been reviewed by Edjigayehu Seyoum and McIntire (1987) and Jones (1973), respectively.

The relative effect of Togo (TRP), Egyptian rockphosphates (ERP), and triplesuperphosphate (TSP) was evaluated using *Medicago sativa* grown in an Andept. Egyptian rockphosphate applied at 43 kg ha$^{-1}$ and TRP applied at 60 kg ha$^{-1}$ gave 92 and 64% the yield, respectively, at that obtained with TSP. Beyond these rates, relative agronomic effectiveness (RAE) values declined (Nnadi and Haque, 1988c). Figure 6 shows the dramatic effects of both TSP and ERP on clovers grown on P-deficient Vertisol for six years. The clovers grew very poorly where no P was applied and reacted similarly to both TSP and ERP at 15 and 30 kg P ha$^{-1}$. TSP was more effective than ERP at 45 and 60 kg ha$^{-1}$ (Haque, unpublished data). Relative efficiency of TSP, 50% and 35% acidulated and unacidulated Minjingu rockphosphate from Tanzania, was compared on *Stylosanthes guianensis* c.v. Cook grown on highland Ultisol. Application of TSP, 50% and 25% acidulated and pure rockphosphate significantly increased dry matter yield at various rates relative to control. All sources were equally effective, demonstrating that Minjingu rockphosphate is reactive and could be used on acid soils (Haque, unpublished data).

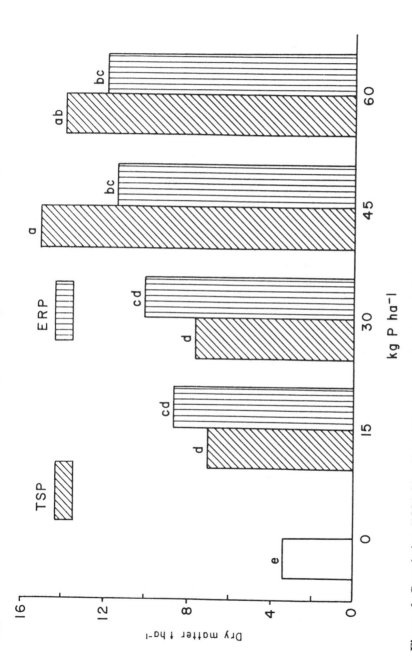

**Figure 6.** Cumulative (1984-89) effect of triple superphosphate and Egyptian rock-phosphate on clover grown on highland Vertisol. (From Haque, unpublished data.)

Many African rockphosphates are low in chemical reactivity and are unsuitable for direct application on short season crops (IFDC, 1984). An alternative choice may be to broadcast the rockphosphate and band a soluble source to provide P while the rockphosphate dissolves. This technique has been successful in Africa with locally available RP of low citrate solubility (Pichot and Roche, 1972).

## E. Correction of Other Nutrient Deficiencies

The role of sulphur (S) in tropical African countries has been reviewed and deficiencies of S have been reported in 23 sub-Saharan African countries (Kanwar and Mudahkar, 1986). Good responses by forage legumes to elemental S and gypsum have been recorded in Uganda (Horrell and Court, 1965; Wendt, 1970), in Kenya (Anon, 1969) and in Nigeria (Haggar, 1971). Forage legumes also positively responded to K application in Swaziland (I'Ons, 1968), Uganda (Wendt, 1970), and Nigeria (Tening et al., 1992).

Micronutrient deficiencies and responses in various crops, forages and livestock nutrition have been reported and reviewed (Kang and Osiname, 1972; 1985; Cottenie et al., 1981; McDowell et al., 1982; Mtimuni, 1982; Faye et al., 1983; McDowell et al, 1984; Haque, 1987; Haque et al, accepted. Fertilizers to correct micronutrient deficiencies are not available in SSA and also would be beyond the buying power of resource-poor farmers. A mineral supplement known as kanwa (44.2 ppm Cu), fed to traditionally-managed cattle in central and northern Nigeria proved an attractive source of plant nutrient. Kanwa application up to 90-100 kg ha$^{-1}$ on old and newly established Verano Stylo increased dry matter yield significantly (Figure 7). In the year of its application every kg of Kanwa up to an optimum level increased dry matter by 19 kg in the old Verano stylo pasture as compared with 60 kg in stands established in the same year. The 100 kg ha$^{-1}$ Kanwa increased the amount of protein from 417 to 712 kg ha$^{-1}$ at a cost of US $10 which seems to be insignificant in comparison to potential benefit of increased protein for livestock.

# III. Role of Crop Residues and Animal Manures

Most farming systems in sub-Saharan Africa continue to rely on organic matter recycling for maintaining soil productivity. Livestock have long played a key role in these processes. The cycling of biomass (natural vegetation, crop residues) through animals (cattle, sheep, and goats) into excreta (manure and urine) that fertilizes the soil is an important linkage between livestock and soil productivity (Figure 8). This linkage is particularly important in more arid environments where livestock play a vital role in stabilizing food output. The viability of these systems requires that animals be kept so they can be available for consumption and sale during years of poor rainfall and crop production.

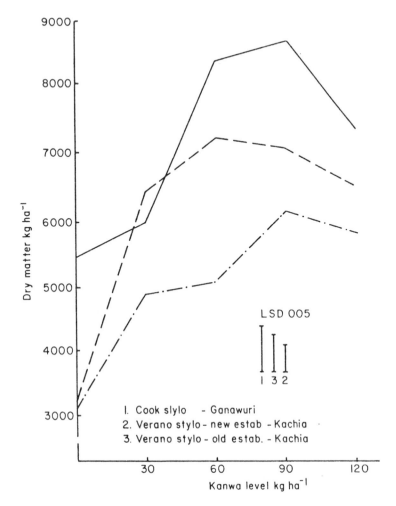

**Figure 7.** Effects of various levels of "kanwa" on stylo dry matter production. (From Mohamed-Saleem and Otsyina, 1987.)

Most biomass in these mixed farming systems must, therefore, be fed first to animals before becoming available as a soil amendment.

Although livestock enhance the sustainability and stability of agricultural production in many farming systems of sub-Saharan Africa, they can also be principal vectors of nutrient loss. Excessive removal of vegetation by grazing can deplete soil organic matter and nutrient reserves and have detrimental effects on soil productivity. Long-term sustainable land management in mixed crop-livestock farming systems necessitates technologies that enhance the beneficial and reduce the competitive relationships between livestock and soil management.

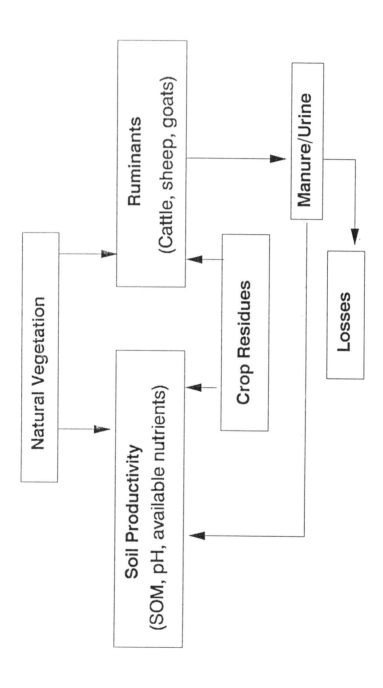

**Figure 8.** Ruminant-soil productivity linkages in mixed farming systems of the Sahel. (From Powell and Williams, 1993.)

## A. Crop Residues for Feed or Soil Conservation

Cereal stovers and other crop residues are vital livestock feeds during the six to eight month dry season, especially in the drier regions (Sandford, 1990). It is in these drier areas, however, that the competition between livestock and soil conservation for crop residues is most acute. High soil temperatures, wind erosion and sand blasting of young plants can pose severe limitations to crop establishment and production in semi-arid environments. When left in the field, cereal stovers in these production systems provide a physical barrier to soil movement, allow soil and organic matter to accumulate, and enhance soil chemical properties (Geiger et al., 1992) and crop yield (Bationo and Mokwunye, 1991).

The continuous removal of biomass (grain, crop residues) from cropland without adequate replenishment can rapidly deplete soil nutrient reserves and jeopardize the sustainability of agricultural production. Nutrient balances (inputs minus offtakes) have become negative for many farming systems in sub-Saharan Africa (Stoorvogel and Smaling, 1990) and grazing animals can contribute to these imbalances. Much greater amounts of biomass and nutrients are removed from cropland by grazing animals than are returned in the form of manure (Table 6). Exceptions to this are situations in Burkina Faso where all residues are harvested and fed to animals in stalls and refused stover, other feeds and manure are collected and returned to fields. Although a large proportion of biomass disappearance from cropland during the dry season can be attributed to removal by grazing livestock, a portion consumed by termites and trampled by animals would remain in the soil surface, decompose and be available for recycling during the subsequent cropping season.

The small manure return relative to the large stover removal can be attributed to the characteristic uneven distribution of manure in the landscape. Animal voidings in grazing systems are usually concentrated around watering points, in resting areas, and along paths of animal movement. The concentration of nutrients in these areas is subject to high losses through soil surface erosion, volatilization, leaching, etc. (Woodmansee, 1979; West et al., 1989). A greater capture and recycling of nutrients voided by animals would require large amounts of manual labor to harvest, store, and feed residues to animals in stalls, technologies that would capture feed refusals, urine, and manure (e.g. compost pits) and additional labour to transport and spread refusals and animal excreta on to fields.

Integrated crop, animal, and soil management strategies are needed to arrest soil degradation in many regions of sub-Saharan Africa. Growing forages to substitute for crop residues, partial rather than total removal of crop residues and an improved balance between feed supplies and animal populations can mitigate the competition between livestock and soil conservation for crop residues in semiarid regions (Unger et al., 1991).

The harvesting pattern of sorghum and millet and the chemical composition of their stover plant parts may offer management possibilities that allow certain

**Table 6.** Cereal stover dry matter (DM), nitrogen (N) and phosphorus (P) removals and manure returns during crop residue grazing in West Africa

| Locations | DM | N | P | Source |
|-----------|-----|-----|-----|--------|
| | | | -kg ha$^{-1}$ | |
| **Stover Removals** | | | | |
| Nigeria | 2470 | 24.5 | 3.9 | 1 |
| Burkina Faso | 1570 | 14.8 | 4.0 | 2 |
| Burkina Faso | 1450 | 13.8 | 1.5 | 2 |
| Niger | 2135 | 16.8 | 1.8 | 3 |
| Niger | 3495 | 33.1 | 3.6 | 3 |
| Niger | 1470 | 13.6 | 1.5 | 4 |
| **Manure returns** | | | | |
| Nigeria | 70-400 | 0.8-4.3 | 0.2-0.8 | 5 |
| Nigeria | 27-161 | 0.3-1.7 | 0.1-0.3 | 1 |
| Burkina Faso | 600-1600 | 7.5-20.0 | 1.5-4.0 | 2 |
| Niger | 175-228 | 3.1-4.1 | 0.8-1.1 | 4 |

(Adapted from Powell and Williams (1993) who used the following sources: (1) Powell and Saleem, 1987; (2) Quiflin and Milleville, 1981; (3) Powell, unpubl.; (4) Fernandez-Rivera, Unpubl.; (5) van Raay, 1975. N and P concentrations to calculate **REMOVALS** for sources (2) and (3), collected from various trials in Niger were 12.6 and 1.3 g/kg for millet leaves and 7.0 and 0.8 for millet stalks, respectively. N and P concentrations to calculate **RETURNS** for sources (2) and (5) were 10.8 and 2.1 g/kg, respectively, taken from source (1); for source (4) were 18.0 and 4.7 g/kg taken from Powell et al., 1993).

stover parts to be harvested for alternative uses (e.g. feed) while returning other parts for soil conservation (Powell et al., 1991; Powell and Fussell, 1993). The upper section of millet stover and non-productive panicles and tillers have the highest feeding value of all stover parts (Table 7). These could be harvested in conjunction with grain leaving lower stover portions for soil conservation. The adoption of such techniques depends, however, on the associated relative costs and benefits. The labor required to selectively harvest cereal stover is being assessed and a series of long-term trials has been initiated to determine animal response to the feeding of and soil response to the land application of millet stover plant parts (ILCA, 1990).

## B. Effects of Animal Manure on Long-Term Soil Management

Animal manures have long played an important role in maintaining cropland productivity in sub-Saharan Africa. Up to 95% of the total nutrients consumed

**Table 7.** Millet uses based on the harvesting pattern and chemical composition of plant parts

| Use | Plant Part | DM | N | P | Digestibility | Crude Protein |
|---|---|---|---|---|---|---|
| | | ------------kg ha⁻¹ ------------ | | | -------g kg⁻¹ ------- | |
| **Food** | | | | | | |
| | Grain | 500 | 10.54 | 2.04 | 747 | 131 |
| | | (382)[1] | (0.87) | (0.18) | (5) | (4) |
| **Feed** | | | | | | |
| | Chaff | 420 | 4.38 | 0.62 | 485 | 65 |
| | | (39) | (0.47) | (0.05) | (7) | (9) |
| | Non-Grain Panicles | 130 (26) | 2.79 (0.53) | 0.43 (0.06) | 518 (10) | 134 (6) |
| | Upper Stover | 380 (22) | 4.45 (0.2) | 0.72 (0.06) | 516 (5) | 73 (4) |
| | Tillers | 190 (21) | 2.32 (0.29) | 0.31 (0.05) | 588 (8) | 804 (6) |
| **Soil Management** | | | | | | |
| | Mid stover | 430 (18) | 3.80 (0.20) | 0.42 (0.04) | 482 (7) | 57 (3) |
| | Lower Stover | 670 (36) | 5.11 (0.37) | 0.61 (0.06) | 446 (11) | 508 (5) |

[1] SE are given in parentheses.
(After Powell and Fussell, 1993.)

by livestock are excreted. The amount and forms of nutrients excreted by animals, and therefore the susceptibility of nutrients to loss when applied to cropland, are highly influenced by animal species, age, diet, and watering regime. Whereas N is voided in both urine and feces, most P is voided in feces (CAB, 1984; Ternouth, 1989).

Seasonal differences in the diet of grazing animals can also greatly influence manure output and its nutrient content. Manure output by grazing cattle during the rainy season (2.2 kg/animal/day) is twice as much as manure output during the dry season (Siebert et al., 1978). Likewise, the N and P concentrations in cattle manure varies considerably by season (Powell, 1986). Wet season manure is more abundant and of higher quality, but it is largely unavailable to cropping, as animals are normal outside the cultivation zone in the wet season (Powell and Williams, 1993).

The total amount and proportion of N excreted in urine and manure depends on animal diet. The urine and manure excreted by animals fed highly digestible diets is more susceptible to N losses than excreta from diets containing greater amounts of roughage. More than twice the amount of N is excreted as urine by animals fed cowpea hay than is excreted by animals fed *Acacia tortilis* pods or no supplement (Coppock and Reed, 1992). Much of this urine N would be lost via ammonia volatilization. Manure from animals fed *A. tortilis* pods contains greater amounts of structural carbohydrates and, therefore, decomposes more slowly in the soil than manure from other diets.

Studies have been conducted to determine the effects of different forages on the amounts and forms of nutrients excreted by sheep and the relationship between diet quality and nutrient voidings and to evaluate the potential impact of these effects on the release of nutrients from urine and manure when used as organic fertilizers. Sheep fed millet and cowpea residues excreted significantly more total and urine N than sheep fed browse leaves. Feeding browse caused a general shift from fecal soluble N (microbial and endogenous N) to fecal insoluble N (undigested plant N). Sheep fed browse leaves voided less urine N and, therefore, produced excreta less susceptible to volatile and leaching N losses. The N released from the various manures would provide 60, 73 and 86% of the N needs of millet. The 8.4-12.2 kg P ha$^{-1}$ released from manure would supply all the annual P requirements of millet (Powell et al., 1993). Selecting feeds that not only satisfy the nutrient requirements of livestock but produce animal excreta less susceptible to losses may improve nutrient cycling in mixed farming systems.

Not all of the nutrients excreted by animals are available for recycling. As mentioned previously, animal voidings in grazing systems are unevenly distributed in the landscape. However, after daytime grazing during the dry season, farmers corral their animals overnight on fields to return both urine and manure to cropland. In more intensively managed systems where animals are stall-fed, only manure may be available for recycling and it must be handled, stored, transported, and spread onto fields. The move from extensive livestock management based on grazing, to more intensive stall feeding (Winrock, 1992) could increase nutrient losses and jeopardize long-term soil productivity if technologies are not available that capture and recycle the nutrients voided by stationary animals.

Farmers annually manure from 20 to 50% of their cultivated areas. Amounts and frequency of manure applications depend on animal and cropping densities, animal type and rainfall. Higher manure applications (and shorter intervals between applications) are generally associated with higher rainfall where cultivation densities are greater and cattle are more important than small ruminants (Table 8). However, the amount and type of manure available for cropping can vary considerably, especially in drier areas. Low and erratic rainfall can cause large fluctuations in livestock numbers and can change the types of livestock that producers keep. The 50 to 80% reduction in animal numbers during the droughts of the early 1970s and mid-1980s reduced manure availability and were perhaps a major reason why yields per unit area declined over the same period (Government of Niger, 1992). The shorter reproductive cycles of sheep and goats relative to cattle allow small ruminants to become the

**Table 8.** Cultivated and manured areas, manure dry matter (DM), nitrogen (N), and phosphorus (P) on farmers' fields in Niger

|  | Zones | | |
|---|---|---|---|
|  | Wet | Mid | Dry |
| Rainfall (mm) | 600 | 425 | 350 |
| Cultivated area (ha/household) | 3.2 | 6.3 | 8.2 |
| % manured | 29 | 52 | 30 |
| Manure DM (kg/ha) | 3800 | 1700 | 1300 |
| % cattle | 52 | 55 | 19 |
| % small ruminant | 48 | 45 | 81 |
| Manure N (kg/ha) | 45 | 23 | 22 |
| Manure P (kg/ha) | 5.7 | 3.0 | 2.7 |

(After Powell and Williams, 1993.)

predominant animal species after drought. A shift in animal species can influence soil management and the nutrient cycles of these farming systems.

The effects of animal type and management and manuring interval on crop yields and nutrient cycling are being investigated in a long-term trial in Niger. Third year results (Table 9) show that millet DM yield where cattle or sheep are corralled (i.e. extensively-managed enterprises where animals graze during the daytime and are corralled on fields at night to deposit manure and urine) during each of the past three years are 75 to 125% greater than yields in plots where only manure was handspread (i.e. intensively-managed enterprises where only manure from stalls is available for recycling). The positive effect of urine is also evident in the residual plots, or where animals were corralled two or three years ago.

The observed positive effects of urine on millet yield were probably associated with urine-N return and other beneficial effects on soil properties. Separate measurements on adjacent plots showed that an average sheep voiding of 62 g deposits an equivalent of 202 kg N ha$^{-1}$ (Powell et al., 1992). Soil pH increased from 5.9 in control plots to 9.5 in urine patches immediately after urine application. Soil pH, ammonium- and nitrate-N and available P and K remained elevated throughout the 130-day cropping period.

Manure application has been shown to be as effective as fallowing in maintaining soil productivity. An analysis of 10- to 20-year trials in Nigeria showed that annual manure applications of 3 t ha$^{-1}$ were sufficient to maintain cereal yields at the same level as a three-year fallow period (Dennison, 1961; Watson and Goldsworthy, 1964). No additional grain response to fertilizer-N was obtained when annual manure applications were 7.5 t ha$^{-1}$ (Abdullahi and Lombin, 1978). Greater weed growth due to manuring can, however, be a particular problem on farmers' fields in central Nigeria (Powell, 1986). Further

**Table 9.** Cattle and sheep manure and urine effects on millet yields in Niger

|        | Yearly manure applicat. (t ha$^{-1}$) | Application interval | | | | | |
|--------|------|------|------|------|------|------|------|
|        |      | every year | | every 2 years | | every 3 years | |
|        |      | Urine application [1] | | | | | |
|        |      | Yes  | No   | Yes  | No   | Yes  | No   |
| Cattle | 3.11 | 5.87 | 2.91 | 6.00 | 2.90 | 2.19 | 2.10 |
|        | 7.10 | 6.90 | 3.40 | 6.00 | 3.52 | 4.07 | 2.54 |
|        | 10.12| 7.26 | 3.24 | 5.47 | 3.79 | 3.63 | 2.73 |
|        | (SEM)| 0.60 | 0.26 | 0.43 | 0.24 | 0.74 | 0.57 |
| Sheep  | 1.48 | 3.75 | 2.84 | 4.53 | 2.11 | 1.67 | 2.08 |
|        | 3.23 | 5.92 | 3.43 | 5.43 | 2.52 | 3.03 | 1.52 |
|        | 4.98 | 5.09 | 3.64 | 5.51 | 3.77 | 2.89 | 1.99 |
|        | (SEM)| 0.45 | 0.23 | 0.30 | 0.12 | 0.35 | 0.43 |

[1] Animals corralled on cropland to apply manure plus urine on "urine application yes" plots, or manure only was applied to obtain "urine application no" plots.
(After Powell, Unpublished.)

north, in more arid Niger, crop response to animal manure is highly influenced by rainfall (Powell and Ikpe, 1992). During a year of low rainfall, total millet production in 40 manured fields of farmers was only 6 to 19% greater than in adjacent non-manured areas and manuring decreased yields significantly in one village. During the following year when rainfall was adequate, millet yields in manured areas were 71 to 98% greater than in non-manured areas.

The positive effects of animal manure on soil surface chemical properties depend largely on the amount of manure applied. Various on-station manuring trials and measurements from farmers fields showed that manuring increases soil pH, soil organic matter from 12 to 100%, total-N from 8 to 86%, and available-P from 33 to 340% over these soil properties in non-manured plots (Table 10). The highest increase in these soil properties due to manuring were derived from on-station trials where average manure application rates appear to be higher than the application levels of farmers. Of particular interest is the large beneficial effect of manuring on available-P in soil. Soils in the region are more P than N deficient (Breman and de Witt, 1983), and manuring appears to greatly offset this deficiency. An accelerated release of P in manure can be of particular importance to soil productivity in savanna soils that are deficient in P (Ruess, 1987).

Sustainable organic matter recycling, and manuring in particular, depends on land and animal management practices that do not deplete the soil nutrient supply in one location in order to maintain or improve soil productivity in

**Table 10.** Effects of manure on soil surface chemical properties in West Africa; samples taken at first crop harvest after manure application

| Manure Application | pH | Organic Matter | Total Nitrogen | Available Phosphorus | Sources |
|---|---|---|---|---|---|
| | | % | -----------mg kg$^{-1}$------------ | | |
| 5.0 | 5.37 | 0.39 | 202 | 10.3 | 1 |
| | (4.98)[1] | (0.29) | (153) | (5.3) | |
| 20.0 | 6.21 | 0.58 | 285 | 22.9 | 1 |
| | (4.98) | (0.29) | (153) | (5.3) | |
| NR‡ | 5.8 | 0.33 | 164 | 9.6 | 2 |
| | (5.1) | (0.26) | (131) | (4.6) | |
| 3.6 | 6.51 | 0.18 | NR[2] | 5.7 | 3 |
| | (6.18) | (0.15) | | (4.3) | |
| 2.1 | 5.04[3] | 0.31 | NR | NR | 3 |
| | (4.86) | (0.28) | | | |
| 3.1 | 5.45 | 0.31 | 150 | 11.0 | 4 |
| | (5.20) | (0.28) | (138) | (6.1) | |
| 10.1 | 5.68 | 0.33 | 169 | 26.8 | 4 |
| | (5.20) | (0.28) | (138) | (6.1) | |

[1] Values in parenthesis are soil properties in non-manured plots; [2] Not recorded.; [3] pH in KCl, otherwise pH in water.
(After (1) Bationo and Mokwunye, 1991; (2) Powell, 1986; (3) Powell, unpubl. from 40 farmers fields in Niger; (4) Brouwer, unpubl. from on station trials in Niger: samples taken from cattle manure with urine plots, Table 9.)

another location. The sustainability of nutrient transfers, whereby livestock graze and gather nutrients during the daytime, and are corralled to fertilize cropland at night, depends on a balance between the feed supply on natural pasture and livestock numbers. While a portion of nutrients voided at night are from grazed cropland, most nutrients come from rangelands, especially from trees and browse (Swift et al., 1989), and during the latter part of the dry season when cropland residues have been exhausted. Sustainable nutrient transfers from rangeland to cropland for various mixed farming systems requires from 4 to 40 hectares of rangeland for each hectare of cropland (Swift et al., 1989, Breman and Traoré, 1986). The climatic and socio-economic changes occurring in many areas of SSA has put more pressure on the natural resource base to produce more food and feed. As more land is cultivated less rangeland is available, thereby increasing the risk of overgrazing. Such intensification of land use without proper management may jeopardise the sustainability of nutrient transfers and decrease crop production.

## IV. Measuring the Sustainability of Improved Crop-Livestock Systems

Growing food imports and accelerating ecological degradation are causes for widespread concern in sub-Saharan Africa. Efforts are being made to develop improved crop-livestock technologies that enhance food production, maintain ecological stability and preserve the natural resource base, i.e. technologies or systems that are economically viable and ecologically sustainable. Despite the vast amount of literature that now exists on the definition of sustainability, there is no information on how to operationalize and measure the concept for use in biological, physical, and economic analyses.

Because of the broad nature of sustainability, most attempts at constructing measures have concentrated on developing partial indices which estimate some aspects of the broader concept. Dumanski (1987), recently reviewed a number of these indices. Physical indices of sustainability include the soil erosion vulnerability index of Pierce et al. (1983), the erosion sustainability index of Lee and Goebel (1986) and of Putman (1986) and the potential land flexibility index of Dumanski (1987).

In one approach, Ehui and Spencer (1990; 1993) merge biological, physical and economic measures into a single index of total factor productivity (TFP). Total factor productivity is defined as "the total value of all output produced by the system during one cycle divided by the total value of all inputs used by the system during one cycle" (Lynam and Herdt, 1989). In normal economic practice the outputs and inputs would be confined to those attributes which are recognized as economic variables purchased inputs, labor costs, the value of harvest, etc. Ehui and Spencer (1990; 1993) have extended this by costing natural resources used within the systems, such as soil nutrients. The costed net productivity factors, are aggregated to give the TFP index. If TFP shows a constant or upward trend over a period of time and does not fluctuate widely, then the system is sustainable. The TFP can be analyzed to determine which factor(s) contribute most or least to sustainability.

A major advantage of this approach is that it can be calculated using only data on the quantities and prices of inputs and outputs. Prices must reflect the true value of resources and outputs to society if the TFP is to give realistic results.

This section reviews the TFP approach as a tool for measuring the sustainability and economic viability of production systems.

### A. Concept of Intertemporal and Interspatial TFP Measures

A major accomplishment of analytical work in productivity measurement is to find the best productivity indicator for more general representation of technology. Ball (1985) and Antle and Capabo (1988) recommend changes in total factor productivity from period or systems s to t as:

$$
\begin{aligned}
Ln(\frac{TFP_s}{TFP_t}) = &\frac{1}{2} \sum_{i=1}^{m} ( r_{is} + r_{i,t} ). Ln(\frac{Y_{is}}{Y_{i,t}}) \\
&-\frac{1}{2} \sum_{j=1}^{n} (w_{js} + w_{j,t} ). Ln(\frac{X_{js}}{X_{j,t}})
\end{aligned}
\qquad (1)
$$

where $y_i$ denotes output quantity for product $_i$; $x_j$ is input quantity j; $r_j$ are output revenue shares; and $W_j$ are input cost shares. Appropriate weighting and aggregation can make a substantial difference in measured productivity.

One limitation of traditional TFP measure is that it does not properly account for natural resource stock and flows. The sustainability of agricultural production is determined by the complex interactions of the biological, physical and socio-economic factors that constitute the basis of all production systems. The agricultural sector utilizes a common pool natural resources (e.g. air, water, and soil nutrients). The stock of these resources affects the production environment, but is in many cases beyond the control of the farmer. For example, soil nutrients are removed by crops, erosion or leaching beyond the crop root zone, or through other processes such as volatilization of nitrogen. Agricultural production can also contribute to the stock of some nutrients, particularly, nitrogen, by leguminous plants. When the stock of resources is reduced, the farmer faces an implicit cost in terms of forgone productivity. Conversely, when the stock of resources is increased during the production process (e.g. via nitrogen fixation) the farmer derives an implicit benefit from the system.

If these implicit costs and benefits are not accounted for when TFP is measured the result will be biased. Ehui and Spencer (1990; 1993) extended the TFP approach to account for natural resource stock and flows.

The approach is based on the intertemporal TFP measure as developed by Denny and Fuss (1983). Intertemporal TFP is defined in terms of the productive capacity of the system over time. However, this productive capacity for a sustainable system includes the unpriced contributions from natural resources and their unpriced production flows. Ehui and Spencer (1990; 1993) conclude that a system will be sustainable if the associated TFP index, which incorporates and values changes in the resource stock and flow, does not decrease.

Unlike sustainability, economic viability is a static concept which refers to the efficiency with which resources are employed in the production process at a given period. A new production system can be said to be more economically viable (or efficient) than an existing one if its TFP is greater at a given point in time after accounting for differences in quantities of inputs and unpriced natural resources used in each system during a season.

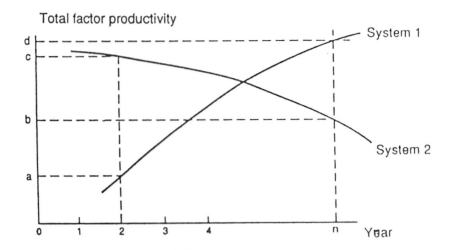

**Figure 9.** Total factor productivity for two hypothetical agricultural systems over time. (From Ehui and Spencer, 1990.)

Figure 9 illustrates the difference between intertemporal and interspatial TFP for two hypothetical systems. System 1 is sustainable since its intertemporal TFP increases for the same period. On the other hand, system 2 is economically more viable than system 1 in year 2 ($c > a$), but it is economically less viable in year n.

## B. Indices of Intertemporal and Interspatial TFP

Whether depletion or replenishment in the resource stock is or is not accounted for will affect the TFP measure. In case of net positive change in resource stock, the index of output quantities must capture the resource flow effect. Disaggregating the output index yield the following TFP measure:

$$Ln(\frac{TFP_s}{TFP_t}) = \frac{1}{2} \sum_{i=1}^{m} [\ r_{si} + r_{ti}\ ].[\ Ln\ \frac{Y_{is}}{Y_{it}}\ ]$$
$$+ \frac{1}{2} \sum_{k=1}^{q} [\ v_{sk} + v_{tk}\ ].[\ Ln\ \frac{Z_{sk}}{Z_{tk}}\ ]$$
$$- \sum_{j}^{n} [\ w_{sj} + w_{jt}\ ].[\ Ln\frac{X_{js}}{X_{jt}}\ ]$$
$$- Ln[\frac{B_s}{B_t}]$$

(2)

where $Z_{sk}$ denotes the resource flow k in time periods. $B_s$ denotes resource abundance in periods; and $V_{sk}$ denotes the revenue share of resource flow Z. Other variables are defined as above. Ehui and Spencer (1990; 1993) provide details about derivation of this equation. The same formula applies for the interspatial TFP measures by identifying s and t with two spatially separated zones.

In case of a net negative change in resource stock, the decrease in resource stock is treated as a cost, the TFP indices now become:

$$Ln(\frac{TFP_s}{TFP_t}) = \frac{1}{2} \sum_{i}^{m} (r_{is} + r_{it})\ Ln\ (\frac{Y_{is}}{Y_{it}})$$
$$- \frac{1}{2} \sum_{k}^{q} (v_{ks} + v_{kt})\ Ln(\frac{Z_{ks}}{Z_{kt}})$$
$$- \frac{1}{2} \sum_{j}^{n} (w_{js} + w_{jt})\ Ln(\frac{X_{js}}{X_{jt}})$$
$$- Ln\ (\frac{B_s}{B_t})$$

(3)

**An example**: These formulas were used to measure the sustainability and economic viability of some tropical farming systems in Nigeria. Four cropping systems, namely A, B, C, and D were evaluated over a two-year period (1986-88). The data and full description of the systems are reported in Lal and Ghuman (1989).

In system A, land was cleared manually and cropped by a local farmer. Yam, melon, and plantains were grown in 1986. In 1988, plantain, melon, and cassava were grown. In all the other systems, the land was cleared by a tractor equipped with a shear blade and cropped by the researchers. In system B, cassava, maize, and cowpea were planted in 1986 and only cassava was planted in 1988. In system C, maize, and cassava were planted in 1986 and rice in 1988. All crops in system C were grown in alleys formed by hedgerows of nitrogen-fixing trees or shrubs. In this system, known as alley cropping, the hedgerows were periodically pruned during the cropping season to prevent shading and reduce competition with food crops (Kang et al., 1989). In system D, plantain was grown during the 1986-88 period. No fertilizer was used in any of the cropping systems.

Input data used included labor, planting materials, and implements. Natural resource study was measured by levels of the major soil nutrients, nitrogen (N), phosphorous (P), and potassium (K). Changes in their abundance levels were accounted for by aggregating these three soil nutrients. Detailed explanations about the computation of output and input indices and their aggregation is reported in Ehui and Spencer (1990).

Intertemporal and interspatial productivity indices for the four cropping systems were calculated using equations (1) and (2) and are reported in Tables 11 and 12. In column I, there is no adjustment for changes in resource stock abundance levels and flows. Column II provides productivity measures allowing for variations in the resource stock only. In column III, full correction is made by accounting for both changes in the resource stock level and the flows.

From column III in Table 11, total factor productivity increased for systems B and C and declined for systems A and D. Systems B and C produced 6.25 and 11.58 times as much output in 1988 as in 1986 using the 1986 input bundle. Systems B and C can, therefore, be said to be sustainable over the two-year interval since after properly accounting for temporal differences in input quality and quantity and resource flows and stocks, they produced more than in the reference year (1986). Systems A and D produced only 0.22 and 0.88 as much output in 1988 as in 1986 using the 1986 input bundle. Thus, A and D can be said to be non-sustainable.

Completely accounting for changes in resource abundance levels and flows substantially alters the productivity measures (Table 11). This is particularly true for system C where the hedgerows trees fix atmospheric N and recycle nutrients and system D where the plantains heavily depleted the soil of its nutrients. Note that in system C, if the N contribution of the trees is not accounted for, the intertemporal productivity index is lower than unity (column 1), leading to the erroneous conclusion that the system is not sustainable. Soil nutrients increased by 31%, representing nearly 30% of the net revenue in 1988. This is important to the value of output which explains the high intertemporal productivity index number in system C.

Similarly, if the depleted resources in system D are not accounted for, the erroneous conclusion would be reached that the system is sustainable. Similar erroneous conclusions are reached when only changes in resource abundance levels are accounted for and flows are ignored (column II). The stock of soil nutrients in this system decreased by 23%, representing about 8% of the total cost of production. Systems A and B are relatively stable because, although

**Table 11.** Intertemporal total factor productivity sustainability indices for four cropping systems under experimental conditions, in southwestern Nigeria. 1986-1988[1]

| Systems[2] | No correction I[3] | Resource stock only II | Resource stock and flows III |
|---|---|---|---|
| System A | 0.20 | 0.19* | 0.22* |
| System B | 6.38 | 6.14* | 6.25* |
| System C | 0.02 | 0.01* | 12.23* |
| System D | 3.27 | 4.23** | 0.88** |

[1] Numbers with one star (*) indicate the case of a net positive change in resource abundance, while those with two stars (**) indicate the case of a net negative change in resource abundance levels; [2] See text for description of systems; [3] In column I there is no correction in soil resource stock and flows in Column II only resource stock use is corrected for. Column III allows for both the resource stock and flows. is no correction in soil resource stock and flows in Column II only resource stock use is corrected for. Column III allows for both the resource stock and flows.
(After Ehui and Spencer, 1990, 1993.)

soil nutrients increased, they represent only 0.7% and 0.1% of the total revenue in each system.

Results in Table 11 confirm that unless variations in the resource flows and stock are fully accounted for, TFP results will be biased. The bias depends on the magnitude of resource flow and stock. While an increased (reduced) resource stock level serves to reduce (increase) the productivity growth rate (column II), the associated change in the resource flow has the opposite effect (column III). A positive change in the resource stock level is a benefit to the farmer and thus contributes to improving the sustainability of the system. When the change in the resource stock is negative, the farmer faces a cost (though it is hidden) which negatively affects the system's sustainability.

The economic viability of cropping systems B, C, and D relative to A are compared in Table 12. In 1986, after accounting for changes in resource abundance and flows, systems B and C are shown to be relatively less productive than the reference base system. The interspatial TFP indices are estimated to be 0.73 and 0.76 for systems B and C, respectively, indicating that these systems use relatively more resources and produce a comparatively lower output than system A. Only system D (in which only plantain was grown) is more productive. In 1988, productivity indices for all the systems show a different pattern. With interspatial TFP indices of 9.26 and 1.12, systems B and C are now found to be more economically viable than system A. Similarly, with a TFP index of 0.14, system D is found to be less economically viable than the reference base system. The changes in productivity measures in 1988 compared

**Table 12.** Interspatial total factor productivity (economic viability) indices for four cropping systems under experimental conditions in southern Nigeria, during 1986 and 1988

| Systems | 1986 | | | | | | 1988 | | | | | |
| | No correction | Resource stock only | | Resource stock and flows | | | No correction | Resource stock only | | Resource stock and flows | | |
| | I | II | | III | | | I | II | | III | | |
| System A | 1 | 1 | | 1 | | | 1 | 1 | | 1 | | |
| System B | 1.73 | 2.02** | | 0.73** | | | 68.50 | 81.34** | | 9.26** | | |
| System C | 5.37 | 6.68** | | 0.76** | | | 0.37 | 0.36* | | 1.12* | | |
| System D | 0.06 | 0.18* | | 2.40* | | | 1.04 | 1.31** | | 0.14** | | |

Note: Refer to footnotes in Table 11 for details on the various systems, the interpretation meaning of the columns as well as the stars (* and **).

(After Ehui and Spencer, 1990, 1993.)

with 1986 are attributable to the changes in soil nutrient status over the two-year period. For example, in system C (where crops are grown in association with leguminous trees), soil nutrients increased by 2.3% in 1988 compared with system A, with a revenue share of about 6%. In system D, where only plantain was grown, chemical fertility was depleted over time. This is reflected in the lower 1988 productivity measure. In system D, soil nutrients decreased by 21% in this system compared to system A representing about 7% of the full cost faced by the farmer in 1988. Soil nutrients decreased by 16% for system B in 1988 representing about 10% of the total cost. As shown in Table 11, when variations in resource stock levels and the flows are not accounted for, productivity measures are biased. The biases depend on the magnitude of changes in resource stock levels.

# V. Research Priorities

The review highlights the potential of various components of improved crop-livestock production strategies for soil management and provides an alternative for resource poor-environments. However, further research in the following areas is necessary to fully exploit this potential:

- Baseline databases in various crop-livestock systems on environmental, agricultural, socio-economic, and livestock variables, their productive potential and constraints need to be defined using decision support systems, simulation modeling, and geographic information systems

- Policy studies on land tenure, use of external inputs, input subsidies, credit availability, and price policies for crop-livestock products

- Characterization and selection of food crops, forage and tree legumes, and grasses to environmental constraints for nutrient-poor environments

- Nitrogen fixation by forage, tree, and grain legumes and the proportion of legume N that is available to subsequent crop or crop intercropped with or undersown with legumes

- The rate of mineralization of N from organic matter, legume roots, mulch and crop residues varying in N, lignin and tannin contents

- Role of various tillage and stubble management systems in the maintenance of soil fertility, water-use efficiency and long-term crop yields

- The amounts of nutrient losses via crop removal, weeds, soil erosion, and other pathways and grains to prepare balance sheets for various crop-livestock systems

- Quantification of legume based crop-livestock systems on soil, water, and nutrient management and conservation

- The capacity of rangeland to support livestock and nutrient harvesting needs to be assessed at various rangeland cropland ratios

- A move from extensive to intensive livestock production will require improved handling, storage and land spreading techniques for manure and urine if nutrient losses are to be minimized. Strategies are required that synchronise manure and urine application and nutrient release with plant nutrient demands

- The complex cultural, technical and socio-economic issues involved in crop-livestock production soil management strategy necessitate an interdisciplinary research approach. An understanding of the key interactions between plants, animals, soils, and agricultural productivity is needed for developing scientifically sound and socially acceptable technologies that improve the role of livestock in natural resource management.

- At ILCA, efforts are being made to expand the methodology (V-B) to crop-livestock systems. This requires that additional data set be established. Additional data set to complement the data set used above includes, among others, weight gain, price per head and per kilogram of animal, stocking rates, crop residue yields and their prices or opportunity costs. In order to have more meaningful results, long-term experiments need to be established which allow for sufficient time for observation of trends in the variables being observed.

- Other approaches are also being explored at ILCA. While the TFP approach may be appropriate for examining trade offs between productivity and natural resource conservation, it is of limited usefulness for examining irreversible ecological processes like species extinction, loss of ancient forest, and salinization of soils. Safe minimum standards approach offer an opportunity for examining these types of questions (Swallow, 1992).

## References

Abate Tedla, Tekalign Mamo, and Getinet Gebeyehu. 1992. Integration of forage legumes into cereal cropping systems in Vertisols of the Ethiopian highlands. *Trop. Agric. (Trinidad)* 69 (1): 68-72.

Abate Tedla, M.A. Mohamed-Saleem, Tekalign Mamo, Alemu Tadesse, and Miressa Duffera. 1993. Grain, fodder and residue management. p. 103-137. In: Tekalign Mamo, Abiye Astatke, K.L. Srivastava, and Asgelil Dibabe (eds.), *Improved management of Vertisols for sustainable crop-livestock production in the Ethiopian highlands: Synthesis Report* 1986-1992. Technical Committee of the Joint Vertisol Project, Addis Ababa, Ethiopia.

Abdullahi, A. and G. Lombin. 1978. Long-term fertility studies at Samaru-Nigeria: Comparative effectiveness of separate and combined applications of mineral fertilizers and farmyard manure in maintaining soil productivity under continuous cultivation in the savanna. Samaru Misc. Pub. No. 75, Ahmadu Bello University, Zaria, Nigeria, 54 pp.

Abiye Astatke, S. Jutzi, and Abate Tedla. 1989. Sequential cropping of Vertisols in the Ethiopian highlands using a broadbed and furrow system. *ILCA Bulletin* 34:15-20.

Abiye Astatke, H. Airaksinen, and M.A. Mohamed-Saleem. 1991. Supplementary irrigation for sequential cropping of Vertisols in the Ethiopian highlands using broad and furrow land management system. *Agric. Water Management* 20 (3):173-185.

Adegboola, A.A. 1964. Forage crop research and development in Nigeria. *Nigerian Agric. J.* 1:34-39.

Adeoye, K.B. 1987. Some aspects of soil management in stylo based pastures in the subhumid zone of Nigeria. Terminal Report to ILCA, Kaduna, Nigeria. 79 pp.

Aduayi, E.A. and I. Haque. 1992. Screening forage and browse legumes germplasm to nutrient stresses: V. Highland leucaena germplasm for tolerance to phophorus deficiency in solution, soil and sand culture. Plant Science Division Working Document No. 18, ILCA, Addis Ababa, Ethiopia. 115 pp.

Anon. 1969. Annual Report for 1968. Part 2. Pasture Research Station, Kitale, Min. of Agric., Kenya. 47 pp.

Antle, J. M. and S.M. Capalbo. 1988. An introduction to recent developments in production theory and productivity measurement. p 17-95. In: S.M. Capalbo and J. M. Antle (eds.), *Agriculture productivity: Measurement and Explanation. Resources for the Future.* Washington D.C.

Attak-Krah, A.N., J.E. Sumberg, and L. Reynold. 1986. Leguminous fodder trees in the farming system: An overview research at the humid zone programme of ILCA in southwest Nigeria. p. 307-329. In: I. Haque, S.C. Jutzi and P.J.H. Neate (eds.), *Potential of forage legumes in farming systems of sub-Saharan Africa.* Proc. of a workshop held at ILCA, 16-19 September, 1985, Addis Ababa, Ethiopia.

Atta-Krah, A.N. and J.E. Sumberg. 1988. Studies with *Gliricidia sepium* for crop-livestock production systems of West Africa. *Agroforestry Systems* 6:67-118.

Atta-Krah, A.N. 1990. Alley farming with Leucaena: effect of short grazed fallows on soil fertility and crop yields. *Expl. Agric.* 26:1-10.

Ball, V.E. 1985 Output, input and productivity measurement in U.S. Agriculture. *Amer. J. Agri. Econ.* 67:475-487.

Bationo, A. and A.U. Mokwunye. 1991. Role of manures and crop residues in alleviating soil fertility constraints to crop production: With special reference to the Sahelian and Sudanian zones of West Africa. p. 217-226. In: A.U. Mokwunye (ed.), *Alleviating Soil Fertility Constraints to Increased Crop Production in West Africa*. Kluwer Academic Publishers. Dordrecht, Boston, London.

Blumenthal, M.Z., V. P. Quach, and P.G.E. Searle. 1988. Effect of soybean population density on soybean yield, nitrogen accumulation and residual nitrogen. *Aust. J. Expl. Agric.* 29:99-106.

Boultwood, J.N. 1964. Two valuable perennial legumes — hourse marmalade (*Desmodium discolor*) and Kuru vine (*Desmodium introtum*). *Rhod. Agric. J.* 61:70-72.

Breman, H. and C.T. de Witt. 1983. Rangeland productivity and exploitation in the Sahel. *Sci.* 221:1341-1347.

Breman, H. and N. Traoré. 1986. Analyse des conditions de l'élevage et proposition de politiques et de programmes. Burkina Faso. OECD, Paris, SAHEL D (86)300. 202 pp.

Brumby, P. 1991. Livestock, food production and land degradation. p. 403-412. In: *Evaluation for sustaibable land management in the developing world*. Volume 2. Technical Papers. Bangkok, Thailand, IBSRAM, Proc. No. 12 (2).

Bumpus, E.D. 1957. Legume nodulation in Kenya. *East Afr. Agric. & For. J.* 22:91-99.

Cobbina, J., K. Mulongoy, and A.N. Atta-Krah. 1988. Soil factors influencing initial growth of *Leucaena* and *Gliricidia* in Southern Nigeria. ILCA Humid Zone Programme Document.

Commonwealth Agricultural Bureaux (CAB), Agricultural Research Council. 1984. *The Nutrient Requirements of Ruminant Livestock*, Supplement No. 1, The Lavenham Press Ltd., Lavenham, Suffolk. 45 pp.

Coppock, L.D. and J.D. Reed. 1992. Cultivated and native browse legumes as calf supplements in Ethiopia. *J. Range Management* 45:231-238.

Cottenie, A., B.T. Kang, L. Kiekens, and A. Sajjapongse. 1981. Micronutrient status. p. 149-163. In: D.J. Greenland (ed.), *Characterization of soils*. Clarendon Press, Oxford, England .

Couto, W. 1983. Effect of excess water in an Oxisol on ammonium, nitrate, iron and manganese availability and nutrient uptake of two tropical forage species. *Plant and Soil* 73:159-166.

Danso, S.K.A. and I. Papastylianou. 1992. Evaluation of the nitrogen contribution of legumes to subsequent cereal. *J. Agric. Sci. (Camb)* 119:13-18.

Davidson, N. and R. Galloway (eds.), 1993. *Productive use of saline land*. Proc. of a workshop held at Perth, Western Australia, 10-14 May, 1991. ACIAR Proc. No. 42. 124 pp.

Dennison, E.B. 1961. The value of farmyard manure in maintaining fertility in northern Nigeria. *Emp. J. Expl. Agric.* 29:116: 330-336.

Dumanski, J. 1987. Evaluating the sustainability of agricultural systems. p. 195-205. In: *Africaland—land development and management of acid soils in Africa II*: IBSRAM Proceedings No. 7 (International Board for Soil Research and Management)., Bangkok, Thailand.

Economic Commission for Africa/Food and Agriculture Organization (ECA/FAO). 1991. Land degradation and food supply: Issues and operations for food self-sufficiency in Africa. Addis Ababa, Ethiopia. 16 pp.

Edjigayehu Seyoum, and J. McIntire. 1987. Literature review and economic analysis of crop response to phosphate rocks in eastern Africa. *ILCA Bull.* 29:12-19.

Ehui, S.K. and D.S.C. Spencer. 1990. Indices for measuring the sustainability and economic viability of farming systems. Resource and Crop Management Program Research Monograph No. 3. International Institute of Tropical Agriculture, Ibadan, Nigeria. 28 pp.

Ehui, S.K. and D.S.C. Spencer 1993. Measuring the sustainability and economic viability of tropical farming systems: a model from sub-Saharan Africa. *Agricultural Economics* 9:279-296.

FAO. 1986. African Agriculture: the next 25 years. Annex II. The Land Resource Base. FAO, Rome, Italy.

Faye, B., C. Grillet, and Abebe Tessema. 1983. Report on survey of trace element status in forages and blood of domestic ruminants in Ethiopia. National Veterinary Institute, Debre Zeit, Ethiopia.

Foy, C.D. 1988. Plant adaptation to acid aluminum toxic soils. *Comm. Soil Sci. & Plant Analysis* 19:959-987.

Francis, C.M. and A.C. Devitt. 1969. The effect of waterlogging on the growth and isoflavone concentration of *Trifolium subterraneum* L. *Aust. J. Agric. Res.* 20:819-825.

Francis, C.M. and M.L. Poole. 1973. Effect of waterlogging on the growth of annual *Medicago* species. *Aust. J. Expl. Agric. and Animal Husbandry* 13:711-713.

Geiger, S.C., A. Manu, and A. Bationo. 1992. Changes in a sandy Sahelian soil following crop residue and fertilizer additions. *Soil Sci. Soc. Am. J.* 56:172-177.

Government of Niger. 1992. La situation du Niger à long terme en graphique. Ministry of Plan. Niamey, Niger. 29 pp.

Gryseels G. and F.M. Anderson. 1983. Research on farm and livestock productivity in the central highlands: Initial results, 1977-80. *ILCA Research Report No. 4*. ILCA, Addis Ababa, Ethiopia.

Haggar, I. 1971. The production and management of *Stylosanthes gracilis* at Shika, Nigeria: 1. Sown pasture. *J. Agric. Sci* 77:427-436.

Hammond, L.L., S.H. Chien, and A.U. Mokwunye. 1986. Agronomic value of unacidulated and partially acidulated phosphate rocks indigenous to the tropics. *Adv. Agron.* 40:89-140.

Haque, I. and S. Jutzi. 1984. Nitrogen fixation by forage legumes in sub-Saharan Africa: Potential and Limitations. *ILCA Bull.* 20: 2-13.

Haque, I., S.C. Jutzi, and P.J.H. Neate (eds.), 1986. *Potentials of forage legumes in farming systems of sub-Saharan Africa.* Proc. of a workshop held at ILCA. 16-19 September, 1985. Addis Ababa, Ethiopia. 575 pp.

Haque, I., L.A. Nnadi, and M.A. Mohamed-Saleem. 1986. Phosphorus management with special reference to forage legumes in SSA. p. 100-119. In: I. Haque; S.C. Jutzi and P.J.H. Neate (eds.), *Potentials of forage legumes in farming systems of sub-Sahara African.* Proc. of a workshop held at ILCA, 16-19 September, 1985. Addis Ababa, Ethiopia.

Haque, I. 1987. Molybdenum in soils and plants and its potential importance to livestock nutrition with special reference to sub-Saharan Africa. *ILCA Bull.* 26: 20-28.

Haque, I. and J.C. Tothill. 1988. Forages and pastures in mixed farming systems of sub-Saharan Africa. p. 107-131. In: *Land management and management of acid soils in Africa. Proc. of the second regional workshop on land management and management of acid soils in Africa.* 9-16 April, 1987. Lusaka and Kasama, Zambia.

Haque, I. 1990. Nitrogen fixation of the forage legumes and their residual effects on wheat growth and yield. p. 40-52. In: M.Gueye, K. Mulongoy, and Y. Dommergues (eds.), *Maximizing biological nitrogen fixation for agriculture and forestry production in Africa. Proc. of 2rd AABNF Conference: Collection Actes de L'ISRA*, Vol. 2. 7-12 November, 1988. Dakar, Senegal.

Haque, I. 1993. Management of soil productivity under population pressure. Consortium for Integrated Natural Resource Management for the Highlands of East and Central Africa. Environmental Sciences Division, ILCA, Addis Ababa, Ethiopia. 15 pp.

Haque, I. 1993. Nutrient management and cycling in legume based crop-livestock systems of cool tropics. p. 7-12. In: ILCA Project Protocols: 1993 portifolio. ILCA. Addis Ababa, Ethiopia.

Haque, I., E.A. Aduayi, and S. Sibanda. Copper in soils, plants and ruminant animal nutrition with special reference to sub-Saharan Africa. *J. Plant Nutrition* 16:2149-2212.

Heichel, G.H. 1987. Legume nitrogen: symbiotic fixation and recovery by subsequent crops. p. 63-80. In: A.R. Helsel (ed.). *Energy in Plant Nutrition and Pest Control.* Elsevier Science Publisher. Amsterdam. The Netherlands.

Horrell, C.R. and M.N. Court. 1965. Effect of the legume *Stylosanthes gracilis* on pasture yields at Serere, Uganda. *J. Br. Grasel Soc.* 20:72-76.

Hubick, K.T., G.D. Farquhar, and R. Shorter. 1986. Correlation between water use efficiency and carbon isotope discrimination in diverse peanut (*Arachis*) germplasm. *Aust. J. Plant Physiol* 13:803-816.

Hulugalle, N.R. 1989. Effect of tied ridges and undersown *Stylosanthes hamata* (L.) on soil properties and growth of maize in the Sudan savannah of Burkina Faso. *Agric., Ecosyst. and Environ.* 25:39-51.

Hulugalle, N.R. and B.T. Kang. 1990. Effect of hedgerow species in alley cropping systems on surface soil physical properties of an oxic Paleustalf in south western Nigeria. *J. Agric. Sci. (Camb.)* 114:301-307.

I'Ons, J.H. 1968. The development of tropical pastures for the Swaziland Midland. In: *Proc. of the Grassland Society of Southern Africa* 3:67-73.

IFDC. 1984. Annual Report, IFDC, Muscle Shoals, Alabama, U.S.A.

ILCA. 1987. Fodder offtake and crop yields in alley farming. p. 62. Annual Report. ILCA. Addis Ababa, Ethiopia.

ILCA. 1988a. Sustainable production from livestock in sub-Saharan Africa: ILCA's programme plans and funding requirements, 1989-1993. ILCA, Addis Ababa, Ethiopia.

ILCA. 1988b. Nutrient management and water use studies in multipurpose trees — highlands. p. 93-94. ILCA Annual Report. Addis Ababa, Ethiopia.

ILCA. 1988c. Effect on crop yield in alley farming of offtake of fodder at different times. p. 38-39. Annual Report, ILCA. Addis Ababa, Ethiopia.

ILCA. 1989a. Alley cropping with grazed fallow using *Gliricidia sepium* on degraded land. p. 50-51. Annual Report, ILCA. Addis Ababa, Ethiopia.

ILCA. 1989b. Evaluation of rhizobium strains in *Sesbania sesban*. p. 81. Annual Report, ILCA, Addis Ababa, Ethiopia.

ILCA. 1990. Stability and sustainability of livestock in mixed farming systems: Research project and experimental protocols. ILCA Semi-arid Zone Programme, Sahelian Center, Niamey, Niger.

ILCA. 1991a. ILCA Annual Programme Report. p. 152. ILCA. Addis Ababa, Ethiopia.

ILCA. 1991b. Annual Report and Program Highlights. p. 11-12 and 20-22. ILCA. Addis Ababa, Ethiopia.

ILCA. 1992. ILCA Annual Programme Report. ILCA. Addis Ababa, Ethiopia.

ILCA. 1993a. ILCA's Long Term Strategy, 1993-2000. ILCA Addis Ababa, Ethiopia. 102 pp.

ILCA. 1993b. Sustainable production from livestock: ILCA's medium term plan, 1994-1998. p. 53-54. ILCA. Addis Ababa, Ethiopia.

Jabbar, M.A., J. Cobbina, and L. Reynold. 1992. Optimum fodder mulch allocation of tree foliage under alley farming in southeast Nigeria. *Agroforestry Systems* 20: 187-198.

Jones, M. J. 1973. A review of the use of rock phosphate as fertilizers in francophone West Africa. Samaru Miscellaneous Paper 43. Institute for Agricultural Research, Samaru, Ahmadu Bello University, Zaria, Nigeria. 10 pp.

Jutzi, S.C., I. Haque, and J.C. Tothill 1985. Prospects low for native pasture improvement on black soils. *ILCA Newsletter* 4 (3):3-4.

Jutzi, S.C., I. Haque, and Abate Tedla 1987. The production of animal feed in Ethiopian highlands: potential and limitations. p. 141-142. In: *IAR Proc.. 1st National Livestock Improvement Conference*. 11-13 February, 1987. Addis Ababa, Ethiopia,,

Kamara, C.S. and I. Haque, 1988. Soil moisture related properties of Vertisols in the Ethiopian Highlands. p. 201-222. In: S.C. Jutzi, I. Haque, J. McIntire, and J.E.S. Stares (eds.), *Management of Vertisols in sub-Saharan Africa*. Proc. of a conference held at ILCA. 31 August to 4 September, 1987. Addis Ababa, Ethiopia.

Kamara, C.S. and I. Haque. 1991. Intercropping maize and forage type cowpeas in the Ethiopian highlands: I. Growth and dry matter yields. II. Soil moisture, soil temperature, solar radiation regimes and water use efficiency. Plant Science Division Working Document No. 14. ILCA. Addis Ababa, Ethiopia.

Kang, B.T. and O.A. Osiname. 1972. Micronutrient investigations in West Africa. Ford Foundation/IITA/IRAT International Seminar on Tropical Soil Research. 22-26 May, 1972. Ibadan, Nigeria. 15 pp.

Kang, B.T., G.F. Wilson, and L. Spikes. 1981a. Alley cropping maize (*Zea mays* L.) and leucaena (*Leucaena leucocephala* LAM) de Witt in southern Nigeria. *Plant and Soil* 163: 165-179.

Kang, B.T., L. Sipkens, G.F. Wilson, and D. Nandju. 1981b. Leucaena (*Leucaena leucocephala* (LAM) de Wit) prunings as a nitrogen source for maize (*Zea mays* L.). *Fertilizer Research* 2:279-287.

Kang, B. T. and O.A. Osiname. 1985. Micronutrient problems in Tropical Africa. *Fertilizer Research* 7:131-150.

Kang, B.T., A.C.B.M. van der Kruijs, and D.C. Couper. 1989. Alley cropping for food production in the humid and sub humid tropics. p 16-26. In: B.T. Kang and L. Reynolds (eds.), *Alley farming in the humid and subhumid tropics*. Proc. of an international workshop. 10 - 14 March, 1986. Ottawa, Ont. IDRC. International Institute of Tropical Agriculture, Ibadan, Nigeria.

Kang, B.T., L. Reynolds and A.N. Atta-Krah. 1990. Alley farming. *Adv. Agron.* 43:315-359.

Kang, B.T. and K. Mulongoy. 1992. Nitrogen contribution of woody legume in alley cropping systems. p. 367-375. In: K. Mulongoy, M. Gueye, and D.S.C. Spencer (eds.), *Biological Nitrogen Fixation and Sustainability of Tropical Agriculture*. John Wiley & Sons, Chichester, England.

Kanwar, J.S. and M.S. Mudahaker. 1986. *Fertilizer Sulphur and Food Production*. Martinus Nijhoff/Dr. W. Junk. Boston, MA.

Kategile, J. A. (ed.). 1985. Pasture improvement research in eastern and southern Africa. Proc. of a workshop held in Harare, Zimbabwe, 17-21 September, 1984. 508 pp.

Keya, N.C.O. and C.L.M. van Eijnathan. 1975. Studies on oversowing of natural grassland in Kenya: effects of seed inoculation and pelleting on the nodulation and productivity of *Desmodium uncinatum* (JOCO) DC. *East Afr. Agric. & For. J.* 40:351-358.

Kouamé, C.N. 1991. Effects of Stylosanthes interplanting on millet grain yield, herbage yield, water use efficiency, and yields of subsequent millet crop. Ph.D. Dissertation. University of Florida, Gainsville, Florida. 119 pp.

Kumar, A. and I.P. Abrol. 1986. *Grasses in alkali soils*. Central Soil Salinity Research Institute. Karnal, India. 95 pp.

Kumarasinghe, K.S., C. Kirda, A.R.A.G. Mohammed, F. Zapata, and S.K.A. Danso. 1992. $^{13}$C isotope discrimination correlates with biological nitrogen fixation in soybean *(Glycine max (L), Merill)*. *Plant and Soil* 130:145-147.

Lal, R. 1987. Managing the soils of sub-Saharan Africa. *Science* 236:1069-1076.

Lal, R., G.F. Wilson, and B.N. Okigbo. 1978. No till farming after various grasses and leguminous cover crops in tropical Alfisol. I. crop performance. *Field Crops Research* 1:71-84.

Lal, R., G.F. Wilson, and B.N. Okigbo. 1979. Changes in properties of an Alfisol produced by various cover crops. *Soil Science* 127 (6):377-382.

Lal, R. and B.S. Ghuman. 1989. Effects of deforestation and land use on soil, hydrology, microclimate and productivity in the humid tropics. Final Report (January 1984–June 1989) of a joint IITA/UNU/Ohio State University Project on Climatic, Biotic and Human Interactions in the Humid Tropics. IITA. Ibadan, Nigeria. (Unpublished).

Lal, R. 1990. Soil erosion and land degradation: the global risks. *Advances in Soil Science* 11:129-172.

Lal, R. 1991. Myths and scientific realities of agroforestry as a strategy for sustainable management for soils in the tropics. *Adv. Soil Sci.* 15:91-137.

Lee, L.K. and J.J. Goebel. 1986. Defining erosion potential on cropland: A comparison of the land capability class-subclass system with RKLS/T categories. *J. Soil and Water Conser.* 41:41 - 44.

Lynam, J.K. and R.W. Herdt. 1989. Sense and sustainability: sustainability as an objective in international agricultural research. *Agric. Econ.* 3:381 - 398.

Malcolm, C. 1992. Use of halophyte forages for rehabilitation of degraded lands. p. 1-21. In: *Int. Workshop on halophytes for reclamation of saline wastelands and as a resource for livestock problems and prospects*. UNEP (Nairobi)/The University of Adelaide, S. Australia. 22-27 November, 1992. Nairobi, Kenya,

Malik, K. A., Z. Aslam, and M. Naqvi. 1986. *Kallar grass: a plant for saline land*. Nuclear Institute for Agriculture and Biology. Faisalabad, Pakistan. 93 pp.

Marshall, T. and A.J. Millington. 1967. Flooding tolerance of some Western Australian pasture legumes. *Aust. J. Expl. Agric. and Animal Husb.* 7:367-371.

McDowell, L.R., J.H. Conrad, and G.L. Ellis. 1982. Research in mineral deficiencies for grazing ruminants. Annual Report, AID Mineral Research Project, University of Florida, Gainesville, USA.

McDowell, L.R., J.H. Conrad, and G.L. Ellis. 1984. Mineral deficiencies and imbalance and their diagnosis. p.67-88. Proc. of the Symposium on Herbivore Nutrition in Sub-tropics and Tropics. Problems and Prospects. Pretoria, South Africa.

McIvor, J.G. 1976. The effect of waterlogging on the growth of *Stylosanthes guyanensis*. *Trop. Grass*. 10 (3):173-178.

McNamara, R.S. 1992. A global population policy to advance human development in the twenty-first century, with special reference to sub-Saharan Africa. Global Coalition for Africa. Kampala, Uganda. 54 pp.

Mohamed-Saleem, M.A. 1985. Effect of sowing time on grain and fodder potential of sorghum undersown with stylo in the subhumid zone of Nigeria. *Trop. Agric. (Trinidad)* 62:151-153.

Mohamed-Saleem, M.A. and R.M. Otsyina. 1986. Grain yields of maize and the nitrogen contribution following *Stylosanthes* pasture in the Nigerian subhumid zone. *Expl. Agric*. 22:207-214.

Mohammed-Saleem, M.A., H. Suleiman, and R.M. Otsyina. 1986. Fodder banks: for pastoralists or farmers? p. 420-437 In: I. Haque; S.C. Jutzi, and P.J.N. Neate (eds.) *Potentials of forage legumes in farming systems of sub-Saharan Africa*. Proc. of Workshop held at ILCA. September 16-19, 1985. Addis Ababa, Ethiopia.

Mohamed-Saleem, M.A. and R.M. Otsyina. 1987. Effect of the naturally occurring salt "kanwa" as a fertilizer on the productivity of Stylosanthes in the subhumid zone of Nigeria. *Fertilizer Research* 13:3-11.

Mtimumi, J.P. 1982. Identification of mineral deficiencies in soil, plant and animal tissues as constraints to cattle in Malawi. Ph.D. Dissertation. University of Florida, Gainesville, FL.

Mugwira, L.M. and I. Haque. 1991. Variability in the growth and mineral nutrition of African clovers. *J. Plant Nutrition* 14 (6):553-569.

Mugwira, L.M. and I. Haque. 1993a. Screening forage and browse legumes germplasm to nutrient stresses: I. Tolerance of *Medicago sativa* L. to aluminum and low phosphorus in soils and nutrient solutions. *J. Plant Nutrition* 16 (1): 17-35.

Mugwira, L.M. and I. Haque. 1993b. Screening forage and browse legumes germplasm to nutrient stresses: II. Tolerance of *Lablab purpureus* L. to acidity and low phosphorus in two acid soils. *J. Plant Nutrition* 16(1): 37-50.

Mugwira, L.M. and I. Haque. 1993c. Screening forage and browse legumes germplasm to nutrient stresses: III. Tolerance of sesbanias to aluminum and low phosphorus in acid soils and nutrient solutions. *J. Plant Nutrition* 16(1):51-66.

Mugwira, L.M. and I. Haque. 1993d. Screening forage and browse legumes germplasm to nutrient stresses: IV. Growth rates of sesbanias as affected by aluminum and low phosphorus in acid soils and nutrient solutions *J. Plant Nutrition*. 16 (1): 67-83

Mulongoy, K. 1986. Potential of *Sesbania rostrata* (berm) as nitrogen source in alley cropping systems. *Biology Agric. and Hort*. 3:341-346.

Mulongoy, K. and B.T. Kang. 1986. The role and potential of forage legumes in alley cropping, live mulch and rotation systems in humid and sub-humid tropical Africa. p. 212-231. In: I. Haque, S. Jutzi, and P.J.H. Neate (eds.), *Potentials of forage legumes in farming systems of sub-Saharan Africa*. Proc. of a workshop held at ILCA. 16-19 September, 1985. Addis Ababa, Ethiopia.

Nnadi, L. A. and I. Haque. 1985. Estimating phosphorus requirements of native Ethiopian clovers using phosphorus sorption isotherms. In: *International conference on soil fertility, soil tillage and post clearing land degradation in the humid tropics*. 21-26 July, 1985. *Ibadan, Nigeria*.

Nnadi, L.A. and I. Haque. 1986. Forage legume cereal systems: improvement of soil fertility and agricultural production with particular reference to sub-Saharan Africa. p. 330-362. In: I. Haque; S.C. Jutzi, and P.J.H. Neate (eds.) *Potentials of forage legumes in farming systems of sub-Saharan Africa*. Proc. of a workshop held at ILCA. 16-19 September, 1985. Addis Ababa, Ethiopia.

Nnadi, L.A. and I. Haque. 1988a. Forage legumes in African crop-livestock production systems. *ILCA Bull*. 30:10-19.

Nnadi, L.A. and I. Haque. 1988b. Root nitrogen transformation and mineral composition in selected forage legumes. *J. Agric. Sci*. 111:513-518.

Nnadi, L.A. and I. Haque. 1988c. Agronomic effectiveness of rockphosphates in an Andept of Ethiopia. *Comm. Soil Sci. & Plant Analysis* 19(1): 79-90.

Nnadi, L.A. and I. Haque. 1990. Performance of legume maize intercrops on an upland soil of Ethiopian highlands. *East Afr. Agric. & For J.* 55 (3):93-1-2.

Norris, D.O. 1966. The legumes and their associated rhizobium. p. 89-105. In: *Tropical Pastures*. Faber and Faber Ltd. London.

Oke, O.L. 1967. Nitrogen fixing capacity of some Nigerian legumes. *Expl. Agric*. 3:315-321.

Otchere, E.O. 1986. The effect of supplementary feeding of Bunaji cattle in the sub-humid zone of Nigeria. p. 351-364. In: R. von Kaufmann, S. Chater and R. Blench (eds.), *Livestock systems research in Nigeria's sub-humid zone*. Proc. of the Second ILCA/NAPRI Symposium. Oct. 29-Nov 2, 1984. Kaduna, Nigeria. ILCA. Addis Ababa, Ethiopia.

Papastylianou, I. and D.W. Puckridge. 1983. Stem nitrate nitrogen and yield of wheat in a permanent rotation experiment. *Aust. J. Agric. Resh*. 34:599-606.

Papastylianou, I. and Th, Samios. 1987. Comparison of rotation in which barley for grain fallows woolypod vetch or forage barley. *J. Agric. Sci. (Camb)* 108:609-615.

Pichot, J. and P. Roche. 1972. Phosphore dans les sols tropicaux. *Agron. Trop*. 27:939-965.

Pierce, F.J., W.E. Larson, R.H. Dowdy, and W.A.P. Graham. 1983. Productivity of soils: assessing long-term change due to erosion. *J. Soil and Water Conservation* 38: 39 - 44.

Powell, J.M. 1986. Manure for cropping: A case study from Central Nigeria. *Expl. Agric.* 22:15-24.

Powell, J.M. and M.A. Mohamed-Saleem. 1987. Nitrogen and phosphorus transfers in a crop-livestock system in West Africa. *Agric. Syst.* 25:261-277.

Powell, J.M., F.M. Hons, and G.G. McBee. 1991. Nutrient and carbohydrate partitioning in sorghum stover. *Agron. J.* 83:933-937.

Powell, J.M. and F.N. Ikpe. 1992. Livestock management and nutrient cycles in semi-arid sub-saharan Africa. In: Livestock sustaining human lives. Information Centre for Low-External-Input and Sustainable Agriculture (ILEIA) *Newsletter* (3) 92:13-14.

Powell, J.M., F.N. Ikpe, and Z.C. Somda. 1992. Nutrient cycling in low external input millet-based mixed farming systems of West Africa. p. 70. Agronomy Abstracts. American Society of Agronomy. Madison, Wisconsin.

Powell, J.M. and L.K. Fussell. 1993. Nutrient and carbohydrate partitioning in millet. *Agron. J.* (in press).

Powell, J.M., S. Fernandez-Rivera, and S. Höfs. 1993. Sheep diet effects on nutrient cycling in semi-arid mixed farming systems of West Africa. *Agric. Ecosys. & Environ.* (under review).

Powell, J.M. and T.O. Williams. 1993. Livestock, nutrient cycling and sustainable agriculture in the West African Sahel. Gatekeeper Series No. SA37, International Institute for Environment and Development (IIED), London.

Putman, J.W. 1986. A conservation policy monograph. *J. Soil and Water Conservation* 41: 406 - 409.

Quilfen, J.P. and P. Milleville. 1981. Résidus de culture et fumure animale: un aspect des relations agriculture-élevage dans le nord de la Haute Volta. *L'Agron. Trop.* 38: 206-212.

Ruess, R.W. 1987. The role of large herbivors in nutrient cycling of tropical savannas. p. 67-91. In: B.H. Walker (ed.), *Determinants of tropical savannas*. The International Union of Biological Sciences, Monograph No. 4, IRL Press, Oxford.

Sanchez, P.A. and G. Uehara 1980. Management considerations for acid soils with high phosphorus fixation capacity. p. 471-514. In: F.E. Khasawneh, E.C. Sample and E.J. Kamprath, (eds.), *The Role of Phosphorus in Agriculture*. ASA/SSA. Madison, Wisconsin, USA.

Sanchez, P.A. and J.G. Salinas. 1981. Low input technologies for managing Oxisols and Ultisols in tropical soils. *Adv. Agron.* 34:279-406.

Sanchez, P.A. and J.J. Nicholaides. 1982. Plant nutrition in relation to soil constraints in developing world. Consultants Report, TAC Secretariat. FAO. Rome, Italy.

Sanchez, P.A. 1982. A legume based pasture production strategy for acid infertile soils of tropical America. In: W. Kusson, S.A. El-Swaify, and J. Mannering (eds.), *Soil erosion and conservation in the tropics*. Special Publication 43:97-120. American Society of Agronomy, Madison, Wiconsin, USA.

Sandford, S. 1990. Integrated cropping-livestock systems for dryland farming in Africa. p. 861-872. In: P.W. Unger, T.V. Sneed, W.R. Jordan, and R. Jansen (eds.), *Challenges in dryland agriculture - A global perspective*, Proc. of the International Conference on Dryland Farming. 15-19 August, 1988. Amarillo/Bushland, TX, Texas Agricultural Experiment Sta., College Station.

Sanginga, N., N. Ayanaba, and K. Mulongoy. 1984. Effect of inoculation and mineral nutrients on nodulation and growth of *Leucaena leucocephala* (Lam) deWit. p. 419-427. In: H. Ssali, (ed.), *Proceedings of the first conference of the African association for biological nitrogen fixation*, (AABNF). 23-27 July, 1984. Nairobi, Kenya.

Sanginga, N., K. Mulongoy, and N. Ayanaba. 1988a. Response of *Leucaena/-Rhizobium* symbiosis to mineral nutrients in southwestern Nigeria. *Plant and Soil* 112:121-127.

Sanginga, N., K. Mulongoy, and A. Ayanaba. 1988b. Nodulation and growth of *Leucaena leucocaphala* (Lam.) de Wit as affected by inoculation and nitrogen fertilizer. *Plant and Soil* 112:12-135.

Sanginga, N., K. Mulongoy, and A. Ayanaba. 1988c. Nitrogen contribution of *Leucaena/rhizobium* symbiosis to soil and a subsequent maize crop. *Plant and Soil* 112:137-141.

Sanginga, N., K. Mulongoy, and M.J. Swift. 1989. Contribution of nitrogen by *Leucaena leucocephala* and *Eucalyptus grandis* to soils and a subsequent maize crop. p. 253-258. In: *Trees for development in SSA*. Proc. of a Regional Seminar held by the International Foundation for Science (IFS). ICRAF House. 15-20 February, 1989. Nairobi, Kenya.

Sanginga, N., K. Mulongoy, and N. Ayanaba. 1989a. Nitrogen fixation of field inoculated *Leucaena leucocephala* (Lam.) de Wit estimated by the $^{15}N$ and the difference methods. *Plant and Soil* 117:269-274.

Sanginga, N., K. Mulongoy, and N. Ayanaba. 1989b. Effectivity of indigenous rhizobia for nodulation and early nitrogen fixation with *Leucaena leucocephala* grown in Nigerian soils. *Soil Biol. & Biochem.* 21:231-235.

Sanginga, N., G.D. Bowen, and S.K.A. Danso. 1991. Intra-specific variation in growth and P accumulation of *Leucaena leucocephala* and *Gliricidia sepium* as influenced by soil phosphate status. *Plant and Soil*: 201-208.

Siebert, B.D.A., R.A. Romero, R.G. Hunter, J.J. Megerrity, Lynch, J.D. Glasgow, and M.J. Breen. 1978. Partitioning intake and outflow of nitrogen and water in cattle grazing tropical pastures. *Aust. J. Agric. Resh.* 29:631-644.

Spain, J.M., C.A. Francis, R.H. Howler, and F. Calvo. 1975. Differential species and varietal tolerance to soil acidity in tropical crops and pastures. p. 308-329. In: E. Bornemiza and A. Alvarado (eds.), *Soil management in Tropical America*. North Carolina State Univ., Releigh, NC, USA.

de Souza, D.I.A. 1969. Legume nodulation and nitrogen fixation studies in Kenya. *East Afr. Agric. & For. J.* 34:299-315.

Stoorvogel, J.J. and E.M.A. Smaling. 1990. Assessment of soil nutrient depletion in Sub-Saharan Africa: 1983-2000. The Winand Staring Centre, Wageningen, the Netherlands.

Sumberg, J.E. 1985. Small ruminant feed production in a farming systems context. p. 41-46. In: J.E. Sumberg and K. Cassaday (eds.), *Sheep and goats in humid West Africa*. Proc. of a workshop held at ILCA. 23-26 January, 1984. Ibadan, Nigeria,

Sumberg, J.E., J. McIntire; C. Okali, and A. Atta-Krah. 1987. Economic analysis of alley farming with small ruminants. *ILCA Bull.* 28-2-6.

Sumberg, J.E. and A.N. Atta-Krah. 1988. The potential of alley farming in humid West Africa: a re-evaluation. *Agroforestry Systems* 6:163-168.

Swallow, B.M. 1992. Potential economic impacts: p. 11-25. In: *Potential for Impact: ILCA looks to the future.* ILCA Working Paper 2. ILCA. Addis Ababa, Ethiopia.

Swift, M.J., P.G.H. Frost, B.M. Campbell, J.C. Hatton, and K. Wilson. 1989. *Nutrient cycling in farming systems derived from savanna: Perspectives and challenges.* p. 63-76. In: M. Clarholm, and D. Berstrom (eds.), *Ecology of Arid Lands.* Kluwer Academic Publishers.

Szabolcs, C. 1992. *Salt affected soils.* Proc. Intern. Symp. on Strategies for Utilizing Salt Affected Lands. 17-25 February 1992. Bangkok, Thailand.

Tarawali, G. 1991. The residual effect of Stylosanthes fodder banks on maize yield at several locations in Nigeria. *Trop. Grass.* 25:26-31.

Tening, A.S., G. Tarawali, M.A. Mohammed-Saleem, K.B. Adeoye, and J.A. I. Omueti. 1992. Management and nutrient requirements of stylosanthes in the pasture and cropping systems in the sub-humid zone of Nigeria. Paper Presented in Int. Workshop on Stylosanthes as a Forage and Fallow Crop held at Kaduna, Nigeria. October 26-30, 1992.

Ternouth, J.H. 1989. Endogenous losses of phosphorus by sheep. *J. Agric. Sci.* 113:291-297.

Thomas, D. 1973. Nitrogen from tropical pasture legumes on the African continent. *Herbage Abstracts* 43 (2):33-39.

Tothill, J.C. 1986. The role of legumes in farming systems of sub-Saharan Africa. p. 162-185. In: I. Haque; S.C. Jutzi, and P.J.H. Neate. (eds.), *Potential of forage legumes in sub-Saharan Africa.* Proceedings of a Workshop held at ILCA. 16-19 September, 1985. Addis Ababa, Ethiopia.

Tothill, J. 1987. Application of agroforestry to African crop-livestock farming systems. *ILCA Bull.* 29:20-23.

Unger, P.W., B.A. Stewart, J.F. Parr, and R.P. Singh. 1991. Crop residue management and tillage methods for conserving soil and water in semi-arid regions. *Soil and Till.* Res. 20:219-240.

van Raay, H.G.T. 1975. Rural Planning in a Savannah Region. Rotterdam. University Press.

Watson, E.R. 1963. The influence of subterranean clover pastures on soil fertility: I. Short term effects. *Aust. J. Agric. Res.* 14:796-807.

Watson, K.A. and P.R. Goldsworthy. 1964. Soil fertility investigations in the middle belt of Nigeria. *Emp. J. Expl. Agric.* 22: 290-302.

Wendt, W.B. 1970. Responses of pasture species in Eastern Uganda to phosphorus, sulphur and potassium. *East Afric. Agric. & For. J.* 36:211-219.

Wendt, W.B. 1971. Effects of inoculation and fertilizers on *Desmodium intortum* at Serere, Uganda. *East Afr. Agric. and For. J.* 36:317-321.

West, C.P., A.P. Mallarino, W.F. Wedin, and D.B. Marx. 1989. Spatial variability of soil chemical properties in grazed pastures. *Soil Sci. Soc. Am. J.* 53:784-789.

Whitman, P.C., M. Seitlheko, M.E. Siregar, A.K. Chudasama, and R.R. Javier. 1984. Short-term flooding tolerance of seventeen commercial tropical pasture legumes. *Trop. Grass.* 18 (2):91-96.

Wilson, G.F. and B.T. Kang. 1980. Developing stable and productive biological cropping systems for the humid tropics. Proc. of a workshop on biological husbandry: A scientific approach to organic farming. 28-30 August, 1980. Wye College, London.

Wilson, G.F., R. Lal, and B.N. Okigbo. 1982. Effects of cover crops on soil structure and on yield of subsequent arable crops grown under strip tillage on an eroded Alfisol. *Soil and Till. Res.* 2:233-250.

Wilson, G.F., B.T. Kang, and K. Mulongoy. 1986. Alley cropping: trees as source of green manure and mulch in the tropics. *Biological Agri. and Horticulture* 3:251-267.

Winrock. 1992. *Assessment of Animal Agriculture in sub-Saharan Africa.* Winrock International USA.

Wondimagegne Shiferaw, H.M. Shelton, and H.B. So 1992. Tolerance of some subtropical pasture legumes to waterlogging. *Trop. Grass.* 26:187-195.

Woodmansee, R.G. 1979. Factors influencing input and output of nitrogen in grasslands. p. 117-134. In: N.R. French (ed.), *Perspectives in Grassland Ecology.* Springer-Verlag, New York.

Wright, G.C., K.T. Hubick, and G.D. Farquhar. 1988. Discrimination in carbon isotopes of leaves correlates with water use efficiency of field grown peanut cultivars. *Aust. J. Plant Physiol.* 15: 815-825.

Yamoach, C.F., A.A. Agboola, G.F. Wilson, and K. Mulongoy. 1986. Soil properties as affected by the use of leguminous shrubs for alley cropping with maize. *Agric. Ecosys. and Envir.* 18:167-177.

# Long-Term Soil Fertility Management Experiments in Eastern Africa

B.R. Singh and H.C. Goma

## I. Introduction

The principal objective of any cropping practice or soil management program is sustained profitable production. Long-term experiments have a special value for providing information to guide agricultural developments. Carefully conducted and fully recorded experiments, where the composition of both soils and crops are determined, provide ideal conditions for measuring the effects of fertilizers, amendments, and cropping system on soil properties and soil productivity.

ISBN 1-56670-076-0

347

The most common traditional farming systems in Eastern Africa are all variations of shifting slash-and-burn cultivation. In these traditional systems of cultivation, recuperation of soil fertility is accomplished by a period of about 15 to 20 years bush fallow which allows crops to be grown without the use of fertilizers or lime for hundreds of years (Woode, 1983). With the need of a more settled farming system farmers are generally aware of the benefits of the use of improved farming methods in these cases, but the use of organic fertilizers and lime is still not great. These systems are, however, reaching their limits due to rapid increase in population and land use. Recent advances in soil management research have shown that, with appropriate use of amendments and fertilizer, high yields of several crops is possible in the highly leached low fertility soils (Vicente-Chandler, 1974, Sanchez et al., 1982, Singh, 1989). However, whether these high yields are sustainable and whether a stable and economically viable crop production over a long period of time can be attained in Eastern Africa is still a debatable question.

Although it has been seen that arable production on tropical Ultisols and Oxisols, be it under traditional shifting cultivation or intensive farming, can lead to a decline in soil fertility due to rapid decrease in soil organic matter, hence nitrogen supply, and the removal of elements such as phosphorus, potassium, calcium, magnesium, and sulphur. It has also been experienced in some areas of Northern Zambia and Mbeya region of Tanzania that yields have actually been declining gradually probably caused by continuous cultivation and more intense use of commercial fertilizers without adequate liming program or unbalanced nutrients management.

Sustainable land management, according to one definition is "a system of technologies that aims to integrate ecological and socioeconomic principles in the management of land for agricultural and other uses to achieve intergenerational equity". In order to make farming systems more sustainable, the various constraining factors must be identified and alleviated on a long-term basis. It is also felt that long-term experiments on soil fertility management often provide more useful information than many short-term trials.

It is argued that intensive use of commercial fertilizer and continuous cropping on low fertility tropical soils lead to soil degradation and decline in productivity. This paper studies the information available on crop yields and soil properties from several long-term experiments conducted in Eastern Africa, with a major emphasis on experiments from Zambia and Tanzania.

## II. Soil Fertility Experiment on a Soil Catena in Tanzania

A soil fertility experiment was initiated in 1981 along a Mlingano Catena which is developed on metamorphic rocks in the Tanga region of northeastern Tanzania. The concept of the catena has been used and found practical in many soil studies including those of soil fertility and management research (Hanegraaf, 1985; Harrop, 1985; Kraus, 1988). The objective of the study was to monitor

**Table 1.** Chemical and physical properties of the Ferralsol and Luvisol in Mlingano, Catena

|  | Ferralsol | | Luvisol | |
|---|---|---|---|---|
|  | Ap<br>0-15 cm | Bws1<br>35-65 cm | Ap<br>0-12 cm | Bt1<br>30-75 cm |
| Clay (%) | 53 | 6.1 | 39 | 53 |
| Silt (%) | 10 | 6 | 15 | 11 |
| Sand (%) | 37 | 3.3 | 56 | 36 |
| pH-$H_2O$ | 6.3 | 5.4 | 7.1 | 6.6 |
| pH-KCl | 5.6 | 4.7 | 6.4 | 5.6 |
| Organic C (%) | 1.8 | 0.3 | 2.0 | 0.4 |
| CEC cmol kg$^{-1}$ | 13.6 | 6.9 | 22.6 | 13.5 |
| Exch. K cmol kg$^{-1}$ | 0.6 | 0.2 | 1.6 | 0.8 |
| Exch. Ca cmol kg$^{-1}$ | 9.3 | 2.1 | 16.6 | 8.0 |
| Exch. Mg cmol kg$^{-1}$ | 1.4 | 0.6 | 4.1 | 2.6 |
| Base Sat. (%) | 83 | 42 | 100 | 70 |
| P - Bray mg kg$^{-1}$ | 5 | 2 | 6 | 1 |

(From Haule et al., 1989.)

crop response and changes in soil properties resulting from continuous monocropping with maize and the addition of N and P fertilizers. Since sulphate of ammonia was used as a N source, its acidifying effect was also studied.

## A. Soils and Climate Characteristics

The soils under catena under study (Mlingano Catena) are developed on metamorphic Precambrian rocks, mainly gneiss. The catena is dominant on the undulating topography at the foot of the Usambara Mountains. The crests of the ridges are broad and slightly convex with slopes between 2 to 14% and are at an altitude of 200 m. Soils on the crests and upper slopes classify as Rhodic Ferralsols and those on the lower slopes as Chromic Luvisols (FAO,1988). The Ferralsols are deep, well drained, clayey soils, and the Luvisols are deep or moderately deep, well drained, and have sandy clay over clay textures. Some important properties are presented in Table 1.

The experiments were conducted on sites, cleared from secondary bush, occupying a crest position (2% slope), a mid-slope position (5% slope, and a lower slope position (9% slope) in the catena described above. The rainfall pattern is bimodal with a long rainy season from March to June and a short rainy season from October to December with an average rainfall of 1150 mm (Haule et al., 1989).

Maize (*Zea mays* L.) was grown with N rates of 20, 40, and 80 kg ha$^{-1}$ and P rates of 8.8, 17.6, 26.4 kg ha$^{-1}$. An absolute control without N and P was also included. Nitrogen and P were applied through sulphate of ammonia and

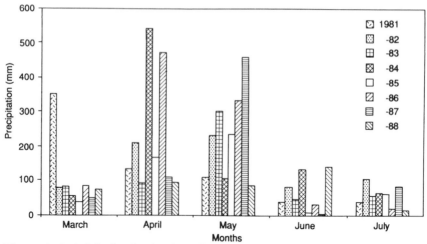

**Figure 1.** Rainfall distribution in different years at Mlingano. (Redrawn from Haule et al., 1989.)

triplesuperphosphate. The experiment was laid out in 3 x 3 factorial plus one control in complete randomized block design. Composite soil samples taken from the topsoil (0 to 20 cm) each year before planting were analyzed for the main soil properties. The rainfall distribution during the main growing months is shown in Figure 1. The rainfall figures show a large variation from season to season and generally low rainfall in June often puts the crop under moisture stress.

## B. Crop Yield

Maize yields of unfertilized plots presented in Table 2 show that the yields depended on soil type and rainfall characteristics of the season. Yields were lowest on the Ferralsols, followed by the intergrade soil, whereas, the highest yields were obtained on the Luvisols. Yields generally declined in the later years. Maize did not respond to P application but its response to N was significant. Highest responses were obtained on the Ferralsol and the highest yield increments occurred at the low application of 20 or 40 kg ha$^{-1}$ (Table 3). In Uganda, Jones (1972) reported that the yields of bean grown on a Ferralsol in the control treatment declined steadily over four years, whereas yields in the treatment receiving fertilizer mixture remained at roughly the same level throughout the period. No such decline in cotton yields in the second rains in the same experiment was seen, but the yield levels in all the years were higher with the use of fertilizers. His results illustrated that the natural fall in soil fertility that occurs during an arable phase, can be prevented by the routine use of fertilizers.

**Table 2.** Unfertilized maize yields in Mlingano Catena

| Year | Ferralsol | Luvisol |
|------|-----------|---------|
| | ----------------------------kg ha$^{-1}$---------------------------- | |
| 1981 | 240 | 3355 |
| 1982 | 1110 | 5780 |
| 1983 | 555 | 2890 |
| 1984 | 990 | 4670 |
| 1985 | 595 | 2965 |
| 1986 | 815 | 2370 |
| 1987 | - | 2445 |
| 1988 | 620 | 2720 |

(From Haule et al., 1989.)

**Table 3.** Mean yield response to nitrogen at Mlingano Catena, 1981-1986; 1988

| | N (20 kg ha$^{-1}$) | N (40 kg ha$^{-1}$) | N (80 kg$^{-1}$) | N (none) |
|---|---|---|---|---|
| | ----------------------Yield (kg ha$^{-1}$)---------------------- | | | |
| Ferralsol | 1680 b[a] | 2170 a | 2040 ab | 705 c |
| Luvisol | 3945 a | 4130 a | 4225 a | 3535 b |

[a] Values followed by the same letter within the same row do not differ significantly at P=0.05.
(From Haule et al., 1989.)

## C. Effect on Soil Properties

Organic matter and pH values declined over the years with stronger decline observed in the last years (Figure 2). Erosion might be held responsible for this as extremely heavy downpours were experienced in the 1986 and 1987 seasons. Similar trend in exchangeable bases were also observed in both soils with minor exceptions in the Luvisol (Figure 3). In general the decrease was more pronounced in the second half of the experimental period. Similar to the results of this study, Jones (1972) working on ferrallitic soils in Uganda found a large decrease in organic matter content only after three years of cultivation. In Tanzania, Anderson (1970) found a decrease of about 30% in organic matter and a drop in pH of 0.9 unit already after five years of maize cultivation on a soil derived from a similar parent material, but having a coarse sandy loam texture. He also showed that the use of ammonium sulphate nitrate reduced the exchangeable Ca, Mg, and K significantly (P=0.001) but it increased exchangeable Mn and electrical conductivity. The strong decrease in organic matter content and other soil parameters as a result of continuous cultivation show that a careful soil management policy has to be followed if long-term cultivation of

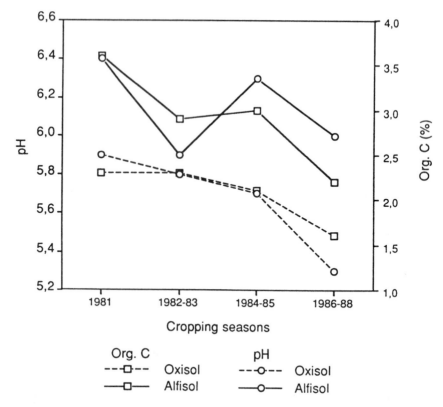

**Figure 2.** Changes in organic carbon and soil pH under continuous cultivation in an Oxisol and an Alfisol in the catena at Mlingano, Tanzania. (Redrawn from Haule et al., 1989.)

Ferralsols is envisaged to avoid mining and rapid degradation of the soil fertility status. The yearly addition of sulphate of ammonia seems to have little effect on the soil pH in the Luvisol but a large decrease was found in the Ferralsol after 8 years of sulphate of ammonia application. This decrease in due course may affect maize yield, but no such decrease was yet experienced in the 1988 crop (Haule et al., 1989).

The results clearly show that long-term studies of this kind provide valuable information on the effects of continuous cultivation on crop yield and soil fertility status and are often more useful than many short-term trials. Ferralsols responded strongly to the addition of N fertilizer, whereas responses on the Luvisol were slight and in general not significant. Responses to added N on the Ferralsol became more pronounced in later years, probably due the decline in the soil fertility status.

Grant (1967), while studying the fertility of Sandveld soil under continuous cultivation in Zimbabwe, reported that under primitive shifting agriculture there

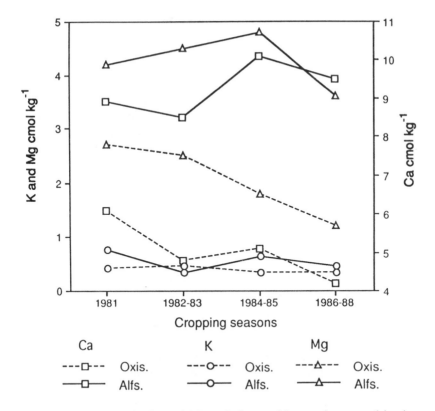

**Figure 3.** Exchangeable Ca and Mg as influenced by continuous cultivation at Mlingano catena.
(Redrawn from Haule et al., 1989.)

is rapid decline in cropping potential and soils are "worked out" in two or three seasons. However, both soil and plant analyses showed that the general fertility of the soil could be maintained under continuous cultivation by the use of fertilizers provided that care was taken to supply some secondary and minor nutrients as well. He found that in the early years of cropping available Mg in the main rooting zone was sufficient for a high yielding crop, but in the fifth season the unmanured plants lacked Mg either because the available Mg was insufficient or because the plants on these plots were unable to draw on the Mg in the deeper layers as their roots were not well developed.

## III. Soil Fertility Experiments in Zambia

### A. Location, Objectives, and Treatments

Earlier it was believed that on newly cleared virgin lands only P fertilizers were required to increase the yields of maize crop in Zambia (McEwan,1932). It was, however, found later that maize responded significantly to both N (Rattray, 1949) and S (Pawson, 1959). In the sixties, therefore, it was decided to initiate a series of long-term fertilizer experiments in the main ecological zones from south to north in Zambia. The results from two contrasting sites from southern and northern provinces are discussed in this paper. The experimental site in the southern province was located on the Magoye Regional Research Station at latitude 16° 00' S and longitude 27° 43' E in the northern province at the Misamfu Regional Research Station at latitude 10° 10' S and longitude 31° 10' E. The Magoye and Misamfu lie at altitudes of 1024 and 1385 m above mean sea level, respectively. The main objective of these trials was to asses the capability of Zambian soils to support continuous maize cultivation under commercial farming conditions. The experiments were further aimed to find out the followings:

(i) Duration of cultivation when P and/or K become limiting factors for maize production,

(ii) Correlation of maize yield response and soil and/or leaf analyses,

(iii) Assess merit of economic levels of fertilizer application over a long-term period, and

(iv) The proper composition of the fertilizer mixture.

The trial sites were cleared by chain stumping using caterpillar tractors from virgin forest and were subjected to a uniformity trial prior to implementation of the experiment. The experiment at Misamfu was initiated in the 1965-66 season and at Magoye in the 1966-67 cropping season.

The rates of N at both sites were 78, 136, and 190 kg ha⁻¹ and those of K were 0 and 9 kg ha⁻¹. At Misamfu, P was applied at the rate of 0 and 10 kg ha⁻¹ and at Magoye S at the rate of 11 and 22 kg ha⁻¹. At Misamfu, Zn was included as an additional treatment and it was applied at the rate of 0 and 50 kg ha⁻¹. Zinc, however, was applied only in 1971, 1972, and 1976 and its residual effect was seen in the rest of the years.

### B. Climate

Northern Zambia, has sub-tropical climate due to its elevation although it lies on tropical latitudes. The dry season, from May to October, consists of a cool and a hot phase, with the former from May to mid-August and the latter from mid August to October.

Rainfall diagram based on data collected by the Misamfu Agrometeorological station is shown in Figure 4. The pattern is similar to most parts of the region.

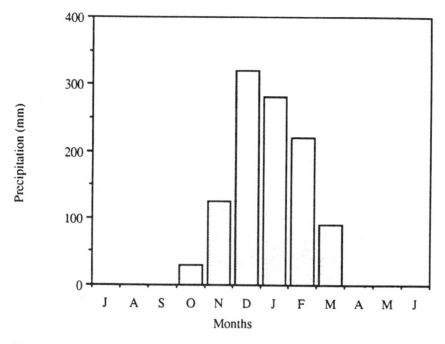

**Figure 4.** Rainfall distribution in the cropping season at Kasama, Zambia. (Redrawn from SPRP Report, 1983-86.)

The figure shows a very good rainfall distribution in the cropping season and at least in 3 months, coinciding with active plant growth, rainfall exceeds the total evepotranspiration. The mean annual rainfall for Kasama is 1360 mm and the wettest months are December and January. During the rainy season the mean monthly maximum temperature is about 16° C. The relative humidity ranges between 63 and 82%.

The cimate at Magoye is relatively drier with the mean annual rainfall of 800 mm mostly in the same period as Misamfu. However, Magoye experiences warmer temperature of about 23° C and greater water stress during the growing season.

## C. Vegetation

The northern Province is covered with vegetation type locally called "Miombo", "Chipya", and "Mateshi" woodlands. The Miombo woodland extends over the greatest part of the plateau. It is also often referred to as *Brachystegia-Julbernardia* woodland after the two tree genera dominating the vegetation. The grassland of the Miombo areas are dominated by *Hyperrhenia* and *Digitaria* species.

Munga woodlands dominated by species of Acacia Combretun,and Terminalia occur in southern province. This vegetation is two storied and is parklike. Munga vegetation is used by agriculturists as an indicator of good arable land.

## D. Soils

The soils of the northern region are strongly leached and derived from non-basic rocks, and the most dominant group consists of mainly Oxisols and Ultisols (USDA Soil Survey Staff, 1975) and are well drained. The experiment at Misamfu was carried out on two benchmark soil series of the region, viz, Kasama sandy loam soil series and Chinsali sandy loam soil series both belonging to Oxisols(Woode, 1987), and that at Katito farm was conducted on Katito soil series belonging to Ultisol order. The soil at Magoye in the southern province belongs to ferric/orthic Luvisol (FAO) or Typic Paleustalf (Soil Survey Staff, 1975). Major characteristics of these soils are presented in Table 4.

## E. Crop Yields

The yield responses to fertilizer at Misamfu were extremely low 10 years after continuous cultivation. In some of the years the treatment effects were, however, overridden by the vagaries of nature. On the Chinsali soil series there was a significant response to nitrogen application in the first three years only (Figure 5a). The Kasama soil series continued to respond to nitrogen fertilization for a further two years, and also showed a small significant response in the 13th, 15th, and 17th cropping seasons for the lower level of application (Woode, 1983).

The Chinsali soil series only showed a response to phosphorus fertilization in five out of the eight years from 1974 onward (Figure 5b), whereas the Kasama soil series showed highly significant responses to phosphorus in most years probably due to lower levels of available P in this series as compared to Chinsali. In almost all the years after the sixth season there was a strong response to K fertilization in both soil series. However, the response was little in the 10th year. A simple regression analysis showed that there was a significant ($P = 0.05$) positive correlation between the rate of application of K fertilizer and the yield response for the Chinsali soil series throughout the trial (Figure 5c). There was no response to zinc application in any of the soil series throughout the experiment. In spite of adequate fertilization, the yield levels declined continuously in both series over the years, which indicates that sustained high yields of monoculture maize on such soils cannot be expected.

At Magoye, linear effect of N on maize yield varied considerably from year to year, without any obvious trend (Table 5). On average for all the years $N_0$ and $N_2$ gave generally the same yield and there was no significant difference. The effect of S on maize yield was significant in the first three years when 0 level of S was used, but the 1969-70 season, the rates of S were changed to 11

**Table 4.** Major characteristics of some of the important soils series of northern Zambia

| Soil series/horizon | Soil depth (cm) | Sand (%) | Clay (%) | pH (CaCl$_2$) | Org. C (%) | Total N (%) | Exchangeable cations —cmol kg$^{-1}$— | | | | CEC Al + bases |
|---|---|---|---|---|---|---|---|---|---|---|---|
| | | | | | | | Al$^{3+}$ | Ca$^{2+}$ | Mg$^{2+}$ | K$^+$ | |
| **Misamfu sandy loam (yellow Typic Haplustox)** | | | | | | | | | | | |
| A | 0 - 10 | 78 | 18 | 4.9 | 1.50 | – | TR | 1.3 | 1.3 | 0.2 | 2.8 |
| AB | 10 -21 | 78 | 19 | 4.2 | 0.81 | – | TR | 1.3 | 0.3 | 0.2 | 1.8 |
| **Misamfu sandy loam (red) Typic Haplustox** | | | | | | | | | | | |
| A | 0 - 16 | 77 | 17 | 4.9 | 1.77 | 0.07 | TR | 1.5 | 0.8 | 0.1 | 2.4 |
| BA | 16 -37 | 68 | 27 | 4.8 | 0.48 | 0.03 | 0.3 | 0.5 | 0.5 | TR | 1.3 |
| **Katito clayey Oxic Paleustult** | | | | | | | | | | | |
| A1 | 0 - 10 | 59 | 36 | 4.5 | 1.20 | 0.07 | 1.3 | TR | 0.3 | 0.2 | 1.8 |
| A2 | 10 - 12 | 56 | 39 | 4.4 | 0.95 | 0.05 | 1.2 | TR | 0.3 | 0.1 | 1.6 |
| **Magoye clayey Typic Paleustalf** | | | | | | | | | | | |
| A | 0 - 8 | 75 | 12 | 5.0 | 0.88 | 0.07 | TR | 1.4 | 1.4 | 0.6 | 3.4 |
| EA | 8 - 21 | 72 | 16 | 4.9 | 0.75 | 0.06 | TR | 1.1 | 1.3 | 0.4 | 2.8 |

(From SPRP, 1983-1986; Wood, 1987.)

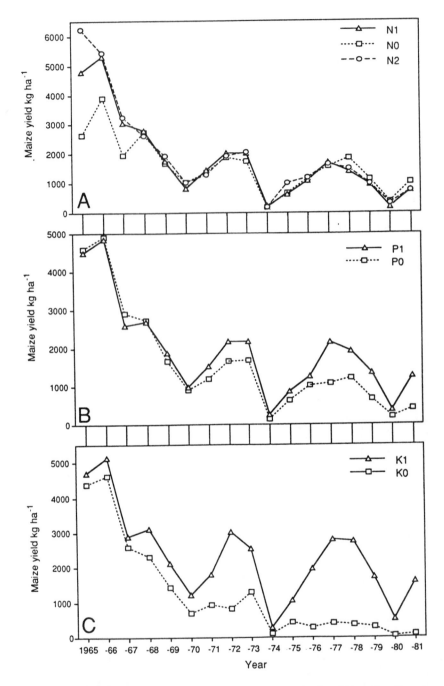

**Figure 5.** Maize response to N, P, and K fertilizers in the Chinsali soil series under continuous cultivation.
(Redrawn from Woode, 1983.)

**Table 5.** The effect of nitrogen on the yield of grain maize in the Luvisol at Magoye

| Year | N rates (kg ha$^{-1}$) | | | Mean | SE | Signifi-cance |
|------|------|------|------|------|------|------|
| | N0 (78) | N1 (136) | N2 (190) | | | |
| | --------------------kg ha$^{-1}$-------------------- | | | | | |
| 1966-67 | 3839 | 3918 | 3936 | 3898 | 103 | <0.001 |
| 1967-68 | 5343 | 5284 | 5074 | 523 | 110 | <0.05 |
| 1968-69 | 5812 | 6412 | 6550 | 6258 | 209 | <0.01 |
| 1969-70 | 4041 | 4517 | 4584 | 4281 | 113 | <0.001 |
| 1970-71 | 4476 | 5283 | 5659 | 5139 | 178 | <0.001 |
| 1971-72 | 4796 | 5846 | 5798 | 5480 | 116 | <0.001 |
| 1972-73 | 6511 | 5964 | 5024 | 5833 | 324 | <0.001 |
| 1973-74 | 5589 | 6768 | 6830 | 6396 | 173 | <0.001 |
| 1974-75 | 5843 | 7335 | 7100 | 6759 | 246 | <0.001 |
| 1975-76 | 5816 | 6163 | 5286 | 5755 | 258 | <0.001 |
| 1976-77 | 5579 | 5597 | 4789 | 5322 | 229 | <0.01 |
| 1977-78 | 4106 | 4592 | 4117 | 4272 | 222 | <0.05 |
| 1978-79 | 3018 | 2824 | 2354 | 2752 | 221 | <0.05 |
| 1979-80 | 3492 | 3442 | 2924 | 3286 | 242 | <0.05 |
| 1980-81 | 4563 | 4861 | 3985 | 4470 | 217 | <0.05 |
| | | | | | | |
| Mean | 4855 | 5254 | 4934 | 5014 | | |
| SE | | | 95 | | | |
| | | | <0.001 | | | |

(From Tveitnes and McPhillips, 1989.)

and 22 kg ha$^{-1}$. Thereafter, no significant response to S was observed. Similarly, no significant response to K was observed in any of the years. In the first nine years of trial, basal P at the rate of about 30 kg ha$^{-1}$ was applied to all plots, but from the 1975-76 season the dummy treatment was used with half the plots receiving 32.5 kg P ha$^{-1}$. From 1975 to 1978 no significant response was obtained but from the 1978-79 season there was significant effect of freshly applied P (Table 6) (Tveitnes and Svads, 1989).

## F. Effect on Soil Properties

### 1. Organic C and Total N

At Misamfu the organic carbon level in both soil series declined continuously showing a rapid decline to begin with and flattening of the curve in the later years (Figure 6). The level of organic matter in the top 0 to 15 cm of the

**Table 6.** The effect of phosphorus on grain yield

| Year | P rates (kg ha⁻¹) | | SE | Significance |
|------|---------|---------|-----|--------------|
|      | PO (0) | P1 (32) | | |
|      | ------------kg ha⁻¹---------- | | | |
| 1975-76 | 5713 | 5797 | 211 | n.s. |
| 1976-77 | 5204 | 5440 | 187 | n.s. |
| 1977-78 | 4166 | 4377 | 180 | n.s. |
| 1978-79 | 2465 | 2999 | 180 | $<0.01$ |
| 1979-80 | 2946 | 3627 | 198 | $<0.01$ |
| 1980-81 | 3560 | 5379 | 177 | $<0.001$ |
| Mean | 4042 | 4636 | | |
| SE | | 146 | | |
| Significance | | $<0.001$ | | |

(From Tveitnes and McPhillips, 1989.)

Chinsali series fell by about 27 t ha⁻¹ from 59 to 32 t ha⁻¹ during the 17 years of cultivation (Woode, 1983). This represents an average loss of 1.6 t ha⁻¹ y⁻¹. This decline in organic matter content is in agreement with observations from other countries (Jones and Wild, 1975; Jones, 1972; Andersen, 1970). Virgin grassland soils traditionally lose organic matter rapidly after they are first cultivated (Allison, 1973; Rasmussen and Collins, 1991). With cultivation, a substantial portion of dry matter production is removed for food or forage and is lost through wind and water erosion.

It was earlier believed that the organic matter content in soils could not go below a certain minimum level. According to Rasmussen and Collins (1991) the organic matter level depends on the rate of residue addition in relation to the rate of residue decomposition and soil erosion. Soil organic matter will continue to change as long as any of the controlling factors continue to change. New equilibrium levels will be highly dependent on farming practices, especially those involving crop residue utilization, crop rotation, and tillage.

The decline in total N content in both soil series was very rapid irrespective of the fertilizer treatment. The application of inorganic fertilizer N had no effect on the total N content of the soil measured. This is probably because fertilizer N not taken up by the crop was lost, either by leaching down the profile or through gaseous losses to the atmosphere. A positive relationship of organic carbon with the total N content is widely reported in the literature. Rasmussen and Collins (1991) showed very positive correlation between organic C and organic N. Organic N comprises more than 99% of the total N present in the soil in the absence of substantial $NO_3$-N accumulation (Allison, 1973). Organic carbon and total nitrogen contents were found to be related to the mean annual rainfall and the clay content of the soil at 295 sites in West Africa (Jones, 1973).

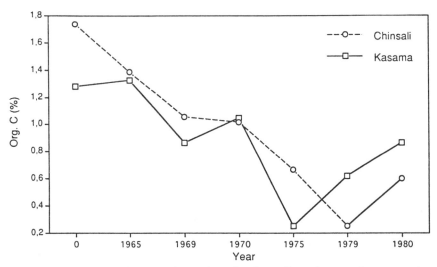

**Figure 6.** Changes in organic carbon in the soil under continuous maize cultivation. Year 0 is value for virgin site.
(Redrawn from Woode, 1983.)

The C:N ratio in this experiment was about 17 under virgin conditions. But it rose to nearly 22 for the Kasama soil series and 19 for the Chinsali soil series and it fell again to levels similar to virgin conditions after 17 years of cultivation (Woode, 1983). It is interesting to note that the C:N ratios in the two soils temporarily rose soon after cultivation and then dropped to original levels (virgin) after 17 years of cultivation. This perhaps is due to faster decline in the N content than that of organic C of the soils, as stated above, when these soils were brought under cultivation and thereby resulted in greater C:N ratio.

Such drastic decreases in the organic C and N contents of the Alfisol at Magoye were not seen, although the contents tended to decrease over the years (Table 7).

## 2. Exchangeable Calcium and Magnesium

The levels of exchangeable Ca and Mg in both soil series at Misamfu rose during the first three years under cultivation and then they declined very rapidly (Figure 7) but only minor changes in the levels of these nutrients were observed in the Alfisol at Magoye (Table 7). This rise in the Ca and Mg levels is similar to those described by Nye and Greenland (1964) when investigating the changes that occurred upon clearing tropical bush. This rise could be due in part to use of calcium ammonium nitrate top-dressing and initial liming with dolomitic limestone and in part due to conversion of unavailable Ca and Mg to an available form after clearing the virgin land (Fauck et al., 1969).

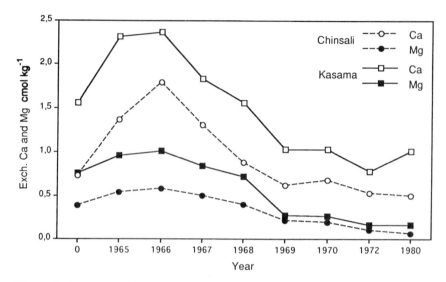

**Figure 7.** Exchangeable Ca and Mg in the soil as affected by continuous maize cultivation. Year 0 is value for virgin site.
(Redrawn from Woode, 1983.)

**Table 7.** Soil chemical properties of the surface soil (0-20 cm) at Magoye

|                          | 1966-67 | 1967-68 | 1968-69 | 1969-70 | 1970-71 | 1971-72 | Mean |
|--------------------------|---------|---------|---------|---------|---------|---------|------|
| pH                       | 5.5     | 5.4     | 5.4     | 4.8     | 4.9     | 4.8     | 5.1  |
| Ex. Ca cmol kg$^{-1}$    | 2.22    | 2.54    | 2.43    | 2.22    | 2.33    | 2.18    | 2.32 |
| Ex. Mg cmol kg$^{-1}$    | 1.43    | 1.84    | 1.77    | 1.66    | 1.63    | 1.43    | 1.63 |
| Ex. K cmol kg$^{-1}$     | 0.39    | 0.39    | 0.38    | 0.31    | 0.30    | 0.22    | 0.33 |
| P mg kg$^{-1}$           | 2.2     | 4.8     | 3.9     | 5.4     | 7.0     | 4.8     | 4.7  |
| Org C %                  | 0.94    | -       | -       | -       | 0.71    | -       | 0.83 |
| Tot N %                  | 0.063   | -       | -       | -       | 0.059   | -       | 0.06 |

(From Tveitnes and Svads, 1989.)

The Mg level in both soil series fell below 0.2 c mol kg$^{-1}$ from 1970 onward and the Ca:Mg ratios steadily rose from 2:1 in 1965 to > 5:1 from 1972 onward. This level of exchangeable Mg in the soil may cause Mg deficiency, especially on acid soils such as these (Young, 1976). Magnesium deficiency has been shown to be aggravated by a high Ca: Mg ratio (>5:1). From these observations Woode (1983) deduced that probably the maximum potential yield of maize was limited by a magnesium deficiency after 1970. In this experiment

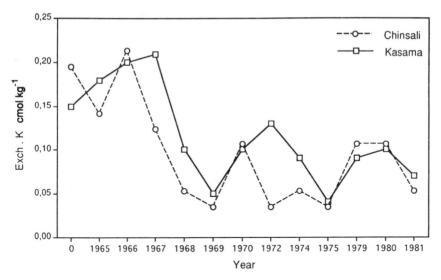

**Figure 8.** Exchangeable K in the soil as affected by continuous maize cultivation. Year 0 is value for virgin site.
(Redrawn from Woode, 1983.)

Mg was lost more rapidly than Ca, leading to an increase in the Ca:Mg ratio. This was observed regardless of the application of dolomitic calcium ammonium nitrate and two applications of dolomitic limestone for the first nine years. It has been reported in the literature that Ca is usually lost more rapidly than Mg, although the opposite trend has also been observed (Jones and Wild, 1975).

3. Exchangeable Potassium

Woode (1983) reported an initial increase in the level of exchangeable K from the virgin state in the Kasama soil series, but not in the Chinsali soil series. During the first four years of cultivation the K level fell below 0.05 me 100 g$^{-1}$ in both soil series (Figure 8). The mean exchangeable potassium values in succeeding years did not rise above 0.10 c mol kg$^{-1}$ with minor exceptions in the Kasama soil series. A significant positive correlation between the level of exchangeable K in the top soil (at the 0.05 and 0.001 c mol kg$^{-1}$ levels for the Chinsali and Kasama soil series, respectively) and the residual potassium fertilizer effect was observed when the first three years data, during which the soil reserve of K was high, were not included in the calculation (Woode, 1983). Tveitnes and McPhillips (1989) found that in the Alfisol at Magoye the level of exchangeable K varied from 0.22 to 0.39 c mol kg$^{-1}$ during the experimental

**Table 8.** Mean available P in the 0-15 cm depth of the Kasama series

| | P rates applied (kg ha⁻¹) | |
|---|---|---|
| | 0 | 10 |
| 0 | 8.4 | 8.4 |
| 1965 | 20.9 | 21.9 |
| 1966 | 10.1 | 11.4 |
| 1967 | 4.8 | 5.6 |
| 1968 | 5.21 | 6.8 |
| 1969 | 9.2 | 12.5 |
| 1970 | 14.9 | 20.9 |
| 1975 | 9.71 | 14.8 |
| 1979 | 11.5 | 12.7 |
| 1980 | 11.6 | 29.8 |
| 1981 | 18.71 | 43.8 |

(From Woode, 1983.)

period (Table 7) and this was perhaps the reason that maize in this experiment did not respond to applied K as reported earlier under crop yield.

In most tropical soils the exchangeable K level of about 0.1 c mol $kg^{-1}$ is considered critical for most crops (Boyer, 1972). In both Chinsali and Kasama soil series the exchangeable K levels were below the critical limit in most of the years after the fifth cropping season (Woode, 1983). This was reflected in the significant yield response to K fertilizer in almost all years after the fifth cropping season. The response was probably due to an absolute deficiency of potassium, not necessarily as a result of a cation imbalance.

## 4. Available Phosphorus

In both Chinsali and Kasama soil series at Misamfu the level of available P increased after the bush was cleared and burnt. It was also observed that these sites showed a net loss of N, a substantial gain in exchangeable bases, and a massive gain in available P when the change in level of each variable was expressed as a percentage of its level under virgin conditions (Woode, 1983). This observation was in close agreement with Nye and Greenland (1960) for soils under similar conditions. From the second up to the fourth cropping season the available P levels in the Kasama soil series fell steeply (Table 8) and the same is true for the Chinsali soil series (results not presented). This fall was probably caused by the high rate of removal by the crop in the first four years when the yields were high (Woode, 1983). However, from the fifth and subsequent years the maize yields were on average lower than the previous years, and P uptake would probably have also been inhibited by the rise in exchangeable Al.

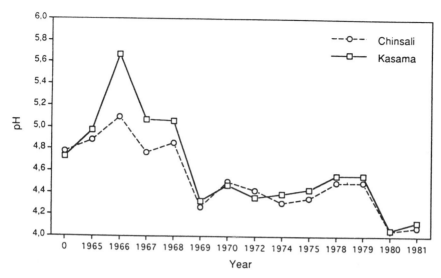

**Figure 9.** Changes in soil pH under continuous maize cultivation. Year 0 is value for virgin site.
(Redrawn from Woode, 1983.)

In the 12th year the amount of annual application of phosphorus fertilizer was raised from 10 kg P/ha to 35 kg P/ha for the plots receiving P. Little effect was seen in the levels of available P until in the 16th and 17th cropping seasons, when the P levels increased substantially, especially in the treated plots (Table 8) (Woode, 1983). In the Alfisol at Magoye, the available P level also increased over time (Table 7) but the increase in this soil was not so large as reported for the Oxisol above.

## 5. Soil Acidity and Aluminum

During the first three years of cultivation, the soil pH rose over the entire trial (Figure 9). This was attributed to the application of calcium ammonium nitrate (CAN) to all plots as top dressing, and to the effect of ashes from the burning of the cleared forest vegetation. This is in line with the results obtained by other researchers (Nye and Greenland, 1960; SPRP Reports, 1987, 1988; Singh, 1987a) working on similar soils following bush clearing.

Woode (1983) observed a higher increase in soil pH during the first three years in the plots receiving higher N levels and related this increase to greater annual lime application in the form of CAN.

Five years after continuous cultivation the soil pH fell sharply (Figure 9) and this fall in pH was linearly correlated ($P=0.001$) to the fall in total exchangeable bases (TEB) and the base saturation percentage (BSP). Woode (1983) observed

**Table 9.** Summary of physical properties (0-15 cm), Chinsali and Kasama series

|  | ------Virgin sites---- | | -Cultivated plots 1975- | |
|  | Chinsali | Kasama | Chinsali | Kasama |
|---|---|---|---|---|
| Bulk Density g cm$^{-3}$ | 1.15 | 1.20 | 1.32 | 1.30 |
| 2 h infiltration, cm h$^{-1}$ | 20.6 | 37.1 | 13.9 | 12.7 |
| 3 h infiltration, cm h$^{-1}$ | 1.73 | 3.10 | 1.10 | 1.01 |
| Total porosity % | 53.5 | 51.0 | 47.2 | 48.1 |
| Clay % | 13 | 18 | 14 | 15 |
| Silt % | 7 | 8 | 8 | 7 |
| USDA texture | SL | SL | SL | SL |
| 0.33 bar water content % | 20.3 | 21.5 | 14.5 | 14.5 |
| 15 bar water content % | 7.5 | 8.3 | 7.4 | 8.8 |
| Available water content % | 12.8 | 13.2 | 7.1 | 5.8 |
| Aggregation % | 93 | 94 | 77 | 87 |

(From Woode, 1983.)

that the fall in pH was not related to the level of N fertilizer applied, but to the overall decline in soil fertility shown by the falling TEB and BSP curves. These results have a general similarity to those reported by Andersen (1970) and Jones (1972) for Tanzania and Uganda, respectively. A similar decrease in pH in the Alfisol at Magoye after some years of cultivation was also observed by Tveitnes and McPhillips (1989).

In the Kasama soil series exchangeable Al(1 N KCl) increased from 0.18 c mol kg$^{-1}$ in the virgin topsoil with a pH of 4.8 (CaCl2) to 0.67 c mol kg$^{-1}$ with a pH of 4.2 after 17 years of cultivation. This type of relationship between Al and pH is reported by many investigators for tropical soils (Kamprath, 1970; Brams, 1971; Brenes and Pearson, 1972; Pearson, 1975; Singh, 1987b). At pH levels of around 4.2 the Al saturation of both soil series was approximately 50%. Excessive soil acidity manifested in low pH, low levels of exchangeable Ca, Mg, and K, and high levels of Al, and or Mn, is detrimental to plant growth (Sanchez, 1976; Foy, 1984) and to soil microbiological processes (Alexander, 1977).

In the 6th and 12th years 1 t/ha and 1.5 t/ha lime were applied respectively, which resulted in higher yields. Since all plots including the control were limed, it is not possible to say with confidence if the apparent response to liming was, in fact real.

## 6. Physical Properties

Perhaps the most important effect of cultivation on the soil physical properties was the reduction in infiltration rates of both soil series (Table 9). This

**Table 10.** Mean values of soil properties,  Chinsali series

| Soil property | --------1965-1969------- | | -------1970-1981-------- | |
|---|---|---|---|---|
| | Topsoil | Subsoil | Topsoil | Subsoil |
| Organic C, % | 1.37 | 0.71 | 0.98 | 0.23 |
| Total N, % | 0.07 | 0.06 | 0.04 | N.D. |
| Ex. Ca, cmol kg$^{-1}$ | 1.27 | 0.34 | 0.58 | 0.26 |
| Ex. Mg, cmol kg$^{-1}$ | 0.44 | 0.18 | 0.14 | 0.07 |
| Ex. K, cmol kg$^{-1}$ | 0.08 | 0.04 | 0.06 | 0.05 |
| CEC, cmol kg$^{-1}$ | 5.5 | 3.9 | 5.6 | 4.7 |
| pH, CaCl$_2$ | 4.8 | 4.4 | 4.3 | 4.2 |
| Available P, mg kg$^{-1}$ | 12.1 | 9.3 | 23.2 | 14.4 |

(From Woode, 1983.)

reduction in infiltration rates may be due to the break down of top soil structural units under the influence of rain drops impact or mechanical cultivation, followed by differential illuviation of the silt and clay fractions. A further possible consequence of this differential illuviation would be a reduction in the crop-rooting depth. Lal and Maurya (1982) reported that the roots of a maize crop grown in an ideal soil with a bulk density similar to that of the trial under review reached a depth of 2.40 m after 8 weeks of growth. The average rooting depth observed in this experiment, however, was only 0.43 m at maturity during the 17th cropping season.

The Kasama soil series had higher infiltration rates than the Chinsali soil series under virgin conditions (Table 8) suggesting that the Kasama soil series had a better structure. However, following the structural degradation of both soil series under cultivation, the Chinsali soil series showed a significantly higher infiltration rate, presumably due to its sandier texture (Woode, 1983).

## 7. Sub-soil Characteristics

In a few selected years sub-soil samples were also collected from 22-45 cm between 1965 to 1969 or from 30-45 cm between 1974-81 from both Chinsali and Kasama soil series and these samples were analyzed for some chemical properties. The mean values of the results for the Chinsali series are presented in Table 9. The table also gives comparison of topsoils and subsoils. The periods mentioned above can be considered to represent high productivity and low productivity periods, respectively. There was decline in organic C, and exchangeable Ca and Mg contents from the high to the low productivity period (Table 10). In a number of properties significant correlation between the mean values of topsoils and subsoils were observed with the highest correlation being observed for soil pH (Woode, 1983).

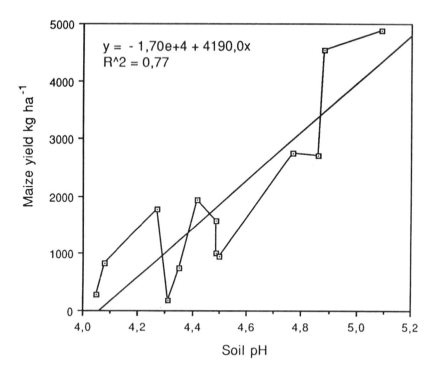

**Figure 10.** Relationship between soil pH and maize yield in the Chinsali soil series under continuous maize cultivation.
(Redrawn from Woode, 1983.)

## G. Relationships between Soil Properties and Crop Yield

Some positive correlations among soil parameters such as between organic C and total N and between exchangeable Ca and Mg and pH were observed. These parameters also showed significant correlations with crop yield. A relationship between soil pH and crop yield for the Chinsali series is shown in Figure 10. However, the correlation coefficients on yearly basis showed no consistent pattern, probably because the correlation's in individual seasons were masked by the treatment effects. Therefore the correlation coefficients for each variable with crop yield over the duration of the experiment were calculated. With the exception of CEC and available P, all the soil variables in the two soil series showed a highly significant positive linear correlation with crop yield (Table 11). Available phosphorus levels were negatively correlated with crop yield in both soil series. This was due to the overall accumulation of available P during the course of the trial, whilst the yield declined due to other limiting factors (Woode, 1983). The accumulation of available P as a result of application of inorganic P fertilizers is widely reported in surface soils of different types and

**Table 11.** Correlation of soil properties with crop yield

| Soil property | Chinsali series | Kasama series |
|---|---|---|
| Organic C | 0.757* | 0.727 n.s. |
| Total N | 0.972** | 0.690 n.s. |
| Ex. Ca | 0.904*** | 0.781* |
| Ex. Mg | 0.908*** | 0.842** |
| Ix. K | 0.656* | 0.603* |
| CEC | -0.068 n.s. | 0.340 n.s. |
| pH | 0.880*** | 0.794*** |
| Available P | -0.418 n.s. | 0.456 n.s. |

*, **, ***, and n.s. indicate significance at 0.05, 0.01, 0.001 levels, and non-significant, respectively.
(From Woode, 1983.)

of different texture classes (Harrison, 1987; Kamprath, 1967; Novias and Kamprath, 1978). These workers also found accumulation of available P in several long-term residual P experiments.

The correlations of both the exchangeable calcium and magnesium levels with yield were highly dependent on the rate of application of potassium fertilizer. The maize yield response was more strongly correlated with exchangeable Ca and Mg levels at the zero potassium application level than where potassium was applied. This suggests that as the potassium application increases, the absolute demand for Ca and Mg increases such that the relative differences of exchangeable Ca and Mg in the soil become less significant to the crop.

In contrast to the Chinsali and Kasama series, the effects of fertilizer on soil properties in the Magoye soil series were inconsistent and less marked and hence they did not show significant relationship with crop yield.

## IV. Long-Term Effects of Liming

### A. Location and Soil Characteristics

The experiments summarized in this section were conducted in northern Zambia, and a summary of the climatic and vegetation data is presented in the previous section entitled "Soil fertility experiments in Zambia". Three bench mark soils of the region; Misamfu red and yellow, and the Katito soil series were used for these experiments.

Both Misamfu yellow and red soil series (*Typic Haplustox*) are deep and well-drained Oxisols. Strictly, these soils lie on the boarder of Oxisols and Ultisols (Woode, 1987). The major differences between yellow and red lies in their Fe composition, hence their color (differences in their ratios of hematite and geothite), and $Al^{3+}$ content in the subsoil; the Misamfu yellow being more acidic in the subsoil compared to the red. Both have good physical properties for crop production but need to be supplied with macro- (N, P, K, and S, and Ca and Mg), and some micro-nutrients (Mo and Zn) for maximum production.

In the Katito soil series (*Oxic Paleustult*), the topsoil is sitting on a clay layer at 20 cm depth which is predominantly kaolinitic with small amounts of geothite, and this makes it prone to water erosion hazards. It is strongly acidic (pH < 4.5), low in bases, and high in aluminum saturation ( > 60 %) throughout the profile although exchangeable Al is relatively low in the top soil layer.

## B. Crop Response to Liming

A series of multilocation lime trials were started in early seventies to study the direct and residual effects of liming on maize and groundnut rotation on sandy loam, Misamfu red soil series. Six rates of lime were selected i.e. 0, 500, 1000, 2000, 4000 kg ha$^{-1}$ for initial and 500 kg ha$^{-1}$ for yearly application. The trial design was a randomized block with 8 replications for each of the test crops of groundnut (*Arachis hypogaea* L.) and Maize (*Zea mays* L.). The cropping sequence for first five years were groundnut/maize or maize/groundnut depending on which crop was grown first in the plots. From the sixth year crops were grown in rotation on all the replications. All plots received a basal dose of 120, 69, 50, and 20 kg ha$^{-1}$ of N, $P_2O_5$, $K_2O$, and S, respectively.

In spite of initial low pH in the soil (4.4) maize response to either freshly applied lime or in a long-term study conducted on the Misamfu red soil series (Oxisol) was not found significant (P < 0.05) during the first four cropping seasons (Figure 11). However, from the 5th to 9th seasons the highest yields were obtained in the treatment with an initial lime application of 4,000 kg ha$^{-1}$ (Tveitnes and Svads, 1989). In the 12th and 14th cropping seasons treatment with the annual application of 500 kg lime ha$^{-1}$ gave the highest yields. Generally, in ten cropping seasons during the fourteen years, the treatments with 4,000 kg lime ha$^{-1}$ applied initially or 500 kg lime ha$^{-1}$ applied annually gave higher yields of maize than other treatments, and these treatments were also able to maintain yield levels between 3,000 to 4,500 kg ha$^{-1}$ over the years. Maize responded to liming even better in the Katito soil series (Ultisol) where the yield differences between the different liming rates in the two first cropping seasons were significantly different (McKenzie et al., 1988; Singh, 1987a; Øygard, 1986). The yield differences between limed and unlimed treatments became larger over the years and the higher rates of liming, for example 8,000 kg lime ha$^{-1}$, maintained yield levels between 4.5 to 6 Mg ha$^{-1}$ all through out. After the fifth cropping season the crop performed very badly or failed completely in unlimed plots (Figure 12). However, in the limed plots the effects of liming in this experiment were still present after 9 years. (McKenzie, et al., 1988)

Groundnut grown in rotation with maize on Misamfu red soil series gave highly significant response to liming (Figure 11). However, the yield difference among lime rates in the first four seasons did not differ significantly (P = 0.05).

Mean: 4415 3922 3902 2730 4464 596 2926 2996 2762 1521
LSD5% n.s   n.s   n.s   n.s  434  215  n.s   586 1103 650

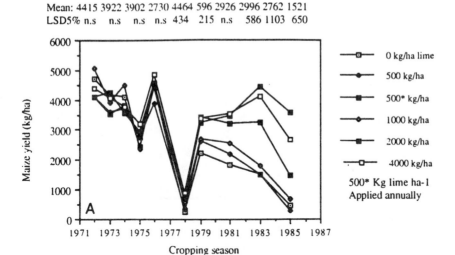

Mean: 1054 1327 1024 912 784   439  1339   851   416
LSD5%: n.s  n.s   n.s  n.s 124   60    88   194   194

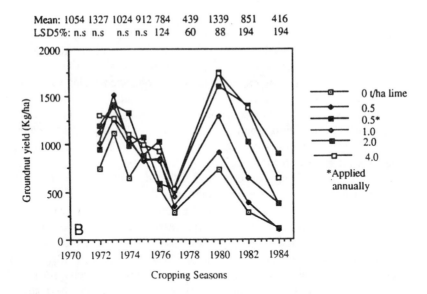

**Figure 11.** Direct and residual effects of liming on the yields of maize and groundnut grown in an Oxisol.
(Redrawn from Tveitnes and Svads, 1989.)

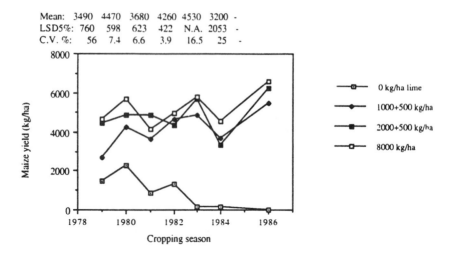

Mean:   3490   4470   3680   4260   4530   3200   -
LSD5%:   760    598    623    422   N.A.   2053   -
C.V. %:    56    7.4    6.6    3.9   16.5     25   -

**Figure 12.** Effects of liming on maize yield on Katito (Ultisol) soil series. (Redrawn from Zam-Can Annual Report, 1986.)

In the 11th and the 13th seasons annual applications of 500 kg ha$^{-1}$ gave the best yield. Tveitnes and Svads (1989) and Singh (1989) reported that the residual effects of 4,000 kg ha$^{-1}$ and 2,000 kg ha$^{-1}$ were as good as 500 kg ha$^{-1}$ applied annually. Although these experiments were conducted under very poor farm management conditions, data available from some good years of experimentation are indicative that lime applied at the rate of 4000 kg ha$^{-1}$ initially or 500 kg ha$^{-1}$ annually along with basal application of deficient plant nutrients could maintain yield levels at economically viable level for at least 2 to 3 times the period it was possible under shifting cultivation.

In a lime x phosphorus experiment conducted on the Misamfu red soil series where different lime sources were applied at the rate of 2 and 4 t ha$^{-1}$ with (33 kg P ha$^{-1}$) and without P, and maize and groundnut crops were grown in rotation Singh (1989) reported that both crops responded highly significantly to P application but no significant direct response to lime applied through different sources over a period of seven years was observed (Figure 13). Absence of response to lime was associated with low initial level of exchangeable Al$^{3+}$ in the soil. The results of soil analysis showed that both lime and application generally increased pH and decreased Al$^{3+}$ saturation to acceptable level in the first year and kept them under tolerable limits in the subsequent years.

The results reported above from Oxisol and Ultisol soils of Zambia show similarity to those reported from similar soil types in other parts of Africa (Foster, 1970; Ofori, 1973) or South America (Abruna et al., 1975; Goedert, 1983). A tremendous range in tolerance of soil acidity and response to lime among tropical crops is reported. In Uganda, Foster (1970) reported no yield response by maize grown on an Oxisol to lime when soil pH was above pH 5.5,

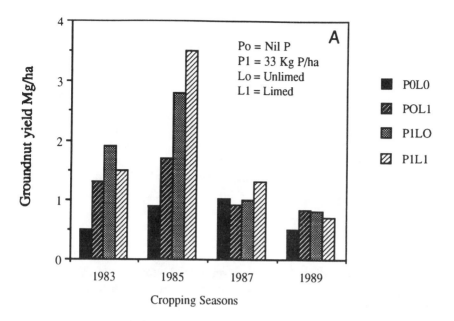

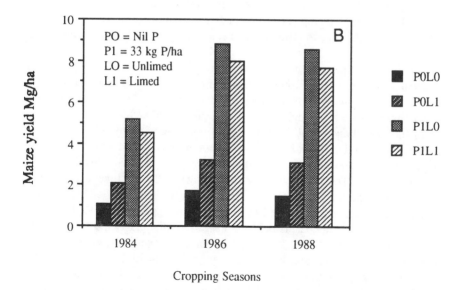

**Figure 13.** Effects of liming and phosphorus on the yields of maize and groundnut grown in rotation in an Oxisol.
(SPRP Annual Reports, 1983-89; Singh, 1989.)

but there were three instances of response at pH around 5.1. On a similar soil, Ofori (1973) reported increase in maize yields from liming, which he attributed to counteracting the effect of $(NH_4)_2 SO_4$ in acidifying these poorly buffered soils. Goedert (1983) summarized the results of one experiment established in 1972 on a Haplutox in which graded doses of lime (0 to 8,000 kg ha[-1]) were incorporated at either 0 to 15 cm or 0 to 30 cm depths. Of the seven crops grown in the experiment up to 1978, the first three, the fifth, and the sixth were maize, the fourth was sorghum, and the seventh was soybean. A marked yield response to lime was observed for all crops. Yields were approximately doubled by an application of 2,000 kg ha[-1] lime. In this experiment also, the long residual effect of liming, even for the smaller rates of application, was evident up to the seventh year.

## C. Effects of Liming on Soil Properties

In spite of the great deal of work done on soils of humid tropics, the nature of the soil acidity and the causes of poor plant growth on these acid soils are not properly understood. Pearson et al. (1977), in their study of comparative response of three plant species in soils of the southeastern United States and Puerto Rico, found different response patterns, which were not clearly related to either crop species or soil category. Similarly, Amedee and Peech (1976) reported that the same crop performed differently on different soils of the same pH value, probably due to differences in concentrations of Al in soil solution at equal pH values.

Although a complete time series of soil samples from the liming experiments reported above are not available, analysis of the soil samples collected at some selected times show some changes in soil properties as a result of liming alone or in combination with P application. The results from the Misamfu yellow soil series (Oxisol) showed that the soil pH did not change when 500 to 1,000 kg lime ha[-1] was applied but at higher rates of liming (2,000 and 4,000 kg ha[-1]) the pH increase was about 0.4 unit (Tveitnes and Svads, 1989). Soil pH dropped in all plots, but the pH levels in the plot supplied with 4000 kg lime ha[-1] ten years earlier and in the plot supplied with 500 kg lime ha[-1] annually were the same after ten years cultivation (Table 12). Similar to the results of this study, Foster (1970) reported that in Oxisols in Uganda whose initial pH values ranged from 4.3 to 5.8, the first 5,000 kg lime ha[-1] increased soil pH about 0.6 unit regardless of the initial pH, but the second increment of 10,000 kg ha[-1] had varied effects ranging from 0.2 to 0.8 unit. Generally, soil pH changes induced by liming tend to be more transient in these tropical soils than in those of temperate regions (Singh, 1987b). The exchangeable calcium in the surface soil more or less followed the pH trend. A general decline in the concentration of Ca, Mg, and K in all plots irrespective of the lime rate (Table 11) from 1972-73 to 1981 was observed. The CEC remained unaffected by liming (Tveitnes and Svads, 1989; Singh, 1987b). In the Katito soil series soil pH increased and Al

**Table 12.** Change in soil properties as result of liming in 0-15 cm in the Misamfu soil series

| Property | 1972-73 Lime rate, kg ha$^{-1}$ | | | | 1981-82 Lime rate, kg ha$^{-1}$ | | | |
|---|---|---|---|---|---|---|---|---|
| | 0 | 500* | 1000 | 4000 | 0 | 500* | 1000 | 4000 |
| pH (CaCl$_2$) | 4.5 | 4.5 | 4.5 | 4.4 | 4.2 | 4.7 | 4.2 | 4.7 |
| Ex. Ca, cmol kg$^{-1}$ | 0.36 | 0.45 | 0.57 | 1.29 | 0.42 | 1.07 | 0.40 | 1.36 |
| Ex. Mg, cmol kg$^{-1}$ | 0.33 | 0.28 | 0.31 | 0.34 | 0.07 | 0.16 | 0.09 | 0.14 |
| Ex. K, cmol kg$^{-1}$ | 0.12 | 0.07 | 0.08 | 0.07 | 0.05 | 0.06 | 0.05 | 0.04 |
| CEC, cmol kg$^{-1}$ | 3.42 | 3.44 | 3.52 | 4.38 | 4.38 | 4.60 | 4.25 | 4.70 |

* Applied every year.
(From Tveitnes and Svads, 1989.)


**Table 13.** Effect of different rates of lime on soil pH (CaCl$_2$) and Al saturation in Katito soil series

| Lime rate Kg ha$^{-1}$ | Depth cm | pH 1979 | pH 1981 | pH 1984 | Al sat., % 1984 |
|---|---|---|---|---|---|
| 0 | 0-15 | 4.5 | 4.0 | 3.8 | 84 |
| | 15-30 | 4.5 | 3.9 | 4.0 | 84 |
| | 30-45 | 4.5 | - | 4.0 | 87 |
| 1000 + 500 annually | 0-15 | 5.2 | 4.5 | 4.8 | 10 |
| | 15-30 | 4.5 | 4.0 | 4.1 | 71 |
| | 30-45 | 4.5 | - | 4.1 | 71 |
| 2000 + 500 annually | 0-15 | 5.2 | 4.7 | 5.0 | 3 |
| | 15-30 | 4.6 | 4.0 | 4.2 | 54 |
| | 30-45 | 4.5 | - | 4.2 | 62 |
| 8000 | 0-15 | 5.9 | 5.6 | 6.5 | 0.3 |
| | 15-30 | 4.5 | 4.1 | 5.7 | 7 |
| | 30-45 | 4.5 | - | 4.7 | 24 |

(From Zam-Can, 1981-84.)


saturation decreased progressively with increased rates of liming (Table 13). The applied lime apparently moved down in the soil profile as indicated by the Al saturation. In 1985, a soil analysis of samples up to 75 cm depth showed that exchangeable Ca and Mg were higher and soluble Al lower at all depths with high lime application rates (Hodgins and Aulakh, 1985). McKenzie et al. (1988) further showed that in this soil series soluble Al in the unlimed plots was more than doubled from 1980 to 1985, but where lime was applied to reduce soluble Al in the subsoil the soluble Al reduced as the liming rates increased. Seven years after liming the highest rate of lime (8,000 kg ha$^{-1}$) reduced soluble Al at 45 to 75 cm depth by as much as 50%, while the corresponding soil pH increased only 0.3 unit. Similarly, Goedert (1983) found that in Oxisol from Brazil, changes in soil pH and in the exchangeable cation pool were closely related to lime rates. Soil pH increased initially but decreased at all rates during the following 5 years. Exchangeable Al and Al saturation increased with time, but became no higher than the initial level. The inverse occurred with exchangeable Ca + Mg.

Available P seemed to have dropped by lime addition in the lime x phosphorus experiment reported above, where as the exchangeable K remained unaffected by liming. The drop in available P is not entirely unexpected because results reported by other workers show that the effect of lime on P availability can vary from beneficial to detrimental. Amarasiri and Olsen (1973) showed that soil P solubility in an Oxisol from Columbia decreased sharply as pH was increased by CaCO$_3$ application up to about pH 6.5, and the limed soil had a higher maximum P adsorption capacity than the unlimed soil. The authors suggested that the freshly precipitated sesquioxides were responsible for the increased inactivation of added P as lime rate increased. Similarly, Heynes (1982) found that liming could increase, decrease, or not affect the phosphate that could be extracted from highly weathered acid soils. Mendoz and Kamprath (1978) found on some Oxisols of Panama that liming to neutralize Al, when combined with low levels of added P, increased plant growth. When Al saturation was below 60% and large amounts of P were added, there was no benefit from liming; and when Al saturation was above 60%, not even large P applications could completely overcome the detrimental effects of Al on plant growth.

A close relationship between soil pH and exchangeable Al (Figure 14A) as well as between soil pH and percent Al saturation (Figure 14B) on Misamfu yellow soil series was observed. The percent Al saturation decreased from around 80 % at pH levels of 4.0 to nearly zero percent as the soil pH approached 5.0. These observations are consistent with the observations of Pearson (1975), who reported that exchangeable Al practically disappeared from both Ultisols and Oxisols in Puerto Rico at pH values around 5.5, but that it was difficult to maintain levels that high, in spite of the rate of lime applied.

## V. Future Research Needs for Long-Term Experimentation

The results presented in this paper were obtained from experiments which suffer from many limitations. The systematic characterization of soils with respect to chemical, physical, and microbiological properties of the soils used prior to the initiation of the experiments is often missing. Most of them have been conducted under very poor management conditions. The dynamics of soil properties under the experimental conditions has not been properly studied because soil samples were not taken or measurements were not done after harvest of each crop.

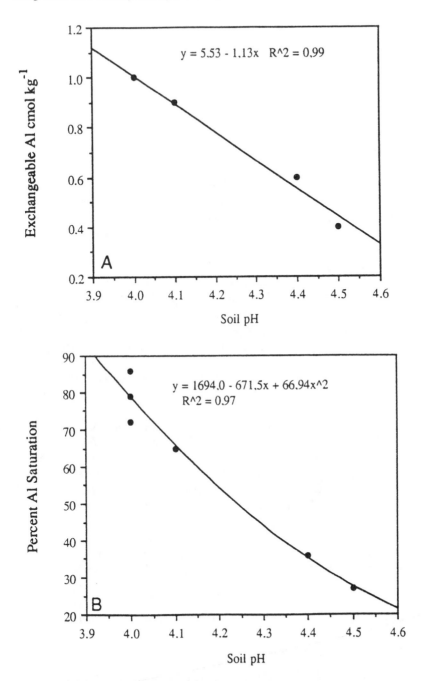

**Figure 14.** Relationship between soil pH and exchangeable Ala and Al saturation in an Oxisol 15 years after liming.
(Redrawn fromTveitnes and Svads, 1989.)

In spite of the limitations mentioned above, the review of the experiments presented in this paper has shown that the sustained crop production in thesesoils is dependent on soil-crop conditions. For example, continuous monoculture of maize on the Oxisols even with appropriate use of lime and plant nutrients cannot be expected to produce sustained high yields, whereas the Alfisols under similar management conditions were able to maintain high yields of maize both in Tanzania and Zambia. In spite of the variable effects of liming on crop yields and soil properties in this paper, the main conclusion seems to be that the addition of lime of appropriate type and in right quantity is essential for sustained crop production in these soils.

Some of the research priorities which require long-term experimentation for better fundamental understanding of what happens to soils when they are brought under continuous cultivation and which of the cropping systems are able to maintain a sustained production in these highly fragile soils of the Savannas in general are:

1. Characterization of soil fertility and study of the dynamics of soil properties, chemical, physical, and microbiological, with time in important soil-crop system is required. Long-term monitoring of the changes in soil properties would enable to predict soil chemical or physical deterioration that could occur under continuous cultivation and to adopt measures to correct them before they actually happen. This type of monitoring could also provide answers about the nutrient recycling, N turnover in systems involving legumes and the efficiency of fertilizer use.

2. Understanding the long-term effects of organic matter in soils dominated by low activity clay (Oxisols and Ultisols) in reducing soil acidity, nutrient economy, and soil conservation is required. Efforts should be directed to identify management systems that will sustain organic matter at reasonably higher levels than are experienced under conventional systems. Maintaining soil organic matter at levels consistent with sustainable land-use management is extremely important in the region.

3. Assessment of long-term productivity of benchmark soils and identification of nutrient deficiency or imbalances is only possible under long-term experimentation.

4. In areas where soils are deteriorated in their chemical or physical properties, research programs are needed for rejuvenation of such soils in order to make them part of a sustainable land-use system.

5. In acid soils understanding of soil Al in relation to nutrient uptake and root development is essential for their long-term use. The comparative role of P and lime in reducing soil acidity and their interactions need to be seen more critically.

6. Subsoil acidity is often a problem in heavy textured soils of the region and where occasional moisture stress is experienced. Therefore it is important to find efficient and economic means to move Ca to the subsoil layers.

## Acknowledgements

The presentation of this paper in the workshop on "Long-Term Soil Management Experiments" held at The Ohio State University, Columbus, Ohio, U.S.A. was supported by the grant from the Research Council Of Norway through the Centre for Sustainable Development of the Agricultural University of Norway. The financial support from the Reserach Council Of Norway is gratefully acknowledged.

The authors also thank Mr. P. Woode for allowing us to use selected data from his thesis.

## References

Abruna, F., R.W. Pearson, and R. Peres-Escolar. 1975. Lime response of corn and beans grown on tropical Oxisols and Ultisols of Puerto Rico. p. 261-262. In: E. Bornemisza and A. Alvardo (eds.), *Soil Management in Tropical America*. N. State Univ. Raleigh, N.C.

Alexander, M. 1977. *Introduction to Microbiology*. John Wiley & Sons, NY.

Allison, F.E. 1973. *Soil organic matter and it's role in crop production*. Dev. Soil Sci. 3. Elsevier, Amsterdam.

Amarasiri, S.L. and S.R. Olsen. 1973. Liming as related to solubility of P and plant growth in an acid tropical soil. *Soil Sci. Soc. Am. Proc.* 37:716-721.

Amedee, G. and M. Peech 1976. Liming of highly weathered soils of the humid tropics. *Soil Sci.* 121:259-266

Anderson,G.D. 1970. Fertility studies of a sand loam in semiarid Tanzania. I. Effects of N, P, and K fertilizers on yields of maize and soil nutrient status. *Expl. Agric.* 6:1-12.

Brams, E.A. 1971. Continuous cultivation of West African soils: Organic matter diminution and effects of applied lime and P. *Plant Soil* 35:401-414

Brenes, E. and R.W. Pearson. 1973. Root response of three gramineae species to soil acidity in an Oxisol and an Ultisol. *Soil Sci.* 116: 295-302.

Boyer, J. 1972. Soil potassium. In: *Soils of the Humid Tropics*. Natl. Acad. Sci. Wageningen.

FAO. 1988. *Soil Map of the World — Revised Legend*. World Soil Resources Report 60 FAO/UNESCO, Rome.

Fauck, R., G. Moreaux, and C. Thomann. 1969. Changes in soils of Sefa (Cassmance, Senegal) after fifteen years of continuous cropping. *Agron. Trep.* (Paris) 24:263-301.

Foster, H.L. 1970. Liming continuously cultivated soils in Uganda. *E. Afr. Agric. For. J.* 36:58-69.

Foy, C.D. 1984. Physiological effects of hydrogen, aluminum, and manganese toxicities in acid soils. p. 57-59. In: Adams, F. (ed.), *Soil Acidity and Liming*, 2nd ed. Amer. Soc. Agron., Madison, WI.

Goedert, W.J. 1983. Management of the Cerrado soils of Brazil: a review. *J.Soil Sci.* 34:405-428.

Grant, P.M. 1967. The fertility of Sandveld soil under continuous cultivation.II. The effect of manure and nitrogen fertilizer on the base status of the soil. *Rhod. Zamb. Mal. J. Agric. Res.* 5:117-128

Hanegraaf, M.C. 1985. Soil fertility related to catenas in Tanzania. Internal Publ. of Inst. Soil Fertility. The Netherlands.

Harrison, A.F. 1987. *Soil Organic Phosphorus. A Review of World Literature.* CAB International.

Harrop, J.F. 1985. Prospects for sequential experimentation for soil fertility studies in Tanzania. In 8th A.G.M. of the E.A. *Soil Sci. Soc.* Dec. 1985. Arusha, Tanzania.

Heynes, R.J. 1982. Effects of liming on phosphorus availability in acid soils, a critical review. *Plant Soil* 68:289-308.

Hodgins, L.W. B.S. Aulakh. 1985. Effect of liming on wheat performance in Northern Zambia. Zam-Can, Mount Makulu, Mimeo.

Haule, K.L., E.I. Kimambo, and J. Floor. 1989. Effect of continuous cultivation and fertilizer applications on yields of maize and soil characteristics of a Ferralsol-Luvisol catena on gneiss in northeastern Tanzania. Soil Fertility Report F5, Mlingano Agric. Res. Inst. Tanga, Tanzania.

Jones, E. 1972. Principles for using fertilizer to improve red ferrallitic soils in Uganda. *Expl. Agric.* 8:315-332

Jones, M.J. 1973. The organic matter content of the savanna soils of West Africa. *J. Soil Sci.* 24:42-53.

Jones, M.J. and A. Wild. 1975. Soil of the West African Savanna. The maintenance and improvement of their fertility. Com. Bureau Soil Tech. Comm. No. 55. Harpendeen, U.K.

Kamprath. E.J. 1967. Residual effects of large application of P on high P fixing soils. *Agron. J.* 59:25-27.

Kamprath. E.J. 1970. Exchangeable Al as a criteria for liming leached mineral soils. *Soil Sci. Soc. Amer. Proc.* 34:252-254.

Kraus A. 1988. Soil of East Africa. Nutrient reserves, availability and requirements. Paper presented at Seminar on Increased Crop Production through Efficient and Balanced Plant Nutrition. Addis Ababa, 1988.

Lal, R. and P.R. Maurya. 1982. Root growth of some tropical crops in uniform columns. *Plant and Soil* 68:193-206.

McEwan, T. 1932. Notes on fertilizer trials. Ann. Bull. Dept. Agric. Zambia.

McKenzie, R.C., D.C. Panny, L.W. Hodgins, B.S. Aulakh, and H. Ukrainnetz. 1988. The effects of liming on an Oxisol in Northern Zambia. *Comm. in Soil Sci. Plant Anal.* 19:1355-1669.

Mendoz, J. and E.J. Kamprath.1978. Liming of Latosols and the effect on P response. *Soil Sci. Soc. Amer. J.* 42:86-88.

Novias, R. and E.J. Kamprath. 1978. Phosphorus supplying capacity of of previously heavily fertilized soils. *Soil Sci. Soc. Am. J.* 42:931-935.

Nye, P. H. and D.J. Greenland. 1960. The soil under shifting cultivation. Commonwealth Bur. Soil Tech. Comm. 51. 156 pp.

Nye, P.H. and D.J. Greenland. 1964. Changes in the soil after clearing tropical forest. *Plant Soil* 21:101-112.

Ofori, C.F. 1973. Decline in fertility status of a tropical forest Ochrosol under continuous cropping. *Exp. Agr.* 9:15-22.

Øyagard R. 1986. Economic aspects of agricultural liming in Zambia. Ph.D. Thesis, Agric. Univ. of Norway. Aas, Norway.

Pearson, R.W. 1975. Soil Acidity and Liming in the humid Tropics. Cornell Intern. Agric. Bull. 30. Cornell University. Ithaca, N.Y.

Pearson, R.W., R. Perez-Escolar, F. Abruna, Z.F. Lund, and E.J. Brenes. 1977. Comparative response of three crop species to liming several soils of the Southeastern United States and Puerto Rico. Univ. Puerto Rico. *J. Agric.* 61:361-382.

Pawson, E. 1952. Fertility relationships of the Upper Valley Soils of Northern Rhodesia with special reference to maize. Thesis. London Univ.

Rasmussen, P.E and P. Collins. 1991. Long-term impacts of tillage, fertilizer, and crop residue on soil organic matter in temperate semi-arid regions.*Adv. Agron.* 45: 93-134.

Rattray, A.G.H. 1949. Annual Report of Experiments. Rhod. *Agric. J.* 46:326.

Sanchez, P.A. 1976. Properties and management of soils in the tropics. John Wiley, New York.

Sanchez, P.A., D.E. Bandy, J.H. Villchica, and J.J. Nicholaides. 1982. Amazon basin soils: management for continuous crop production. *Science* 216:821-827.

Singh, B.R. 1989. Evaluation of liming materials as ameliorants of acid soils in high rainfall areas of Zambia. *Norwegian J. Agric. Sci.* 3: 1321.

Singh, B.R. 1987a. Amendments for soil acidity amelioration. In: Lathem, M. (ed.), *Africaland — Land Development and Management of Acid Soils in Africa*. Lusaka, Zambia.

Singh, B.R. 1987b. Effect of liming on soil properties of Oxisols and Ultisols. In: Woode, P. (ed.), XII[th] International Forum on Soil Taxonomy and Agrotechnology Transfer. Lusaka, Zambia.

Soil Survey Staff. 1975. Soil Taxonomy, A basic system of soil classification for mapping and interpretation soil surveys. USDA. Washington D.C.

SPRP Research Reports 1983-89. Research Branch, MAWD. Kasama, Zambia.

Tveitnes S. and H. Svads. 1989. The effect of lime on maize and groundnut yields in the high rainfall areas of Zambia. *Nor. J. Agric. Sci.* 3:173-180.

Tveitnes S. and J.K. McPhillips. 1989. Maize yield response to N, P, and S in fertilizer under continuous cultivation in the southern province of Zambia. *Nor. J.Agric. Sci.* 3:181-189.

Vicente-Chandler, J., R. Caro-costas, J. Figarella, S. Silva, and R.W. Pearson. 1974. Bull. 223. Univ. of the Philippines Regional Agric. Expt. Station.

Woode, P.R. 1983. Changes in soil characteristics in a long-term fertilizer trial with maize in Northern Zambia. Vols.I and II. M.Sc. Thesis, University of Aberdeen, Scotland.

Woode, P.R. 1987. XII[th] International Forum on Soil Taxonomy and Agrotechnology Transfer. Lusaka, Zambia.

Young, A. 1976. *Tropical Soils and Soil Survey*. Camb. Univ. Press.

# C. Temperate and Mediterranean Climate

# The Effects of Long Continued Applications of Inorganic Nitrogen Fertilizer on Soil Organic Nitrogen — A Review

M.J. Glendining and D.S. Powlson

## I. Introduction

The use of inorganic nitrogen (N) fertilizer has been widespread for many years, although the rapid upsurge of its use only began in the 1950s. This paper reviews the effects of long–term applications of mineral N on soil N content, and in particular, on mineralizable soil N. There are two main reasons for interest in this subject.

First, mineralized soil N can make an important contribution to a crop's N requirements, even in soils with a relatively low organic matter content. For example, it has been calculated that, on average, 61 kg N ha$^{-1}$ is mineralized each year from an old arable soil in SE England with a total N content of 0.1% in the topsoil (**Broadbalk** plot 8, Powlson et al., 1986). Reliable predictions of the optimum fertilizer N requirement of a particular crop in a particular field

ISBN 1-56670-076-0

require an estimate of the amount of N made available to the crop by mineralization. Such predictions should reduce wasteful over- and under-applications of
fertilizer N, which are undesirable for both economic and environmental
reasons.

Second, although an increased supply of mineralized soil N may be of benefit
to subsequent crops, it may increase the quantity of N lost from agricultural soil
to the wider environment via leaching and gaseous emissions. There is
considerable concern about increasing concentrations of nitrate in potable waters
and the environmental consequences of nitrates in rivers (Department of
Environment, 1986; Keeney et al., 1987; House of Lords, 1989). Long–term
changes in the supply of mineralized soil N will have a profound effect on
nitrate leaching. They will also influence the amount of nitrate in soil at risk to
denitrification with consequent emission of nitrous oxide, a gas that is thought
to account for 5-10% of current global warming (Jenkinson, 1990a).

For many years it has been recognized that long–term applications of *organic*
N fertilizers lead to an increase in soil organic matter content, and thus to an
increased amount of mineralizable N in the soil (Woodruff, 1949; Stevenson,
1965). For example, the annual application of 35 t ha$^{-1}$ of farmyard manure
(FYM) to the **Hoosfield** Continuous Barley Experiment at Rothamsted since
1852 has more than doubled total soil N content (Jenkinson and Johnston, 1977).
The greater mineralization of N in these plots is reflected in average yield
increases in spring barley of about 100%, compared to the plots given only
inorganic fertilizer, when no additional fertilizer N is applied. It is also reflected
in the quantity of inorganic N (mainly nitrate) in the soil profile at risk to
leaching. Measurements made during the winter of 1986-87 showed that the
FYM plot contained at least 80 kg ha$^{-1}$ more inorganic N than the plot receiving
only inorganic fertilizers: total leaching over winter was estimated to have been
100 kg N ha$^{-1}$ greater from the FYM plot (Powlson et al., 1989).

Applications of *inorganic* N fertilizers are widely assumed to be of benefit to
the current crop only, with little or no effect on the supply of N to the next
crop. According to Cooke (1976), this idea seems to have originated from early
results from the **Broadbalk** Continuous Wheat Experiment at Rothamsted.
Drainage water collected from plots receiving inorganic N fertilizer contained
no more nitrate at the end of winter than plots given no fertilizer N. In a lecture
in 1891, Warrington said, "The very similar amount of nitrate passing away
during the winter is a striking illustration of the well-known fact that ammonium
salts leave no residue in the soil, which is of use to the next season's crop".
These views long dominated thinking on N residues (Cooke, 1976).

If inorganic fertilizer N is applied to a crop at an excessive rate, or too late
in the growing season for effective uptake, inorganic residues may remain in the
soil after harvest. In a temperate maritime climate, such as in northwest Europe,
such residues will be subject to leaching during winter and are unlikely to be of
benefit to subsequent crops. In a climate having a drier winter, they persist and
may be taken up in the next growing season. If applied at an appropriate rate
and time, in accordance with the growth pattern and expected yield of the crop,

inorganic N residues derived from fertilizer are often negligible. For example, Macdonald et al. (1989), using [15]N-labelled fertilizer, never found more than 5 kg ha[-1] labelled inorganic N in soil after winter wheat crops given fertilizer N rates up to 234 kg N ha[-1] on three soil types in SE England. Chaney (1990) also found that nitrate in soil after harvesting wheat crops only increased substantially if the amount of fertilizer N applied was more than that required for maximum grain yield. Although in the short–term, fertilizer N appears to have no effect on the N–supplying capacity of the soil (unless residues of inorganic N remain from the previous application), the central hypothesis of this paper is that in the long–term, continued applications of fertilizer N will indirectly increase soil organic N levels, and thus the quantity of mineralizable N.

The quantity of organic N in a soil is the net result of inputs of organic matter (e.g. organic manures and crop residues), and outputs of mineralized N, removed from the soil by plant uptake, leaching, and gaseous losses (and soil erosion if this occurs). This can be simply represented as:

$$dN/dt = - K_1(t).N + K_2 + K_3(t).Y(t)$$

where $N$ is soil organic nitrogen content, $K_1$ is a decomposition coefficient that may change with time (t), $K_2$ a non-crop addition term (e.g. organic manures), $K_3$ a crop addition term, influenced by $Y$, plant yield at time $(t)$ (Russell, 1975). An equilibrium level of soil organic N will be established when total annual inputs are equal to total annual outputs. There are feedback relationships between plant yield and soil organic N levels - soil organic N content can affect plant yield, and, through the return of plant residues to the soil, plant yield can affect soil organic N levels. The use of inorganic fertilizer N, through its influence on crop yield, may thus indirectly influence inputs of organic N, and ultimately equilibrium levels of soil organic N. Fertilizer N may increase both the amount of crop residue (roots, root exudates, stubble, etc.), and its N content. The N concentration of the residue will also influence its rate of mineralization and thus the rate of turnover of organic N (Jenkinson, 1984). There may also be direct immobilization of fertilizer N into the soil organic matter. Any other factors which affect crop yield, such as genotype and crop protection, may also affect inputs of organic N and thus soil equilibrium levels of N.

Such effects are small and may take many years to show up against the large background of organic N in the soil. Thus all the information in this review is obtained from long–term field experiments, the only satisfactory way of monitoring slow changes in soil fertility.

## A. Scope of the Review

This review is not intended to be a complete compilation of all long-term soil fertility experiments (see Steiner and Herdt (1993) for a comprehensive directory of non-European long-term agronomic experiments). Rather, results are presented from a number of selected experiments, so as to include a wide range of environments, soil types, cropping systems, years of fertilizer application, and rates and types of inorganic nitrogen fertilizer addition. As there are few appropriate long-term experiments in the tropics, the review is predominately concerned with temperate agriculture, although wherever possible examples from tropical agriculture are included.

The following criteria were used when selecting the experiments, although occasionally results from other experiments are included, for comparison.

1) The review is restricted to the effects of *inorganic* nitrogen fertilizers. For information and reviews on the effects of *organic* sources of N, see Stevenson (1982); Power and Doran (1984); and Arden–Clarke and Hodges (1988).

2) Long–term experiments which include legumes are not considered. This includes all long-term grassland experiments, as legumes are nearly always present, particularly in unfertilized treatments. This excludes many well known long–term experiments, e.g. the Jordan soil fertility plots, Pennsylvania, USA (White, 1955) and the Ley-Arable experiments at Rothamsted and Woburn (Johnston, 1973) and rotation experiments that include legumes. Greenland (1971) gives information on changes in soil N status under pasture. Legumes can fix large amounts of atmospheric N (e.g. the equivalent of 150–200 kg N ha$^{-1}$ in *Trifolium repens*/grass swards – Frame and Newbould, 1984). However, inputs of biologically fixed N are unlikely to be the same under different fertilizer treatments and thus it is difficult to separate the effect of inorganic fertilizer N from other factors. Fixation may be decreased if inorganic N is applied, for example by reducing soil pH (as in the Jordan Experiment, White, 1955), or from greater suppressive effects of companion species or by direct inhibition of fixation (Frame and Newbould, 1984). Alternatively, small amounts of inorganic N can encourage legume growth, particularly in monocultures, and increase the amount of N fixation.

3) *All* treatments should receive adequate supplies of all other nutrients, otherwise plant growth and N uptake may be restricted, thus reducing the return of organic N to the soil in the plant residues. Many long–term experiments do not include appropriate control plots (i.e. plots receiving phosphorus (P), potassium (K), and all other necessary nutrients, with the exception of nitrogen, subsequently referred to as PK control plots). If this is the case, results are compared with unfertilized treatments, although this is obviously unsatisfactory.

4) All experiments should be adequately limed to prevent acidification, particularly from the use of ammonium–based fertilizers. Acidification can be a major problem especially in tropical soils, irrespective of the source of N (Nnadi and Arora, 1985). This can lead to serious restriction of plant growth, and to a decline in soil fertility. Although an important consequence of the long–term usage of inorganic N fertilizers, acidification is outside the scope of this review. For more information, see Cooke (1967) and Wild (1988).

5) All fertilizer treatments have been applied continuously for at least seven years although some changes in soil N content have been detected even earlier than this and are mentioned in the text. A distinction has been drawn between 'long-term' and 'medium-term' treatments, arbitrarily defined as more than 40 years, and seven to 40 years, respectively.

## II. Results

### A. Total N Content of the Soil

Relatively few long- and medium-term experiments provide information on the effects of inorganic N additions on mineralizable soil N, although many give details of the effects on total soil N content, subsequently referred to as total N. Such information is of relatively limited value in predicting the effect on mineralization, for the following reasons.

Cultivated topsoils contain huge amounts of N (commonly 2 to 5 t N ha$^{-1}$, of which more than 95% is present in organic forms), which are not available to plants until mineralized. Much of this organic N is very resistant to mineralization, with a very long turn-over time (Jenkinson and Rayner, 1977), and changes in total N may not be apparent for many years. The large background of relatively inert organic N may obscure changes in the smaller fractions of more readily mineralizable N. There can also be problems in detecting small changes in total N, as the % N content of soil is very low, commonly between 0.06% and 0.5% of cultivated topsoils. Measurements of total N can provide useful information on the long-term effects of major treatments which have a profound effect on soil organic N content — for example, when comparing organic and inorganic manures, or continuous grassland and arable cropping. However, it is much more difficult to detect long-term changes in total N resulting from less extreme treatments — for example, the use of inorganic N fertilizers. Furthermore, measurements of changes in *total* N give no information on changes in the different fractions of soil N.

Tables 1 and 2 show total soil N content after many years of fertilizer N application in a number of long-term and medium-term field experiments. Strictly, soil N content on a % dry soil basis should be corrected for changes in bulk density brought about by the different treatments and presented as kg N ha$^{-1}$ in a specified layer of soil. However, the necessary bulk density measurements are seldom reported, so to allow comparison between experiments, all results have been presented as the percentage of N in dry soil. If available, the initial soil N content at the start of the experiment has been included to show any changes in soil N over the course of the experiment. Any differences between the NPK and PK treatments (i.e. with and without fertilizer N, all other nutrients supplied) are expressed as a percentage of the PK treatment, or unfertilized (none) treatment if no PK treatment is available (in brackets).

**Table 1.** Total soil Nitrogen content in long-term field experiments, after more than 40 years of inorganic fertilizer nitrogen application. % N in air-dry soil

| Site details (further details in numbered footnotes) | Years of fertilizer treatment[a] | | | kg N ha⁻¹ y⁻¹ | Total soil N content[b] % of dry soil | | | | Difference[c] | |
| --- | --- | --- | --- | --- | --- | --- | --- | --- | --- | --- |
| | None | PK | NPK | | Initial | None | PK | NPK | % | kg N ha⁻¹ [d] |
| 1) Broadbalk, Rothamsted, UK. Continuous w.wheat | 136 | 136 | 136 | 48 | 0.113 | 0.102 | 0.104 | 0.113 | +9 | +220 |
| | | " | " | 96 | | | | 0.124 | +19 | +480 |
| | | | " | 144 | | | | 0.126 | +21 | +530 |
| | | | 20 | 192 | | | | 0.123 | +18 | +460 |
| 2) Hoosfield, UK. | 124 | 124 | 124 | 48 | 0.130 | 0.101 | 0.105 | 0.098 | -7 | -170 |
| 3) Barnfield, UK. | 113 | 113 | 113 | 96 | - | 0.092 | 0.088 | 0.096 | +9 | +190 |
| 4) Halle, Germany. | | | | | | | | | | |
|   Continuous rye | 84 | 84 | 84 | 40 | 0.095 | 0.072 | 0.074 | 0.080 | +8 | +140 |
|   Continuous rye | 108 | 108 | 108 | " | | 0.073 | 0.082 | 0.080 | -2 | -50 |
|   Maize (since 1961) | " | " | " | " | | 0.074 | 0.077 | 0.082 | +6 | +120 |
|   Rotation (since 1961) | " | " | " | " | | 0.076 | 0.074 | 0.079 | +7 | +120 |
| 5) Dehérain, Grignon, France. | 60 | 58 | 60 | 87 | 0.204 | 0.124 | 0.120 | 0.138 | +15 | +430 |
| | 84 | - | 84 | " | | 0.098 | - | 0.123 | (+26) | (+600) |
| 6) Askov, Denmark.   Sand | 79 | 79 | 79 | 60 | 0.073 | 0.053 | 0.062 | 0.066 | (+24) | (+310) |
|   Loam | " | 39 | " | " | 0.146 | 0.108 | 0.121 | 0.119 | (+10) | (+260) |
| 7) Bad Lauchstädt, Germany. | 76 | 76 | 76 | 70 | 0.160 | - | 0.136 | 0.151 | +11 | +360 |

**Table 1.** continued

| Site | Treatment | | | | | | | | | | |
|---|---|---|---|---|---|---|---|---|---|---|---|
| 8) Coimbatore, India. | Dry | - | 66 | 66 | 125 | NA | - | 0.028 | 0.029 | +4 | +20 |
|  | Irrigated | - | 39 | 39 | " | NA | - | 0.034 | 0.038 | +12 | +100 |
| 9) Pendleton, Oregon, USA. | | 45 | - | 45 | 45 | 0.095 | 0.079 | - | 0.083 | (+5) | (+100) |
|  | | 55 | - | 55 | | | 0.079 | - | 0.087 | (+10) | (+190) |
| 10) Weihenstephen, Germany. | $CaCN_2$ | - | 53 | 53 | 72 | NA | - | 0.096 | 0.105 | +9 | +220 |
|  | $(NH_4)_2SO_4$ | - | " | " | | | | | 0.108 | +13 | +290 |
|  | limed $(NH_4)_2SO_4$ | - | 39 | 39 | | | | | 0.107 | +11 | +260 |
|  | $NaNO_3$ | - | 53 | 53 | | | | | 0.100 | +4 | +100 |
|  | $Ca(NO_3)_2$ | - | " | " | | | | | 0.099 | +3 | +70 |
| 11) Stackyard Field, Woburn, UK. Continuous cereals | | - | 11 | 11 | 46 | 0.156 | - | 0.125 | 0.134 | +7 | +220 |
|  | | - | " | " | 92 | | | | 0.142 | +14 | +410 |
| cereals | | - | 50 | 50 | 36 | | - | 0.104 | 0.103 | -1 | -20 |
|  | | - | " | " | 50 | | | | 0.106 | +2 | +50 |
| 12) Uman, Ukraine. Orchard | 0-20 cm | 50 | 50 | 50 | 120 | NA | 0.149 | 0.157 | 0.160 | +2 | +70 |
|  | 20-40 cm | | | | " | " | 0.124 | 0.128 | 0.131 | +2 | |
|  | 40-100 cm | | | | " | " | 0.082 | 0.086 | 0.088 | +2 | |
| 13) Skierniewice, Poland. Continuous potatoes | | - | 46 | 46 | 37 | NA | - | 0.080 | 0.083 | +4 | +70 |
|  | | - | 50 | 50 | " | | - | 0.080 | 0.082 | +3 | +50 |
| Continuous rye | | - | 46 | 46 | " | | - | 0.090 | 0.095 | +6 | +120 |
|  | | - | 50 | 50 | " | | - | 0.084 | 0.088 | +5 | +100 |

**Table 1.** continued

| Site | | | | | PK | None | NPK | [c] | [d] |
|------|---|---|---|---|------|------|------|-----|------|
| 14) Sanborn Field, USA. | 50 | - | 50 | 72 | 0.170 | 0.100 | 0.100 | (0) | (0) |
| 15) Lyberetski, Russia. | 49 | - | 49 | 35 | NA | 0.091 | 0.091 | (0) | (0) |
| 16) Bernburg, Germany. | | | | | | | | | |
| Continuous potatoes 0-25 cm | 42 | 42 | 49 | NA | NA | 0.116 | 0.123 | +6 | +170 |
| 25-50 cm | " | " | " | | - | 0.103 | 0.095 | -8 | - |
| Continuous rye 0-25 cm | " | " | 29 | | - | 0.125 | 0.133 | +6 | +190 |
| 25-50 cm | " | " | " | | - | 0.113 | 0.112 | -1 | - |

NA = not available.

[a] Fertilizer treatments: 'None' = unfertilized, 'PK' = No N, but adequate supplies of all other nutrients, 'NPK' = inorganic N plus all other nutrients.

[b] Soil depth 0-20 cm, unless stated otherwise.

[c] Difference between NPK and PK treatments or, where no PK available, between None and NPK (figures in brackets).

[d] Difference in kg N ha$^{-1}$ calculated from standard soil bulk density of 2.4 $\times$ 10$^6$ kg ha$^{-1}$ and standard soil depth of 0-20 cm.

1) Glendining and Powlson (1991). Started 1843, fertilizer treatments established in 1852. Silt loam, 19-27% clay. Initial N content extrapolated from Jenkinson (1977) assuming a soil weight of 2.91 $\times$ 10$^6$ kg ha$^{-1}$ and 3.3 t N ha$^{-1}$ (0-23 cm). See also Tables 3, 5, 6, and 9 and Fig.1a. No replicate plots. Sections 1 and 9. Straw removed. Soil depth 0-23 cm.

2) Jenkinson and Johnston (1977). Rothamsted. Continuous s.barley. Started 1852. Silt loam. Straw removed. Initial N estimated by Jenkinson and Johnston (1977). See also Table 6 and Fig.1b. No reps. Soil depth 0-23 cm.

3) Warren and Johnston (1962); Avery et al. (1972). Rothamsted. Started 1843. Root crops, mainly mangolds. Silty clay loam. No reps. Soil depth 0-23 cm. 48 kg N ha$^{-1}$ applied 1845-1860. See also Table 6.

4) Kolbe and Stumpe (1969); Garz et al. (1982); Garz and Hagedorn (1990). Started 1878. "Ewiger Roggenbau" Sandy loam, 13% clay, loess. Initial N content estimated from total C content, assuming a C:N ratio of 13 (Garz et al., 1982). Continuous rye 1878-1961. Cropping changes (continuous maize and potato/rye rotation) introduced in 1961. Fertilizer treatments separated by sunken concrete walls. See also Tables 3, 8, and 9 and Fig.1d. No reps.

5) Morel et al. (1984); R. Chaussod (pers. comm.); Houot et al. (1989). Started 1878. Agrudalf-loam, 28% clay, Brown colluvial soil. W.wheat/sugar beet rotation, straw and tops removed. PK treatment started 27 years after NPK. See also Tables 3 and 9 and Fig.1c. Soil depth 0-30 cm at 60-year sampling, 0-20 cm at 84-year sampling. 8 reps.

6) Kofoed and Nemming (1976); Kofoed (1982). Started 1894. Sand 4% clay. Loam 11% clay. Rotation (including grass/clover ley) PK treatment in loam soil started 40 years later than other treatments. Initial N content = unfertilized treatment after 18 years. NPK after 18 years = 0.081 (sand), 0.171 (loam). 4 reps. See also Fig.1e.

7) Körschens et al. (1982). Started 1902. Rotation (no legumes). Humic loess loam. 'PK' = Mean of PK and unfertilized treatments. 'NPK' = Mean of NPK, NK, NP, and N treatments. Initial value from Eich et al. (1982). See also Tables 6 and 8. No reps.

8) Muthuvel et al. (1977; 1979). Permanent Manurial Expts., Tamil Nadu Agric. Univ. Calcerous red loam. Sorghum and cotton. 'Old' expt. started 1907 (not irrigated). 'New' expt. started 1934 (irrigated). See also Table 8. No reps.

9) Rasmussen et al. (1980; 1989). Silt loam. W.wheat/fallow rotation, straw incorporated. Estimated from Fig.7, assuming soil bulk density of $3.79 \times 10^6$ kg ha$^{-1}$ (0-30 cm). No PK applied, N applied cropped years only. 34 kg N ha$^{-1}$ 1931-66. 4 reps.

10) Bosch and Amberger (1983). München. 3-course rotation. Sandy-silt loam. See also Tables 3, 7, and 8. 4 reps, 0-25 cm.

11) Johnston, 1975; Mattingly et al., 1975. Loamy sand/sandy loam. 1876-1926. Mean of Continuous W.Wheat and Continuous S.Barley Experiments, and two forms of N ($NH_4^+$-N and $NO_3^-$-N). No reps. Soil depth 0-23 cm. See also Fig.1f.

12) Kopytko and Gérkiyal (1983). Clay loam, dark grey forest soil. Apple orchard. Humus content declined by 3 and 9% over 50 years in the NPK and unfertilized treatments respectively. See also Table 8. Probably no reps.

13) Dobransky (1976). Loamy sand, 8% clay. Started 1923. Limed regularly. See also Table 8. 5 reps, soil depth NA.

14) Upchurch et al. (1985); Smith (1942). Missouri. Mexico silt loam. Continuous wheat. Initial soil N = virgin sod. % N calculated from soil bulk density of $2.24 \times 10^6$ kg ha$^{-1}$. No reps. Soil depth 0-18 cm.

15) Shevtsova (1966). Moscow Region, Derno-podsolic loam. Continuous flax. See also Tables 3 and 7. Soil depth NA.

16) Wabersich (1967). Halle, Black earth. Started 1910. NPK mean of N as $NO_3^-$ and $NH_4^+$-based fertilizers. No reps.

**Table 2.** Total soil nitrogen content in medium-term field experiments, after between 7 and 40 years of inorganic fertilizer nitrogen application. % N in air-dry soil

| Site details (further details in numbered footnotes) | Years of fertilizer treatment | | | $kg\ N\ ha^{-1}\ y^{-1}$ | Total soil N content % of dry soil | | | | Difference | |
|---|---|---|---|---|---|---|---|---|---|---|
| | None | PK | NPK | | Initial | None | PK | NPK | % | $kg\ N\ ha^{-1}$ |
| 1) Kansas Ag. Expt. Stn, USA. | - | 35 | 35 | 23 | 0.139 | - | 0.133 | 0.114 | -14 | -460 |
| 2) Ak-Karak, Kazakhstan. | 35 | - | 35 | 125 | NA | 0.058 | - | 0.065 | (+12) | (+170) |
| 3) Mironov Expt. Stn. Kiyev, Ukraine. | 31 | - | 31 | 40 | NA | 0.203 | - | 0.210 | (+3) | (+170) |
| | | | " | 65 | | | | 0.198 | (-3) | (-120) |
| 4) Melfont, Canada. | | | | | | | | | | |
| Continuous spring wheat | 31 | - | 31 | 52 | 0.55 | 0.535 | - | 0.529 | (-1) | (-100) |
| Fallow/wheat/wheat rotation | " | | " | 14 | | 0.493 | | 0.503 | (+2) | (+160) |
| 5) Indian Head, Canada. | | | | | | | | | | |
| Continuous spring wheat | 30 | - | 30 | 46 | 0.2 | 0.194 | - | 0.213 | (+10) | (+420) |
| Fallow/wheat rotation | " | | " | 6 | | 0.181 | | 0.187 | (+3) | (+130) |
| Fallow/wheat/wheat rotation | " | | " | 6 | | 0.182 | | 0.199 | (+9) | (+370) |
| 6) Halle, Germany. Arable rotation | 30 | - | 30 | 50 | 0.123 | 0.115 | - | 0.118 | (+3) | (+70) |
| | | | | 100 | | | | 0.118 | (+3) | (+70) |
| 7) Lyberetski Expt. Stn, Moscow, Russia. | - | 27 | 27 | 70 | NA | | 0.042 | 0.045 | +7 | +70 |
| N as NaNO₃ | | | | | | | | | | |
| N as (NH₄)₂SO₄ | | | " | " | | | | 0.047 | +12 | +120 |

**Table 2.** continued

| | | | | | | | | | |
|---|---|---|---|---|---|---|---|---|---|
| 8) "36 Parcelles", France. | | | | | | | | | |
|   without straw | 19 | - | 19 | 70 | 0.147 | 0.121 | 0.128 | (+6) | (+170) |
|   straw composted with N | " | - | " | | | 0.135 | 0.139 | (+3) | (+100) |
|   without straw | 26 | - | 26 | | | 0.110 | 0.117 | (+6) | (+170) |
|   straw composted with N | " | - | " | | | 0.139 | 0.145 | (+4) | (+140) |
| 9) Tocklai Expt. Stn., Assam, India. Tea. | 26 | - | 26 | 45 | 0.120 | 0.074 | 0.082 | (+11) | (+190) |
| | | | " | 90 | | | 0.083 | (+12) | (+220) |
| | | | " | 135 | | | 0.087 | (+18) | (+310) |
| 10) La Estanzuela, Uruguay. | 23 | | 23 | 36 | 0.181 | 0.128 | 0.148 | +16 | +480 |
| 11) Southern Sweden. Mean of 6 sites, 3 N rates. | 20 | 14 | 14 | 50 | 0.18 | 0.156 | 0.162 | +4 | +140 |
| | | | | 100 | | | 0.168 | +8 | +290 |
| | | | | 150 | | | 0.173 | +11 | +410 |
| Each site, highest N rate only | | | | | | | | | |
|   Västraby | - | 22 | 22 | 150 | 0.20 | 0.158 | 0.173 | +10 | +360 |
|   Örja | | | " | " | 0.13 | 0.104 | 0.123 | +18 | +460 |
|   Fjärdingslöv | | | " | " | 0.18 | 0.135 | 0.152 | +13 | +410 |
|   Ekebo | | | " | " | 0.20 | 0.179 | 0.204 | +14 | +600 |
|   Orup | | | " | " | 0.22 | 0.186 | 0.195 | +5 | +220 |
|   S. Ugglarp | | | " | " | 0.15 | 0.125 | 0.144 | +15 | +460 |
|   Mean | | | " | " | 0.18 | 0.148 | 0.165 | +12 | +410 |
| 12) Martonvasar, Hungary. | 21 | - | 21 | 120 | NA | 0.161 | 0.172 | (+7) | (+260) |
| 2 sites. Maize/wheat rotation | " | | " | 160 | | 0.260 | 0.258 | (-1) | (-50) |

**Table 2.** continued

| | | | | | | | | | | |
|---|---|---|---|---|---|---|---|---|---|---|
| 13) Gross Kreutz, Germany. | 21 | - | 21 | 67 | NA | 0.065 | - | 0.060 | (-8) | (-120) |
| 14) Scottsbuff, Nebraska, USA. Continuous maize | - | 20 | 20 | 45 | 0.05 | - | 0.051 | 0.055 | +8 | +100 |
| | | | " | 90 | | | | 0.056 | +10 | +120 |
| | | | " | 135 | | | | 0.056 | +10 | +120 |
| | | | " | 180 | | | | 0.056 | +10 | +120 |
| 15) Broom's Barn, UK.  0-25 cm | 19 | - | 19 | 75 | 0.102 | 0.094 | - | 0.099 | (+5) | (+120) |
| 0-25 cm | | | " | 150 | 0.102 | | | 0.104 | (+11) | (+240) |
| 25-50 cm | 19 | | 19 | 75 | NA | 0.073 | - | 0.072 | (-1) | - |
| 25-50 cm | | | " | 150 | NA | | | 0.082 | (+12) | - |
| 16) Lethbridge, Canada. Continuous spring wheat | 18 | 13 | 18 | 45 | 0.149 | - | 0.145 | 0.166 | +15 | +500 |
| Wheat/fallow/fallow rotation | " | " | " | " | 0.120 | - | 0.140 | 0.154 | +10 | +340 |
| 17) Saria, Burkina Faso. Sorghum without crop residues | 9 | - | 9 | 40 | NA | 0.023 | - | 0.020 | (-13) | (-70) |
| | 17 | - | 17 | " | | 0.030 | | 0.030 | (0) | (0) |
| | 19 | - | 19 | " | | 0.018 | | 0.019 | (+6) | (+20) |
| with crop residues | 9 | - | 9 | 40 | | - | | 0.028 | | - |
| | 17 | - | 17 | " | | - | | 0.035 | | - |
| | 19 | - | 19 | " | | - | | 0.018 | | |

**Table 2.** continued

| | | | | | | | | | | |
|---|---|---|---|---|---|---|---|---|---|---|
| **18) Fundulea, Romania.** | | | | | | | | | | |
| Continuous w.wheat | 18 | - | 18 | 100 | NA | 0.107 | - | 0.130 | (+21) | (+550) |
| Wheat/maize rotation | " | - | " | " | NA | 0.120 | - | 0.147 | (+23) | (+23) |
| **19) N. Nigeria. Savanna.** | | | | | | | | | | |
| Continuous cotton since 1961 | - | 15 | 15 | 12 | NA | - | 0.026 | 0.029 | +12 | +70 |
| | | " | " | 24 | | | | 0.029 | +12 | +70 |
| | | 18 | 18 | 12 | | | 0.025 | 0.025 | 0 | 0 |
| | | " | " | 24 | | | | 0.026 | +4 | +20 |
| **20) Hermitage Res. Stn., Australia. Wheat and barley** | - | 13 | 13 | 12 | NA | - | 0.151 | 0.153 | +1 | +50 |
| | | " | " | 24 | | | | 0.160 | +6 | +220 |
| | - | 20 | 20 | 12 | | | 0.136 | 0.141 | +4 | +120 |
| | | " | " | 24 | | | | 0.145 | +7 | +220 |
| **21) Swift Current, Canada.** | | | | | | | | | | |
| Continuous spring wheat | - | 10 | 10 | 32 | 0.178 | 0.175 | 0.174 | 0.204 | +17 | +720 |
| | - | 15 | 15 | " | | | 0.178 | 0.196 | +10 | +430 |
| Fallow/wheat/wheat rotation | - | 10 | 10 | " | | | 0.179 | 0.179 | 0 | 0 |
| | - | 15 | 15 | " | | | 0.184 | 0.184 | 0 | 0 |
| **22) Sofia, Bulgaria.** 0-20 cm | 13 | 13 | " | 60 | NA | - | | 0.177 | (+1) | (+50) |
| Wheat/maize rotation | | " | " | 120 | | | | 0.177 | (+1) | (+50) |
| | | " | " | 180 | | | | 0.179 | (+2) | (+100) |
| 20-40 cm | 13 | " | " | 60 | | 0.169 | | 0.164 | (-3) | - |
| | | " | " | 120 | | | | 0.171 | (+1) | - |
| | | " | " | 180 | | | | 0.167 | (-1) | - |

**Table 2.** continued

| | | | | | | | | | |
|---|---|---|---|---|---|---|---|---|---|
| 23) Columbia, Oregon, USA. Wheat/fallow rotation | 0-7.5 cm | 11 | 22.4 | NA | – | – | 0.111 | | |
| | " | " | 44.8 | | | | 0.117 | | |
| | " | " | 67.2 | | | | 0.125 | | |
| | " | " | 89.6 | | | | 0.127 | | |
| | 0-22.5 cm | " | 22.4 | | | | 0.101 | | |
| | " | " | 44.8 | | | | 0.104 | | |
| | " | " | 67.2 | | | | 0.107 | | |
| | " | " | 89.6 | | | | 0.110 | | |
| 24) Guelph, Canada. | Stover retained | 9 | 112 | NA | 0.08 | – | 0.09 | +13 | +240 |
| | | " | 224 | | | – | 0.09 | +13 | +240 |
| | Stover removed | 9 | 112 | | 0.12 | – | 0.09 | -25 | -720 |
| | | " | 224 | | | – | 0.10 | -17 | -480 |
| 25) Lind, USA. | Dry | 9 | 34 | NA | – | 0.058 | 0.059 | (+2) | (+20) |
| | | " | 67 | | | | 0.063 | (+9) | (+120) |
| | | " | 135 | | | | 0.063 | (+9) | (+120) |
| | | " | 270 | | | | 0.067 | (+16) | (+220) |
| | Irrigated | 9 | 34 | NA | – | 0.055 | 0.055 | (0) | (0) |
| | | " | 67 | | | | 0.063 | (+15) | (+190) |
| | | " | 135 | | | | 0.069 | (+25) | (+340) |
| | | " | 270 | | | | 0.072 | (+31) | (+410) |
| 26) Samaru, Nigeria. Savanna soil, fine sandy loam | | 9 | 26 | NA | – | 0.042 | 0.046 | (+10) | (+100) |
| | | " | 52 | | | | 0.047 | (+12) | (+120) |

**Table 2.** continued

| | | | | | | | | | | |
|---|---|---|---|---|---|---|---|---|---|---|
| 27) Barrackpore, India. Jute-rice-wheat rotation | 8 | - | 8 | 150 | 0.075 | 0.061 | - | 0.066 | (+8) | (+120) |
| | | | " | 300 | | | | 0.067 | (+10) | (+140) |
| | | | " | 450 | | | | 0.069 | (+13) | (+190) |
| 28) Lethbridge, Canada. | | | | | | | | | | |
| Continuous spring wheat | 8 | - | 8 | 45 | NA | 0.17 | - | 0.18 | (+6) | (+240) |
| Wheat/fallow rotation | " | | " | " | | 0.15 | - | 0.15 | - | (0) |
| Wheat/wheat/fallow rotation | " | | " | | | 0.12 | - | 0.13 | (+8) | (+240) |
| 29) Sugarcane Res. Stn., India. Continuous sugarcane | 7 | | 7 | 132 | 0.11 | 0.033 | - | 0.036 | (+9) | (+70) |

NA = Not available.   See Table 1 for definitions of fertilizer treatments, differences and soil depth.

1) Dodge and Jones (1948).  Prairie soil, silty clay loam.  Cont. wheat.  Initial N after 4 years of treatments.  No reps, 0-18 cm.

2) Shevtsova (1966).  Tashkent.  Grey soil, Serozem.  Continuous cotton.  See also Tables 4 and 5.  Soil depth NA.

3) Shevtsova (1966).  Light loam, black earth Chernozem.  Continuous maize.   See also Tables 4, 7 and 8.  Soil depth NA.

4) Campbell et al. (1991b); Zentner et al. (1990).  Silty clay loam, thick black Chernozem.  No K applied.  LSD (P<0.10) = 0.0202.  Actual bulk density (1.2 x $10^6$ kg ha$^{-1}$) used to calculate differences.  See also Table 8.  4 reps, 0-15 cm.

5) Campbell et al. (1991a;c); Zentner et al. (1987).  Saskatchewan.  Heavy clay, thin black Chernozem.  No K applied, LSD (P<0.10) = 0.0192.  Bulk density (1.63 x $10^6$ kg ha$^{-1}$) used for differences.  See also Tables 4, 8 and 9.  4 reps, 0-15 cm.

6) Stumpe and Hagedorn (1982).  Sandy loam brown Chernozem.  Potato/cereal/maize/s.beet; 1975-1981 mean. Soil depth NA.

7) Shevtsova (1966).  Sandy loam, Derno-podsolic.  4-course rotation (legumes?).  See also Tables 4, 5 and 8.  Soil depth NA.

8) Morel et al. (1984); Houot et al. (1989).  Grignon.  Bare fallow since 1959.  Loam (26% clay).  7 t ha$^{-1}$ yr$^{-1}$ wheat straw. Composted straw: C = 1300, N = 58 kg ha$^{-1}$ yr$^{-1}$.  Composted with N:C = 1100, N = 68 kg ha$^{-1}$ yr$^{-1}$.  See also Tables 4 and 9.  Soil depth 0-25 cm (19 yrs) and 0-20 cm (26 yrs).  6 and 1 reps, respectively.

9) Gokhale (1959).  Acid soil, loamy sand, alluvial origin.  Pruned annually, litter left on site.  7 reps, soil depth 0-23 cm.

10) Diaz (pers. comm.). Colonia. Silty clay loam. Arable rotation, no legumes. Mean N rate (range 0-100). No K applied. Initial N = mean of 1964-66. Other values = mean of 1984-86. 3 reps.

11) Jansson (1983); Bjarnason (1989); Ivarsson and Bjarnason (1988). Barley/oilseed/wheat/s.beet rotation. Mean of 4 PK levels. Initial soil N in 1957. Soil bulk density = $2.5 \times 10^6$ kg ha$^{-1}$. 2 reps at each site. Taken from linear regression of changes in total soil N between 1960 and 1984 (fertilizer treatments started in 1962). See also Tables 4 and 7.

12) Hargitai (1982). Forest residue Chernozem brown earth. 5 and 4 reps, respectively. Soil depth NA.

13) Asmus et al. (1982). Loam, yellow earth. Rye/oat/potato rotation. 40 kg N ha$^{-1}$ 1959-65; 80 kg N ha$^{-1}$ 1966-79. 0-30 cm.

14) Anderson and Peterson (1973). Fine sandy loam. No K applied. 2 reps, soil depth 0-30 cm.

15) Last et al. (1985). Suffolk. Sandy loam. S.beet/w.wheat/s.barley rotation. 2 reps. See also Table 4.

16) Janzen (1987). Alberta. Sandy loam, dark brown Chernozem. Straw incorporated. Initial N 14 years before start. No K applied. 'PK' = mean of P and none. 'NPK' = mean of NP and N. See also Table 4. No reps, 0-10 cm.

17) Pichot et al. (1981). Sand. Started in 1960. Crop residues applied alternate years (since 1969).

18) Ionescu et al. (1986). Podzolic reddish-brown soil. 5 reps.

19) Bache and Heathcote (1969); Jones (1971). Light brown loamy fine sand. Mean of 3 P, K + FYM rates, each replicated 3 times. N as $(NH_4)_2SO_4$ lowered pH considerably. Differences not significant. Soil depth 0-15 cm.

20) Dalal (1989); Dalal et al. (1991). Queensland. Cracking clay. LSD ($P < 0.05$) = 0.008 (13 yr) and 0.006 (20 yr). Same trend in organic C. See also Tables 4 and 9. Mean of 2 tillage and 2 residue treatments with 4 reps, 0-10 cm.

21) Biederbeck et al. (1984); Campbell et al. (1983). Saskatchewan. Loam. Straw retained. No K applied. Initial soil N estimated from samples taken 50 m away. See also Tables 4 and 9. 3 reps, soil depth 0-15 cm.

22) Petrova and Gospodinov (1986). Weakly leached Chernozem. See also Table 4. 4 reps.

23) Rasmussen and Rohde (1988). Silt loam. N applied alternate years, 11 times in 22 years. No PK applied. 3 reps.

24) Ketcheson and Beauchamp (1978). Loam. Continuous maize. See also Tables 4 and 6. Soil depth NA, 4 reps.

25) El-Haris et al. (1983). Washington. Silt loam. Wheat/fallow rotation. See also Table 4. 3 reps, soil depth 0-15 cm.

26) Jones (1971). Mean of cont. cotton and cotton/sorghum/groundnut rotation and 2 mulching levels, plough layer.

27) Mandal et al. (1984; 1985). W. Bengal. New Gangetic Alluvial sandy loam. 3 crops a year, 4 reps, 0-22 cm.

28) Dormaar and Pittman (1980). Alberta. Dark brown Chernozem. No PK applied, N to fallow. No reps, 0-13 cm.

29) Singh (1964). Muzaffarnagar, Uttar Pradesh. Sandy loam. N as $(NH_4)_2SO_4$, soil pH 7.8. 2 reps, soil depth NA.

Obviously, an increase of, say, 0.01% N is proportionally greater in soils with a low background total N content. Thus, in order to compare experiments with different total N contents, we have also expressed any change in terms of kg N ha$^{-1}$. As few authors give details of soil bulk density, we used a standard value of 2.4 x 10$^6$ kg ha$^{-1}$ to a standard depth of 0 to 20 cm, the mean (range 2.2 to 2.5 x 10$^6$) quoted in the experiments in Tables 1 and 2 (excluding the site at **Melfont**, Canada, with the low value of 1.2 x 10$^6$ kg ha$^{-1}$ 0-15 cm, Campbell et al., 1991b; Table 2, expt. 4).

In general, the results presented in Table 1 indicate that long and very long–term applications of inorganic N fertilizer have increased total N compared to plots receiving no fertilizer N. The increase is generally small, with only the **Broadbalk, Dehérain, Bad Lauchstädt** and **Weihenstephen** experiments (Table 1, expts. 1, 5, 7, 10) showing increases of more than 10% compared to the PK control, equivalent to an extra 250 to 530 kg N ha$^{-1}$ (0-20 cm, standard bulk density). Most other experiments show increases of 5-10%. However, N fertilizer appeared to have no effect at **Hoosfield, Uman, Sanborn Field** and **Lyberetski** (Table 1, expts. 2, 12, 14, 15).

N application rates were modest or low at most sites, and in general the greatest increase in total N occurred at the highest N rate. The largest increase was around 20%, or 500 kg N ha$^{-1}$ (0-23 cm), after applying 144 kg N ha$^{-1}$ to the **Broadbalk** experiment for 135 years (since 1852, Glendining and Powlson, 1991). As the straw is removed each year, this increase must arise from greater returns of organic N in roots, root exudates, stubble, chaff, and from fallen leaves, plus any direct immobilization of fertilizer N into soil organic matter. It appears to have been established within 30 years — in 1881 the plot contained 28% more total N than the PK treatment (Figure 1a). As the experiment contains no true replication, it could be argued that this was due to soil differences present before the experiment began. However, a further plot given 192 kg N ha$^{-1}$ since 1968 (previously 48 kg N ha$^{-1}$) has shown a 9% increase in total N over the past 20 years, in comparison with the plot which has continued to receive 48 kg N ha$^{-1}$, further evidence that moderate applications of inorganic N at this site can produce measurable changes in total N within 20 to 30 years.

Smaller applications of N to **Broadbalk** (48 and 96 kg N ha$^{-1}$; Table 1, expt 1) have produced smaller, but measurable, increases in total N. In contrast, the annual application of 48 kg N ha$^{-1}$ to the adjacent **Hoosfield** experiment (Table 1, expt. 2), also since 1852, has had no effect, with some evidence of *less* total N in the NPK than PK plot from 1913 onward (Figure 1b). Both experiments are on the same soil series, with a similar initial total N content. Above-ground crop yields (winter wheat on **Broadbalk**, spring barley on **Hoosfield**) were consistently greater from the NPK than from the PK treatment on both experiments (Warren and Johnston, 1967; Garner and Dyke, 1969), although fertilizer N may have increased organic N inputs from roots more under the longer growing winter wheat. There may be more soil movement on **Hoosfield** (Warren and Johnston, 1967) which would tend to even out any differences between the treatments (see Discussion). Although the application of fertilizer N to the

**Hoosfield** experiment has not led to a measurable increase in *total* N, mineralizable soil N has increased substantially (Jenkinson and Johnston, 1977).

In around half of the experiments in Table 2, medium-term (7-40 years) applications of inorganic N fertilizer increased total N by 10% or more, compared to control treatments given no N fertilizer. This is equivalent to an extra 120 to 720 kg N ha$^{-1}$ (standard bulk density, 0-20 cm). Five other experiments gave increases of 5 to 10%. In many cases the NPK treatment was compared with an unfertilized control. However, reserves of P, K, and other nutrients may be sufficient for much of the experiment, particularly in the shorter trials. Total soil C content generally followed the same trend as total N, usually slightly higher in plots receiving inorganic N fertilizer for many years. In some experiments, differences were apparent after as little as 10 years — for example, Biederbeck *et al*, (1984) reported that total N under continuous spring wheat at **Swift Current** (Table 2, expt. 21) receiving inorganic N at 32 kg N ha$^{-1}$ for 10 years was 17% greater than in control plots receiving only PK fertilizer. This difference was only observed in the 0 to 15-cm soil depth, under continuous cropping (straw retained); it was not detected in a fallow/wheat/wheat rotation. Smith (1962) reported that applications of 45 to 135 kg N ha$^{-1}$ yr$^{-1}$ (as urea or $(NH_4)_2SO_4$) to continuous tea at various sites in East Africa significantly increased soil organic matter content (generally well correlated with total N) after only 5 years, due to the greater return of crop residues (leaves, roots, etc.). In addition, soil pH declined by around 0.4 pH units at all sites where $(NH_4)_2SO_4$ was used. Djokoto and Stephens (1961) summarized 30 medium-term (3-9 years) experiments in Ghana, under various rotations with a range of soils. Fertilizer N (applied as $(NH_4)_2SO_4$) significantly increased total N, at all but two sites. At these two there was a larger decline in soil pH (around 0.5 pH units) than at the other sites (0.3 pH units or less).

There are a number of exceptions. For example, a prairie soil in Kansas growing continuous wheat for 35 years apparently lost more total N when fertilizer N was applied (Dodge and Jones, 1948, Table 2, expt. 1). Although this experiment contained no replicates, total N in both plots was very similar at the start of the experiment (0.140 and 0.137% N). Erosion was undoubtedly an important factor in the loss of soil N (and C), but was not measured.

N fertilizer is applied every year to the bare fallow "**36 Parcelles**" experiment at Grignon (with and without the addition of wheat straw) (Morel et al., 1984, Table 2, expt 8). After 19 years, total N had declined under all treatments, but the application of N fertilizer appeared to reduce this loss. As no plants are allowed to grow, presumably the inorganic N is incorporated directly into the soil organic N pool, via the microbial biomass.

At one dryland site (El–Haris et al., 1983; Table 2, expt. 25) the increase in total N after 9 years at up to 270 kg N ha$^{-1}$ was mainly due to an accumulation of *inorganic* N. Total N content was very low (0.06%), and at the highest N rate, inorganic N accounted for 13% of total N. Under irrigation, there was a significant increase in the *organic* N content at the highest N rate, as well as the mineral N content. The small increase in total N in a heavily fertilized soil in

West Bengal (Mandal et al. 1984, 1985; Table 2, expt. 27) may also have been due to an accumulation of inorganic N.

## B. Temporal Changes in Total N

Figures 1(a-f) show how total N has changed over the course of six of the oldest long-term experiments. Initial soil N was measured at the **Dehérain** and **Stackyard** experiments, and estimated (dotted line) at **Broadbalk, Hoosfield** and **Halle** (see Table 1, expts. 5, 11, 1, 2, 4 respectively for references). Total N has remained fairly stable for many years at **Broadbalk** (Figure 1a) and **Hoosfield** (Figure 1b) with some evidence of a decline on **Hoosfield** after the 1950s, particularly in the NPK plot. This could be due to deeper ploughing, diluting the topsoil with subsoil low in organic matter, but this explanation is not supported by the rate of P accumulation (Jenkinson and Johnston, 1977).

In contrast, all treatments at **Dehérain** (Figure 1c) lost total N throughout the experiment, even when managed in the same way for over 100 years. This may be due to a relatively high initial N content (0.204% compared to values of around 0.1% at the other sites) due to different soils and previous cropping. The **Broadbalk** and **Hoosfield** sites are thought to have been in continuous arable cropping for many years before the experiments began, whereas lucerne had been grown for several years before the start of the experiment at **Dehérain** (Guerillot, 1935). It may be many more years before an equilibrium soil N content is established at **Dehérain**, as in 1987 the winter wheat/sugar beet rotation practised for many years was replaced with continuous maize.

At the **Halle** experiment (Figure 1d) total N remained constant between the 1930s and 1960s, then declined, probably due to the introduction of two new cropping systems (continuous maize and a potato/rye rotation). Previously, continuous rye had been grown on the whole site since 1878. At the **Askov loam** site total N decreased during the first 20 to 30 years and then reached an equilibrium value of a little over 0.1% N (Figure 1e). There was no corresponding initial decline at the **Askov sand** site, with a constant % N of about 0.07 for about 40 years, but a small decrease during the 1950s; the reasons for this are not clear. This very well-known experiment is included in this review, even though there is a grass/clover ley one year in four, because it has two contrasting soil types, and has been sampled so regularly.

Total N was lost throughout the 50 years of the **Stackyard** experiment (Figure 1f), declining by around a third under all treatments from the initial value of about 0.15%. Originally this was a grassland site, probably ploughed in the 1830s (Johnston, 1975). After 11 years there was some evidence that the NPK plots contained more total N than the PK plot, but this difference had disappeared by 1927. Soil acidity developed very rapidly, particularly on the plots given ammonium-based fertilizers, reducing yields (and presumably organic N inputs to the soil) from over 3 t ha$^{-1}$ (grain, 85% DM) in 1877 to 1886 to less

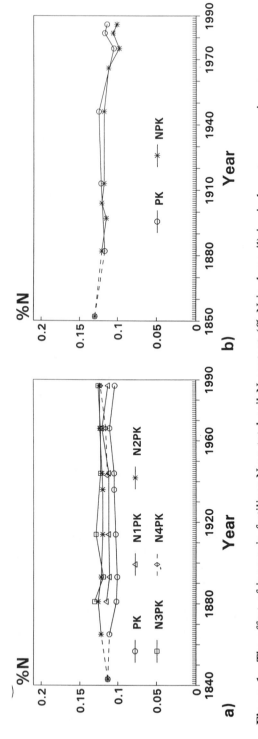

**Figure 1.** The effect of inorganic fertilizer N on total soil N content (% N in dry soil) in six long-term experiments.

a) Broadbalk Continuous Winter Wheat, Rothamsted. Started 1843. Sections 1 and 9 only (continuous wheat) 1966 and 1987. Fertilizer N applied (kg ha$^{-1}$ yr$^{-1}$) N1 = 48, N2 = 96, N3 = 144 since 1852; N4 = 192 since 1968, previously N1. (From Glendining and Powlson, 1991.)

b) Hoosfield Continuous Spring Barley, Rothamsted. Started 1852. 48 kg N ha$^{-1}$ yr$^{-1}$ applied to NPK until 1966, both plots then received 70 kg N ha$^{-1}$ yr$^{-1}$. (From Jenkinson and Johnston, 1977; and P.R. Poulton, unpublished.)

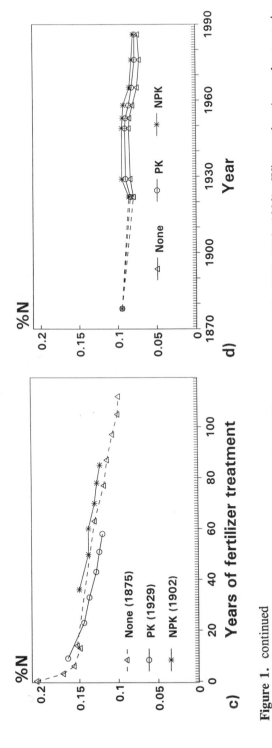

**Figure 1.** continued

c)  Dehérain, Grignon, France. Started 1875. (NPK treatment started in 1902, PK in 1929). Winter wheat/sugar beet rotation until 1987. 87 kg N ha$^{-1}$ yr$^{-1}$ to NPK. (From Morel et al., 1984; and R. Chaussod, personal communication.)

d)  Halle 'Eternal Rye', Germany. Started 1878. Continuous rye until 1961, then mean of continuous rye, continuous maize and potato/rye rotation. 40 kg N ha$^{-1}$ yr$^{-1}$ to NPK. (From Garz et al., 1982; and Garz and Hagedorn, 1990.)

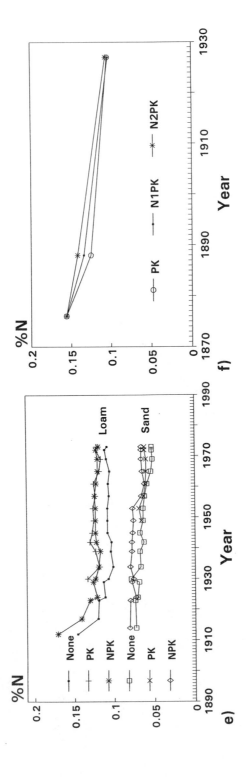

**Figure 1.** continued

e) Askov, Denmark. Started 1894. Loam soil = 11% clay, Sand soil = 4% clay. 4 year winter cereal/root crop/summer cereal/grass clover ley rotation. 60 kg N ha$^{-1}$ yr$^{-1}$ to NPK. (From Kofoed and Nemming, 1976; and Kofoed, 1982.)

f) Stackyard, Woburn, England. 1876-1926. Mean of continuous winter wheat and continuous spring barley experiments. N1 = 36, N2 = 50 kg N ha$^{-1}$ yr$^{-1}$ to NPK. (From Mattingly et al., 1975.)

than 1 t ha$^{-1}$ by 1917 to 1926 (Johnston, 1975). Soil erosion may also have contributed to the decline in total N (Mattingly et al., 1975).

## C. Modelling Changes in Soil N Content

A comprehensive review of models which describe and predict changes in soil organic N is outside the scope of this paper. Lathwell and Bouldin (1981), Russell (1981), Jenkinson (1990b), and Rasmussen and Collins (1991) summarize the different approaches used. Relatively few models operate on the years-to-centuries timespan, most describe decomposition of organic matter in the early weeks or months of decay. A number of authors (e.g. Lucas et al., 1977; Russell, 1981; Bjarnason, 1989; Wolf and Van Keulen, 1989; Jenkinson, 1990b) have modelled the long-term effect of inorganic fertilizer N on soil organic matter dynamics (N and C) with data from long-term experiments. Russell (1981) used data from a 10-year continuous sorghum experiment at **Narayen**, Australia, with an initial soil N content of 0.3% to predict changes in total N over 100 years. Total N was predicted to decline to 0.055% N in the absence of fertilizer N, but to only 0.095% N with the application of 100 kg N ha$^{-1}$ yr$^{-1}$. Although these predictions are based on only 10 years of data, and crop yield changes have been related entirely to nitrogen, with no account of other factors such as rainfall, temperature, etc., they demonstrate the potential of inorganic fertilizer N to influence total N content.

The Rothamsted multicompartmental C turnover model (Jenkinson and Rayner, 1977; Jenkinson, 1990b) was derived from data on the decomposition of labelled plant material in different soils, followed for 10 years in the open under Rothamsted conditions, radiocarbon dating of **Broadbalk** soil and long-term (10-100 years) changes in the soil organic matter content of some of the 'Classical' experiments at Rothamsted. It was able to simulate the measured increase in soil C in **Broadbalk** soil due to NPK fertilizer, compared to the plot never given fertilizer, by using plausible values for the different annual inputs of organic C (and presumably N).

Wolf and Van Keulen (1989) also tested their model against data from long-term experiments: the **Bad Lauchstädt** experiment in Germany, started in 1902 (arable rotation); the **Broadbalk** experiment in England; and a 21-year flooded rice experiment at **Aomori**, Japan. Agreement between observed and computed values was generally good, with inorganic fertilizer N increasing soil N content at **Bad Lauchstädt** and **Broadbalk**. In contrast, plots given inorganic N at **Aomori** lost more total N than plots given no fertilizer N (35 and 20 kg N ha$^{-1}$ yr$^{-1}$, respectively). The reasons for this are unclear.

### D. Mineralizable Soil N

Mineralizable soil N is defined in this review as the fraction of soil organic N that is mineralized and made available to crops within a year. Changes in this fraction have been measured by a range of biological and chemical methods, all of which have advantages and drawbacks (for reviews of the techniques available see Bremner (1965) and Keeney (1982)).

Measurements of the inorganic N released during laboratory incubations give a direct measurement of the net amount of organic N mineralized by the soil, i.e., gross mineralization minus immobilization (Tables 3 and 4). Incubations can be aerobic or anaerobic, short–term (1–6 weeks) or long–term (up to 30 weeks or more). Several North American authors (Janzen, 1987; Biederbeck et al., 1984; El–Haris et al., 1983) measure the 'potentially' mineralizable or labile soil N pool. This is based on the long–term incubation method of Stanford and co–workers (Stanford and Smith, 1972; Stanford et al., 1974; Stanford, 1977), which estimates N mineralization from a soil sample over an extended period (up to 30 weeks), with the inorganic N being removed periodically by leaching during the incubation. The N mineralization potential, $N_o$, is estimated from the cumulative amount of N mineralized. This method tends to give higher values than the conventional aerobic incubation, and under field conditions probably not all of this N would be mineralized in a year.

Care should be taken when interpreting the results of incubations – Stanford and co–workers showed that the results of short–term incubations (e.g. 1–2 weeks) do not necessarily reflect the long–term N mineralization capacity of the soil. The true rate of N mineralization is often only established once the effects of sample handling, recently added crop residues, etc. have been overcome. For example, drying or freezing the soil prior to incubation greatly alters the results obtained in the short term. Part of the soil microbial biomass is killed, resulting in an initial flush of N mineralization, as killed organisms are decomposed.

An alternative approach has been to measure mineralizable soil N indirectly, as crop uptake of 'soil' N (Tables 5-7). This N will be from three sources (a) non–fertilizer inputs, such as in rain, seed, dry deposition of ammonia, ammonium compounds and oxides of nitrogen, and biological fixation; (b) residues of inorganic fertilizer N from the previous year; (c) mineralization of soil organic N. If it can be assumed that non–fertilizer inputs of N are the same for all fertilizer treatments, and residues of inorganic N from the fertilizer applied to the previous crop are negligible (as is often the case), then differences in the uptake of N by the crop can be assumed to reflect differences in N mineralization. Fertilizer N may lead to a small increase in non-fertilizer inputs, principally due to a greater crop surface area, and thus more effective capture of N derived from dry deposition. This increase was estimated as 10 kg N ha$^{-1}$ to the crop on the **Broadbalk** plot given 144 kg N ha$^{-1}$ compared to the PK plot (K.W.T. Goulding, *pers. comm.*). If residues of fertilizer N are thought to be significant, an unfertilized crop may be grown, and discarded, before the main crop is grown (e.g. Mattsson, 1988, Table 7, expt. 4).

**Table 3.** The effects of long-term applications of inorganic nitrogen fertilizer on net nitrogen mineralized during incubation

| Site details (further details in numbered footnotes) | N fertilization | | Incubation conditions | | N mineralized during incubation[a] mg N kg⁻¹ soil day⁻¹ | | | % difference |
| --- | --- | --- | --- | --- | --- | --- | --- | --- |
| | Years | kg ha⁻¹ yr⁻¹ | Temp °C | Days | None | PK | NPK | |
| 1) Broadbalk, Rothamsted, UK. Continuous w. wheat | 129 | 48 | 25 | 20 | – | 0.265 | 0.285 | +8 |
| | " | 96 | " | " | | | 0.360 | +36 |
| | " | 144 | " | " | | | 0.420 | +58 |
| | 13 | 192 | " | " | | | 0.460 | +74 |
| | 130 | 48 | 25 | 20 | – | 0.264 | 0.294 | +11 |
| | 14 | 192 | " | " | | | 0.641 | +143 |
| 2) As (1)   (NH₄)₂SO₄ in spring | 105 | 96 | 25 | 28 | 0.36 | 0.36 | 0.43 | +19 |
| (NH₄)₂SO₄ in autumn | " | " | " | " | | | 0.36 | 0 |
| NaNO₃ in spring | 73 | " | " | " | | | 0.36 | 0 |
| 3) Halle, Germany. Continuous rye (since 1878) | 104 | 40 | 30 | 14 | 0.43 | – | 0.80 | (+86) |
| Rotation (since 1961) | 104 | " | " | " | 0.37 | – | 0.40 | (+8) |
| Continuous rye (since 1878) | 107 | " | " | 252 | 0.15 | – | 0.20 | (+33) |
| 4) Dehérain, France. | 83 | 87 | 28 | 28 | 0.19 | – | 0.36 | (+89) |
| Wheat/sugar beet rotation | | | " | 168 | 0.17 | – | 0.22 | (+29) |
| 5) As (4) | 83 | 87 | 28 | 168 | 0.16 | – | 0.19 | (+19) |

Table 3. continued

| | | | | | | | | |
|---|---|---|---|---|---|---|---|---|
| 6) As (4). Continuous maize. | 86 | 87 | 28 | 28 | 0.21 | – | 0.32 | (+52) |
| (previously wheat/sugar beet) | 87 | " | " | " | 0.31 | – | 0.38 | (+23) |
| 7) Weihenstephan, Germany. CaCN$_2$ | 53 | 72 | 35 | 210 | – | 2.43 | 2.71 | +12 |
| (NH$_4$)$_2$SO$_4$ | " | " | " | " | | | 2.74 | +13 |
| limed (NH$_4$)$_2$SO$_4$ | 39 | " | " | " | | | 2.36 | –3 |
| NaNO$_3$ | 53 | " | " | " | | | 2.27 | –7 |
| Ca(NO$_3$)$_2$ | " | " | " | " | | | 2.21 | –9 |
| 8) Lyberetski, Russia. | 49 | 35 | 28 | 30 | 1.25 | – | 1.18 | (–6) |

See Table 1 for definitions of fertilizer treatments and % difference. [a] Total inorganic N, i.e. NO$_3^-$ and NH$_4^+$, unless stated otherwise. Aerobic incubation with fresh soil (0-20 cm), unless stated otherwise.

1) Shen et al. (1989). 50% water-holding capacity. Soil frozen before incubation. Section I. See also Tables 1, 5, 6 and 9. SE means = 0.0382 (129 years) and 0.0301 (130 years) based on 3 replicate samples, no field reps. Soil depth 0-23 cm.

2) Gasser (1962). Soil frozen before incubation. Mean of 7 sampling dates, Sept. 1957 - Sept. 1958 and 3 Sections (II, III and V). No errors given. Soil depth 0-18 cm.

3) Garz et al. (1982); Garz and Hagedorn (1990). Rotation previously continuous rye. See also Tables 1, 8, and 9. Short-term anaerobic incubation. Long-term incubation after Stanford and Smith (1972). No reps.

4) Chaussod (1987). See also Tables 1 and 9.

5) Houot et al. (1989). 85% field capacity.

6) Chaussod (pers. comm.). Soil depth 0-30 cm.

7) Bosch and Amberger (1983). After Stanford and Smith (1972). See also Tables 1, 7, and 8. 4 reps, soil depth not given.

8) Shevtsova (1966). Moscow Region. Continuous flax. 60% capillary moisture capacity. NO$_3^-$-N only. See also Tables 1 and 7. Soil depth and number of reps not given.

**Table 4.** The effects of medium-term applications of inorganic nitrogen fertilizer on net nitrogen mineralized during incubation

| Site details (further details in numbered footnotes) | N fertilization | | Incubation conditions | | N mineralized during incubation[a] mg N kg⁻¹ soil day⁻¹ | | | % difference |
|---|---|---|---|---|---|---|---|---|
| | Years | kg ha⁻¹ yr⁻¹ | Temp °C | Days | None | PK | NPK | |
| 1) Ak-Karak, Kazakhstan. | 35 | 125 | 28 | 30 | 0.970 | – | 0.550 | (-43) |
| 2) Mironov, Kiyev, Ukraine. Continuous maize | 31 | 40 | 28 | 30 | 0.417 | – | 0.683 | (+64) |
| | " | 65 | " | " | | | 0.703 | (+69) |
| 3) Indian Head, Canada. Continuous spring wheat | 30 | 46 | 35 | 112 | 3.17 | – | 4.85 | (+53) |
| Fallow/wheat rotation | " | 6 | " | " | 2.15 | – | 2.80 | (+31) |
| Fallow/wheat/wheat rotation | " | " | " | " | 2.48 | – | 3.29 | (+33) |
| 4) Lyberetski,   N as $NaNO_3$ Russia.   N as $(NH_4)_2SO_4$ | 27 " | 70 " | 28 " | 30 " | – | 0.74 | 1.00 0.67 | +35 -9 |
| 5) S.Sweden. Annuals, mostly cereals | 27 | 80 | 24 | 10 | | 0.64 ± 0.04 | 0.81 ± 0.06 | +27 |
| 6) "36-Parcelles", Without straw France.   With straw   With composted straw | 27 " " | 70 " " | 28 " " | 28 " " | 0.37 0.60 0.45 | – – – | 0.37 0.59 0.58 | (0) (-2) (+29) |
| 7) As (6)   Without straw   With composted straw | 26 " | 70 " | 28 " | 28 " | 0.22 0.31 | – – | 0.35 0.45 | (+59) (+45) |

**Table 4.** continued

| | | | | | | | | |
|---|---|---|---|---|---|---|---|---|
| 8) As 6) Without straw | 26 | 70 | 28 | 168 | 0.08 | - | 0.10 | (+25) |
| With composted straw | " | " | " | " | 0.17 | - | 0.17 | (0) |
| 9) 6 sites in S. Sweden. | | | | | | | | |
| Västraby | 22 | 150 | 25 | 119 | - | 0.415 | 0.485 | +17 |
| Örja | " | " | " | " | - | 0.551 | 0.781 | +42 |
| Fjärdingslov | " | " | " | " | - | 0.537 | 0.642 | +20 |
| Ekebo | " | " | " | " | - | 0.539 | 0.662 | +23 |
| Orup | " | " | " | " | - | 0.631 | 0.615 | -3 |
| S.Ugglarp | " | " | " | " | - | 0.558 | 0.726 | +37 |
| Mean | | | | | | 0.539 | 0.652 | +21 |
| 10) Bet Dagan, Israel. 0-20 cm | 22 | 225 | 27 | 77 | 0.20 | - | 0.45 | (+125) |
| Irrigated cotton 20-40 cm | " | " | " | " | 0.17 | - | 0.31 | (+82) |
| 40-60 cm | " | " | " | " | 0.16 | - | 0.23 | (+44) |
| 60-90 cm | " | " | " | " | 0.13 | - | 0.17 | (+31) |
| 90-120 cm | " | " | " | " | 0.12 | - | 0.16 | (+33) |
| 11) Broom's Barn 0-25 cm | 19 | 75 | 30 | 14 | 1.40 | - | 1.43 | (+2) |
| Suffolk, UK. 0-25 cm | " | 150 | " | " | | - | 1.64 | (+17) |
| 25-50 cm | " | 75 | " | " | 0.86 | - | 1.00 | (+16) |
| 25-50 cm | " | 150 | " | " | | - | 1.07 | (+24) |
| 12) Lethbridge, Canada. | | | | | | | | |
| Continuous spring wheat | 18 | 45 | 35 | 126 | - | 1.12 | 1.48 | +32 |
| Wheat/fallow/fallow rotation | " | " | " | " | - | 0.88 | 1.09 | +24 |

**Table 4.** continued

| Site | | | | | | | | | |
|---|---|---|---|---|---|---|---|---|---|
| 13) Swift Current, Canada. | | 16 | 32 | 21 | 10 | - | 1.27 | 1.73 | +36 |
| | | " | " | 35 | 126 | - | 1.47 | 1.83 | +24 |
| 14) Sofia, Bulgaria. | | 16 | 50 | NA | NA | - | | 31.9 | (+5) |
| | | " | 100 | | | 30.4 | | 46.7 | (+54) |
| | | " | 150 | | | | | 51.8 | (+70) |
| 15) Hermitage, Australia. | | 13 | 69 | 40 | 7 | - | 2.34 | 2.99 | +28 |
| 16) Lind, USA. | Dry | 9 | 34 | 25 | 84 | 0.887 | - | 1.237 | (+39) |
| | | " | 67 | " | " | | | 1.337 | (+51) |
| | | " | 135 | " | " | | | 2.343 | (+164) |
| | | " | 270 | " | " | | | 3.786 | (+327) |
| | Irrigated | 9 | 34 | 25 | 84 | 1.069 | - | 1.645 | (+54) |
| | | " | 67 | " | " | | | 2.123 | (+99) |
| | | " | 135 | " | " | | | 2.961 | (+177) |
| | | " | 270 | " | " | | | 4.152 | (+288) |
| 17) Guelph, Canada. | Stover retained | 9 | 112 | NA | 14 | - | 1.79 | 1.89 | +6 |
| | | " | 224 | | " | | | 1.99 | +11 |
| | Stover removed | 9 | 112 | NA | 14 | - | 2.10 | 2.13 | +1 |
| | | " | 224 | | " | | | 1.94 | -8 |

NA = not available.   See Table 1 for definitions of fertilizer treatments and % difference.   [a] Total inorganic N, i.e. $NO_3^-$ and $NH_4^+$, unless stated otherwise.   Aerobic incubation with fresh soil (0-20 cm), unless stated otherwise.

1) Shevtsova (1966). Tashkent. Continuous cotton. 60% capillary moisture capacity. $NO_3^-$-N only. No further details. Two times more inorganic N in NPK-treatment than unfertilized before incubation. See also Tables 2 and 5.

2) Shevtsova (1966). As (1). See also Tables 2, 7, and 8.

3) Campbell et al. (1991c). Saskatchewan. Soil-sand mixture, leached with 0.01 $M$ $CaCl_2$. See also Tables 2, 8, and 9. 4 reps, soil depth 0-15 cm.

4) Shevtsova (1966). 4-course rotation. 60% capillary moisture capacity. See also Tables 2, 5, and 8. No further details.

5) Schnürer et al. (1985). Sandy clay loam, residues removed. Soil at 80% DM. 4 reps. See also Table 9.

6) Chaussod (pers. comm). Grignon. Bare fallow. Field moist soil, no reps. See also 7), 8) and Tables 2 and 9.

7) Chaussod (1987). As 6).

8) Houot et al. (1989). As 6). Fresh soil at 85% field capacity.

9) Bjarnason (1989). Long-term incubation, air-dried soil rewetted to 40% water-holding capacity. Mineralization in first 2 weeks discounted. Mineralization significantly greater in NPK treatments ($P<0.05$). 2 reps. See also Tables 2 and 7.

10) Hadas et al. (1986a; 1989). 50% clay. $NO_3^-$-N in soil profile, in the field. No crop or N, covered to prevent leaching. Temperature = mean soil temperature. Manure applied to NPK treatment 1 year in 7. 4 reps.

11) Last et al. (1985). Anaerobic incubation, $NH_4^+$-N (after Waring and Bremner, 1964). See also Table 2. 2 reps.

12) Janzen (1987). Alberta. Straw incorporated. Periodic leaching (0.001$M$ $CaCl_2$, under suction). 100% relative humidity. C mineralization followed similar trend. No K applied. See also Table 2. No reps, soil depth 0-10 cm.

13) Biederbeck et al. (1984). Field moist soil, stored 0°C. 3-day pre-incubation (21°C) before 10 d incubation; 126 d incubation periodically leached. C mineralization showed same trend. See also Tables 2 and 9. 3 reps, soil depth 0-7.5 cm. LSD ($P < 0.05$) = 0.53 (10 d), 0.29 (126 d).

14) Petrova and Gospodinov (1986). mg N kg⁻¹ soil. Incubation conditions not known. See also Table 2. Soil depth 0-30 cm.

15) Dalal (1989). Queensland. Cracking clay. Anaerobic incubation, $NH_4^+$-N (Keeney, 1982). 60-90 soil layer, no differences in soil from 0-10 and 10-20 cm depths (data not given). See also Table 2. 4 reps.

16) El-Haris et al. (1983). Washington. Air-dry soil, rewetted. Periodic leaching (0.01 $M$ $CaCl_2$, under suction, after Stanford and Smith, 1972). 3 reps, 0-15 cm soil depth. See also Table 2. LSD ($P < 0.05$) = 0.1657 (dry), 0.1501 (irrigated).

17) Ketcheson and Beauchamp (1978). Soil sampled in spring. Mineralization rates not significantly different. Significantly more $NO_3^-$-N in soil before incubation in NPK-treated soils (residual fertilizer N?). See also Tables 2 and 6. 4 reps, soil depth NA.

**Table 5.** Crop uptake of unlabelled N when [15]N-labelled fertilizer was applied, as a measure of the effect of long-term applications of inorganic N on the supply of mineralized soil N

| Site details (further details in numbered footnotes) | Usual N fertilization | | Year | Crop N uptake kg ha$^{-1}$ | | | |
|---|---|---|---|---|---|---|---|
| | Years | kg ha$^{-1}$ yr$^{-1}$ | | None | PK | NPK | % difference |
| 1) Broadbalk, Rothamsted, UK. Continuous w.wheat Section 1 | 129 | 48 | 1980 | - | 30.1 | 39.9 | +33 |
| | " | 96 | | | | 61.0 | +103 |
| | " | 144 | | | | 74.3 | +147 |
| | " | 192 | | | | 75.6 | +151 |
| Uptake of unlabelled N | 13 | 48 | 1981 | - | 30.9 | 50.0 | +62 |
| | 130 | 96 | | | | 72.6 | +135 |
| | " | 144 | | | | 73.2 | +137 |
| | 14 | 192 | | | | 71.7 | +132 |
| 2) Ak-Karak, Kazakhstan.  Unlabelled N | 35 | 125 | 1961 | 37.6 | - | 37.6 | (0) |
| Labelled N | | | | 128.6 | | 144.0 | (+12) |
| 3) Lyberetski, Russia.  Unlabelled N | 27 | 70 | 1961 | - | 48.6 | 63.0 | +30 |
| Labelled N | | | | | 124.6 | 128.8 | +3 |

See Table 1 for definitions of fertilizer treatments and % difference.

1) Powlson et al. (1986). [15]N-labelled fertilizer applied at the usual rate. SE means = 2.07 (129 years) and 4.41 (130 years) based on 3 replicate samples, no field reps. Uptake of unlabelled N by weed (determined in 1981 only): PK = 12.4, NPK (48 kg N) = 6.4 kg N ha$^{-1}$. Weeds negligible in other treatments. See also Tables 1, 3, 6, and 9.

2) Shevtsova (1966). Tashkent. Pot experiment using soil from the field experiment. N uptake by barley, mg N kg$^{-1}$ soil. [15]N applied to NPK and control. Number of replicates not given. See also Tables 2 and 4.

3) Shevtsova (1966). As (2). N as NaNO$_3$. See also Tables 2, 4, and 8.

**Table 6.** The effects of long- and medium-term applications of inorganic fertilizer N on crop uptake of soil N, in the absence of the usual fertilizer N. a) Field experiments

| Site details (further details in numbered footnotes) | Usual N fertilization | | Year | DM yield t ha⁻¹ | | | | N uptake kg ha⁻¹ | | | |
|---|---|---|---|---|---|---|---|---|---|---|---|
| | Years | kg ha⁻¹ yr⁻¹ | | None | PK | NPK | % diff | None | PK | NPK | % diff |
| 1) Broadbalk, UK. | 131 | 144 | 1983 | - | 1.30 | 2.25 | +73 | - | 38.6 | 52.0 | +35 |
| 2) As (1) Plots 3 (None) 5 (PK) and 16 (NPK) | 13 | 192 | 1865 | 0.93 | 1.03 | 2.38 | +131 | | | | |
| | | | 1866 | 0.87 | 0.94 | 1.25 | +33 | | | | |
| | | | 1867 | 0.60 | 0.65 | 1.01 | +55 | | | | |
| | | | 1868 | 1.18 | 1.27 | 1.64 | +29 | | | | |
| | | | 1869 | 0.95 | 1.06 | 1.12 | +6 | | | | |
| | | | 1870 | 1.07 | 1.35 | 1.35 | 0 | | | | |
| | | | 1871-77 | 0.69 | 0.81 | 0.81 | 0 | | | | |
| | | | 1878-83 | 0.79 | 0.93 | 0.86 | -8 | | | | |
| 3) Hoosfield, Rothamsted, UK. Continuous s.barley | 115 | 48 | 1968 | - | 3.01 | 3.09 | +3 | - | NA | NA | |
| | | | 1969 | - | 1.43 | 2.53 | +77 | - | NA | NA | |
| | | | 1970 | - | 1.70 | 2.74 | +61 | - | 17.03 | 32.58 | +91 |
| | | | 1971 | - | 2.03 | 3.69 | +82 | - | 22.76 | 33.61 | +48 |
| | | | 1972 | - | 1.28 | 3.03 | +137 | - | 13.43 | 27.76 | +107 |
| | | | 1973 | - | 1.21 | 1.87 | +55 | - | NA | 15.97 | |

**Table 6.** continued

| | yrs | N | Year | | | | | | | | |
|---|---|---|---|---|---|---|---|---|---|---|---|
| 4) Bad Lauchstädt, Germany. | | | | | | | | | | | |
| Potatoes fresh wt | 78 | | 1979 | – | 17.6 | 23.5 | +34 | | | | |
| W. wheat 85% DM | | 70 | 1980 | – | 4.7 | 6.2 | +32 | | | | |
| Sugar beet fresh wt | | | 1981 | – | 12.3 | 39.0 | +217 | | | | |
| 5) Exhaustion Land, Rothamsted, UK. | 43 | | 1902 | 2.07 | 1.74 | 3.84 | +121 | 34.4 | 28.3 | 87.5 | +209 |
| | | | 1903 | 0.71 | 0.76 | 1.62 | +113 | 12.1 | 13.2 | 27.9 | +111 |
| Wheat and potatoes | | 96 | 1904 | 1.06 | 1.06 | 1.53 | +44 | 17.3 | 16.2 | 25.2 | +56 |
| 6) Limburgerhof, Germany. | 20 | 89 | 1979 | – | 1.62 | 1.90 | +17 | | | | |
| | | 155 | 1979 | – | – | 2.84 | +75 | | | | |
| 7) Guelph, Canada. | | | | | | | | | | | |
| Stover retained | 9 | | 1972 | – | 3.1 | 4.0 | +29 | | | | |
| Stover removed | | 224 | " | – | 4.3 | 5.0 | +16 | | | | |
| 8) Barnfield, Rothamsted, UK. | 6 | | 1853 | 1.28 | 1.33 | 1.48 | +11 | | | | |
| | | | 1854 | 1.13 | 1.21 | 1.31 | +8 | | | | |
| | | 48 | 1855 | 1.14 | 1.21 | 1.29 | +7 | | | | |

NA = not available. See Table 1 for definitions of fertilizer treatments and % difference.

1) Powlson et al. (1986). Rothamsted. Continuous w. wheat. Plot 8, Section 1. Grain DM yield. N uptake by crop and weed. $^{15}$N-labelled fertilizer applied at usual rate in 1982. No N in 1983. NPK N uptake included 2.2 kg ha$^{-1}$ labelled N. See also Tables 1, 3, 5 and 9.

2) Johnston (1970); Lawes and Gilbert (1885). Grain DM yield. NPK applied to plot 16 1852-64, then unfertilized (no N or PK) 1865-83. Treatments to plots 3 and 5 unchanged 1852-1883. 2 replicate plots.

3)  Jenkinson and Johnston (1977); P.R. Poulton (pers. comm).   Grain and straw yield (85% DM) and N uptake.   NPK applied 1852-1967.   No N 1968-73.   See also Table 1.   No reps.

4)  Körschens et al. (1982).   Arable rotation.   NPK 1903-78.   No N 1979-81.   'NPK' = mean of NPK, NP, NK+N treatments, 'PK' = mean of PK and unfertilized treatments.   Estimated from graph.   See also Tables 1 and 8.   No reps.

5)  Johnston and Poulton (1977).   Grain DM yield, grain and straw N uptake.   No N or PK 1902-4, barley and oats grown.

6)  Lang and Sturm (1983).   Arable rotation, no legumes.   Grain DM yield.   NPK 1959-78.   No N 1979.   Estimated from Figure 3.   Mean annual N rate.   Number of reps not given.

7)  Ketcheson and Beauchamp (1978).   Maize.   Grain DM yield.   No N 1972.   Significantly more $NO_3^-$-N in NPK soil — residual fertilizer?   See also Tables 2 and 4.   Four reps.

8)  Johnston (1970); Lawes and Gilbert (1857).   Turnips and swedes 1845-52, barley 1853-55.   NPK 1845-50.   No N 1851 or 52, no NPK 1853-55.   Grain DM yield.   No reps.   See also Table 1.

**Table 7.** The effects of long- and medium-term applications of inorganic fertilizer N on crop uptake of soil N, in the absence of the usual fertilizer N. b) Pot experiments, with soil from field experiments

| Site details (further details in numbered footnotes) | Usual N fertilization | | DM yield t ha⁻¹ | | | | N uptake mg kg⁻¹ | | | |
|---|---|---|---|---|---|---|---|---|---|---|
| | Years | kg ha⁻¹ yr⁻¹ | None | PK | NPK | % diff | None | PK | NPK | % diff |
| 1) Weihenstephan, Germany.   CaCN₂ | 53 | 72 | | | | | - | 15.1 | 17.7(18.3) | +17(+21) |
| (NH₄)₂SO₄ | " | " | | | | | | | 18.0(34.5) | +19(+129) |
| limed (NH₄)₂SO₄ | 39 | " | | | | | | | 19.2(16.9) | +27(+12) |
| NaNO₃ | 53 | " | | | | | | | 12.2(12.7) | -19(-16) |
| Ca(NO₃)₂ | " | " | | | | | | | 13.8(9.6) | -9(-36) |
| 2) Lyberetski, Russia. | 49 | 35 | 2.51 | - | 3.33 | (+33) | 18.8 | - | 32.6 | (+73) |
| 3) Mironov, Ukraine. Continuous maize. | 31 | 40 | 1.72 | | 2.59 | (+50) | 12.7 | | 24.5 | (+93) |
| | " | 65 | | | 2.68 | (+56) | | | 20.7 | (+63) |

See Table 1 for definitions of fertilizer treatments and % differences.

1) Bosch and Amberger (1983). Ryegrass in pot experiment. pH 5.8-6.0, except unlimed (NH₄)₂SO₄ = 4.85. Figures in brackets = pH adjusted to 5.9-6.6, (NH₄)₂SO₄ to 5.3. See also Tables 1, 3, and 8. 4 reps, soil depth sampled 0-15 cm.

2) Shevtsova (1966). Moscow. Continuous flax. Oats in pot experiment. No N applied, only P and K. No further details. See also Tables 1 and 3.

3) Shevtsova (1966). Kiyev. As (2). See also Tables 2, 4, and 8.

**Table 7.** continued

4) Four sites in S. Sweden.

| | | | | | | | | |
|---|---|---|---|---|---|---|---|---|
| Fors | 21 | 25.2 | 3.8 | 4.1 | +8 | - | 9.3 | (8.4) | -10 |
| | " | 50.7 | | 3.6 | -5 | | | (7.3) | -22 |
| | " | 75.7 | | 3.5 | -8 | | | 7.7 | -17 |
| Kungsägen | 21 | 25.2 | 1.6 | 2.4 | +50 | - | 7.0 | (9.6) | +37 |
| | " | 50.5 | | 3.3 | +106 | | | (10.7) | +53 |
| | " | 75.7 | | 2.7 | +69 | | | 9.6 | +37 |
| Ekebo | 27 | 49.3 | 2.6 | 1.8 | -31 | - | 12.0 | (8.5) | -29 |
| | " | 98.5 | | 2.1 | -19 | | | (10.1) | -16 |
| | " | 147.8 | | 3.7 | +42 | | | 12.1 | - |
| Orja | 27 | 49.3 | 2.1 | 2.2 | +5 | - | 7.4 | (8.6) | +16 |
| | " | 98.5 | | 2.5 | +19 | | | (8.3) | +12 |
| | " | 147.8 | | 2.8 | +33 | | | 9.8 | +32 |

4)  Mattsson (1988). Arable rotation (no legumes). Barley crop grown in pot experiment, no N, only PK. DM yield gm pot$^{-1}$. Pot = 4.5-5.5 kg soil. N uptake figures in brackets calculated from mean soil weight, N1 and N4 treatments. No significant differences between N treatments, except at Ekebo, DM yield LSD (P < 0.05) = 0.5. 3 replicate pots, 2 replicate plots. Soil depth 0-20 cm. See also Tables 2 and 4.

N released on incubation is generally closely correlated with uptake of N from soil by plants in glasshouse studies (see Keeney, 1982, for references). However, the 21% increase in N mineralization due to fertilizer N in a long-term laboratory incubation reported by Bjarnason (1989; Table 4, expt. 9) was not always accompanied by greater N uptake in a pot experiment (Mattsson, 1988; Table 7, expt 4). Nitrogen uptake was found to be significantly correlated with both soil C content and the earlier N application rate, but not with the N application rate alone. Not all of the N mineralized and made available to plants may actually be taken up by the crop: some may be lost by leaching or denitrification. Thus this indirect method may only gives a partial measure of mineralized soil N. Sampling and mixing if the soil is used in pot experiments (Table 7) may lead to a flush of mineralization, thus giving apparently higher mineralization rates than occur in the field.

Crop uptake of soil N can be measured by applying $^{15}$N-labelled fertilizer, to differentiate between soil-derived and fertilizer-derived N (Table 5). Differences in the uptake of *unlabelled* N are assumed to reflect differences in the amount of mineralized soil N that is available. However, in certain circumstances, fertilizer N can stand proxy for soil N that otherwise would have been immobilized, or lost gaseously, or (incompletely) taken up by plants. If $^{15}$N–labelled fertilizer is applied, the effect of immobilization is, for example, to cause an apparent increase in the amount of unlabelled (i.e. mineralized) N taken up by a crop, described as a positive apparent 'added nitrogen interaction' (ANI) (Jenkinson et al., 1985). To avoid this possibility, an alternative approach is to measure crop uptake of N in the absence of the usual fertilizer N application (Tables 6 and 7). However, this method also has drawbacks — crop growth in the absence of fertilizer N may be so poor that root growth, and thus the uptake of mineralized N, may be limited (a 'real' ANI).

A third approach to measuring changes in the supply of mineralizable soil N, and one adopted in many of the experiments included in this review, is to use a chemical extraction procedure (Table 8). This method is attractive, as it is generally more rapid and convenient than the biological methods. However, it seems unlikely that a chemical extractant will be able to simulate the action of micro-organisms, in mineralizing (and immobilizing) plant–available forms of soil N (Bremner, 1965; Jenkinson, 1984). Mild extractants, such as 0.01 M CaCl$_2$ and hot water generally show the closest relationship with biological measurements of mineralizable soil N, and may indicate changes in the more readily available soil N fractions. Intermediate and strong extractants, such as alkaline permanganate and 6 $M$ HCl generally show little relationship with mineralizable N determined by incubation. For example, Hadas et al. (1986b) obtained a better estimate of mineralizable N from total soil N content than a number of chemical extractants.

The application of fertilizer N for many years increased mineralizable soil N measured by the various methods in Tables 3-7 in many of the long- and medium-term experiments. The increases were generally substantial (20% or more, compared with the PK treatment) and tended to be greater than any

**Table 8.** The effects of long-term applications of inorganic fertilizer N on 'available' soil N, measured by various chemical methods

| Site details (further details in numbered footnotes) | Usual N fertilization | | Method and extractant | mg N kg⁻¹ soil | | | % diff |
|---|---|---|---|---|---|---|---|
| | Years | kg ha⁻¹ yr⁻¹ | | None | PK | NPK | |
| 1) Halle, Germany. 0-20 cm | 100 | 40 | Cornfield (1960) | 36 | - | 50 | (+39) |
| Continuous rye    20-40 cm | " | " | 1 M NaOH (28°C)[a] | 28 | - | 32 | (+14) |
| 0-20 cm | 107 | 40 | Hot water extractable N[b] | 20.4 | - | 24.3 | (+19) |
| 2) Bad Lauchstädt, Germany. Arable rotation | 85 | 70 | Hot water extractable N | 31 | 35 | 44 | +26 |
| 3) Skierniewice, Poland.       Continuous    0-25 cm | 60 | 60 | N extracted by Electro-Ultra-Filtration | - | 20 | 31 | +55 |
| potato         50-70 cm | " | " | | - | 11 | 15 | +36 |
| Continuous    0-25 cm | " | " | | - | 21 | 45 | +114 |
| rye           50-70 cm | " | " | | - | 3 | 8 | +167 |
| Rotation      0-25 cm | " | " | | - | 12 | 35 | +192 |
| 4) Coimbatore, India.              Dry | 66 | 125 | Digested with 0.32% KMnO₄ and 2.5% NaOH | - | 58.8 | 75.6 | +29 |
| Irrigated | 39 | " | | - | 58.8 | 81.2 | +15 |
| 5) Weihenstephan, Germany. 3-course rotation | 53 | 72 | 6 M HCl 'Non hydrolysable' N | - | 845 | 916 | +8 |
| | | | | - | 111 | 123 | +11 |

**Table 8.** continued

| | | | | | | | |
|---|---|---|---|---|---|---|---|
| 6) Uman, Ukraine. | 50 | 120 | 'Easily hydrolysable' N | 123 | 145 | 167 | +15 |
| Apple orchard 20-60 cm | " | " | | 108 | 116 | 117 | +1 |
| 60-100 cm | " | " | | 94 | 90 | 99 | +10 |
| 0-20 cm | " | " | 'Less easily hydrolysable' N | 240 | 264 | 361 | +37 |
| 20-60 cm | " | " | | 153 | 206 | 258 | +25 |
| 60-100 cm | " | " | | 110 | 143 | 167 | +17 |
| 0-20 cm | " | " | 'Non-hydrolysable' N | 1084 | 1120 | 1018 | -9 |
| 20-60 cm | " | " | | 759 | 781 | 728 | -7 |
| 60-100 cm | " | " | | 573 | 536 | 525 | -2 |
| 7) Mironov, Kiyev, Ukraine. | 31 | 40 | Tyurin and Kononova (1934) H₂SO₄ | 55.5 | - | 60.4 | (+8) |
| Continuous maize | " | 65 | | | | 62.7 | (+13) |
| 8) Melfont, Canada. | | | Hydrolyzed in 6 M HCl for 24 hrs | | | | |
| Continuous spring wheat | 31 | 52 | | 1587 | - | 1524 | (-4) |
| Fallow/wheat/wheat | " | 14 | | 1448 | - | 1383 | (-5) |
| 9) Indian Head, Canada. | | | Hydrolyzed in 6 M HCl for 24 hrs | | | | |
| Continuous spring wheat | 30 | 46 | | 677 | - | 748 | (+10) |
| Fallow/wheat rotation | " | 6 | | 626 | - | 639 | (+2) |
| Fallow/wheat/wheat | " | 6 | | 623 | - | 690 | (+11) |
| 10) Lyberetski, Russia. | | | Tyurin and Kononova (1934) H₂SO₄ | | | | |
| N as NaNO₃ | 27 | 70 | | - | 53.3 | 61.9 | +16 |
| N as (NH₄)₂SO₄ | " | " | | - | - | 68.8 | +29 |

See Table 1 for definitions of fertilizer treatments and % difference.

1) Garz et al. (1982); Garz and Hagedorn (1990). [a] Described as 'hydrolysable' N. [b] Soil dried at 70°C. No reps. See also Tables 1, 3 and 9.

2) Körschens et al. (1990). Converted from kg N ha$^{-1}$ (0-30 cm), assuming bulk density of 3.6 x 10$^6$ kg ha$^{-1}$. Mean of 4 arable crops (no legumes). See also Tables 1 and 6. 4 reps.

3) Mercik and Németh (1985). Inorganic N and low molecular weight organic molecules (amino acids, amides, etc.). See also Table 1. 5 reps.

4) Muthuvel et al. (1977). See also Table 1. No reps, soil depth 0-20 cm. Extraction method of Subbiah and Asija (1956).

5) Bosch and Amberger (1983). Method of Aldag and Kickuth (1973). 'Non hydrolyzable' N = total N minus N extracted with 6 M HCl. Mean of 5 fertilizer N treatments. See also Tables 1, 3 and 7. 4 reps, soil depth not given.

6) Kopytko and Gérkiyal (1983). See also Table 1. 10 replicate soil samples. Extraction methods not given.

7) Shevtsova (1966). Described as 'easily hydrolysable' N. See also Tables 2, 4, and 7.

8) Campbell et al. (1991b). Hydrolyzable amino compounds (amino acids and amino sugars). LSD (P<0.10) = 96. See also Table 2. 4 reps, soil depth 0-15 cm.

9) Campbell et al. (1991d). As for (8). LSD (P<0.10) = 64. See also Tables 2, 4, and 9. 4 reps, soil depth 0-15 cm.

10) Shevtsova (1966). Moscow. 4 course rotation. Described as 'easily hydrolyzable' N. See also Tables 2, 4, and 5.

increases in total soil N. For example, 150 kg N ha$^{-1}$ applied to six sites in Southern Sweden for 22 years increased N mineralized during incubation by 21% (Table 4, expt. 9), with a much smaller increase (12%; Table 2, expt. 11) in total N (Bjarnason, 1989). Carbon mineralization followed the same trend: 25% more $CO_2$-C produced during the 17-week incubation. Medium-term applications of N (16-18 years) at **Lethbridge** (Janzen, 1987; Table 4, expt. 12) and **Swift Current** (Biederbeck et al., 1984; Table 4, expt. 13) increased mineralized N by 24-36%, with much smaller increases in total N (up to 17%; Table 2, expts. 16 and 21). In part of the **Hoosfield** experiment 48 kg N ha$^{-1}$ was applied for 115 years and then stopped. Crop DM yield and N uptake in the absence of any fertilizer N over the next seven years was almost twice that from the plot never given fertilizer N (Jenkinson and Johnston, 1977), with no decline in the difference with time (Table 6, expt. 3), although there was no increase in total soil N (Table 1, expt. 2).

'Available' soil N, measured by various chemical extractants (Table 8), also tended to be greater when fertilizer N had been applied for many years. At **Skierniewice** (Mercik and Nemeth, 1985; Table 8, expt. 3) 60 years of N fertilization led to a large increase in the amount of N extracted by electro-ultrafiltration: 55-192%, depending on crop rotation, compared to increases in total N of around 5% (Table 1, expt 13). This technique extracts inorganic N and low molecular weight organic molecules, and has been used to predict mineralizable soil N (Appel and Mengel, 1992; Ziegler et al., 1992). However, crop yields in this experiment are now very low, and air pollution from nearby industrial areas may well be influencing crop growth and soil processes (M. Fotyma, *pers. comm.*). At **Uman** (Kopytko and Gérkiyal, 1983; Table 8, expt. 6) 120 kg N ha$^{-1}$ applied to apple trees for 50 years had virtually no effect on total soil N, measured to 1 metre (Table 1, expt 12), but increased the proportion of N in two fractions described as hydrolysable which together constituted about 30% of total soil N.

At **Broadbalk**, **Halle** and **Dehérain** N mineralization during incubation has been measured several times (Table 3, expts. 1-6). The results are quite variable, but in general, the effect of fertilizer N appears to be greatest when measured in short-term incubations, suggesting it is increasing the most readily mineralizable fractions of organic N. For example, in one case with the **Dehérain** experiment (expt 4), the increase in mineralization in a soil receiving 87 kg N ha$^{-1}$ for 83 years was 89% during a 28-day incubation but only 29% over 168 days (expt. 4, Chaussod, 1987). Other short-term incubations gave smaller differences (expt. 6). At **Broadbalk**, Shen et al. (1989) reported a 58% increase in N mineralization during incubation, as a result of applying 144 kg N ha$^{-1}$ for 130 years (Table 3, expt. 1). The soil had previously been frozen, which probably led to a flush of mineralization in this short-term (20 day) incubation. Gasser (1962; Table 3, expt. 2), also incubating previously frozen soil, found much smaller effects.

Other results from **Broadbalk** do suggest that long-term fertilizer N applications have increased the amount of soil N which is mineralized and available to the crop each year, and that this increase is greater than the maximum increase in total soil N of 20%. When $^{15}$N-labelled fertilizer was applied (Table 5, expt. 1), uptake of unlabelled N by crop plus weeds (where present) indicated that at least 30 kg ha$^{-1}$ more N was mineralized from the plot given the highest long-term N rate, compared to the plot never given N (Powlson et al., 1986). Inorganic residues of fertilizer N from the previous crop were negligible (Macdonald et al., 1989) and there was no evidence of any ANI (Powlson et al., 1986). Presumably some mineralized N was lost (by leaching or denitrification) before it could be taken up by the crop. When the usual fertilizer N (144 kg ha$^{-1}$) was withheld from part of the experiment in 1983 (Table 6, expt. 1), N uptake was 35% greater than from the plot never given N, further evidence of greater mineralization.

The use of fertilizer N for 15 years or less resulted in large increases in mineralized N at many sites (Table 4). El-Haris et al. (1983; Table 4, expt. 16) reported very large increases in potentially mineralizable N as calculated from the Stanford and Smith (1972) method after only 9 years of NPK fertilizer application: increases were up to 300% although this was in comparison with unfertilized controls. The **Broadbalk** plot given 192 kg N ha$^{-1}$ for only 13 years mineralized as much N as a plot given 144 kg N ha$^{-1}$ for 129 years (Table 3, expt. 1 and Table 5, expt. 1), although it contained less total N (Table 1, expt. 1; Powlson et al., 1986; Shen et al., 1989). Some early results from **Broadbalk** (Table 6, expt. 2; Johnston, 1970) also show that the application of 192 kg N ha$^{-1}$ for just 13 years increased N mineralization. This N rate was applied to plot 16 between 1852 and 1864, no further N, P, or K was then applied for the next 19 years. In 1865, the first year without N, the yield on plot 16 was more than twice that on the PK control, but may well be attributable to large inorganic residues from the previous year, as this relatively high N rate was always surplus to the crops' requirements, and 1864 was particularly dry. Over the next three years grain yields were 0.31 - 0.37 t ha$^{-1}$ greater than in the plot never given N, suggesting greater mineralization. Thereafter, yields were very similar. Presumably the additional N in the soil resulting from 13 years of N inputs was in a short-lived fraction.

At **Weihenstephen** (Bosch and Amberger, 1983) inorganic N was applied as four different forms, all supplying 72 kg N ha$^{-1}$, with only a modest effect on total N (Table 1, expt 10), mineralized N (Table 3, expt. 7; Table 7, expt. 1) or N extracted with 6 $M$ HCl (Table 8, expt. 5). Surprisingly, the addition of FYM at 10 t ha$^{-1}$ yr$^{-1}$ led to similar increases (total soil N content of 0.111% N, a mineralization rate of 2.67 mg N kg soil$^{-1}$ day$^{-1}$). The ammonium-based fertilizers showed the greatest increase compared to the control.

In contrast to all the other sites, N mineralization at **Ak-Karak** (Shevtsova, 1966), as measured in a laboratory incubation, was substantially less in fertilized soil than unfertilized (Table 4, expt. 1). However, the fertilized soil contained twice as much inorganic N at the start of the incubation and total soil N was greater from the NPK-treated soil (Table 2, expt. 2).

At **Guelph** (Ketcheson and Beauchamp, 1978) and **Lyberetski** (continuous flax, Shevtsova, 1966) N fertilization had little effect on total N or N mineralized during incubation, but crop uptake of soil N was much greater (Table 6, expt. 7 and Table 7, expt. 2). This was probably due to inorganic residues of fertilizer N from the previous crop, rather than increased mineralization.

Also included in Table 4 (expt. 10) are results from an experiment in Israel (Hadas et al., 1989) in which N mineralization was measured *in situ* as the accumulation of inorganic N in the soil profile. The bare soil was covered with black polythene to prevent leaching and minimize evaporation. After 11 weeks there was almost twice as much inorganic N in the soil profile of the plots normally given NPK (225 kg N ha$^{-1}$) as the unfertilized plots (there was no PK treatment). Most of the difference was in the top 40 cm. However, part of the increase in mineralization may be due to the application of manure to the NPK plots every seventh year (Hadas et al., 1986a) rather than a direct effect resulting from inorganic N application.

## E. Soil Microbial Biomass N

A few authors give details of the effects of long- and medium-term applications of inorganic fertilizer N on microbial biomass N (Table 9). The soil microbial biomass is both the agent of change in soil, mediating immobilization and mineralization, and a repository of considerable quantities of N in a form that is much more readily mineralizable than the N in most of the soil organic matter (Jenkinson and Ladd, 1981). Although accounting for only a small fraction of the soil organic matter (1–3% of soil organic C, Jenkinson and Ladd, 1981), biomass measurements may reveal changes brought about by soil management long before such changes can be detected in total organic N or C content (e.g. Powlson and Jenkinson, 1981; Powlson et al., 1987; Saffigna *et al.*, 1989).

At the first six sites in Table 9, plots given N fertilizer for 13 years or more contained at least 20% more microbial biomass N than plots given no fertilizer N. At **Halle** (expt. 2; Garz and Hagedorn, 1990) and **Dehérain** (expt. 3; Chaussod, 1987) the difference was 70% or more, but as at **Indian Head** (expt 4; Campbell et al., 1991a), part or all of this increase may have been due to the application of PK fertilizers, which were not applied to the control plots.

In contrast, there was no significant difference in microbial biomass N content of plots given N and P fertilizer and those given only P at **Swift Current** (expt. 8; Biederbeck et al., 1984). Bacterial numbers were greatest under the P treatment, but N fertilization led to a significant increase in N mineralization (24% in a long-term incubation, Table 4, expt. 13). This suggests that microbial *activity* was greater under the NP treatment, and that the P treatment contained a large but comparatively inactive microbial population. Bjarnason (1988, data not shown) found that N fertilizer applied for 22 years to six sites in Southern Sweden had no effect on microbial biomass C content (generally well-correlated with biomass N, but not measured in this experiment), even though N

**Table 9.** The effects of long-term applications of inorganic fertilizer N on microbial biomass N

| Site details (further details in numbered footnotes) | Usual N fertilization | | Year | Biomass Nᵃ mg kg⁻¹ | | | |
|---|---|---|---|---|---|---|---|
| | Years | kg ha⁻¹ yr⁻¹ | | None | PK | NPK | % diff |
| 1) Broadbalk, Rothamsted, UK. Continuous w.wheat | 129 | 48 | 1980 | - | 52.9 | 59.9 | +13 |
| | " | 96 | | | | 67.8 | +28 |
| | " | 144 | | | | 63.4 | +20 |
| | 13 | 192 | | | | 64.3 | +22 |
| | 130 | 48 | 1981 | | 46.0 | 70.5 | +53 |
| | 14 | 192 | | | | 82.0 | +78 |
| 2) Halle, Germany. | | | | | | | |
| Continuous rye (since 1878) | 104 | 40 | 1981 | 6.6 | - | 11.2 | (+70) |
| Potato/rye rotation (since 1961) | " | " | | 5.2 | - | 8.8 | (+69) |
| Continuous rye (since 1878) | 107 | " | 1984 | 4.9 | - | 10.5 | (+114) |
| 3) Dehérain, Grignon, France. | 83 | 87 | 1985 | 25.0 | - | 48.8 | (+95) |
| 4) Indian Head, Canada. | | | | | | | |
| Continuous spring wheat | 30 | 46 | 1987 | 58.9 | - | 76.1 | (+29) |
| Fallow/wheat rotation | " | 6 | | 48.5 | - | 46.0 | (-5) |
| Fallow/wheat/wheat rotation | " | " | | 54.6 | - | 54.0 | (-1) |
| 5) S. Sweden. | 27 | 80 | 1982 | 58±5 | | 76±2 | +31 |
| 6) "36 Parcelles", Grignon, France. | | | | | | | |
| Without straw | 26 | 70 | 1985 | 23 | - | 19 | (-17) |
| Straw composted with N | " | " | | 27 | - | 46 | (+70) |

**Table 9.** continued

| | | | | | | | |
|---|---|---|---|---|---|---|---|
| 7) Hermitage, Australia. Wheat/barley | 20 | 23 | 1988 | 65.1 | - | 67.2 | +3 |
| | " | 69 | | | | 70.3 | +8 |
| 8) Swift Current, Canada. Continuous spring wheat | 16 | 32 | 1982 | 86 | - | 65 | -24 |
| Wheat/fallow rotation | " | " | | | | 62 | -28 |

See Table 1 for definition of fertilizer treatments and % difference. [a] Microbial biomass N measured by the Chloroform Fumigation/Incubation method of Jenkinson and Powlson (1976). Soil depth sampled 0-20 cm, unless stated otherwise.

1) Shen et al. (1989). Section 1. Soil frozen before incubation, 0-23 cm. $K_N = 0.57$. See also Tables 1, 3, 5, and 6. SE means = 3.57 (129 years) and 3.02 (130 years) based on 3 replicate samples. No field reps.

2) Garz et al. (1982; Garz and Hagedorn (1990). Rotation previously continuous rye. See also Tables 1, 3, and 8. $K_N$ value not given. No reps.

3) Chaussod (1987). Modified method of Chaussod et al. (1986). $K_N = 0.34$. See also Tables 1 and 3. No reps.

4) Campbell et al. (1991a). Soil frozen before incubation, pre-incubated 3 d @ 21°C. $K_N = 0.4$. LSD (P<0.10) = 13.5. See also Tables 2 and 4. 4 reps, soil depth 0-15 cm.

5) Schnürer et al. (1985). Annuals, usually cereals. $K_N = 0.4$. See also Table 4. 4 reps.

6) Houot et al. (1989). Bare fallow. As for (3). See also Tables 2 and 4. No reps.

7) Dalal et al. (1991). Field moist soil, pre-incubated 7 d @ 22°C. $K_N = 0.5$. LSD (P<0.05) = 4.2. Biomass N and N mineralized during anaerobic incubation correlated, $r = 0.72$. See also Tables 2 and 4. Soil depth 0-10 cm, 4 reps.

8) Biederbeck et al. (1984). Field moist soil, stored 0°C, preincubated 21°C for 3 d. $K_N = 0.40$. See also Tables 2 and 4. Soil depth 0-7.5 cm, 3 reps. LSD (P<0.05) = 35.

mineralization increased by 21%. Insam et al. (1991) measured biomass C in three long-term (59-77 years) experiments in Alabama. Overall, the fully fertilized (NPK) plots contained significantly more biomass C than unfertilized plots, but there was no difference between NPK and PK plots (data not shown).

A possible explanation for these apparently conflicting results is that measurement of microbial biomass give the size of the microbial population, not its activity (Jenkinson, 1988). The rate of N mineralization will give a better indication of microbial biomass activity. Furthermore, there may be too much variation (either field spatial heterogeneity or lack of analytical precision) to detect the subtle differences arising from past inorganic fertilizer treatments. At the **Broadbalk** and **Halle** experiments, where biomass N was measured at least twice, there was considerable year-to-year variation in the effect of fertilizer N on biomass N. The biomass N content of the **Broadbalk** plot given 192 kg N ha$^{-1}$ measured at harvest each year between 1980 and 1983 was 64, 82, 70, and 75 mg N kg$^{-1}$ soil respectively, a variation of 28% (Shen et al., 1989), with no consistent variation with time.

Soil microbial biomass values were particularly low at the **Halle** and "**36 Parcelles**" sites (Table 9, expts. 2 and 6), both in absolute terms and as percentages of total soil N; this presumably reflects the low soil organic matter contents at **Halle**, and the very low organic inputs to the bare-fallow experiment at Grignon. The application of straw composted with N approximately doubled microbial biomass N content in the "**36 Parcelles**" experiment. There was more biomass N under continuous cropping than in rotations including fallow years (Table 9, expts. 8 and 4; Biederbeck et al., 1984; Campbell et al., 1991a).

## III. Discussion

In most agricultural systems, production is limited by the availability of inorganic N, and the application of fertilizer N will increase both root and shoot DM production. The response of the shoot tends to be greater, resulting in an increase in the ratio of shoot to root (Welbank et al., 1974; Russell, 1977). Below-ground residues show a more complicated response to N, especially when other factors, e.g. water (Campbell et al., 1977), other nutrients and light (Russell, 1977) interact.

Increasing amounts of fertilizer N have been shown to increase both the amount and concentration of N in wheat straw and stubble (Powlson et al., 1986), pruning litter from tea bushes (Gokhale, 1959), and roots of spring wheat (Campbell et al., 1977), winter wheat (Hart et al., 1986), and apple trees (Kopytko and Gerkiyal, 1983) and of root exudates of wheat (Ushakov et al., 1985). Dormaar and Pittman (1980) reported an 18% increase in below-ground organic N residues (> 1.0 mm) when 45 kg N ha$^{-1}$ was applied to continuous spring wheat. If the C:N ratio of the crop residue is decreased when fertilizer N is applied, this may increase net N mineralization as the residues decompose, at least in the short-term, as less N is immobilized (Jenkinson, 1984).

The main hypothesis of this review is that the long-term use of inorganic N fertilizer, through its influence on crop residue production and composition, will increase soil organic N content, and thus the amount of mineralized soil N. Results from the majority of long-term experiments included in this paper support this hypothesis.

Increases in total soil N content tend to be small, typically around 10%. The **Broadbalk** continuous wheat experiment at Rothamsted, England, the longest running soil fertility experiment, showed one of the largest increases in total soil N of 20% or 500 kg N ha$^{-1}$ (0-23 cm) after applying 144 kg N ha$^{-1}$ for 136 years (Table 1, expt. 1). Many experiments gave measurable increases in total N after 20-30 years, especially when N application rates were high, or large amounts of above-ground crop residues were incorporated, or if the background soil N content was low. The application of 192 kg N ha$^{-1}$ to **Broadbalk** for just 20 years has increased total soil N by 9% (0.01% N), even though the straw is removed each year.

At most sites in Tables 1 and 2 total N has decreased since the start of the experiments, although the application of inorganic N often slowed the decline (e.g. Bjarnason, 1989, Table 2, expt. 11). Eventually a new equilibrium will be established, although this may take many years — for example, the **Dehérain** experiment had not reached an equilibrium after 100 years (Figure 1c). Subsequent changes in management or the environment will lead to the establishment of a new equilibrium.

Changes in the supply of mineralizable soil N were estimated using several methods (Tables 3-8). The increases in this fraction were substantial (20% or more) at many sites, and tended to be greater than any increases in total soil N. For example, N mineralization during incubation at six sites in Sweden given 150 kg N ha$^{-1}$ for 22 years was 21% greater than in plots given only PK fertilizer, whereas total soil N increased by only 12% (Bjarnason, 1989; Table 4, expt. 9 and Table 2, expt. 11). This suggests that the small amount of additional organic N returned to the soil as a result of using nitrogen fertilizer is in fractions that turn over more rapidly than the N in other organic matter fractions. Shen et al. (1989) showed, using $^{15}$N labelling, that recently added residues of wheat to the **Broadbalk** experiment were about seven times more mineralizable than the other fractions of soil organic matter.

A clear trend observed in the experiments in this review is that readily mineralizable fractions of soil N build up proportionally faster than total N. The converse will also be true: thus environmental or management practices which lead to a loss of total organic N will almost certainly result in greater losses of 'active' organic N fractions. This has been observed for biomass in situations where total soil organic N content is decreasing (Powlson and Jenkinson, 1976).

Results from many sites demonstrate the difficulty of obtaining absolute values for differences in mineralization arising from past fertilizer treatments (e.g. **Broadbalk**). The results of a laboratory incubation on a soil sample taken at a particular time of year may not reflect differences occurring over the whole year in the field. *In situ* field incubations of soil cores may be useful in this respect (Hatch et al., 1990) but, even so, the disturbance that is inherent in obtaining samples for incubation may still influence the results.

The changes in mineralizable N observed were not always accompanied by similar changes in microbial biomass N (Table 9). Major changes in inputs of organic matter (e.g. as a result of straw incorporation, Powlson et al., 1987) are generally accompanied by changes in soil microbial biomass content. However, the more subtle changes in inputs arising from past inorganic fertilizer treatments may have little effect on the *amount* of microbial biomass, although it may be more active, as demonstrated by increased mineralization of soil N per unit of biomass N. Furthermore, seasonal and annual variation in microbial biomass content may mask any small underlying changes.

Soil erosion may be a major cause of the loss of soil N. In addition to direct removal of the soil, both wind and water erosion is known to be selective — the N content of the eroded material may be much higher than in the soil that remains (Neal, 1944; Allison, 1973). The use of fertilizer N may prolong crop cover, and thus protect soil from erosion for a longer period of the year (Biederbeck et al., 1984). Again, if fertilized crops leave more trash on the soil surface than unfertilized, there may be less erosion.


## A. Limitations of Long-Term Experiments

Long-term experiments have a number of limitations, mainly arising from experimental design and sampling. The major ones pertinent to this review are:

(a) Lack of starting samples. Frequently soil samples were not taken until the experiment was well under way. Observed differences may have been present *before* the fertilizer treatments were imposed as a result of spatial variability in soil properties.

(b) Lack of replication and randomization, particularly in the older experiments, so that no true measure of error can be made.

(c) No true control treatments - in many cases it has been necessary to compare NPK treatments with unfertilized plots, rather than those receiving PK fertilizer.

(d) Modifications to the experiment over time — for example, changes to N application rates, cropping, pest and disease control, etc. There will always be a conflict between updating the management of an experiment, so that it remains relevant to modern agriculture, and maintaining the original treatments, which soon become obsolete. If the original experimental area is large enough, the best solution is to do both, as has occurred at some of the sites at Rothamsted (Jenkinson, 1991).

(e) Incomplete description of the experiments and sampling protocol — soil type and depth sampled, number of replicates, fate of crop residues, and whether all treatments receive adequate P, K, other nutrients and lime. Most N fertilizers acidify the soil, and in a few experiments this may not have been counteracted by liming, allowing acidity to decrease yields.

(f) The movement of soil between plots, principally by cultivation, but also due to water, wind, and soil fauna. Concrete partitions sunk into the ground prevent soil movement by cultivation at **Halle** (Kolbe and Stumpe, 1969, Table 1, expt. 4), but at most other sites there are no barriers to soil movement. Sibbesen (1986) simulated soil movement at 21 long-term (50 years or more) experiments, based on a model of changes in soil P at **Askov**. Assuming the same rate of soil movement as at **Askov**, he calculated that less than a third of the topsoil now present in the central quarter of each plot originated from topsoil present at the start of the experiments. Soil movement will tend to even out any differences between plots due to long-term fertilizer N applications, and thus underestimate the impact such effects could have on the field scale. The extent of any effect depends on many factors, including size of plot, soil type, number, method and direction of cultivation. For example, soil movement was thought to be greater at the **Hoosfield** spring barley experiment than on the adjacent **Broadbalk** winter wheat experiment, as more cultivations were required to control weeds with the spring-sown crop before the advent of chemical weedkillers (Warren and Johnston, 1967).

## B. Factors Influencing the Effect of Fertilizer N on Soil N Content

The above limitations should be borne in mind when interpreting the results from long-term field experiments. Furthermore, it is not possible to directly compare individual experiments, due to the range of crops grown, soil types, climates, N application rates, length of experiment, etc. But when considered as a whole, it is possible to draw some general conclusions.

In soils with a low available N content (generally equivalent to a low total N content) fertilizer N will tend to increase the production of root and other crop residues more than in soils with a greater available N content, where less root is required to provide the nutrients the crop requires (Hart et al., 1986; Gregory, 1988). Around 50 kg N ha$^{-1}$ was applied to continuous spring wheat for 30 years at two sites in Saskatchewan, with very different total N contents (0.5% N at **Melfont**, 0.2% N at **Indian Head**, Campbell et al., 1991a-c). Fertilizer N more than doubled estimated crop residue N (including straw) at **Indian Head** (12 to 32 kg N ha$^{-1}$ yr$^{-1}$), but had less of an effect at the more fertile **Melfont** (24 to 34 kg N ha$^{-1}$ yr$^{-1}$). Total soil N increased by 310 kg N ha$^{-1}$ (0 to 15 cm), or 10% at **Indian Head**, with no measurable increase at **Melfont**. Results from other experiments (Hargitai, 1982, Table 2, expt. 12; Kofoed, 1982, Table 1, expt. 6; Bjarnason, 1989, Table 2, expt. 11) also suggest that fertilizer N will have a greater effect on total N in soils with a lower background total N content. The actual increase in total N (in kg N ha$^{-1}$) due to N fertilization at five sites in Sweden (0.10 to 0.18% N, Bjarnason, 1989) was negatively correlated with total N (r = -0.87*). Results from one site, Ekebo, were omitted from the correlation, due to the abnormally high C:N ratio of 17.5 (normally 10-12, Allison, 1973). Obviously, it is also more difficult to detect relatively small changes in total N against a high background value.

N application rates to many of the experiments are low by modern standards and have sometimes been changed over the course of the experiment. The **Broadbalk** experiment (Table 1, expt. 1) is one of the few with a range of unchanged rates. Evidence from this and other sites (e.g. Jansson, 1983, Table 2, expt 11; Rasmussen and Rohde, 1988, Table 2, expt. 23) indicates that the greater the rate of N applied, the greater the increase in total and mineralized soil N. At other sites (e.g. Anderson and Peterson, 1973, Table 2, expt. 14; Ketcheson and Beauchamp, 1978, Table 2, expt. 24; Stumpe and Hagedorn, 1982, Table 2, expt. 6; Petrova and Gospodinov, 1986, Table 2, expt. 22) the rate of N applied appears to have no effect, or to affect only N mineralization. This may be because the experiment has only been running for a few years. Another possibility is that at these sites small applications of fertilizer N are sufficient to produce all the 'support tissue' (root, straw, etc.) that the crop requires; above this rate, additional N will have little impact on residue production.

It is difficult to draw any conclusions about the effects of the form of fertilizer N applied. Unless the experimental sites are known to have been limed regularly, the differences in soil pH which can arise, particularly from the use of ammonium sulphate, may effect biological activity and thus N mineralization (e.g. at **Stackyard**, Mattingly et al., 1975, Table 1, expt. 11). At **Lyberetski** (Table 4, expt. 4) N mineralization increased when N was applied as NaNO$_3$, but not as (NH$_4$)$_2$SO$_4$ — this was probably a pH effect.

In general, the effect of fertilizer N is greatest in the topsoil (e.g. Kopytko and Gerkiyal, 1983; Table 8, expt. 6). Rasmussen and Rohde (1988; Table 2, expt. 23) report that N fertilization affected total N content primarily in the upper 22.5 cm of the soil, with only minor effects in the 22.5 to 45 cm depth, and the greatest effect in the top 7.5 cm. This is not surprising because most crop residue is concentrated in the plough layer, although some roots are found well below this depth (e.g. Barraclough and Leigh, 1984). Dalal (1989; Table 4, expt. 15), working on a vertisol reported an increase in N mineralized during aerobic incubation in the 60 to 90 cm depth only, speculating that this may be due to organic matter enriched in N from fertilizer applications leaching, falling, or washing down through cracks.

N fertilization tends to have a greater impact on total and mineralizable soil N under conditions which maximise the return of organic N in crop residues - for example, when straw and other above-ground residues are incorporated or retained on the surface (Ketcheson and Beauchamp, 1978, Table 2, expt. 24; Pichot et al., 1981, Table 2, expt. 17), when other nutrients are applied (Janzen,

1987), when crops are irrigated (Muthuvel et al., 1977, Table 1, expt. 8; El-Haris et al., 1983, Table 2, expt. 25), and grown continuously (Biederbeck et al., 1984, Table 2, expt 21; Janzen, 1987, Table 2, expt. 16).

## IV. Conclusions

These results have a number of agricultural and environmental consequences. Soil organic matter has declined in many agricultural systems of the world (Biederbeck et al., 1984; Rasmussen and Collins, 1991), with serious effects on soil physical and biological properties. In the more productive regions, it may be possible to slow or halt this decline by increasing the return of organic N in crop residues, through the judicious use of fertilizer N, combined with good crop management practices. For example, fertilizer N applied for just 10 years, in combination with continuous wheat (straw retained), has increased total soil N content by 17% at a temperate semi-arid site in Canada, previously under an unfertilized wheat/fallow rotation for 45 years (Biederbeck et al., 1984; Table 2, expt 21).

A greater supply of mineralized soil N will obviously benefit subsequent crops, and should be taken into account when recommending the optimum rate of fertilizer N to apply. However, the actual increases tend to be modest, for example, an extra 30 kg N ha$^{-1}$ mineralized and taken up by the crop from the **Broadbalk** plot given 144 kg N ha$^{-1}$ for 135 years, compared to the plot which never receives fertilizer N. Furthermore, one of the main results of this review is that the response of total and mineralizable soil N to fertilizer N is varied, with relatively large effects at some sites, and little or no effects at others. Current models for the long-term turnover of organic matter in the soil (e.g. Jenkinson et al., 1990b; Parton *et al.*, 1988) are able to predict gross changes, for example due to differences between existing climates, on the basis of measurements made elsewhere. The more subtle indirect effect of fertilizer N may be harder to predict quantitatively.

More mineralizable N will also increase losses of N from the soil system, via leaching and gaseous emissions. In temperate maritime climates the main source of nitrate leached from arable land is from the mineralization of organic N in the autumn, when warm soils become moist, crop uptake of N is low or non-existent, and conditions may be ideal for leaching (Jenkinson, 1986; Macdonald et al., 1989). In areas of relatively low through drainage, this extra N may increase the nitrate concentration of percolating water above the 1980 EC Drinking Water Quality Directive limit of 11.3 mg N l$^{-1}$. For example, in the **Broadbalk** experiment, which has an average through drainage of 200 mm yr$^{-1}$, only 23 kg N ha$^{-1}$ yr$^{-1}$ would need to be leached to bring the drainage water above this limit. Thus, although in the short-term, direct leaching losses of N from fertilizer may be very low, if applied at the appropriate rate and time (Macdonald et al., 1989; Chaney, 1990), in the long-term, much of the N lost from old arable soils may be indirectly derived from fertilizer N, through its ability to build up soil organic matter. Similarly, any nitrate produced from mineralization may be exposed to denitrification if soil conditions become conducive to this process. This nitrate will then become one of the sources of nitrous oxide.

An inevitable consequence of increased soil fertility is an increased risk of nutrient leakage to aquifers or the atmosphere, with the possibility of adverse environmental consequences. Long-term experiments are one of the few ways in which the full consequences of different agricultural activities can be quantified and compared, taking account of both benefits and adverse impacts. Availability of such data is essential if rational decisions are to be made when planning the strategies required to meet the growing worldwide demand for agricultural produce, but in ways that are socially and environmentally acceptable.

## Acknowledgments

This work was partially funded by a grant from the Commission of the European Communities. The authors thank David Jenkinson who made many invaluable suggestions, also Paul Poulton for helpful discussion; the staff of Rothamsted Library; Nora Gray, Rémi Chaussod, Birgit Hütsch and numerous visitors to the Soil Science Department for help with translation; and Christine Jaggard for typing the manuscript.

## References

Aldag, R., and R. Kickuth. 1973. Stickstoffverbindungen in Böden und ihre Beziehung zur Humusdynamik. *Z. Pflanzenernähr. Bodenkde.* 136: 193-202.

Allison, F.E. 1973. Soil organic matter and its role in crop production. *Developments in Soil Science 3*. Elsevier, Amsterdam.

Anderson, F.N., and G.A. Peterson. 1973. Effects of continuous corn (*Zea mays* L.), manuring and nitrogen fertilization on yield and protein content of the grain and on the soil nitrogen content. *Agron. J.* 65: 697-700.

Appel, T., and K. Mengel. 1992. Nitrogen uptake of cereals grown on sandy soils as related to nitrogen fertilizer application and soil nitrogen fractions obtained by electro-ultrafiltration (EUF) and $CaCl_2$ extraction. *Eur. J. Agron.* 1: 01-09.

Arden-Clarke, C., and R.D. Hodges. 1988. The environmental effects of conventional and organic/biological farming systems. II. Soil ecology, soil fertility and nutrient cycles. *Biol. Ag. Hort.* 5: 223-287.

Asmus, F., U. Volker, and H. Koriath. 1982. Wirkung langjährig unterschiedlicher Düngung auf Pflanze und Boden in einem Dauerversuch auf Tieflehm-Fahlerde. *Tag.-Ber., Akad. Landwirtsch.-Wiss. DDR, Berlin.* 205:79-86.

Avery, B.W., P. Bullock, J.A. Catt, A.C.D. Newman, J.H. Rayner, and A.H. Weir. 1972. The soil of Barnfield. *Rothamsted Expt. Stn Ann. Rep. for 1971* Part 2:5-37.

Bache, B.W., and R.G. Heathcote. 1969. Long-term effects of fertilizers and manure on soil and leaves of cotton in Nigeria. *Expt. Agric.* 5:241-247.

Barraclough, P.B., and R.A. Leigh. 1984. The growth and activity of winter wheat roots in the field: the effect of sowing date and soil type on root growth of high-yielding crops. *J. Agric. Sci., Cambs.* 103:59-74.

Biederbeck, V.O., C.A. Campbell, and R.P. Zentner. 1984. Effect of crop rotation and fertilization on some biological properties of a loam in southwestern Saskatchewan. *Can. J. Soil Sci.* 64:355-367.

Bjarnason, S. 1988. Turnover of Organic Nitrogen in Agricultural Soils and the Effects of Management Practices on Soil Fertility. Dissertation, Swedish University of Agricultural Sciences, Uppsala.

Bjarnason, S. 1989. The long-term soil fertility experiments in southern Sweden. III. Soil carbon and nitrogen dynamics. *Acta Agric. Scand.* 39:361-371.

Bosch, M., and A. Amberger. 1983. Einfluß langjähriger Düngung mit verschiedenen N-Formen auf pH-Wert, Humusfraktionen, biologische Aktivität und Stickstoffdynamik einer Acker-Braunerde. *Z. Pflanzenernaehr. Bodenk.* 146:714-724.

Bremner, J.M. 1965. Nitrogen availability indexes. p. 1324-1345. In: Black, C.A. et al. (eds.), *Methods of Soil Analysis*, Part 2. Agronomy Monograph No. 9, Am. Soc. Agron., Madison, Wis.

Campbell, C.A., V.O. Biederbeck, R.P. Zentner, and G.P. Lafond. 1991a. Effect of crop rotations and cultural practices on soil organic matter, microbial biomass and respiration in a thin Black Chernozem. *Can. J. Soil Sci.* 71:363-376.

Campbell, C.A., K.E. Bowren, M. Schnitzer, R.P. Zentner, and L. Townley-Smith. 1991b. Effect of crop rotations and fertilization on soil organic matter and some biochemical properties of a thick Black Chernozem. *Can. J. Soil Sci.* 71:377-387.

Campbell, C.A., D.R. Cameron, W. Nicholaichuk, and H.R. Davidson. 1977. Effects of fertilizer N and soil moisture on growth, N content, and moisture use by spring wheat. *Can. J. Soil Sci.* 57:289-310.

Campbell, C.A., G.P. Lafond, A.J. Leyshon, R.P. Zentner, and H.H. Janzen. 1991c. Effect of cropping practices on the initial potential rate of N mineralization in a thin Black Chernozem. *Can. J. Soil Sci.* 71:43-53.

Campbell, C.A., D.W.L. Read, R.P. Zentner, A.J. Leyshon, and W.S. Ferguson. 1983. First 12 years of a long-term crop rotation study in southwestern Saskatchewan - Yields and quality of grain. *Can. J. Plant Sci.* 63:91-108.

Campbell, C.A., M. Schnitzer, G.P. Lafond, R.P. Zentner, and J.E. Knipfel. 1991d. Thirty year crop rotations and management practices effects on soil and amino nitrogen. *Soil Sci. Soc. Am. J.* 55:739-745.

Chaney, K. 1990. Effect of nitrogen fertilizer rate on soil nitrate nitrogen content after harvesting winter wheat. *J. Ag. Sci., Cambs.* 114:171-176.

Chaussod, R. 1987. Relation entre le type de sol, les charactéristiques de la biomasse microbienne et la minéralisation d'azote. p. 19-31. In: C. Egoumenides (ed.), *Statut Organique du Sol et Mineralization d'Azote*. Proc. of the 1987 meeting of Group d'Etude de la Matiére Organique des Sols, Montpellier, France.

Chaussod, R., B. Nicolardot, and G. Catroux. 1986. Mesure en routine de la biomass microbienne des sols par la methode de fumigation au chloroforme. *Sci. Sol.* 24:201-211.

Cooke, G.W. 1967. The control of soil fertility. The English Language Book Society. Crosby Lockwood, London.

Cooke, G.W. 1976. Long-term fertilizer experiments in England: The significance of their results for agricultural science and for practical farming. *Ann. Agron.* 27:503-536.

Cornfield, A.H. 1960. Ammonia released on treating soils with $N$ sodium hydroxide as a possible means of predicting the nitrogen-supplying power of soils. *Nature* 187: 260-261.

Dalal, R.C. 1989. Long-term effects of no-tillage, crop residue and nitrogen application on properties of a Vertisol. *Soil Sci. Soc. Am. J.* 53:1511-1515.

Dalal, R.C., P.A. Henderson, and J.M. Glasby. 1991. Organic matter and microbial biomass in a Vertisol after 20 years of zero-tillage. *Soil Biol. Biochem.* 23:435-441.

Department of the Environment. 1986. Nitrate in water. A report by the Nitrate Coordination Group. Department of the Environment Central Directorate of Environment Protection. Pollution Paper 26. HMSO, London, UK.

Djokoto, R.K., and D. Stephens. 1961. Thirty long-term fertilizer experiments under continuous cropping in Ghana. II. Soil studies in relation to the effects of fertilizers and manures on crop yields. *Emp. J. Expt. Ag.* 29:245-258.

Dobransky, B. 1976. The application of mineral and organic fertilizers on two crop rotations (Pologne-Skierniewice 1923). *Ann. Agron.* 27:625-642.

Dodge, D.A., and H.E. Jones. 1948. The effect of long-time fertility treatments on the nitrogen and carbon content of a Prairie soil. *J. Am. Soc. Agron.* 40:778-785.

Dormaar, J.F., and U.J. Pittman. 1980. Decomposition of organic residues as affected by various dryland spring wheat-fallow rotations. *Can. J. Soil Sci.* 60:97-106.

Eich, D., E. Bahn, and E. Buhtz. 1982. Ertragsentwicklung und Entwicklung der Gehalte an organischer Substanz und Nährstoffen im Statischen Versuch Lauchstädt. *Tag.-Ber., Akad Landwirtsch.-Wiss. DDR, Berlin.* 205:37-48.

El-Haris, M.K., V.L. Cochran, L.F. Elliott, and D.F. Bezdicek. 1983. Effect of tillage, cropping and fertilizer management on soil nitrogen mineralization potential. *Soil Sci. Soc. Am. J.* 47:1157-1161.

Frame, J., and P. Newbould. 1984. Herbage production from grass/white clover swards. p. 15-35. In: D.J. Thomson (ed.), *Forage Legumes* Occ. Symp. 16, British Grassland Soc.

Garner, H.V., and G.V. Dyke. 1969. The Broadbalk Yields. *Rothamsted Expt. Stn. Ann. Rep. for 1968* Part 2:26-49.

Garz, J., and E. Hagedorn. 1990. Der Versuch 'Ewiger Roggenbau' nach 110 Jahren. p. 9-30. In: H.-J. Liste (ed.), *110 Jahre Ewiger Roggenbau.* Kong. Tag.-Ber. Martin-Luther-Univ., Halle-Wittenberg, Halle (Salle) 1990.

Garz, J., A.P. Scerbakov, and M. Rossbach. 1982. Der Stickstoff im Boden des Versuches Ewiger Roggenbau (Halle) und seine Pflanzenverfügbarkeit. *Tag.-Ber., Akad. Landwirtsch.-Wiss. DDR Berlin.* 205:195-202.

Gasser, J.K.R. 1962. Effects of long-continued treatment on the mineral nitrogen content and mineralizable nitrogen of soil from selected plots of the Broadbalk experiment on continuous wheat, Rothamsted. *Plant Soil* 17:209-220.

Glendining, M.J., and D.S. Powlson. 1991. The effect of long-term applications of inorganic nitrogen fertilizer on soil organic nitrogen. p. 329-338. In: W.S. Wilson (ed.), *Proc. Symp. Advances in Soil Organic Matter Research: the Impact on Agriculture and the Environment*, Sept. 1990. University of Essex, England.

Gokhale, N.G. 1959. Soil nitrogen status under continuous cropping and with manuring in the case of unshaded tea. *Soil Sci.* 87:331-333.

Greenland, D.J. 1971. Changes in the nitrogen status and physical condition of soils under pastures, with special reference to the maintenance of the fertility of Australian soils used for growing wheat. *Soil Fert.* 34:237-251.

Gregory, P.J. 1988. Growth and functioning of plant roots. p. 113-167. In: Wild, A. (ed.), *Russell's Soil Conditions and Plant Growth*, 11th edition. Longman Scientific, Harlow, UK.

Guerillot, J. 1935. L'histoire et les buts de la Station Agronomique de Grignon. *Ann. Agron.* Année 5:610-621.

Hadas, A., S. Feigenbaum, A. Feigin, and R. Portnoy. 1986a. Distribution of nitrogen forms and availability indices in profiles of differently managed soil types. *Soil Sci. Soc. Am. J.* 50:308-313.

Hadas, A., S. Feigenbaum, A. Feigin, and R. Portnoy. 1986b. Nitrogen mineralization in profiles of differently managed soil types. *Soil Sci. Soc. Am. J.* 50:314-319.

Hadas, A., A. Feigin, S. Feigenbaum, and R. Portnoy. 1989. Nitrogen mineralization in the field at various soil depths. *J. Soil Sci.* 40:131-137.

Hargitai, L. 1982. Änderungen des Humuszustands und der Stickoffrerhältnisse im Boden von Dauerversuchen. *Tag.-Ber., Akad. Landwirtsch.-Wiss. DDR, Berlin.* 205: 203-210.

Hart, P.B.S., J.H. Rayner, and D.S. Jenkinson. 1986. Influence of pool substitution on the interpretation of fertilizer experiments with $^{15}$N. *J. Soil Sci.* 37:389-403.

Hatch, D.J., S.C. Jarvis, and L. Philipps. 1990. Field measurement of nitrogen mineralization using soil core incubation and acetylene inhibition of nitrification. *Plant Soil* 124:97-107.

Houot, S., J.A.E. Molina, R. Chaussod, and C.E. Clapp. 1989. Simulation by NCSOIL of net mineralization in soils from the 'Dehérain' and '36 Parcelles' fields at Grignon. *Soil Sci. Soc. Am. J.* 53:451-455.

House of Lords. 1989. *Nitrate in Water.* Report by the Select Committee on the European Communities. HMSO, London.

Insam, H., C.C. Mitchell, and J.F. Dormaar. 1991. Relationship of soil microbial biomass and activity with fertilization practice and crop yield of three Ultisols. *Soil Biol. Biochem.* 23:459-464.

Ionescu, F.L., M. Nicolescu, M. Coifan, I. Dincă, E. Banită, G.H. Stefanic, V. Stratula, and L. Pop. 1986. (Effect of rotation on winter wheat crop on the reddish-brown slightly podzolic soil in the central zone of Oltenia). *Analele Institutilui de Cercetări Pentru Cereale di Plante Tehnice Fundulea* 8:267-297 (in Romanian, English summary).

Ivarsson, K., and S. Bjarnason. 1988. The long-term soil fertility experiments in Skåne, Southern Sweden. I. Background, site description and experimental design. *Acta. Agric. Scand.* 38:137-143.

Jansson, S.L. 1983. (Twenty-five years of soil fertility studies in Sweden). *Instit. markvetenskap Avd. växtnäringslära* Rapport 151. Uppsala, Sweden (in Swedish, English summary).

Janzen, H.H. 1987. Effect of fertilizer on soil productivity in long-term spring wheat rotations. *Can. J. Soil Sci.* 67:165-174.

Jenkinson, D.S. 1977. The nitrogen economy of the Broadbalk experiments. I. Nitrogen balance in the experiments. *Rothamsted Expt. Stn. Ann. Rep. for 1976*, Part II:103-109.

Jenkinson, D.S. 1984. The supply of nitrogen from the soil. p. 79-92. In: *The Nitrogen Requirements of Cereals* MAFF/ADAS Ref. Book 385. HMSO, London, UK.

Jenkinson, D.S. 1986. Nitrogen in UK arable agriculture. *J. Roy. Ag. Soc. Eng.* 147:178-189.

Jenkinson, D.S. 1988. Determination of microbial biomass carbon and nitrogen in soil. p. 368-386. In: J.R. Wilson (ed.), *Advances in Nitrogen Cycling in Agricultural Ecosystems*, CAB International, Wallingford, Oxon, UK.

Jenkinson, D.S. 1990a. An introduction to the global N cycle. *Soil Use and Manage.* 6:56-61.

Jenkinson, D.S. 1990b. The turnover of organic carbon and nitrogen in soil. *Phil. Trans. R. Soc. Lond. B.* 329:361-368.

Jenkinson, D.S. 1991. The Rothamsted long-term experiments: Are they still of Use? *Agron. J.* 83:2-10.

Jenkinson, D.S., R.H. Fox, and J.H. Rayner. 1985. Interactions between fertilizer nitrogen and soil nitrogen - the so-called 'priming' effect. *J. Soil Sci.* 36:425-444.

Jenkinson, D.S., and A.E. Johnston. 1977. Soil organic matter in the Hoosfield Continuous Barley Experiment. *Rothamsted Expt. Stn. Ann. Rep. for 1976*, Part II: 87-101.

Jenkinson, D.S., and J.N. Ladd. 1981. Microbial biomass in soil: Measurement and turnover. p. 415-471. In: A.E. Paul and J.N. Ladd (eds.), *Soil Biochemistry*. Dekker, New York.

Jenkinson, D.S., and D.S. Powlson. 1976. The effects of biocidal treatments on metabolism in soil - V. A method for measuring soil biomass. *Soil Biol. Biochem.* 8:209-213.

Jenkinson, D.S., and J.H. Rayner. 1977. The turnover of soil organic matter in some of the Rothamsted Classical experiments. *Soil Sci.* 123:298-305.

Johnston, A.E. 1970. The value of residues from long-period manuring at Rothamsted and Woburn. II. A summary of the experiments started by Lawes and Gilbert. *Rothamsted Expt. Stn. Ann. Rep. for 1969* Part 2:7-21.

Johnston, A.E. 1973. The effects of ley and arable cropping systems on the amounts of soil organic matter in the Rothamsted and Woburn Ley-Arable Experiments. *Rothamsted Expt. Stn. Ann. Rep. for 1972* Part 2:131-159.

Johnston, A.E. 1975. Experiments made on Stackyard Field, Woburn 1876-1974. I. History of the field, details of the cropping and manuring and the yields in continuous wheat and barley experiments. *Rothamsted Expt. Stn. Ann. Rep. for 1974* Part 2:29-44.

Johnston, A.E., and P.R. Poulton. 1977. Yields on the Exhaustion Land and changes in the NPK content of the soils due to cropping and manuring, 1852-1975. *Rothamsted Expt. Stn. Ann. Rep. for 1976* Part 2:53-85.

Jones, M.J. 1971. The maintenance of soil organic matter under continuous cultivation at Samaru, Nigeria. *J. Agric. Sci., Cambs.* 77:473-482.

Keeney, D.R. 1982. Nitrogen Availability Indices. p. 711-733. In: A.L. Page, R.H. Miller, and D.R. Keeney (eds.), *Methods of Soil Analysis, Part 2*. Agronomy Monograph No. 9 (2$^{nd}$ Edn), Am. Soc. Agron. Madison, WI, USA.

Keeney, D.R., J.S. Schepers, J.E. Blodgett, G.R. Hallberg and P.F. Pratt. 1987. Nitrate contributions to groundwater by agricultural practices. *Proc. 1986 Soil Sci. Soc. Am. Workshop*, New Orleans, La, USA.

Ketcheson, J.W. and E.G. Beauchamp. 1978. Effects of corn stover, manure and nitrogen on soil properties and crop yield. *Agron. J.* 70:792-797.

Kofoed, A.D. 1982. Humus in long-term experiments in Denmark. p. 241-258. In: D. Boels, D.B. Davies, and A.E. Johnston (eds.), *Land Use Seminar on Soil Degradation*, Oct. 1980. Wageningen, The Netherlands.

Kofoed, A.D., and O. Nemming. 1976. Askov 1894: Fertilizers and manure on sandy and loamy soils. *Ann. Agron.* 27:583-610.

Kolbe, G., and H. Stumpe. 1969. Neunzig Jahre "Ewiger Roggenbau". *Albrecht-Thaer-Archiv* 13:933-949.

Kopytko, P.G., and Z.V. Gérkiyal. 1983. Content of humus and nitrogen in a dark gray forest soil in an orchard after long-term fertilization. *Soviet Soil Sci.* 15: 52-60. Translated from *Pochvovedenie* (1983) 12:64-72.

Körschens, M., D. Eich, and C. Weber. 1982. Der Statische Düngungsversuch Lauchstädt nach Erweiterung der Versuchsfrage 1978. *Tag.-Ber., Akad. Landwirtsch.-Wiss. DDR, Berlin* 205:49-56.

Körschens, M., E. Schulz, and R. Behm. 1990. Heißwasserlöslicher C und N im Boden als Kriterium für das N-Nachlieferungsvermögen. *Zentralbl. Mikrobiol.* 145:305-311.

Lang, H., and H. Sturm. 1983. Nachwirkung langjährig durchgeführeter N-Düngen und ihre Bewertung für die Ertragsfähigkeit eines IS-Bodes. *Landwirtsch. Forsch.* 36:332-342.

Last, P.J., D.J. Webb, R.B. Bugg, K.M.R. Bean, M.J. Durrant, and K.W. Jaggard. 1985. Long-term effects of fertilizers at Broom's Barn, 1965-82. *Rothamsted Expt. Stn. Ann. Rep. for 1984* Part 2: 231-249.

Lathwell, D.J., and D.R. Bouldin. 1981. Soil organic matter and soil nitrogen behaviour in cropped soils. *Trop. Agric.* 58:341-348.

Lawes, J.B., and J.H. Gilbert. 1857. On the growth of barley by different manures continuously on the same land; and on the position of the crop in rotation. *J. Roy. Ag. Soc. Eng.* 18:454-531.

Lawes, J.B., and J.H. Gilbert. 1873. Report of experiments on the growth of barley for twenty years in succession on the same land. *J. Roy. Ag. Soc. Eng.* 9:5-178.

Lawes, J.B., and J.H. Gilbert. 1885. On the growth of wheat for the second period of 20 years in succession on the same land. *J. Roy. Ag. Soc. Eng.* 20: 97 pp.

Lucas, R.E., J.B. Holtman, and L.J. Connor. 1977. Soil carbon dynamics and cropping practices. p. 333-351. In: W. Lockerelz (ed.), *Agriculture and Energy.* Washington Univ., St Louis, MO, June 1976. Academic Press, New York.

Macdonald, A.J., D.S. Powlson, P.R. Poulton, and D.S. Jenkinson. 1989. Unused fertilizer nitrogen in arable soils — its contribution to nitrate leaching. *J. Sci. Food Ag.* 46:407-419.

Mandal, B.C., A.B. Roy, M.N. Saha, and A.K. Mandal. 1984. Wheat yield and soil nutrient status as influenced by continuous cropping and manuring in a jute-rice-wheat rotation. *J. Ind. Soc. Soil Sci.* 32:696-700.

Mandal, B.C., M.N. Saha, and A.B. Roy. 1985. Effect of long-term fertilizer use on the available NPK status in soil under rotational cropping with jute-rice-wheat sequence over ten years. *Ind. J. Ag. Sci.* 55:421-426.

Mattingly, G.E.G., M. Chater, and A.E. Johnston. 1975. Experiments made on Stackyard Field, Woburn, 1876-1974. III. Effects of NPK fertilizers and farmyard manure on soil carbon, nitrogen and organic phosphorus. *Rothamsted Expt. Stn Ann. Rep. for 1974* Part 2:61-77.

Mattsson, L. 1988. Four Swedish long-term experiments with N, P and K. 3. After-effects of N-fertilization. *Swedish J. Ag. Res.* 18:13-19.

Mercik, S. and K. Németh. 1985. Effects of 60-year N, P, K and Ca fertilization on EUF-nutrient fractions in the soil and on yields of rye and potato crops. *Plant Soil* 83:151-159.

Morel, R., T. Lasnier, and S. Bourgeois. 1984. Les essais de fertilisation de longue durée de la station agronomique de Grignon. Dispositif Dehérain et des 36 Parcelles. Resultats expérimentaux (1938-1982). INA-PG, INRA, Paris, France.

Muthuvel, P., P. Kandaswamy, and K.K. Krishnamoorthy. 1977. Availability of NPK under long-term fertilization. *Madras Agric. J.* 64:358-362.

Muthuvel, P., P. Kandaswamy, and K.K. Krishnamoorthy. 1979. Organic carbon and total N content of soils under long-term fertilization. *J. Ind. Soc. Soil Sci.* 27:186-188.

Neal, O.R. 1944. Removal of nutrients from the soil by crops and erosion. *J. Am. Soc. Agron.* 36:601-607.

Nnadi, L.A., and Y. Arora. 1985. Effects of fertilizer N on crop yields and soil properties in the major Savanna soils, excluding vertisols. p. 223-234. In: B.T. Kang and J. van der Heide (eds.), *N management in farming systems in humid and subhumid tropics*. Inst. Soil Fert. and Int. Inst. Trop. Ag.

Parton, W.J., J.W.B. Stewart, and C.V. Cole. 1988. Dynamics of C, N, P and S in grassland soils: a model. *Biogeochem.* 5:109-131.

Petrova, M., and M. Gospodinov. 1986. (Effect of long-term systematic fertilizing on the nitrogen status of weakly leached Chernozems) *Pochvoznan-ie, Agrokhimiya i Rastitelna Zashchita* 21: 18-26 Sofia, Bulgaria (in Bulgarian, English summary).

Pichot, J., M.P. Sedogo, J.F. Poulain, and J. Arrivets. 1981. Evolution de la fertilite d'un sol ferrugineux tropicale sous l'influence des fumures minerales et organiques. *L'Agron. Tropicale* 36:122-133.

Power, J.F., and J.W. Doran. 1984. Nitrogen use in organic farming. p. 585-598. In: R.S. Hauck (ed.), *Nitrogen in Crop Production*. Am. Soc. Agron., Madison, Wis., USA.

Powlson, D.S., P.C. Brookes, and B.J. Christensen. 1987. Measurement of soil microbial biomass provides an early indication of changes in total soil organic matter due to straw decomposition. *Soil Biol. Biochem.* 19:159-164.

Powlson, D.S., and D.S. Jenkinson. 1976. The effects of biocidal treatments on metabolism in soil - II. Gamma irradiation, autoclaving, air-drying and fumigation with chloroform or methyl bromide. *Soil Biol. Biochem.* 8:179-188.

Powlson, D.S., and D.S. Jenkinson. 1981. A comparison of the organic matter, biomass, adenosine triphosphate and mineralizable nitrogen contents of ploughed and direct-drilled soils. *J. Ag. Sci., Cambs.* 97:713-721.

Powlson, D.S., P.R. Poulton, T.M. Addiscott, and D.S. McCann. 1989. Leaching of nitrate from soils receiving organic or inorganic fertilizers continuously for 135 years. p. 334-345. In: J.A. Hansen and K. Henriksen (eds.), *Nitrogen in Organic Wastes Applied to Soils*. Academic Press, London.

Powlson, D.S., G. Pruden, A.E. Johnston, and D.S. Jenkinson. 1986. The nitrogen cycle in the Broadbalk Wheat Experiment: recovery and losses of $^{15}$N-labelled fertilizer applied in spring and inputs of nitrogen from the atmosphere. *J. Ag. Sci. Cambs.* 107:591-609.

Rasmussen, P.E., R.R. Allmaras, C.R. Rohde, and N.C. Roager. 1980. Crop residue influences on soil carbon and nitrogen in a wheat-fallow system. *Soil Sci. Soc. Am. J.* 44:596-600.

Rasmussen, P.E. and H.P. Collins. 1991. Long-term impacts of tillage, fertilizer and crop residue on soil organic matter in temperate semiarid regions. *Adv. Agron.* 45:93-134.

Rasmussen, P.E. and C.R. Rohde. 1983. Long-term changes in soil C, N and pH produced by $NH_4$-N fertilization. *Agron. Abstr.* p.178.

Rasmussen, P.E. and C.R. Rohde. 1988. Long-term tillage and nitrogen fertilization effects on organic nitrogen and carbon in a semiarid soil. *Soil Sci. Soc. Am. J.* 52:1114-1117.

Russell, J.S. 1975. A mathematical treatment of the effect of cropping system on soil organic nitrogen in two long-term sequential experiments. *Soil Sci.* 120:37-44.

Russell, J.S. 1981. Models of long-term soil organic nitrogen change. p. 222-232. In: M.J. Frissel and J.A. van Veen (eds.), *Simulation of Nitrogen Behaviour of Soil-Plant Systems*. Wageningen, The Netherlands.

Russell, R.S. 1977. Plant root systems: Their function and interaction with the soil. McGraw-Hill, London.

Saffigna, P.G., D.S. Powlson, P.C. Brookes, and G.A. Thomas. 1989. Influence of sorghum residues and tillage on soil organic matter and soil microbial biomass in an Australian vertisol. *Soil Biol. Biochem.* 21:759-765.

Schnürer, J., M. Clarholm, and T. Rosswall. 1985. Microbial biomass and activity in an agricultural soil with different organic matter contents. *Soil Biol. Biochem.* 17: 611-618.

Shen, S.M., P.B.S. Hart, D.S. Powlson, and D.S. Jenkinson. 1989. The nitrogen cycle in the Broadbalk Wheat Experiment: $^{15}$N-labelled fertilizer residues in the soil and in the microbial biomass. *Soil Biol. Biochem.* 21:529-533.

Shevtsova, L.K. 1966. (Effect of long-continued use of manure and mineral fertilizer on the content of humus and nitrogen in various soils). p. 169-188. In: *Conference on Fertilizer and Soil Fertility, USSR*. Moscow, USSR, Kolos (in Russian).

Sibbesen, E. 1986. Soil movement in long-term field experiments. *Plant Soil* 91:73-85.

Singh, A. 1964. Effect of long-term applications of organic and inorganic sources of nitrogen on the yield of sugar cane and on soil fertility. *Empire J. Expt. Agric.* 32:205-210.

Smith, A.N. 1962. The effect of fertilizers, sulphur and mulch on East African tea soils II. The effect on the base status and organic matter content of the soil. *East African Agric. For. J.* 28:16-21.

Smith, G.E. 1942. Sanborn Field. 50 years of field experiments with crop rotations, manure and fertilizers. *Missouri Agric. Exp. Stn. Bull.* 458:1-61.

Stanford, G. 1977. Evaluating the N-supplying capacities of soils. p. 412-418. In: *Proc. Int. Seminar Soil Environ. Fertil. Manage. Intensive Agric.* Comm. IV, ISSS, Tokyo.

Stanford, G., J.N. Carter, and S.J. Smith. 1974. Estimates of potentially mineralizable soil N based on short-term incubations. *Soil Sci. Soc. Am. Proc.* 38:99-102.

Stanford, G., and S.J. Smith. 1972. Nitrogen mineralization potentials of soils. *Soil Sci. Soc. Am. Proc.* 36:465-472.

Steiner, R.A., R.W. Herdt (eds.), 1993. A global directory of long-term agronomic experiments. Vol.1. Non-European Experiments. The Rockefeller Foundation, New York. (in press).

Stevenson, F.J. 1965. Origin and distribution of nitrogen in soil. p. 1-42. In: W.V. Bartholomew and F.E. Clarke (eds.), *Soil Nitrogen* Agronomy Monograph No. 10 Am. Soc. Agron. Madison, WI.

Stevenson, F.J. 1982. Origin and distribution of nitrogen in soil. p. 1-42. In: F.J. Stevenson (ed.), *Nitrogen in Agricultural Soils.* Agronomy Monograph No. 22. Am. Soc. Agron. Madison, WI.

Stumpe, H., and E. Hagedorn. 1982. Die Wirkung von Stallmist und Mineraldüngung auf Pflanzenertrag und Humusgehalt des Bodens in einem 32 jährigen Dauerversuch in Halle. *Tag.-Ber., Akad. Landwirtsch.-Wiss. DDR, Berlin* 205:65-72.

Subbiah, B.V., and G.L. Asija. 1956. A rapid procedure for the estimation of available nitrogen in soils. *Curr. Sci.* 25:259-260.

Tyurin, I.V., and M.M. Kononova. 1934. Determination of the nitrogen requirement of soils. *Chem. Abstr.* 29: 4119 (1935) (Translated from Dokuchaev Soil Inst. 10:49-56).

Upchurch, W.J., R.J. Kinder, J.R. Brown, and G.H. Wagner. 1985. Sanborn Field. Historical Perspective. *Missouri Agric. Exp. Stn. Res. Bull.* 1054.

Ushakov, V.Y., A.Y. Zhezher, and A.I. Yuzhakov. 1985. (The effect of the soil temperature regime on the growth characteristics and functioning of wheat root systems under various nutritional levels during early stages of development). Sibirskii-Vestnik-Sel skokhozyaistvennoi-Nauki. 1:22-27 (in Russian, English summary).

Waberisch, R. 1967. Der Bernburger Dauerdüngungsversuch 2. Mitteilung: Die Kohlenstoff-und Stickstoffverhältnisse. *Albrecht-Thaer-Archiv* 11:859-869.

Waring, S.A., and J.M. Bremner. 1964. Ammonium production in soil under waterlogged conditions as an index of nitrogen availability. *Nature* 201:951-952.

Warren, R.G., and A.E. Johnston. 1962. Barnfield. *Rothamsted Expt. Stn. Ann. Rep. for 1961*: 227-247.

Warren, R.G., and A.E. Johnston. 1967. Hoosfield Continuous Barley. *Rothamsted Exp. Stn. Ann. Rep. for 1966*: 320-338.

Welbank, P.J., M.J. Gibb, P.J. Taylor, and E.D. Williams. 1974. Root growth of cereal crops. *Rothamsted Expt. Stn. Ann. Rep. for 1973* Part 2:26-66.

White, J.W. 1955. Outstanding lessons from the Jordan Soil Fertility Plots. *Penns. State Univ. Coll. Agric. Bull.* 613:12-18.

Wild, A. 1988. Plant nutrients in soil:nitrogen. p. 652-694. In: A. Wild (ed.), *Russell's Soil Conditions and Plant Growth* 11th Edn. Longman Scientific, UK.

Wolf, J., and H. van Keulen. 1989. Modelling long-term crop response to fertilizer and soil nitrogen. II. Comparison with field results. *Plant Soil* 120:23-38.

Woodruff, C.M. 1949. Estimating the nitrogen delivery of soil from the organic matter determination as reflected by Sanborn Field. *Soil Sci. Soc. Am. Proc.* 14:208-212.

Zentner, R.P., K.E. Bowren, W. Edwards, and C.A. Campbell. 1990. Effects of crop rotations and fertilization on yields and quality of spring wheat grown on a Black Chernozem in north-central Saskatchewan. *Can. J. Plant Sci.* 70:383-397.

Zentner, R.P., E.D. Spratt, H. Reisdorf, and C.A. Campbell. 1987. Effect of crop rotation and N and P fertilizer on yields of spring wheat grown on a Black Chernozemic clay. *Can. J. Plant Sci.* 67:965-985.

Ziegler, K., K. Németh, and K. Mengel. 1992. Relationship between electro-ultrafiltration (EUF) extractable nitrogen, grain yield and optimum fertilizer rates for winter wheat. *Fert. Res.* 32:37-43.

# Long-Term Trials on Soil and Crop Management at ICARDA

## Hazel C. Harris

## I. Introduction

Agriculture evolved in west Asia over 10,000 years ago in the area known as the Fertile Crescent. The region is the center of origin of a number of cereal and legume species of temperate agriculture, and of the domestication of sheep and goats. White (1963, 1970) and Watson (1974) reviewed the development of agriculture from these origins through the period of Roman colonization, when the Near East and North Africa were major providers of grain to the remainder of the Empire, and on through first Arab, then Turkish and finally European domination to independence in the twentieth century.

There is a long written record, dating back to Greek and Roman times, of the use of crop rotations involving both cereal-legume and cereal-fallow sequences in the farming systems which evolved (White, 1970; I. Papastyilanou, pers. comm.). The inclusion of fallow throughout history indicates that the major

ISBN 1-56670-076-0

factor limiting productivity has always been, as it remains today, the availability of water in the uncertain rainfall regime which characterizes the region. But there was also recognition by the Greeks more than 2000 years ago that legumes contributed something to the systems that improved the performance of cereals.

The background of soil conditions and management, unfortunately, does not appear to be documented in this literature. Circumstantial evidence that significant soil erosion has occurred can be seen at sites of cities dating back 1800 to 2000 years where the foundation stones of buildings are exposed. Doorsteps are often 60 cm or more above the soil surface of today suggesting that substantial soil loss may have occurred. In north west Syria, hillsides and mountains which are recorded as carrying woodlands dominated by *Quercus* spp within the last 200 years, are now bare rock. What seems to be unknown is when this soil loss occurred; some may have happened slowly but consistently through centuries, or it may have been quite recent. Also unknown to us is where the eroded soil has been deposited; it may be that it remains in the surrounding valleys and plains where today's farming is concentrated.

Developments in the last four or five decades have brought rapid change. National policies have encouraged industrialization, created pricing structures for agricultural products which favor urban populations at the expense of farmers, and neglected infrastructure development in rural areas. Widely, agricultural production as a proportion of GDP has declined, agriculture has stagnated, and there has been migration from farming communities to cities (Tully, 1989). At the same time populations have expanded rapidly and *per capita* food production has declined, the affluence of urban dwellers has increased creating greater demand, especially for animal products, and food imports have escalated. This has put pressure on the farming systems, with expansion of cropping to ever less suitable land, and increases in the rate of use of scarce, and possible irreplaceable, water resources. Animal populations have increased on areas of natural grazing with deterioration of vegetation, most particularly in arid steppe lands. Cereal-fallow rotations have been replaced with continuous cereal cropping, especially of barley as a livestock feed to replace degraded pastures.

The factor having perhaps the greatest physical impact is the introduction of tractors and mechanization of many farming operations. This has allowed the expansion of cultivation to areas which are marginal for cropping because they are too dry, or too steep, or the soil is shallow and stony. Tillage, in particular, has been intensified both in the severity of the tillage practiced and in its frequency. Land in areas with less than 250 to 200 mm of annual rainfall, which previously has never been cultivated, is being cropped annually with barley, the stubble of which is grazed. We suspect that destruction of perennial plants through cultivation, and powdering of the soil surface by the action of animal hooves during grazing are causing increased wind erosion, but this is not easy to quantify. Changes in land tenure have lead to fragmentation of ownership, and often the existence of small holdings in narrow strips up and down hillsides. This confines the direction of tillage to that of the slope and makes it difficult to conceive of ways to introduce water erosion control measures.

All of these changes raise grave concerns for the maintenance of the resource base and for the continued productivity of the farming systems of the region. It is against this background that ICARDA is working with national agricultural institutions in an endeavor to increase food production and the welfare of the farming community. Given the characteristics of the climate of the region, the components of the farming systems remain very much as they have always been. Our aim, therefore, has to be to endeavor to increase the potential productivity of the individual crop and pasture plants, and to devise management strategies that will restore stability to these ancient systems.

# II. The Environment of West Asia and North Africa

Since it differs appreciably from tropical regions on which this meeting largely focuses, it seems appropriate to briefly describe the environment of the region within which ICARDA works.

## A. Soils

The predominant agricultural soils of the region are derived from limestone residuum, and hence are calcareous (FAO, 1978, 1979; Kassam, 1981, 1988). Cooper et al. (1987) have discussed the significance of this from an agronomic viewpoint, with particular reference to the effects of high pH on the availability of phosphorus and micronutrients. Matar et al. (1992) have considered the chemistry of these soils at a more fundamental level. In general, soil organic matter levels are low, and phosphate and nitrogen deficiencies are widespread through the region. Responses to micronutrients have been recorded but are not common in rainfed agriculture. Boron toxicity has recently been recognized as a problem of cereal production in some areas of west Asia.

## B. Climate

The climates of west Asia and north Africa (broadly, the Mediterranean region) have been described in numerous publications (eg. de Brichambaut and Wallen, 1963; Unesco/FAO, 1963; Kassam, 1981, 1988), and summarized in an agronomic context by, for example, Cooper et al. (1987). They are characterized by cool to cold wet winters and warm to hot arid summers. Locally, conditions are modified considerably by continental (in west Asia) or maritime (in north Africa) influences, and by topography (Figure 1). Precipitation, whether as rain, or snow in highland areas of west Asia, is variable in space and unreliable in time, and often deficient in amount. Broadly, coastal areas are wettest, and the amount decreases rapidly with distance inland. On average, rain commences in the autumn (September-October), reaches a peak in January or February and

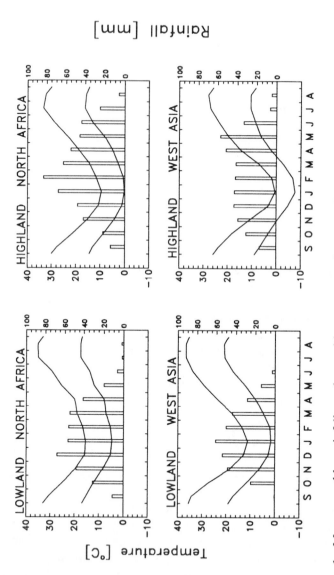

**Figure 1.** Mean monthly rainfall, and monthly mean maximum and minimum temperatures in contrasting environments; north African lowland and highland; and west Asian lowland and highland.

decreases rapidly until April. This pattern changes in the east where in parts of Turkey, Iraq and Iran, the peak is delayed to April-May and some rain can be expected in June and July. However, throughout the region, year to year variability in this distribution is experienced. The first rains may be delayed by as much as two or three months, and a similar uncertainty attaches to the time the rains end.

Winter temperatures are mildest in the north African lowlands, are somewhat more severe but similar in the highlands of north Africa and the lowlands of west Asia, and are very severe, with extended periods of snow cover, in the plateaux of west Asia (eg., Turkey, northern Iran). Summer temperatures are hottest in the lowlands of west Asia, are modified by maritime influences in much of north Africa, and are least extreme in the west Asian highlands.

Although the patterns of rainfall differ across the region, the influence of temperature on the development rate of crops means that almost everywhere flowering takes place after the peak rainfall season. Crops thus depend in large measure on water stored in the soil to complete their life cycle and almost universally suffer some degree of water stress, often severe, during seed/grain development. Yields are strongly governed by how much water can be stored during the rainy season, and soil depth, as it determines water-holding capacity, can be an important feature. The amount of water stored can also be influenced strongly by the nature of the rainfall. Often a significant proportion of the rain comes as small events (5 mm or less) which make little contribution to soil storage. At Tel Hadya, our main research station, for example, approximately 10 percent of rain occurs as events of $\leq 2$ mm, and a further 15 to 25 percent in falls of 2.1 to 5 mm. Class A pan evaporation on raindays averages about 2 mm, so rain in these categories has limited effectiveness.

## C. Farming Systems

The farming systems of the region are based largely on cereals and livestock, chiefly sheep and goats. Wheat is the main cereal of wetter areas ($> ca$ 350 mm) where it is grown in rotation with either fallow, or a range of other crops including barley (*Hordeum vulgare*), faba bean (*Vicia faba*), chickpea (*Cicer arietinum*), and lentil (*Lens culinaris*) as winter sown species, or melon (*Citrullus* spp.), sunflower (*Helianthus annuus*), cotton (*Gossypium* spp.), maize (*Zea mays*), sorghum (*Sorghum bicolor*), sesame (*Sessamum indicum*), or a variety of vegetables sown in spring or summer. Barley, used mainly for feed in systems where livestock form the major output, occupies the drier rainfall zones and is being extended into ever more arid areas to replace the now degraded vegetation of native steppelands. It is rotated with fallow, or increasingly grown in continuous barley systems. In both the wheat-based and barley-based systems fallows may be kept weed-free, or may be the traditional weedy fallows of north Africa where they provide winter and spring grazing. The functioning of these systems has been described in detail by Cooper et al. (1987) and Tully (1989).

The work described in this paper is carried out at Tel Hadya, in northwest Syria. This paper is not an exhaustive coverage of long-term trials conducted by ICARDA, but rather it focuses on three trials in which the work is of most relevance to this meeting.

As with all agronomic work, it must be recognized that there are elements of site- and season-specificity in the results. They are thought to be of general relevance to lowland west Asia, but cannot be extrapolated beyond that environment with any degree of confidence. However, the questions addressed in the experimental program have wide relevance and, as well as the experimental role, the program serves to raise awareness of, and interest in, issues of management of soil and crops within the systems of the region.

## III. Trial Description

Data presented in the following sections are drawn from several trials, only one of which as yet can be described as long-term by the criterion set for this meeting. Others are on-going, and it is hoped that they will be maintained for a sufficient length of time to yield data of the type needed to provide a sound basis for assessing the sustainability of the practices being tested.

The trials which are discussed here are carried out at the main research station of ICARDA, Tel Hadya, with an average annual rainfall of 330 mm. The soils have been classified as Calcixerollic Xerochrepts (J. Ryan, pers. comm.). Typically, they swell when wet and develop deep wide cracks when dry. These soils are very friable and offer few problems of management, except that they become very sticky when wet, which can cause drill blockages during sowing, especially where crop residues are retained. Their depth varies, depending partly on landscape, from 40 - 50 cm to > 2 m. Some depth variability exists in all of the trials described below. The soils are deficient in available phosphorus and nitrogen; organic matter contents are $\leq 1\%$, and the $pH_{sat}$ is approximately 8.

All the trials involve crop rotations with either two- or three-crop sequences. It should be clear from the description of the climate that the unimodal pattern of rainfall limits cropping to one crop on any one plot each rainfall 'year' (September to August). A two-course rotation therefore occupies two rainfall years (or seasons), and a three-course rotation involves three years. However, because variability in rainfall from season to season results in wide fluctuations in yields, all crops are grown each season, i.e.,"phased entry" is practiced.

Crops are established after the first rain in the autumn, except where treatments dictate otherwise. The other exception is the frost-sensitive water melon which, is sown on a sparse grid (2 x 3 m) in the spring to grow on water stored by fallowing during the winter. This crop is sown only when there is at least 1 m of wet soil in early April; otherwise the plots are fallowed. General management follows recommendations derived from other work for fertilizer applications, seeding rates and depths, and weed and pest control. Of the crops common to most trials, cereals are combine harvested, and, except where residue management treatments are included, straw is baled or collected loose and stubbles are grazed. Chickpea is also combine-harvested and residues returned, but lentil is hand-harvested, all above-ground biomass being removed.

## A. Timing of Tillage

This experiment was established in the 1978-79 season, and therefore qualifies as a long-term trial (Trial A). Its original purpose was to study weed control methods in a wheat-lentil rotation, one crop being grown each year, i.e., no phased entry. After seven years the weed control studies were ended, and in 1985-86 the trial was redesigned: tillage treatments were retained; weed control treatments, superimposed on the tillages as split plots, were replaced by uniform control; and phased entry was introduced by splitting the plots across the previous treatments. Plot size is 30 x 30 m.

Two tillage treatments are included, 'conventional tillage' involving deep discing (25 cm) followed by appropriate secondary tillage to prepare a seed bed, and zero till. Conventional tillage represents practices widely used by farmers on similar soils in the wetter zones (> *ca* 300 mm) of Syria and Jordan. Experimentally, it is imposed at three times:

Early (CTE): all operations, including sowing, carried out before the first rain.

Middle (CTM): discing carried out before rain, secondary tillage and sowing after the first rain.

Late (CTL): all operations after the first rains.

Zero till Early (ZTE) and Middle (ZTM) are direct drilled at the same time as the corresponding conventional treatments.

## B. Tillage Methods Trial

Established in the 1985-86 season to test the hypothesis that tillage can be reduced without prejudice to crop yields, this trial com (B) pares two deep tillages (disc and chisel) with minimum (tyned cultivator) and zero till. It is conducted within three course rotations of bread wheat (*Triticum aestivum*)-chickpea-water melon and durum wheat (*Triticum turgidum* var. durum)-lentil-watermelon. The second year of the third cycle of the rotations has just been completed.

Tillage sequences are based on farmers' practice, and deep tillages are carried out before the legume and water melon phases. Because light tillage for weed control is used during the growth of watermelon, no further preparation is needed for establishment of the following wheat crop. Basic plot size is 50 x 300 m, each tillage treatment being 12.5 x 300 m. Plots are split in the other direction at 75 m intervals to include auxiliary treatments as questions arise, but these will not be further considered.


## C. Productivity of Cropping Systems

As ICARDA began to develop 'new' technology it was felt that it should be applied on-station to evaluate the biological potential of cropping systems in the environment. Consequently, two-course rotations of durum wheat with: fallow (Wf), wheat (Ww), lentil (Wl), chickpea (Wc), watermelon (Ws), vetch (*Vicia sativa*) (Wv), and medic pasture (*Medicago* spp.) (Wm) were established in 1983-84 and grown for two seasons without agronomic inputs. In 1985-86 management according to recommendations was introduced, including the use of new cultivars as proven material becomes available, and auxiliary treatments involving four levels of nitrogen application on the wheat (0, 30, 60, 90 kg ha$^{-1}$) and three intensities of grazing of wheat stubble (heavy, moderate, and zero) were superimposed in a split-strip design. Main plot size is 36 x 150 m, nitrogen and residue sub-plots being 36 x 37.5 m and 12 x 150 m, respectively. Medic pastures and vetch are grazed, medic for as long as possible at 10 ewes (and lambs) ha$^{-1}$, and vetch at $\approx$25 to 30 lambs ha$^{-1}$ during the spring. Wheat stubble is grazed by temporarily fencing sub-plots and introducing a large flock of sheep for one or two days. This will be referred to as Trial C.

Data recorded in the trial include: biomass, grain and seed yields of wheat and the food legumes; grazing days and animal production from grazed legumes; soil water balance of three rotations, Wf, Wc, and Wm, at two levels of nitrogen application, 0 and 90 kg ha$^{-1}$; soil nitrogen concentrations and organic matter at the end of summer (occasionally to date, but regular measurements are planned for the future); nitrogen uptake by the grain crops; residues after grazing and immediately before planting; and, quantities, costs, and prices of inputs and outputs.

The fourth cycle of the rotations under the higher level of management has just been completed, but data for the last season are not yet available, so results for 7 years are reported here.

More detailed descriptions of these experiments can be found in Annual Reports of the Farm Resource Management Program (ICARDA, 1985, 1989, 1990, 1991).


# IV. Experimental Results and Discussion

Because of the overriding importance of water as the factor most limiting productivity in the environment, the major focus of the experimental program has been quantification of water balances. Only minimal work has been undertaken on other physical parameters of the soil.

## A. Seasonal Conditions

### 1. Rainfall

Rainfall in the seasons 1985-86 to 1991-92 is shown as cumulative rain through the season in Figure 2. These data illustrate year-to-year variability in the time of start and end of the rains as well as variability in seasonal totals. The data also portray the aridity of the summer. Of the seven seasons, three had average rainfall (330 mm), one was wet, and one was somewhat dry and two extremely dry.

Small rain events early in the season, interspersed with long dry spells, are ineffective. The preferred time of planting is mid-November, but in two of the seven years, crop establishment was delayed until the end of December, and in 1990-91 until the end of January. The most common time for the end of the rains is in March, but in 1988-89 no effective falls were received after the end of December. The length of the historical weather record for the area is relatively short, but we estimate that the wettest season was about a 1-in-30 years event. The probability of termination as early as December defies estimation and this, it seems, can be regarded as a singular happening. The period from 1985 to 1992 spans a substantial amount of the variability in the historical rainfall data, although the record does contain lower seasonal totals than have been experienced in these seven seasons.

### 2. Temperature

Temperatures in the same seasons are summarized as thermal time (base temperature 0°C) in Figure 3. Variations in the rate of accumulation of thermal time have resulted in as much as five weeks difference in the time of anthesis and four weeks in harvest maturity of wheat from the hottest to the coolest seasons. There have been, on average, 39 days per season (range 17 to 57) with minimum temperature $\leq 0°C$, with an extreme low of -9.9°C. Frosts, not uncommonly, occur as late as mid-March (occasionally early April) when, if they are preceded by warm weather, crop development may be advanced and damage to flowering organs severe. A minimum temperature of -8.9°C on March 17, 1989, for example, caused extensive damage to both cereal and legume crops.

Rainfall and temperatures are, of course, not independent. Severe frosts tend to accompany and exacerbate drought, as is to be expected of semiarid regions where radiation frosts result from cloudless conditions. The risk of late frosts, and the termination of rain, in the main, by the end of March mean that there is a narrow flowering window for crops if frost damage is to be avoided and drought stress during seed/grain filling minimized.

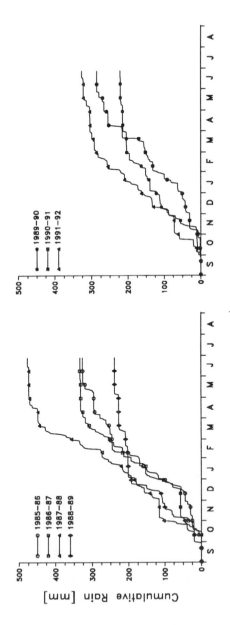

**Figure 2.** Cumulative rainfall during six cropping seasons (September to August) at Tel Hadya, NW Syria.

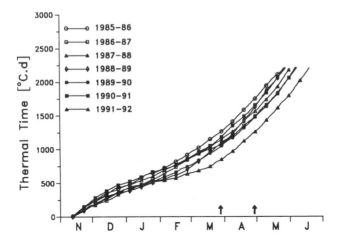

**Figure 3.** Temperature regimes at Tel Hadya, NW Syria, expressed as thermal time (°C.d) from mid-November. (A base temperature of 0°C was used to calculate thermal time. Arrows indicate the approximate time of anthesis of wheat in the warmest and coolest seasons.)

## 3. Soil Water

The variability of the seasonal rainfall is reflected in the total soil water at the time of maximum recharge of the profile in each of six seasons (Figure 4); the data are taken from the wheat phase of a wheat-chickpea rotation. The solid lines show the air-dry profile (left) and the drained upper limit (right). The repeatability of the determination of air dryness can be judged from the scatter about the line at depths greater than 60 cm. Below about 45 cm this profile is equivalent to the lower limit of plant-extractable water (Ritchie, 1980); in the upper (0 to 30 cm) horizons the lower limit is approximately 0.18 cc cc$^{-1}$ (27 mm/15 cm). The drained upper limit was determined under a bare fallow in the wet 1987-88 season.

Clearly, the greatest limitation at this site is the supply of water and not the potential storage capacity of the soil. Even with a much shallower soil the profile would have filled in only one of these seasons. By inference, much higher average rainfall would be required for drainage beyond the root zone to happen regularly on the high clay soils. This has implications for how we think about nitrogen dynamics, fertilizer strategies, and indeed breeding strategies for drought adaptation.

The pattern of recharge and discharge of the 'labile' water in the whole profile (0-180 cm) or the upper profile (0-45 cm) illustrates (Figure 5) that during the summer there are several months when the soil is dry. During the winter, when recharge occurs, soil temperatures are quite low. Thus, there is possibly a

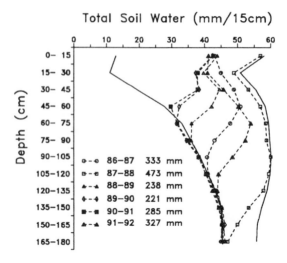

**Figure 4.** Total soil water (mm per 15-cm soil depth) at the time of maximum soil profile recharge, and total seasonal rainfall, in six seasons at Tel Hadya, NW Syria. Total rainfall for each season is also shown. Solid lines are air dry (left) and drained upper limit (right) profiles.

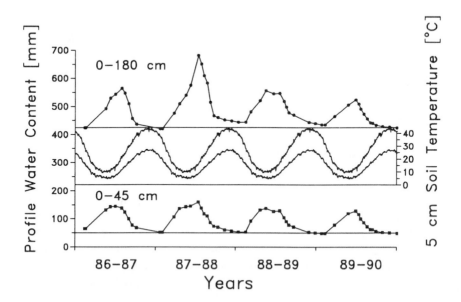

**Figure 5.** Total soil water in the profile (0-180 cm) and upper profile (0-45 cm) in two cycles of a crop rotation, and mean daily maximum and minimum temperatures in the 0-5 cm soil layer at Tel Hadya, NW Syria.

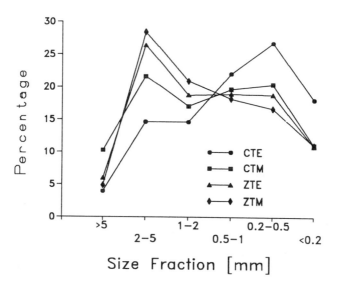

**Figure 6.** The distribution of aggregate size fractions in soil subjected for nine years to three tillage treatments.

relatively short period each season when both moisture and thermal conditions are highly suitable for biological activity, and microbially mediated changes within the soil probably can be expected to occur slowly, especially near the surface.

## B. Tillage

### 1. Timing of Tillage

The effect of the timing of tillage (Trial A) on soil structure is shown in Figure 6. The soil was sampled during the summer when the moisture content of the top 10 cm was between 5.5 and 7.5 percent by weight. The samples were dry sieved mechanically for 10 minutes, using a Endecott sieve shaker, to separate aggregates into six size classes. Corrections for the presence of gravel were made in the three larger fractions which were dispersed in a solution of sodium meta-phosphate, dried and re-sieved. Where all tillage is carried out when the soil is dry (CTE) there has been a decrease in the proportion of aggregates in the 1 to 2 and 2 to 5 mm fractions and a concurrent increase in the proportion at 0.2 to 0.5 mm and ≤ 0.2 mm. Trends are similar but less clear where only the primary tillage is done in dry soil (CTM).

Infiltration was measured in the same trial using a rainfall simulator (Asseline and Valentin, 1977) on plots from which lentil had been hand-harvested and all

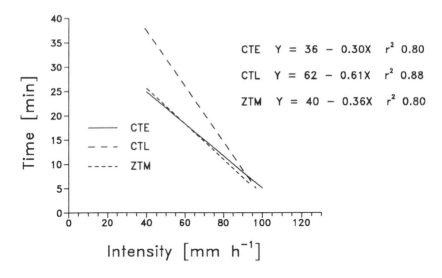

**Figure 7.** Time for surface saturation at a range of simulated rainfall intensities on untilled soil surfaces following nine years of tillage treatments (see text).

plant material therefore had been removed. The work was carried out during the summer, prior to tillage, when the soil was at similar water contents to those above. A late storm had caused the formation of a weak crust, and this was gently and uniformly broken with fingertips before the simulation began. The data recorded were the time for surface saturation to occur under a range of rainfall intensities from 35 to 100 mm h$^{-1}$ (Figure 7). The results appear compatible with the differences found in the laboratory between the conventional tillages, but reasons for the slow infiltration in the zero till plots remain obscure.

These results are exploratory only. Much more detailed work is needed in this trial before it will be possible to draw conclusions on mechanisms associated with the apparent effects of the tillage practices on soil structure.

## 2. Tillage and Water Balance

Three arguments are advanced for widespread use of deep tillage with either disc or moldboard plows in farming systems in the region: that more water is stored because of increased infiltration; deep burial of weed seeds allows better weed control; and deep plowing provides for disruption of a compaction layer in the soil. We have addressed the first of these by studies of the water balance of the four tillage treatments in Trial B.

Soil water is monitored gravimetrically from 0 to 15 cm and by neutron scattering techniques in the remainder of the profile; neutron probe access tubes

**Table 1.** Maximum quantity of water (mm) stored during the rainfall season in four tillage treatments at Tel Hadya, NW Syria

| Season | ZT | DD | CP | DF |
|--------|-----|-----|-----|-----|
| Chickpea | | | | |
| 86-87 | 150 | 142 | 146 | 146 |
| 87-88 | 248 | 209 | 226 | 269 |
| 88-89 | 113 | 111 | 92 | 98 |
| 89-90 | 84 | - | 87 | - |
| 90-91 | 80 | 61 | 76 | 71 |
| 91-92 | 168 | 130 | 142 | 145 |
| | | | | |
| Lentil | | | | |
| 86-87 | 143 | 137 | 137 | 141 |
| 87-88 | 267 | 210 | 220 | 236 |
| 88-89 | 114 | 90 | 98 | 105 |
| 89-90 | 97 | 93 | 82 | 83 |
| 90-91 | 72 | 55 | 60 | 75 |
| 91-92 | 129 | 115 | 136 | 135 |
| | | | | |
| Fallow | | | | |
| 88-89 | 92 | 77 | 81 | 88 |
| 89-90 | 91 | 89 | 80 | 86 |
| 90-91 | 92 | 82 | 89 | 92 |
| 91-92 | 140 | 126 | 110 | 121 |

being permanently installed to a depth of 180 cm. It is characteristic of neutron probe data that minor variation in the soil profile from point to point causes variability in the estimates of soil water content unless calibration is carried out for each measurement point. With permanently installed access tubes calibration of each tube is not possible. However, where we have permanent tubes, we find that most of the spatial variation is very repeatable over time (ICARDA, 1988), and can therefore largely be eliminated by considering, not absolute water content, but change in water content with time. This is referred to as 'stored soil water' and is calculated as the change of soil water status from a first reading for a season taken at the end of the summer and, so far as possible, before the first rain of the season. Negative values of stored soil water reflect water stored in the soil at the time of the first reading.

The maximum amount of water stored in the profile under growing chickpea and lentil crops and under the winter fallow which precedes watermelon (Table 1) shows no clear trends. If anything, storage following deep discing tends to be least (10 of 16 cases), but quantities are similar amongst the other three tillages. Where trends are apparent, the quantities involved are, in the main, small and have yet to translate into a difference in crop yields. Certainly, in the conditions of the trial, there is no advantage of deep tillage with either a disc or chisel plow in terms of water storage.

A more consistent trend is for more water to be stored under the legume crops than under fallow, although, again, the amounts are small. This can be attributed to the fact that the fallowed plots remain rougher through the major part of the rainfall season and therefore have a greater surface area from which evaporation can take place.

In this trial there has been no effect of tillage on weed populations (data not shown). This may be related to the weed flora which is almost entirely broadleafed species. Where there are reports from the region that weed control is improved by moldboard plowing the dominant species were grasses, in particular, *Bromus* species ( Durutan *et al.*, 1991).

## C. Water Relations of Rotations

### 1. Seasonal Water Balances

Patterns of soil water recharge and discharge during rainfall (crop) seasons are shown for four two-course rotations in Figure 8. Data for the rotations of wheat with fallow, chickpea, and medic are taken from trial C, and those for wheat with lentil from trial A. Positive values of stored soil water illustrate storage of rainfall after the first reading, while negative values indicate water present in the profile at the time of the first reading, which is subsequently lost by evaporation from the soil surface or used in transpiration. In general, this is water stored at depth in the soil throughout the summer and it can be expected that its loss will be by transpiration during crop growth.

Data for two cycles of the rotations illustrate the pattern observed for the different crop sequences (Figure 8). Fallow efficiency, calculated as the proportion of the rainfall in one season stored through the following summer, has varied from about 10 to 40%. The most important factor determining the efficiency is the depth to which the soil is wetted; water needs to be stored below about 60 cm to be carried through the summer. The depth of wetting is determined, in turn, not only by the quantity but also by the distribution of the rainfall. A few substantial falls are more effective than numerous small rain events. When, as in 1987-88, water was stored to the full depth of the profile the fallow provided almost 50% of the water used by wheat in the drought of 1988-89.

Lentil is a short season crop normally harvested about one month earlier than either wheat or chickpea. It is therefore less exposed to the severe evaporative demands which can occur late in the cropping season (May and early June) and does not dry the soil to the same extent as other crops. Water remaining in the soil (Figures 8a, 8c) is frequently retained through the summer and becomes available to wheat during the next phase (Figures 8b, 8d), providing some degree of buffering against season to season variability in rainfall. Chickpea and medic

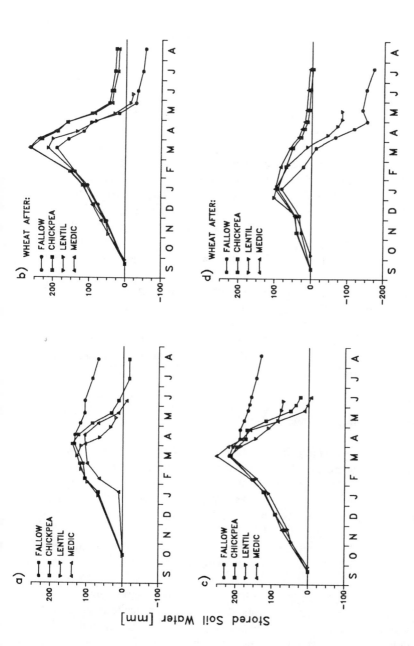

**Figure 8.** Patterns of soil water recharge and depletion (mm) under legumes in a wet (a) and average (c) rainfall season and under the following season's wheat crop (b,d) at Tel Hadya, NW Syria. Positive values show storage and use of current season rainfall; negative values indicate use of water stored from the previous season.

pastures, in contrast, dry the soil throughout the 180 cm of the measured profile to at least the same extent as wheat, so that in rotations with these legumes each phase is entirely dependent on the rainfall of a given season.

## 2. Wheat Grain Yield and Water Use Efficiency

Seven years' data for grain yield of wheat reflect these patterns of water use (Table 2). The three crop sequences which rely on current season rainfall only, wheat after chickpea, wheat after medic, and, by inference, continuous wheat, consistently yield least. After lentil, or vetch which has a similar growth habit, yields are about 80 percent of those after fallow or the sparse watermelon crop. However, when the fallow efficiency is taken into consideration, and water lost through soil evaporation during the fallow is included in the estimation of water-use efficiency, the Wf rotation in general shows a lower efficiency at both levels of N fertilization than either Wc or Wm (Table 3). Wheat-lentil is omitted from this table as the trial in which soil water measurements are made uses a different cultivar.

One consequence of five years of the continuous wheat treatment was decimation of crops in 1987-88 by larvae of the wheat ground beetle, *Zabrus tenebroides*. Crops were re-sown late and yielded poorly. This was the only crop sequence affected, and the event sounds a warning for governments of the region whose policies of price incentives on wheat are encouraging farmers toward continuous wheat cropping.

## D. Soil Nitrogen and Rotations

Differences among the rotations in the concentration of total nitrogen in the top 20 cm of soil were apparent by the time the first measurements were made in the late summer of 1989 (Table 4). Levels of nitrogen are greater where a legume is included in the sequence, the order being Wm > Wv ≈ Wl > Wc. The data also indicate an increase in the soil nitrogen level with increasing levels of N fertilizer application (Table 5), but since both measurements were preceded by seasons of low rainfall these differences may be due to residual fertilizer nitrogen not used by drought-affected crops. They therefore need to be treated with some caution.

It is tempting to conjecture that all of the legumes are contributing nitrogen to these systems. The evidence appears strongest for the medic pasture, and increases following medic have recently been reported from another trial on Tel Hadya (White et al., 1993). However, other work indicates that the food legumes fix, on average, only about the quantity of N that is contained in the seed, or approximately seventy percent of their requirements (Beck et al., 1991). Where the crops are hand-harvested, as is almost always the case for lentil and often the case for chickpea, negative N balances have been recorded. Data for

**Table 2.** Grain yield (t ha$^{-1}$) of durum wheat (cv. Cham 1) in seven two-course rotations for the crop seasons 1985-86 to 1991-92 at Tel Hadya, NW Syria

| Season | Crop Rotations | | | | | | | s.e. | n |
|---|---|---|---|---|---|---|---|---|---|
| | Wf | Ww | Wl | Wc | Ws | Wv | Wm | | |
| **Grain** | | | | | | | | | |
| 85-86 | 2.57 | 1.38 | 2.04 | 2.00 | 2.71 | 2.13 | 1.32 | 0.162 | 12 |
| 86-87 | 1.92 | 1.01 | 1.65 | 1.35 | 1.96 | 1.62 | 1.18 | 0.098 | 12 |
| 87-88 | 4.11 | 0.87[1] | 3.91 | 2.94 | 3.61 | 3.93 | 3.24 | 0.245 | 36 |
| 88-89 | 1.80 | 0.31 | 0.94 | 0.40 | 1.04 | 0.98 | 0.29 | 0.118 | 36 |
| 89-90 | 1.14 | 0.55 | 0.73 | 0.62 | 1.60 | 0.74 | 0.50 | 0.074 | 36 |
| 90-91 | 1.56 | 0.97 | 1.25 | 0.76 | 1.64 | 1.17 | 0.93 | 0.086 | 36 |
| 91-92 | 2.70 | 1.92 | 2.21 | 2.19 | 2.77 | 2.31 | 2.24 | 0.123 | 36 |
| **Mean** | 2.26 | 1.00 | 1.82 | 1.46 | 2.19 | 1.84 | 1.39 | 0.051 | 252 |

[1] This result omitted from the statistical analysis.

**Table 3.** Water use (WU; mm), grain yield (GY; t ha$^{-1}$), and water use efficiency (WUE; kg ha$^{-1}$ mm$^{-1}$) of durum wheat (cv. Cham 1) at two levels of nitrogen fertilization in three rotations at Tel Hadya, NW Syria

| Year | N | ------Wheat/F------ | | | ------Wheat/C------ | | | ------Wheat/M------ | | |
|------|------|------|------|------|------|------|------|------|------|------|
|      |      | WU | GY | WUE | WU | GY | WUE | WU | GY | WUE |
| 86-87 | 0 |     | 1.71 |     | 308 | 1.19 | 3.7 | 307 | 1.01 | 3.3 |
|       | 90 |    | 2.04 |     | 322 | 1.49 | 4.6 | 329 | 1.25 | 3.8 |
| 87-88 | 0 | 686 | 2.99 | 4.4 | 414 | 1.90 | 4.8 | 419 | 2.38 | 5.7 |
|       | 90 | 760 | 4.86 | 6.4 | 453 | 3.81 | 8.4 | 460 | 3.79 | 8.3 |
| 88-89 | 0 | 660 | 1.36 | 2.1 | 183 | 0.34 | 1.8 | 178 | 0.36 | 2.0 |
|       | 90 | 689 | 2.10 | 3.0 | 178 | 0.44 | 2.5 | 184 | 0.25 | 1.4 |
| 89-90 | 0 | 412 | 0.94 | 2.3 | 221 | 0.72 | 3.3 | 208 | 0.62 | 3.0 |
|       | 90 | 428 | 1.26 | 2.9 | 223 | 0.63 | 2.8 | 206 | 0.55 | 2.7 |
| 90-91 | 0 | 391 | 1.14 | 4.2 | 246 | 0.78 | 3.2 | 250 | 0.90 | 3.6 |
|       | 90 | 387 | 1.54 | 5.7 | 253 | 0.78 | 3.1 | 244 | 0.91 | 3.7 |
| 91-92 | 0 | 458 | 2.02 | 4.4 | 289 | 1.78 | 6.2 | 268 | 2.27 | 8.5 |
|       | 90 | 497 | 3.09 | 6.2 | 299 | 2.44 | 8.1 | 278 | 2.24 | 8.0 |

**Table 4.** Total nitrogen concentration (ppm) in the top 20 cm of soil in seven crop sequences at Tel Hadya, NW Syria

| Rotation | 1989 | 1991 |
|----------|------|------|
| Wf | 686 | 688 |
| Ww | 698 | 678 |
| Wl | 737 | 718 |
| Wc | 716 | 696 |
| Ws | 668 | 651 |
| Wv | 739 | 738 |
| Wm | 806 | 792 |
| s.e. | 10.5 | 12.8 |

**Table 5.** Total nitrogen concentration (ppm) in the top 20 cm of soil at four levels of applied nitrogen fertilizer (kg ha$^{-1}$) at Tel Hadya, NW Syria

| N Applied | 1989 | 1991 |
|-----------|------|------|
| 0 | 703 | 685 |
| 30 | 701 | 692 |
| 60 | 732 | 718 |
| 90 | 749 | 740 |
| s.e. | 8.8 | 10.5 |

these rotations may therefore be reflecting only a slower rate of decline in soil N than in those with fallow and watermelon. More time is required before long-term trends can be detected and the question resolved.

In the shorter term, basic studies on biological nitrogen fixation, the fate of applied fertilizer nitrogen and nitrogen and carbon cycling have been commenced within the trial in collaboration with the Soil Science Department of the University of Reading. To date, these do not include the rotations where the legumes are grazed, but given the importance of animals within the farming systems of the region (Cooper et al., 1987) and the key role which forage and pasture legumes could play in the systems of the future, it is desirable that they be included as soon as possible.

# V. Future Needs

Future work falls into three categories. The first is to intensify the level of existing studies of the effects of tillage practices and crop residue management on soil physical properties and the structural integrity of the soils of the higher rainfall areas. To date this has been limited by available resources, especially human, but with increasing recognition of the urgent need to improve soil management it is hoped that this can be remedied.

The second is to extend this work to other parts of the region. A start has already been made by colleagues in Jordan, Algeria, Morocco, and Turkey, but these efforts deserve greater support than they currently command.

The third is to expand work on soil management to other soils and agro-ecologies where the problems, current and potential, are greater. Soils of the drier areas tend to contain less clay and more silt, to form crusts which impede water infiltration and crop establishment, and to have very low organic matter contents. Their structure is fragile, and they are subject to erosion by both wind and water. Farmers are often poorer and have fewer options for changed management than those in the wetter areas.

Two trials, already established in northwest Syria but in only their second or third year, address primarily the question of wind erosion in the drier areas. One is designed to test the feasibility of some retention of stubble and/or the timing of tillage as measures to limit wind erosion after the surface structure of the soil has been destroyed by the action of animal hooves during the grazing of barley stubble. The second is designed to explore the use of hedges of native saltbush (*Atriplex halimus*) to both reduce wind erosion and provide additional grazing in the late summer and autumn when a major feed gap exists. Both include comparisons of continuous barley with a barley-forage legume rotation which small-plot trials indicate will increase the productivity of systems in the area.

# References

Asseline, Y. and C. Valentin. 1977. Construction et mise au point d'un infiltrometre a aspersion. *Series Hydrologie* 15:321-344.

Beck, D.P., J. Wery, M.C. Saxena, and A. Ayadi. 1991. Dinitrogen fixation and nitrogen balance in cool-season food legumes. *Agron. J.* 83:334-341.

Brichambaut, G. de and G. Wallen. 1963. A study of agroclimatology in the semi-arid and arid zones of the Near East. *Technical Note 56*. WMO, Geneva, Switzerland.

Cooper, P.J.M., P.J. Gregory, D. Tully, and H.C. Harris. 1987. Improving water use efficiency of annual crops in the rainfed farming systems of west Asia and north Africa. *Expl. Agric.* 23:113-158.

Durutan, N., M. Guler, M. Karaca, K. Meyveci, A. Avcin, and H. Eyuboglu. 1991. Effect of various components of the management package on weed control in dryland agriculture. p 220-234. In: Hazel C. Harris, P.J.M. Cooper and M. Pala (eds.), *Soil and Crop Management for Improved Water Use Efficiency in Rainfed Areas*. Proc. of an international workshop held in Ankara, Turkey, May 15-19, 1989. ICARDA, Aleppo, Syria.

FAO (Food and Agriculture Organization of the United Nations). 1978. Report on the agro-ecological zones project, Vol 1. Methodology and results for Africa. *World Soils Resources Report 48/1*. FAO, Rome, Italy.

FAO. 1979. Report on the agro-ecological zones project. 2. Results for South West Asia. *World Soils Resources Report, 48/2*. FAO, Rome, Italy.

ICARDA. 1985. Annual Report 1984. Aleppo, Syria.

ICARDA. 1988. Farm Resource Management Program. Annual Report for 1987. Aleppo, Syria.

ICARDA. 1989. Farm Resource Management Program. Annual Report for 1988. Aleppo, Syria.

ICARDA. 1990. Farm Resource Management Program. Annual Report for 1989. Aleppo, Syria.

ICARDA. 1991. Farm Resource Management Program. Annual Report for 1990.

Kassam, A.H. 1981. Climate, soil and land resources in North Africa and West Asia. *Plant and Soil* 58:1-29.

Kassam, A.H. 1988. Some agroclimatic characteristics of high-elevation areas in north Africa, west Asia, and south-east Asia. p. 1-32. In: J.P. Srivastava, M.C. Saxena, S. Varma, and M. Tahir (eds.), *Winter Cereals and Food Legumes in Mountainous Areas*. Proc. of an international symposium 10-16 July, 1988, Ankara, Turkey. ICARDA, Aleppo, Syria.

Matar, A., J. Torrent and J. Ryan. 1992. Soil and fertilizer phosphorus and crop responses in the dryland Mediterranean zone. *Adv. in Soil Sci.* 18:81-146.

Ritchie, J.T. 1981. Water dynamics in the soil-plant-atmosphere system. *Plant and Soil* 58:81-96

Tully, D. 1989. Rainfed farming systems of the Near East region. In: C.E. Whitman, J.F. Parr, R.I. Papendick, and R.E. Meyer (eds.), *Soil, Water, and Crop/Livestock Systems for Rainfed Agriculture in the Near East Region*. Proc. of the workshop at Amman, Jordan, January 18-23, 1986. USDA, Washington, D.C.

Unesco/FAO (United Nations Educational, Scientific and Cultural Organization). 1963. Bioclimatic map of the Mediterranean zone. *Arid Zone Research 21*. 58pp. UNESCO, Paris, France.

Watson, A. 1974. The Arab agricultural revolution and its diffusion. 700-1100. *J. Econ. Hist.* 34, No 1.

White, H. 1963. Roman agriculture in north Africa. *Nigeria Geog. J.* 6:39-49.

White, H. 1970. Fallowing, crop rotation and crop yields in Roman times. *Agric. Hist.* 44:281-290.

White, P.F., N.K. Nersoyan, and S. Christiansen. 1993. Nitrogen cycling in dry Mediterranean zones: changes in soil N and organic matter under several crop/livestock production systems. *Aust. J. Agric. Res.* (in press).

# D. Synthesis and Future Priorities

# An Integrated Approach to Soil Management Experiments

E. Pushparajah and H. Eswaran

## I. Introduction

Long-term experiments on soil management for food crops in the tropics are limited, and when they exist are often conducted by IARC's independently or in association with NARS. On the other hand considerable data on the management of tree crops, in particular rubber (*Hevea brasiliensis*), can be found, especially in Malaysia. Nevertheless, a number of papers reporting on the results of experiments which have been conducted for periods ranging from two to up to eight years with arable crops are readily available. Generally, these address specific issues such as the use of lime, fertilizer inputs, green manuring, use of crop residues, tillage, alley cropping, and soil conservation measures. At times two or more input factors are considered in the investigations. However, many

ISBN 1-56670-076-0
©1995 by CRC Press, Inc.

of the reports appear site specific and often do not include details of the characteristics of the site and/or the weather conditions under which the trials have been conducted.

Nevertheless, the results available are valuable and form a useful base or serve as indicators of new areas of investigations aimed at sustained production systems.

This paper uses the published results of some of the soil management experiments in Southeast Asia and in Africa to highlight the importance of different soil management practices. Subsequently, the need for a reexamination of the approaches to future long-term soil management experiments are forwarded.

## II. Role of Long-Term Experiments

The time lag on the effect of management practices on soils, and consequently on yield, varies with the soil and the type of crop. On sandy acid soils in the subhumid to semiarid areas of Thailand, the yield of cassava dropped from an initial 30 t ha$^{-1}$ to about 15 t ha$^{-1}$ by the fifteenth year (Kubota et al., 1982). However, with a crop of maize on a similar soil in Africa, the yield decline under subsistence crop was rapid; the low yield of about 400 kg of maize in the third year was about 30% of that obtained in the first year after clearing of fallow vegetation (Godo and Yeboua, 1993). On the other hand, in some soil situations degradation and loss in productivity even with cereal crop is slow to set in. Chaiyasit and Sawatdee (1992) working with an Ustic Kandihumult found soil loss of 120, 69, and 224 t ha$^{-1}$ in the first, second, and a third year of upland rice cropping under farmers' practice. Despite such losses of top soil the yield of 1.10 t of upland rice obtained in the third year was the same as that observed in the first year, even when no inputs were used. This implies that from the point of view of decline in productivity under subsistence farming, trials would have to be extended for longer periods.

At the same time it is now realized that to ensure sustained and productive farming at least some inputs are important components of soil management, and these inputs are considered as they will also influence the period over which a trial needs to be conducted.

## III. Major Constraints to Land Management

The major constraints to sustainable soil management include acidity and fertility, erosion, tillage, and water. The needs for managing these attributes have often been addressed individually. These are briefly discussed below.

## A. Management of Acidity and Fertility

There are many reports on the need for liming and the correction of acidity, with some of the trials being continuous for over long periods. Table 1 illustrates the effect of liming and other amendments on yield of soybean on a Kaolinitic Paleustult in Thailand (Tawonmas et al., 1984). Use of lime alone gave a total yield of 5.6 t ha$^{-1}$ over a six-year period; this was double the yield of the control (2.8 t ha$^{-1}$). The use of a high rate of compost at 50 t ha$^{-1}$ gave a total yield of 7.5 t ha$^{-1}$; however, it is not easy to obtain such large amounts of compost in practical farming situations. The results, however, suggest that not only pH and Al but Mn may also be a contributing factor to soil acidity. Another suggestion is that organic matter may also contribute to the alleviation of soil acidity.

Sujadi (1984) suggested that on Ultisols which are always low in P, correction for the deficiency may be more important than liming (Table 2). The use of 2 t ha$^{-1}$ of lime increased the yield of maize from 200 to 700 kg ha$^{-1}$ while 40 kg of P gave a yield of 1500 kg.

When the residual effects with soybean were assessed, similar results were obtained. The yield due to the residual lime was 300 kg while that due to residual P alone was 700 kg.

Pichot et al. (1981) as reported by Pieri (1987) considered an integrated approach to soil fertility management. The results of an investigation over 15 years clearly emphasized the need for combined use of organic manure and inorganic fertilizers to sustain and enhance yield in semiarid Africa. However, cattle manure which was used in the trial is not always available and even where it is, transport and distribution to the farms pose problems. Godo and Yeboua (1993) in Bedici, Cote d'Ivoire used crop and weed residue as a source of organic matter. They confirmed that even with high inputs of mineral fertilizers, the yield of maize declined from about 4 t ha$^{-1}$ in the first year to just over 1 t ha$^{-1}$ in the third year, a decline of 72%. However, where crop residues and NPK fertilizers were used, the yield remained almost stable. At the same time withdrawing inorganic fertilizers also led to nonsustainable yield. Godo and Yeboua (1993) suggested that the nonsustainablity due to use of inorganic fertilizers alone, could be due to the rapidly declining organic C and base saturation.

However, how can such results be applied to areas where crop residue is in critical demand as fodder and fuelwood? The potential answer seems to be in the introduction of alley cropping, which also contributes to soil conservation.

## B. Soil Conservation

Djoko-Santoso et al. (1992) working with an Epiaquic Kandiudult on a 15% slope in Indonesia showed that alley cropping and the use of mulch from the hedge of *Flemengia congesta* enhanced crop productivity. Paningbatan et al. (1992) have drawn attention to the fact that close care needs to be given to the

**Table 1.** Effect of liming and compost on changes in top soil properties and soybean yield on a kaolinitic Paleustult in Thailand

| Fertility parameter | Treatments | | | | | | | |
|---|---|---|---|---|---|---|---|---|
| | ---Control---- | | ----Lime---- | | --Compost-- | | -Compost + lime- | |
| | 1977 | 1981 | 1977 | 1981 | 1977 | 1981 | 1977 | 1981 |
| pH | 4.5 | 4.3 | 4.5 | 4.9 | 4.5 | 4.7 | 4.5 | 5.1 |
| Al (mg kg$^{-1}$) | 284 | 248 | 295 | 168 | 303 | 216 | 264 | 104 |
| Mn (mg kg$^{-1}$) | 24 | 125 | 21 | 10 | 22 | 9 | 22 | 10 |

| | Yield (kg ha$^{-1}$) | | | |
|---|---|---|---|---|
| 1977 | 0.32 | 0.85 | 1.75 | 1.60 |
| 1978 | 0.65 | 0.87 | 1.20 | 1.25 |
| 1979 | 0.07 | 0.46 | 0.52 | 0.65 |
| 1980 | 0.90 | 1.42 | 1.70 | 1.85 |
| 1981 | 0.30 | 0.85 | 1.00 | 1.05 |
| 1982 | 0.55 | 1.10 | 1.27 | 1.35 |
| Total 6 yrs | 2.79 | 5.55 | 7.44 | 7.75 |

(From Towanmas et al., 1984.)

**Table 2.** Effect of P in acid soils

| | -------------Yield (kg ha$^{-1}$)------------- | |
|---|---|---|
| Treatments | Maize | Soybean |
| Nil | 200 | 60 |
| Lime (2 t ha$^{-1}$) | 700 | 300 |
| P (40 kg ha$^{-1}$) | 1500 | 700 |
| P + lime | 1800 | 950 |

(From Sujadi, 1984.)

hedgerow of trees/shrubs as the competition from the roots and shading can considerably reduce the yield of crop rows close to the hedge rows. The results of Djoko-Santoso et al. (1992) showed that the use of fertilizers on the crop which promotes vigour of the crop also helps to reduce soil loss and runoff (Table 3). The results also showed that incorporation of crop residue led to high soil loss even when fertilizers were applied to the crop, thus indicating the need to consider tillage, and surface soil management.

## C. Soil Tillage

Soil tillage is often overlooked. It is often claimed that it is difficult to reduce the pH of subsoils and thus the restriction of roots to the surface soils is due to

**Table 3.** Soil loss as affected by conservation systems and fertilizer inputs

| Conservation treatments | Soil loss (t ha⁻¹)[a] by levels of fertilizer inputs[b] | | |
|---|---|---|---|
| Alley crop + mulch | 52 | 2 | 1.4 |
| Residue burnt, no alley | 290 | 152 | 89 |
| Residue incorporated, no alley | 272 | 125 | 94 |

[a]From June 1990 - June 1992. Crops were rice and peanuts.
[b]Fertilizer input (kg ha⁻¹) by crop.

| Crop | ---Level 1--- | | ------------Level 2----------- | | | |
|---|---|---|---|---|---|---|
| | N | P | N | P | K | Lime |
| Upland rice | 42 | 30 | 90 | 40 | 25 | - |
| Peanuts | 23 | 20 | 45 | 40 | 25 | 2000 |

(From Djoko-Santoso et al., 1992.)

the high Al content in the subsoils. We feel that the development of a plough pan or an impenetrable layer is often the major barrier. De Blick (personal communication, 1988) showed that in an untilled soil, cassava tubers are confined to the surface 10 cm. Tillage is necessary to ensure deeper rooting. On the other hand on an Oxic Paleustult in Thailand, Tawonmas et al. (1984) showed that under some circumstances tillage may reduce the yield of maize. Unfortunately, the study does not provide information on soil properties (chemical and physical) or on rooting depth. Rooting is an important determinant of the volume of soil exploited — for nutrients and soil moisture; the latter is often overlooked too.

## D. Soil Water

In Kenya, an extensive series of fertilizer trials under rainfed conditions failed to provide conclusive results as there was wide variability in rainfall, and the seven seasons covered were unfortunately not typical of the long-term variation. Keating et al. (1992) reexamined the data using a modified CERES - maize model and were able to formulate specific recommendations. This emphasizes the need to relook at older trials and possibly use models to enhance interpretation.

Muchena and Ikitoo (1993), working on a Typic Pellustert showed that drainage increased the yield of a sensitive crop, maize, from 1930 to 3610 kg ha⁻¹ while the yield of sunflower, less sensitive to excess moisture, was raised from 1860 to 1990 kg ha⁻¹ per season of cropping.

The need to consider moisture even in the humid tropics is evident. In Malaysia for example, at critical stages of growth, during the growing season, rains may fail for a period of two to three weeks. This can result in reduced crop or at times crop failure. Lim and Maesschalck (1980) showed that conservation of moisture by mulching enhanced soil moisture and crop yield. Similar results

**Table 4.** Effect of water management on the yield on Vertisols in Chisumbanje, Zimbabwe

| | Yield (kg ha$^{-1}$) | | | |
|---|---|---|---|---|
| | ------------Maize------------ | | ----------Cotton------------ | |
| Season | Control | WH* | Control | WH* |
| 1982-83 | 1330 | 2160 | 1040 | 2190 |
| 1983-84 | 57 | 79 | 82 | 186 |
| 1984-85 | 2790 | 2530 | 2500 | 2440 |
| 1985-86 | 1820 | 2110 | 1560 | 1580 |
| 1986-87 | 510 | 630 | 260 | 280 |
| 1987-88 | 1590 | 3780 | 2590 | 3540 |
| 1988-89 | 510 | 1070 | 1320 | 1730 |
| 1989-90 | 1090 | 1290 | 480 | 1060 |
| Mean | 1170 | 1710 | 1230 | 1620 |

* WH = Water harvesting using tied-furrows.
(From Nyamudeza et al., 1991.)

were seen in the semiarid areas of Thailand (Petchawee, 1988). Unfortunately, the effect of interaction of nutrients and soil moisture is often overlooked.

In semiarid Zimbabwe, water conservation by land shaping on Vertisols enhanced yield (Table 4). This was achieved by concentrating water in the furrows with the ridges acting as the "catchment". Though the data reported here is for Vertisols (Nyamudeza et al., 1991), similar results have been observed on Alfisols.

In water management it is not always a question of water conservation, as is evident for example on Vertisols in humid and subhumid tropics of Africa.

With appropriate water management introduced, the response to inputs and yield change, e.g., if an acid soil has impeded drainage, pH during the growing season increases and there is no response to lime. Similarly, with better water management, responses to fertilizers will become more prominent. Thus long-term fertilizer trials have to consider, and where possible, incorporate evolving changes in soil and water management. To be able to utilize the results of long-term trials, there is a need to consider the characterization of the experimental sites.

## IV. An Integrated Approach

### A. Consideration in Designing Experiments

Experimental design influences the means by which we seek answers to questions of interest. If a design is well thought of and properly implemented a minimum of statistical analyses is needed. When poorly designed, the statistical salvage operation needed can be complicated and time consuming.

In an experimental design, plot size, replications, randomization, and blocking are key components. The plot size should be in relation to the objective, the soil, and the treatment. For example a trial involving mechanical tillage will need a bigger plot size than a simple fertilizer trial. In Vertisols the size of gilgais may dictate the minimum width of a plot.

In many existing trials, these do not seem to have been taken into consideration. In fact there are many other aspects which need to be taken into consideration at the planning stage. These include: choice of treatments (including proper control), the site of the experiment, the actual field layout, the budget for the duration of the trial, personnel, equipment required, ongoing management, methods of data collection, data analysis, and reporting. Often many of these aspects are overlooked at the crucial planning stage.

## B. Experimental Site Characterization

Many of the earlier medium- and long-term experiments do not provide details of site characteristics or variability. When working on acid soils, variability in soil pH induced by clearing methods is often overlooked. A recent scrutiny of an intended long-term trial initiated in 1990 showed considerable variability in soil units even within a plot. Thus, where long-term experiments exist, it may be useful to try to characterize the site and reinterpret the data to obtain greater efficiency.

Assuming that the plots are uniform, in some investigations (e.g. soil conservation) is the measure or observation appropriate? Often at the bottom of the plot the total runoff and soil loss is measured. Does this truly reflect, the actual soil movement within the plot? Figure 1 shows soil movement from the top and accumulation at other points. Similar phenomena occur in farm units and landscapes and ought to be characterized and then monitored.

There are reports of sustainable paddy cultivation in Sri Lanka, Uganda, etc. even without fertilizer inputs. How is this possible? Figure 2 shows that in the upper slopes, tree crops, e.g. rubber, tea, coffee, are cultivated. These cash crops receive fertilizers. Soil and nutrient movement from the upper slopes to the valley bottoms may be acting as a constant source of nutrient inputs for the paddy. If these crops stopped using inorganic fertilizers, will the paddy in the lower areas be sustainable without fertilizer inputs? Somewhat similar problems occur in semiarid Africa; but there is often an added factor of livestock integrated into the system. The upper "forest" regions also act as grazing lands during the cropping season, and the cattle feed on the crop residues after harvest. Thus what happens on the upper plateau affects the mid and lower slope etc. It is imperative that long-term experiments on soil management recognize these interacting factors. Additionally, emphasis must be placed on the integration of soil, water, and nutrients as these factors interact in nature.

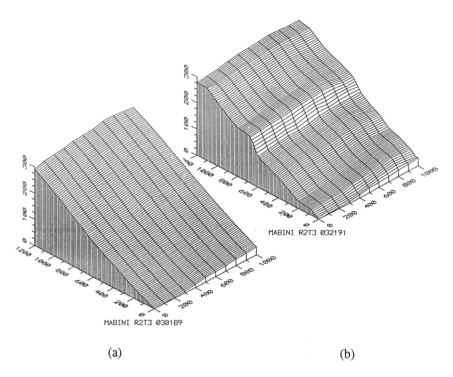

(a)                                                                          (b)

**Figure 1.** Three dimensional surface maps of an erosion plot showing the initial surface (a) and the formation of landstep after 2 years of alley cropping (b). (After E. Paningbatan, The University of the Philippines at Los Baños,1992; private communication.)

## C. Monitoring Properties

The objective of monitoring properties or variables is to measure a change in a given phenomenon or process. Eswaran et al. (1993) have highlighted that the first step in monitoring is the point of reference; at the beginning, or at any point during a time frame and at the end of the process (or evaluation). The indicators used include those which are directly measurable, qualitative, and proxy.

Direct measurement is the most desirable as the data can be tested statistically and compared with the process in another location. On the other hand, in some instances, particularly in dealing with perceptions of individuals, qualitative measurements are useful. In the absence of constraints or lack of facilities a secondary or proxy assessment is useful (e.g. compaction in a soil on a small farm can be assessed through changes in the ease of tillage).

A guide on site characterization and monitoring to assist researchers involved in soil management experiments is available (IBSRAM, 1991) and is currently being revised.

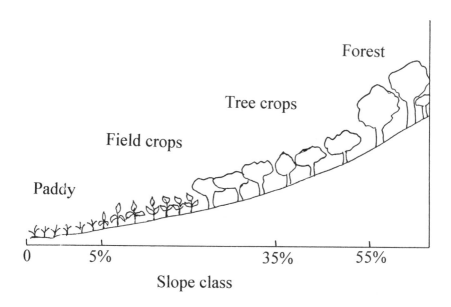

**Figure 2.** Land use in relation to toposequence of soils in N. Thailand.

## D. Archiving Samples

Soil and plant samples in most of the short-, medium-, and long-term experiments reported in literature have been sampled and analyzed regularly or periodiocally, and at times infrequently. In most cases the samples are analyzed for select indicators which are then considered essential, and generally the samples are discarded. Subsequently at a later date when there are suggestions that an indicator not previously assessed may be an important factor, the absence of earlier soil or plant samples prevents the researcher from following up on the investigations.

It is therefore suggested that in all medium- to long-term experiments soil and plant samples be obtained at least once a year, even though analysis of the samples may not be performed at each sampling. The samples collected should be dried, inventoried and stored for later use, if the need arises.

## E. Documenting Information

Data collection should follow a well-designed format amenable for entry into an appropriate data base. The data should include not only measurable criteria but also qualitative and proxy indicators and field observations. Indicators such as the extent of bird damage, post harvest management etc., which may appear unimportant at a particular stage may be an important component to explain variability in trends over a longer period.


## V. Long-Term Experiments to Address Sustainability

Over 35% of the arable land in the tropics is acid and low in inherent fertility (mainly Ultisols and Oxisols). Traditionally, in the tropics, Alfisols have been the preferred soils while Ultisols, Oxisols, and Entisols are also used. However, Vertisols which are chemically more fertile soils, and cover over 300 million ha in the world are often overlooked particularly in Africa where they account for over 100 million ha or 6% of the arable land. According to the concept of sustainability (Eswaran and Virmani, 1990) Vertisols are able to sustain production over a much longer period than even the Alfisols. Thus, in Vertisols declining productivity may be evident only after a considerable period of cultivation with minimal inputs, the period of experimentation to assess sustainability may need to be longer.

Existing and earlier field trials on soil management have provided extremely valuable information. A closer reexamination of the uniformity of the sites followed by a reinterpretation of the data as appropriate, merits consideration. However, the investigations have often tended to consider discrete components. Unfortunately, in nature no single factor acts on its own. Thus there is a need to integrate soil, water, and nutrients in research on soil management. As these are influenced by the positions in a landscape, the watershed needs to be also considered. Such investigations should also take into account the human/social factor. It is emphasized that such investigations are not only costly but time consuming. The use of models is a critical first step to integrate existing data to arrive at tentative appropriate packages.

It is suggested that future long-term experiments in the tropics should be representative of resource management domains, and could usefully address the management of acid soils, potentially acid soils such as the sandy soils in the Sahel, Vertisols, degraded soils (rehabilitation) and soils on sloping and hilly lands.

Logistics, institutional commitment, and other practical considerations are the normal deterrents for the establishment of long-term experiments. A long-term experiment is a commitment for 10 to 20 years and so the objectives must be clearly established at the onset. Few guidelines exist for the establishment, design, and management of long-term experiments. The need for an integrated approach, monitoring of variables, and a clearly defined commitment are obvious. Due to these and other reasons, a most appropriate way appears to be through international collaboration and through a network mode.

# References

Chaiyasit, A., and B. Sawatdee. 1992. Management of sloping lands for sustainable agriculture in Thailand. p. 217-253. In: *Technical Report on the Management of Sloping Lands for Sustainable Agriculture in Asia, Phase 1, 1988-1991 (IBSRAM/*ASIALAND*)*. IBSRAM Network Document No. 2. Bangkok: IBSRAM.

Djoko-Santoso, Sri Adiningsih, and Effendi Suryatna. 1992. Management of sloping lands for sustainable agriculture in Indonesia. p. 29-76. In:*Technical Report on the Management of Sloping Lands for Sustainable Agriculture in Asia, Phase 1, 1988-1991(IBSRAM/*ASIALAND*)*. IBSRAM Network Document No. 2. Bangkok.

Eswaran, H. and S.M. Virmani. 1990. The soil component in sustainable agriculture. p. 45-87. In: *Sustainable Agriculture in the Semi-Arid Tropics:A Compendium*. Patancheru, India, ICRISAT.

Eswaran, H., E. Pushparajah, and C. Ofori. 1993. Indicators and their utilization in a framework for evaluation of sustainable land and management. Preprint:*International Workshop on Sustainable Land Management for the 21st Century*, June 20-26, 1993. Lethbridge, Alberta, Canada.

Godo, G.H., and Y. Yeboua. 1993. Organo-inorganic fertilization and the maintenance of crop yields in southern Cote d'Ivoire. Unpublished Report. IDEFOR/DPO, Abidjan.

IBSRAM (International Board for Soil Research and Management). 1991. Methodological guidelines for IBSRAM's soil management networks. IBSRAM. Bangkok. 45 pp.

Keating, B.A., M.N. Siambi, and B.M. Wafula. 1992. The impact of climatic variability on cropping research in semi-arid Kenya between 1955 and 1985. In: M.E. Probert (ed.), *A search for strategies for sustainable dryland cropping in semi-arid eastern Kenya. Proc. of the symposium held in Nairobi, Kenya,* Dec. 1990, 16-25. ACIAR Proc. No. 41. Canberra. ACIAR.

Kubota,T., P. Verapattananirund, P. Piyapongse, and S. Petchawee. 1982. Improvement of the moisture regime of upland soils by soil management. p. 315-372. In: *Proc. of the International Symposium on Distribution, Characteristics, and Utilization of Problem Soils*. Japanese Society of Soil Science and Plant Nutrition. Tropical Agricultural Series No. 15.

Lim, K.H., and G.G. Maesschalck. 1980. The effect of lalang mulching on soil moisture conservation and temperature control and their influence on the performance of 28 cowpea varieties under Malaysian conditions. p. 345-358. In: *Proc. of the symposium on legumes in the tropics.* Serdang, Malaysia. Universiti Pertanian.

Muchena, F.N. and Ikitoo, E.C. 1993. Management of Vertisols in semiarid areas of Kenya: the effect of improved methods of surplus soil/surface water drainage on crop performance. p. 25-37. In: *Report of the 1992 Annual Meeting on* AFRICALAND—*Management of Vertisols in Africa.* IBSRAM Network Document No. 2. Bangkok.

Nyamudeza, P., O. Mandiringana, T. Busangavanye, and E. Jones. 1991. The development of sustainable management of Vertisols in the lowveld area of southeast Zimbabwe. *IBSRAM Newsletter* 20:9.

Paningbatan, E., A. Maglinao, L.A. Calanog, and G.M. Huelgas. 1992. Management of sloping lands for sustainable agriculture in the Philippines. p. 159-216. In: *Technical report on the management of sloping lands for sustainable agriculture in Asia, Phase 1, 1989-1991 (IBSRAM /*ASIALAND). IBSRAM Network Document No. 2. Bangkok.

Petchawee, S. 1988. Soil organic matter for improving soil productivity. p. 320-330. In: C. Pairintra, K.Wallapapan, J.F. Parr, and C.E. Whitman. (eds.), *Soil water and crop management systems for rainfed agricculture in northeast Thailand.* USAID. Washington, D.C.

Pieri, C. 1987. Management of acid tropical soils in Africa. p. 41-62. In: *Management of acid tropical soils for sustainable agriculture.* IBSRAM Proc. No. 2. Bangkok.

Sujadi, M. 1984. Problem soils in Indonesia and their management. p. 58-73. In: *Ecology and Management of Problem Soils in Asia.* Taiwan: Food and Fertilizer Technology Center for the Asian Pacific Region.

Tawonmas, D., S. Panicakul, S. Ratnarat, and W. Manangsul. 1984. Problem of laterite soils in Thailand. p. 50-57. In: *Ecology and Management of Problem Soils in Asia.* Taiwan: Food and Fertilizer Technology Center for the Asian Pacific Region.

# Need for Long Term Soil Management Research in Africa

C.S. Ofori

## I. Introduction

Increased food production to feed increasing population remains one of the most important challenges in Africa today. Rate of population growth on the continent is presently the highest in the world and yet food production during the past decade has hardly shown any improvement but rather a decline. Although arable land utilization in many of the countries has not reached the limit of further expansion as in, for example, Asia, the soil resources in the continent are generally of low-fertility level and in many cases degraded. The rapid population growth has put further pressure on land and marginal lands are taken into

ISBN 0-87371-889-5

cultivation without the necessary inputs being applied or without the appropriate soil management practices.

Although the predominant land use system — the slash-and-burn system — has undergone some changes, these are not fast enough to keep up with the demand for increased food production resulting from population pressure. The priority given to export crops in the past has resulted in more attention being directed to research and technology development for these crops at the expense of food crops. Research on food crops has suffered from various aspects, such as lack of funding, inadequate institutional development, as well as insufficient training of research-calibre manpower. Further, program development concentrated more on short term solutions rather than a good balance between this and medium- and long-term goals.

A country's agricultural development, let alone that of the whole continent, cannot be planned without medium- to long-term considerations. The main goal is to direct attention to sustainable agricultural production and this can hardly be achieved without first of all identifying the resource base — the soils — their potential and constraints and by developing technology packages that will enhance and sustain their productivity. It is well-recognized that the soil by itself does not change the production system without the development of the other production factors. It is, however, an important baseline on which to build.

The changing patterns of land use, food production, and farming systems in Africa have made it imperative to re-examine some of the research needs, especially on soils, so as to be able to develop management practices that will improve and sustain productivity. In this short communication, some of the issues for the research agenda are highlighted, and attention is drawn to some institutional aspects necessary for technology package development.

## II. Soil Fertility Investigations — Historical Perspective

### A. Soil Surveys

Soil surveys in various countries of Africa were initiated in the 1950s and continue to date. But intensive activities were carried out in the 1960s after most of the countries gained independence. In many countries, surveys were mostly at reconnaissance level with semi-detailed and detailed surveys for only specific projects. With assistance from ORSTOM (France) and other overseas development agencies, a number of West African countries produced soil maps at detailed and semi-detailed scales for planning and natural resource development. A number of soil survey projects linked with institution building and human resource development were undertaken by FAO in many of the African countries. Most of the institutions established have developed into national soils institutes of their respective countries.

## B. Soil Fertility Improvement

Soil management and fertilizer use developed independently from soil surveys. It is significant to note that farming systems at the time (i.e. slash-and-burn land use, predominantly) influenced the programs developed for the improvement of agriculture and increased crop production. The thrust of these programs was mainly on important export crops, such as coffee, tea, but was also on food crops — mainly the annuals: maize, rice, millet, sorghum, yams.

The transition from slash-and-burn to permanent cultivation systems influenced the development of soil management programs focusing attention mostly on soil fertility experimentation to solve immediate soil nutrient replenishment with emphasis on increasing crop yields in the short and medium term. Little attention was paid to long-term evolution of soil fertility and productivity maintenance.

An additional factor influencing the trend of soil fertility program development is the fact that the majority of farmers whose productivity is to be increased were smallholders with limited financial resources to purchase such inputs as fertilizers. Results of experiments from 3 to 4 years were deemed adequate to advise farmers to improve their practices. Such experiments were initiated in many of the African countries by the research divisions of their Ministries of Agriculture. For example, many such experiments were initiated in Ghana in 1949 and early 1950s (Nye, 1949; Djokoto and Stephens, 1960) in Nigeria, Uganda, and many other countries. Many soil fertility experiments were also initiated in the French-speaking African countries.

The experiments initiated at the time were not planned to monitor in-depth nutrient cycling with cropping, neither was much emphasis placed on residual effects of fertilizers applied.

## C. Long-Term Experiments

Long-term soil management experiments initiated in African countries included those in former Belgian Congo and Kade, Ghana experiments in shifting cultivation; 30 long-term experiments on annual crops ($2^5$, NPK-lime-mulch) initiated in Kade, Ghana in 1949 (Nye, 1951); tillage experiments in Senegal in 1962; annual crops in Northern Zambia in 1962; rotation and tillage experiments in Zaria, Moor Plantation and Ife University (Nigeria). From the International Institute for Tropical Agriculture (IITA, Ibadan), a watershed management experiment was initiated in the 1970s (Lal, 1992). These experiments made significant contributions to the knowledge on tillage, and soil and water management, while soil fertility and residue management experiments conducted by Kang at IITA, from 1972-1981, filled part of the gap on nutrient cycling.

Experiments on the effect of ploughing on soils and crop yields in Senegal were initiated in the late 1940s with results reported by Tourte (1951), and Charreau and Nicou (1971). In 1962 another experiment was initiated in Senegal with the objective of studying traditional tillage (using local implements) and conventional tillage/ploughing effects pm yield of millet and groundnuts.

Extensive research programs were developed by the French research institutions in many of the French-speaking countries in West Africa including the Cameroons, particularly in the field of soil fertility and fertilizer use from the mid 1960s. The Belgians also initiated several experiments through INEAC (Institut National Etudes Agronomique Congo), aimed at finding solutions to the land use system of shifting cultivation in Zaire, Burundi, and Rwanda.

## III. The Issues

### A. Soil Degradation

Farming systems and land use have changed radically in Africa as a result of population pressure on land and intensification of agriculture to increase food production per unit area. Soil management practices, however, have not kept pace with this development, neither is the use of such inputs as fertilizers increased to compensate for nutrient uptake in the crops and nutrient lost through soil erosion. Decline in soil fertility has been the major pattern exacerbating the already low fertility levels of these severely weathered soils.

Vast areas have been degraded through inappropriate land use systems and have been stripped of vegetation by land clearing, overgrazing and fuelwood collection. Soils of these areas have been subjected to severe water and wind erosion and depleted of organic matter and nutrients by continued cropping with inappropriate cultural practices.

In order to address some of these issues of natural resource mismanagement in Africa, it is necessary to identify the fundamental problems which appear to be basic to Africa's development crisis as well:

- high rate of population increase,
- declining agricultural productivity, and
- economic starvation and poverty.

One thing that emerges, is the fact that with few exceptions, agricultural research in most African countries has not produced the appropriate technologies needed to increase productivity of land and labour.

### B. Management of Natural Resources

Soil erosion and declining soil fertility together constitute a major threat to agricultural development and sustainable management of natural resources in Africa. An overview of the most serious soil and vegetation degradation problems in Africa is given in Table 1.

**Table 1.** Summary of most serious soil and vegetation degradation problems in Africa by regions

| Region | Arable land | Grazing land | Forest land |
|---|---|---|---|
| Mediterranean and arid North Africa | - Declining soil fertility<br>- Wind and water erosion<br>- Salinization on irrigated lands | - General degradation of vegetation in quality and quantity<br>- Wind and water erosion | - Degradation of vegetation as the deficit in fuelwood and timber increases<br>- Water erosion on degraded forest lands |
| Sudano-Sahelian Africa | - Decline in nutrien levels in the soil<br>- Decline in soil physical properties<br>- Wind and Water erosion | - General degradation of vegetation in quality and quantity<br>- Wind erosion in sub-humid areas | - Degradation of vegetation |
| Humid and sub-humid West Africa | - Decline in nutrient levels in the soil<br>- Decline in soil physical properties<br>- Water erosion | - Degradation of vegetation<br>- Wind erosion in sub-humid areas | - Degradation of vegetation |
| Humid Central Africa | - Degraded soil physical properties<br>- Degraded soil chemical properties | | - Degradation of vegetation |
| Sub-humid and mountain East Africa | - Water erosion<br>- Degradation of soil physical properties<br>- Degradation of soil chemical properties | - Degradation of vegetation in quality and quantity<br>- Water erosion | - Degradation of vegetation<br>- Water erosion |
| Sub-humid and semiarid Southern Africa | - Water erosion<br>- Degradation of soil physical and chemical properties | - Degradation in quality and quantity of vegetation<br>- Wind and water erosion | - Degradation of vegetation<br>- Erosion |

(From FAO, 1986.)

Fertilizer use in Africa remains the lowest per ha arable land in the world. If food production in Africa is to show significant increases, it is necessary to move to more intensive cultivation with higher yielding varieties and increased use of fertilizer inputs, which at the same time does not compromise the long-term productivity of the land. This objective must be the ultimate goal, though admittedly it will take some time and will need careful policy guidance to achieve.

In developing appropriate technology packages, it is necessary to generate data from long-term monitoring of the changes that occur in the resource base - the soil — under given climatic conditions. Short-term experimental data, though useful and necessary, do not provide a sufficiently sound basis for sustainable resource base development or for assuring long-term productivity.

## IV. Research Agenda

Past and ongoing research programs on natural resource management, carried out by national and international institutions, have been beneficial and are still targeted at some of the pressing problems of soil management in the continent. It is important to note, however, that most of these research agendas have the objective of finding immediate and medium-term solutions to soil management problems so as to increase the present levels of production. Long-term effects of resource use and management that should provide the solid basis for sustainable use are yet to be adequately addressed.

Notwithstanding the commendable research work carried out by past and present soil scientists in Africa, results of these research efforts have regretably not been adequately disseminated to the farmers through the extension service mechanism in the various countries. Although there has not been a major breakthrough such as the Green Revolution in Asia, it is true to say that an efficient extension system on the continent could have greatly improved the food production situation.

With the goal of sustainable agricultural production in mind, effective research agendas for agriculture in Africa must address the development of technologies needed to increase land productivity and labor. Some of the technologies critical to sustainable productivity have now begun to receive the attention they merit in African farming systems research. These include soil and water conservation, appropriate tillage systems, crop residue management, agroforestry and crop rotations that could increase soil fertility. In addition to the above, attention should be concentrated not only on the present and future management of resources, but also on programs that tackle the immediate consequences of mismanagement of natural resources — that is programs on rehabilitation of degraded lands. It is estimated that 22.1% of Africa's vegetated land (769 million hectares) suffers from soil degradation in one form or another (Oldeman et al., 1990).

## A. Priority Areas

The topics listed below are by no means exhaustive and not necessarily in order of priority but should be regarded as guidelines in identifying priorities for research and extension.

## 1. Soil erosion and Soil Fertility Decline

Soil erosion and declining soil fertility together constitute a major threat to agricultural development and sustainable natural resource management in Africa. It is crucial to evolve programs on management practices that protect the soil against erosion, salinity, and compaction and provide for cost-effective and energy efficient ultilization of the productive capacity. FAO (1990) recently initiated a program on the "Conservation and Rehabilitation of African Lands." It seeks to develop national soil conservation and rehabilitation programs. Other programs that may have impact on resource management include the FAO/UNEP program on the development of National Soils Policies for some of the countries in the region.

### a. Soil Fertility Decline: Studies on Nutrient Dynamics and Effect on Crop Yields in Various Agro-Ecological Zones

Most of Africa's present agricultural practices could best be described as "exploitative" rather than "balanced" or "generative" agriculture. The continent (south of the Sahara) presently has the lowest fertilizer use in the world with an average of 10 kg nutrients/ha (N, $P_2O_5$ and $K_2O$), the major part of which is used on export or industrial crops. The present low yields of food crops obtained by the farmers could be substantially increased as shown in the extensive demonstrations and experiments carried out under FAO's Fertilizer Program (Table 2) (FAO, 1986).

In a recent study on nutrient balance conducted by the Winand Staring Centre, Wageningen on behalf of FAO, it is estimated that at present level of agricultural production about 80 kg/ha of nutrients is taken from the soil (mainly uptake in crops removed from the land + nutrients lost through erosion), whereas nutrient application amounts to only 12 percent (10 kg/ha) of the total. It is therefore evident that the present agricultural production is based mainly on nutrient exploitation resulting in nutrient deficit.

In the subhumid and humid regions the effect of increasing soil acidity and $Al^{3+}$ saturation and the loss of CEC on crop production are a few of the issues that need to be tackled in long-term experiments.

**Table 2.** Results from FAO's Fertilizer Program

| Country / Zone | Crop | National average | Improved practices | Improved practices + fertilizer |
|---|---|---|---|---|
| | | Yield (kg ha$^{-1}$) | | |
| Algeria (North Africa and the Mediterranean zone) | Millet | 670 | 2760 | 3590 |
| Burkina Faso (Sudano-Sahelian Africa) | Millet | 430 | 520 | 1160 |
| Cameroon (Humid Central Africa) | Rice | 840 | 1360 | 2500 |
| Burundi (Sub-humid and mountain E. Africa) | Bean | 630 | 580 | 1470 |
| Ethiopia (Sub-humid and mountain E. Africa) | Maize | 1100 | 2010 | 4100 |

(From Fao, 1986.)

## 2. Effect of Tillage on the Development and Management of Tropical Soils

From the late 1960s to date, a number of studies on tillage have been undertaken with a view to developing appropriate tillage techniques in the various agro-ecological zones and on different soils [Nicou and Chopart (1979), Lal (1973, 1974, 1979a 1979b), Karnegitter (1967), Ofori and Nandy (1969), Chopart and Nicou (1976)].

## 3. Organic Matter Dynamics

Sandy soils covering 20% of the arable land have low organic matter content, and this is crucial in managing the P supplying power of these soils as well as N supply, and to maintain the CEC.

In the humid and sub-humid regions there is rapid decomposition of organic matter following cultivation until a new equilibrium level is reached. Organic matter management is closely linked to soil fertility maintenance. Two significant sources of organic material additions are available, namely, crop residues and root remains through crop rotations.

The understanding of organic matter dynamics in these soils is crucial to the development of strategies for soil fertility maintenance and improvement of soil structure and stability for sustainable production in the tropics. Long-term studies in different agro-ecological zones provide the best opportunities for generating the basic data required.

## 4. Residue Management

Although residue management is closely linked to organic matter management, this subject needs separate attention as crop residues, especially in the savanna agro-ecosystems, have other economic values such as for fodder, and therefore, may not be entirely available for soil improvement or conservation. Research on residue management should also be linked to crop rotations and soil conservation through biological practices.

## 5. Biology of Tropical Soils

Soil fertility research should not only be concerned with soil chemical and physical aspects, but greater emphasis should also be directed to soil biology. The role of earthworms and termites in the soil in relation to organic matter dynamics cannot be overlooked.

# V. Rehabilitation of Degraded Soils

The present focus on the management of natural resources is justified, but it is necessary to emphasize that several million hectares of arable land in Africa are already degraded through mismanagement. Programs and strategies for their rehabilitation should, therefore, parallel those for conservation and management. The natural resources can best serve sustainable development when already degraded soil resources are rehabilitated and resources under utilization are carefully managed.

Rehabilitation programmes on degraded soils should be multi-disciplinary involving soil scientists, foresters, animal scientists, and socio-economists.

# VI. Institutional Framework

There are several underlying reasons for the lack of long-term soil management studies in Africa. These can be found in:
- Government policy
- Human resources (during the Colonial era and after independence)
- Funding
- Research Agendas
- Institutions.

## A. Government Policy

Government policy toward agricultural research greatly influences research agendas and goals. It is understandable that much more emphasis was placed on short- and medium-term solutions in the past to find answers to urgent soil management problems involving fertilizer use and fertility maintenance. With sustainability issues now at the forefront of resources development and with the present awareness of environmental concerns, governments must direct greater effort to long-term research that may provide lasting solutions to the problems of resource management and also devote adequate resources to finding solutions to the present problems.

## B. Human Resources

Lack of long-term soil management experiments in Africa cannot be fully understood without consideration of research development from the Colonial era to post independence. The rapid turnover of research scientists during the Colonial era and the consequent interest to achieve set goals during the short-term stay of scientists did not encourage long-term planning. There was a general emphasis on short-term studies and on publishing as many scientific papers as possible in the short time available, for further career advancement.

This situation did not change after independence when nationals took charge, in fact it was in some cases exacerbated by the deteriorating economic situation with little or no support for research and with the virtual exodus of scientists from various research institutions.

## C. Funding

Funding for agricultural research on food crops in Africa has never been adequate.

Crops for export enjoyed better status, being the main source of foreign exchange earnings in many cases. However, there is no doubt that for a better balance it would have been more desirable to support agricultural production in the food sector in particular.

When funding becomes a major problem, policy makers hardly see the need for maintaining long-term experiments, the results of which will not be immediately available to them.

## D. Research Agendas

The relatively small impact of agricultural research on food production and resource management in Africa cannot be wholly ascribed to lack of government policy and support, funding, or availability of expertise. To date, agricultural research systems in Africa seem not to have generated improved crop production technologies relevant to small farmers that could make a significant impact if actively promoted by the extension services.

To address sustainability and resource management issues, a carefully planned research agenda needs to be worked out with present, medium- and long-term objectives in view. Long-term experiments are costly and investments made in them should be cost effective. It is, therefore, imperative to plan such experiments very carefully so that results are generated at various stages of the experiment as it progresses. In this way the researcher's interests will be kept up, the ministry funding the work remains involved and intermediate results generated can be applied to increase agricultural production.

### E. Institutions

Long-term experiments cannot be well maintained to achieve their objectives without institutional stability. This issue regrettably has not been adequately examined in developing strategies for finding some of the technological solutions to improve agricultural production in Africa. Strengthening of the National Agricultural Research Systems (NARS) is one of the most needed and valuable kinds of assistance to Africa in order to develop its human resources and institutions in the area of agricultural research and resource management with the long-term goal of developing sustainable production.

## VII. Conclusion

The present decline in agricultural productivity in Africa cannot be remedied on a piecemeal basis, neither should the strategy be limited to present and medium term solutions. Agriculture should be the primary 'engine of growth' in Africa: therefore, a carefully planned, integrated medium- and long-term strategy will be necessary in order to increase food production with the increasing population growth, but at the same time ensuring environmental stability. This is even more pressing when sustainability issues come to the forefront, as is happening today in natural resource development and management. Agricultural research in the past has not made the desired impact partly because of lack of development of appropriate farmer-oriented technology packages.

The small farmer production systems form the backbone of African agriculture. But unlike similar small-farmer systems in Asia, the African scenario lacks an adequate energy input, which still comes predominantly from human labor, with little or no draught animal input for farm operations.

Large areas on the African continent are already degraded and these must be rehabilitated as well as further degradation of existing land arrested and prevented. Soil management practices have to be developed for the different agro-ecological zones taking into account the socio-economic settings.

It is imperative to put together a well-coordinated, medium- and long-term research agenda which will address some of the priority areas in soil management with the objective of providing technology packages and options necessary for resource management. Long-term experiments are essential and provide the necessary vehicle for such development. But the objective cannot be achieved without the appropriate government policy support, human resources development, adequate funding, and sustainable institutions. The way ahead is to improve soil productivity, arrest soil degradation, to develop appropriate technology packages for the wise use of marginal lands, and to ensure sustainable resource use as well as maintain environmental stability.

# References

Charrea, C. and R. Nicou. 1971. L'amélioration du profil cultural dans les sols sableux et sablo-argileux de la zone tropicale séche Quest-Africaine et ses incidences agronomiques. *Agron. Trop.* 26:903-908 et 1183-1247.

Chopart, J.L. et R. Nicou. 1976. Influence du labour sur le développement radiculaire de différentes plantes cultivées au Sénégal. Coonséquences sur leur alimentation hydrique. *Agron. Trop.* 31 (1) 7-28.

Djokoto, R.K. and D. Stephens. 1961. Thirty long-term fertilizer experiments under continuous cropping in Ghana. 1. Crop yields and responses to fertilizers and manures. *Emp. J. Exp. Agric.* 29:181-195.

FAO. 1986. *African Agriculture: The Next 25 Years.* FAO, Rome.

FAO. 1990. *Conservation and Rehabilitation of Africa Lands.* FAO, Rome.

Karnegitter, A. 1967. Zero cultivation and other methods of reclaiming Pueraria fallowed land for flood cultivation in the forest zone of Ghana. *The Trop. Agriculturist* 128:51-73.

Lal, R. 1973. Effects of seedbed preparation and time of planting of maize in Western Nigeria. *Experimental Agric.* 9:304-313.

Lal, R. 1974. No-tillage effects on soil properties and maize (*Zea mays* L.) production in Western Nigeria. *Plant and Soil* 40:129-143.

Lal, R. 1979a. Importance of tillage systems in soil and water management in the tropics. p. 25-32. In: R. Lal (ed.), *Soil Tillage and Crop Production.* IITA, Proc. Ser. 2, Ibadan, Nigeria.

Lal, R. 1979b. Influence of six years of no-tillage and conventional plowing on fertilizer response f maize on an Alfisol in the tropics. *Soil Sci. Soc. Am. J.* 43:399-403.

Lal, R. 1992. Tropical agricultural hydrology and sustainability of agricultural systems — A ten year watershed management project in Southwestern Nigeria. IITA and Ohio State University, Columbus, Ohio. 303 pp.

Nicou, R. 1974. Contribution à l'étude et à l'amelioration de la porosité des sols sableux et sable argileux de la zone tropicale sèche. Conséquences agronomique. *Agro. Trop.* 29 (11):1100-1127.

Nicou, R. and J.L. Chopart. 1979. Water management in sandy soils of Senegal. p. 248-257. In: R. Lal (ed.), *Soil Tillage and Crop Production*. IITA Proc. Ser. 2, Ibadan, Nigeria.

Nye, P.H. 1951. Studies on the fertility of Gold Coast Soils: 1. General account of experiments. *Emp. J. Exp. Agric.* 19:217-223.

Oldeman, L.R., V.W.P. van Engelen, and J.H.M. Pulles. 1990. The extent of human-induced soil degradation, Annex 5 of L.R. Oldeman, R.T.A. Hakkeling, and W.G. Sombroek. *World Map of the Status of Human-Induced Soil Degradation: An Explanatory Note,* rev. 2nd ed. International Soil Reference and Information Centre. Wageningen, The Netherlands.

Ofori, C.S. and W. Nandy. 1969. The effect of method of soil cultivation on yield and fertilizer response of maize grown on a forest ochrosol. *Ghana J. Agric. Sci.* 2:19-24.

Toute, R. 1951. Preparation du sol et enfouissement de la végétation naturelle comme engrais vert. Leur influence sur les rendement du mil au Sénégal. *Ann. CNRA*, Bambey, 120-125.

# The Tropical Soil Productivity Calculator — A Model for Assessing Effects of Soil Management on Productivity

J.B. Aune and R. Lal

## I. Introduction

Despite the impressive progress made in basic sciences, it remains difficult to predict management-induced changes in soil properties and their effect on productivity. Yet, these soil properties are one of the base on which future life on earth depends.

ISBN 1-56670-076-0

The population growth in low income countries except China and India is 2.5% (World Bank, 1993). The agricultural food production will have to keep pace with this growth rate if future food supplies are to be sufficient.

The outlook for improving per capita food production is especially somber for Africa, where per capita food production has declined during the 1980s (FAO, 1991). If food production is to be increased, the agricultural resource base will have to be enhanced. However, the resource base may be shrinking. Stoorvogel and Smaling (1990) estimated that the loss of N, P, and K from soil were 10, 2 and 8 kg ha$^{-1}$ yr$^{-1}$ respectively in sub-Saharan Africa. Furthermore, the increased agricultural production should, from an environmental standpoint, be increased by augmenting land productivity rather than expanding the cultivated area. Clearing new land for cultivation is often associated with deforestation, $CO_2$ emissions, reduced wild life habitat, and increased soil erosion and degradation. Therefore, an important strategy is to invest in enhancing soil productivity, and it should be directed toward areas where potential for future return is high.

Models can serve as a powerful tool for analyzing long-term effects of soil management on productivity, environmental quality, and in development planning. Table 1 lists some of the models that have been developed along with their merits and limitations.

The most advanced model in terms of predicting crop yield is probably the CERES models because of the time step of one day. This time step allows for simulation of the effect of different weather conditions on crop yield. The main emphasis in the CENTURY model is on C, N, and P dynamics, with less emphasis on yield estimation. The QUEFTS model estimates yield on the basis of soil chemical properties, without considering soil moisture regime. The SCUAF model does not predict yield but estimates how yield may change over time as a result of changes in soil properties. The USLE (Universal Soil Loss Equation) is a sub-routine of the SCUAF model.

One of the main factors limiting crop yields in the humid tropics is soil acidity. Sanchez and Logan (1991) estimated that soil acidity including Al-toxicity is a constraint affecting 39% of the geographical area in the tropics and subtropics. As soil pH decreases with cultivation and fertilizer use, the cultivated area affected by acidity also increases. None of the models listed in Table 1 has a built-in routine to compute the effect of acidity on crop productivity. This inherent limitation is a serious constraint of these models for application to soils of the tropics.

Another factor that limits the use of most models to tropical regions is their high data requirements. Many of the data required are not readily available. The data requirement is especially high for the CERES models. Furthermore, soil erosion is not dealt with adequately in any of the models.

The objective of this study, therefore, was to develop a new model designed to predict productivity in relation to fertilizer use (N, P, and K), liming and subliming, tillage and sub tillage, vehicular traffic, residue management, and soil erosion. Furthermore, this model is designed to be simple to use but comprehen-

**Table 1.** Overview of models that can be used to predict sustainable land use in the tropics

| Model | Changes in soil properties | Yield prediction | Time step | Merits | Limitation | Reference |
|---|---|---|---|---|---|---|
| Ceres models | N, H₂O | Yes | Day | Yield estimation | High data requirem., no pH | Ritchie et al. 1989 |
| Century | C, N, P, S | Yes | Month | C, N, P, and S dynamics | Uncertain yield function, no pH | Parton et al. (1988) |
| QUEFTS | | Yes | Year | Simplicity | No water, no pH | Janssen et al. (1990) |
| SCUAF | C and N | Yield changes | Year | Long-term effects | No water, no pH | Young et al. (1990) |

sive enough to account for main factors that influence crop growth and yield in the tropics. The model also can be used to provide data for economical analysis.

## II. Theory and Methods

### A. Model Outline and Data Sources

General features of the model are shown in equation 1:

$$Y = (\text{Potential yield} * (F_1 * F_2 * F_3 * \ldots . F_n)^{B1}) * B2 \qquad \text{(Eq. 1)}$$

where y is yield Mg ha$^{-1}$, $F_1 \ldots F_n$ are indices (range 0-1) of different productivity parameters, and B1 and B2 are constants. The relative value of f-functions determines the magnitude of yield reductions with reference to the potential yield. The potential yield is site- and crop-specific and reflects the yield that can be obtained when soil properties are not limiting productivity. Potential yield is then determined by climatic factors, crop species, and cultivars.

The F-factors are functions which describe the relationship between soil properties and relative productivity. These factors have a value ranging from 0 to 1, the value of 1 indicates that the soil property is not a limiting factor. The f-factors included in Eq. 1 are soil organic carbon (SOC), acidity (pH and Al-saturation), N, available P, exchangeable K, soil bulk density, rooting depth, and weed infestation. These factors are chosen because of their importance in determining productivity of Oxisols, Ultisols, and Alfisols (Stewart et al., 1991). Such functions have been developed before (Kayombo and Lal, 1986; Kang and Osimane 1979), but most of them are site-specific. An attempt is, therefore, made to combine results from many locations and develop a generic function for each soil property. As crops differ in growth requirements, F-factors developed are crop-specific. The procedure for developing these functions is outlined as follows:

The constant B1 is required to use the model across different yield levels. Magnitude of the B1 factor is especially important for predicting yield when f-factors are sub-optimal (low yield). The same B1 constant can be used across locations. The constant B2 is used to fit the correct yield level and is, therefore, site-specific. It is a constant that accounts for f-factors not included in the model and other unexplained variations.

Equation 1 has resemblance to the Mitscherlich-Baule model presented in Equation 2:

$$Y_i = m \, (1-k_1 e^{-B_1 x_{1i}}) \, (1-k_2 e^{-B_2 x_{2i}}) \ldots (1-k_n e^{-B_1 x_{ni}}) \qquad \text{(Eq. 2)}$$

where, m is the asymptotic plateau, $x_1$, $x_2$, $x_n$ are different growth factors and k and B are constants. However, Equation 1 differs from Mitscherlich-Baule equation by assuming that the response to all growth factors does not follow the law of diminishing returns. This assumption is justified by the fact that the relationship between relative yield and pH, Al-saturation, and bulk density do not follow the law of diminishing returns. Another difference lies in the introduction of the exponent B1 in Equation 1. The advantage of selecting a multiplicative rather than an additive formula is that the yield maximum at any location is determined by the most critical yield-limiting factor.

## B. Data Source

The data used to establish f-factors in the Tropical Soil Productivity Calculator were obtained from experiments conducted on Oxisols, Ultisols and, Alfisols in different ecoregions of the tropics (Table 2). Data from these soil orders were grouped together, realizing the fact that Alfisols are relatively less leached and less acid than Oxisols and Ultisols. The proposed model accounts for this difference because the acidity factor is close to one in Alfisols.

## C. Methods for Establishing F-Factors

Different methods were used to establish relationships between soil properties and productivity. Equations relating relative yield to soil acidity, N, P, K, bulk density, and rooting depth were developed from data obtained from several experiments on liming, fertilizer rates, and traffic-induced soil compaction. The relative yield in each of the experiments was obtained by dividing the yield in each treatment with yield in the treatment with the highest yield. A new data set combining different experiments was thus created. Four data points were selected from each experiment to overcome the problem that the number of treatments tested varied among experiments. This normalizing procedure resulted in an equal effect of each experiment on the shape of the curve. A regression analysis was then undertaken across locations, relating soil property to the relative yield.

Equations relating yield to soil acidity and Al saturation were developed from the data of several liming experiments. For other soil properties, however, four data points were selected from the curve that best fitted the data. The reason being, that in many experiments available results were presented graphically rather than in a tabular form.

The data from several long-term experiments conducted in the tropics were used to evaluate the effects of SOC on productivity. This involved a two-step procedure:

1. A linear regression analysis over years was conducted between relative yield and corresponding soil properties in each of the long-term experiments.

2. A new data set was created by selecting four carbon levels and their corresponding relative yields from each experiment. Regression analyses were done on the new data set developed.

**Table 2.** Data sources for computing f-factors

| Soil property | Soil type | Country | Reference |
|---|---|---|---|
| Soil organic carbon (%) | Alfisols | Burkina Faso | Pichot et al., 1981 |
| | Alfisols | India | Acharya et al., 1988 |
| | Oxisols | Ivory Coast | Chabalier, 1986 |
| | Ultisols | Nigeria | Osuji, 1986 |
| | Ultisols | Nigeria | Mbagwu, 1990 |
| | Alfisols | Nigeria | Lal, 1992 |
| | Ultisols | Senegal | Siband, 1972 |
| | Oxisols | Zambia | Woode, 1983 |
| pH and Al-saturation (%) | Oxisols | Brazil | Martini et al., 1974; Smyth et al., 1992 |
| | Oxisols | Colombia | Spain et al., 1975 |
| | Ultisols | Ghana | Lathwell, 1979 |
| | Oxisols, Ultisols | Indonesia | Tropsoil, 1989; Wade et al., 1982 |
| | Ultisols | Nigeria | Friessen et al., 1982; IITA, 1986 |
| | Oxisols | Peru | NSCU, 1976, 1978, 1980; Tropsoil, 1989; Smyth et al., 1990 |
| | Ultisols | Panama | Manrique, 1987 |
| | Oxisols, Ultisols | Puerto Rico | Abruna et al., 1978; Lathwell, 1979; Kamprath, 1984; Rivera et al., 1985 |
| | Ultisols, Oxisols | Zambia | Singh, 1989; Tveitnes et al., 1989 |
| Nitrogen | Alfisols, Oxisols, Ultisols | Africa | IFDC, 1985 |
| | Oxisols, Ultisols | Ghana | Grove, 1978 |
| | Alfisols | Nigeria | IITA, 1981, 1982, 1983 |
| | Oxisols, Ultisols | Puerto Rico | Grove, 1978 |
| Phosphorus | Oxisols, Ultisols, Alfisols | Africa | IFDC, 1985 |
| | Oxisols | Brazil | Smthh et al., 1990; Lins et al., 1985 |
| | | India | Nair et al., 1988 |
| | Oxisols, Ultisols | Indonesia | Wade et al., 1988 |
| | Alfisols | Nigeria | Kang et al., 1979 |

**Table 2.** continued--

| Soil property | Soil type | Country | Reference |
|---|---|---|---|
| Potassium | Oxisols, Ultisols | Brazil | Cox et al., 1992a,b; Ritchie, 1979 |
| | Oxisols, Ultisols | Indonesia | Wade et al., 1988; Gill et al., 1986 |
| | Ultisols | Peru | NCSU, 1978-79; TropSoils, 1989; Cox et. al., 1992a,b |
| Bulk density | Alfisols | Nigeria | Braide, 1992; Kayombo et al., 1991; Onofiok, 1989 |
| | Ultisols | Peru | Alegre et al., 1991 |
| | Oxisols | Tanzania | Kayombo, 1992 |
| Rooting depth | Oxisols | Brazil | Lathwell, 1979; Sanchez, 1977 |
| | Alfisols, Ultisols | Nigeria | Lal, 1976; Mbagwu, 1984a,b |
| | Ultisols | Peru | Alegre et al., 1986 |
| Weeds | Alfisols, Oxisols Ultisols | Africa | Adobundu, 1980; IITA, 1977, 1979, 1982, 1983 |
| | Alfisols | India | FAO, 1982 |
| | Oxisols | S. America | TropSoils, 1989; Doll, 1977 |

The data chosen for analyses were from those experiments wherein SOC was the main factor accounting for changes in productivity. Using all long-term experiments would have lead to erroneous results because in some experiments the effect of carbon on productivity was masked by other factors influencing soil productivity, e.g. Al-saturation.

# III. F-Factors: Functional Relationships between Soil Properties and Agronomic Productivity

Mathematical functions between soil properties (topsoil 0-20 cm) and relative yield were established for different crops (Table 3), and functions which best fitted the data were selected. The critical limit, the level of soil property corresponding to 80% of the maximum yield, was calculated for each property.

## A. Soil Organic Carbon

The direct effect of SOC on productivity in the tropics is difficult to quantify. Few attempts have been made to establish a quantitative relationship between SOC and productivity. The SOC influences productivity through: (i) supplying

**Table 3.** Critical limits of soil properties, and relationships between soil properties and relative yields for different crops

| Crop | F-factor | Critical limit | $R^2$ |
|------|----------|---------|-------|
| All crops, SOC % | $F_1=0.27+0.67.SOC\%-0.16.(SOC\%)^2$ | 1.05% | 0.37 |
| Maize | $F_{2(pH)}=1-82.5.e^{-(1.21*pH)}$ | 5.0 | 0.62 |
| (*Zea mays*) | $F_{2(Al)}=0.98-7.1.10^{-3}.Al$ | 23.5 | 0.50 |
| | $F_3=0.41+9.2.10^{-3}.N-3.6.10^{-5}. N^2$ | | |
| | $F_4=1-0.95.e^{-(0.204*P)}$ | 7.6 | 0.88 |
| | $F_5=1-0.79.e^{-(1.66*K)}$ | 0.83 | 0.43 |
| | $F_6=-2.25+5.44.r_b-2.27.r_b^2$ | 1.5 | 0.64 |
| | $F_7=1-2.54.e^{-(0.11* D)}$ | 22 | 0.61 |
| | $F_8(W_0)=0.6\ 4,\ F_8(W_1)=0.82$ | 22 | |
| Soybean | $F_{2pH}=-6.41+2.42.pH-0.197.pH^2$ | 5.1 | 0.68 |
| (*Glycine max*) | $F_{2\ Al}=1.05-0.012.Al$ | 20.8 | 0.72 |
| | $F_4=1-1.03.e^{-(0.15*P)}$ | 10.6 | 0.78 |
| | $F_5=1-2.11.e^{-(3.32*K)}$ | 0.7 | 0.80 |
| | $F_6=2.02-0.87.r_b$ | 1.39 | 0.71 |
| | $F_7=1-2.22.e^{-(0 .109* D)}$ | 23 cm | 0.83 |
| | $F_8(W_0)=0.51,\ F_8(W_1)=0.75$ | | |
| Groundnut | $F_2=0.973+5.2.1\ 0^{-3}.Al-1.8.10^{-4}.Al^2$ | 48.6 | 0.64 |
| (*Arachis* | $F_4=1-1.03.e^{-(0.15*P)}$ | 10.6 | 0.78 |
| *hypogea*) | $F_5=1-2.11.e^{-(3.32*K)}$ | 0.7 | 0.80 |
| | $F_7=1-2.22.e^{-(0.109* D)}$ | 22 | 0.83 |
| | $F_8(W_0)=0.51,\ F_8(W_1)=0.75$ | | |
| Cowpea | $F_{2\ (PH)}=-17.93+7.25.pH-0.69.pH^2$ | 4.7 | 0.80 |
| (*Vigna* | $F_{2\ (Al)}=0.99-4.9.10^{-3}.Al$ | 38.8 | 0.31 |
| *ungiuculata*) | $F_4=1-1.03.e^{-(0.15*P-ppm)}$ | 10.6 | 0.78 |
| | $F_5=1-2.11.e^{-(3.32*K)}$ | 0.7 | 0.80 |
| | $F_6=2.02-0.87.r_b$ | 1.39 | 0.71 |
| | $F_7=1-2.2\ 2.e^{-(0.109*D)}$ | 22 | 0.83 |
| | $F_8(W_0)=0.51,\ F_8(W_1)=0.75$ | | |
| Sweet potato | $F_2=0.91+3.7.10^{-3}.Al-2.18.10^{-4}.Al^2$ | 32.4 | 0.83 |
| (*Ipomea batata*) | $F_3=0.524+0.0159.N-1.28.10^{-4}.N^2$ | 20.8 | |
| | $F_4=1-0,653.e^{-(0.138*P)}$ | 8.6 | 0.45 |
| | $F_6=-2.69+5.81.r_b-2.3\ 4.r_b^2$ | 1.46 | 0.72 |
| Cassava | $F_2=0.94+2.7.10^{-3}.Al-6.6.10^{-5}.Al^2$ | 70.8 | 0.57 |
| (*Manihot* | $F_3=0.83+2.1.10^{-3}.N-7.1.10^{-6}.N^2$ | | |
| *esculenta*) | $F_4=1-0,653.e^{-(0.138*P )}$ | 8.6 | 0.45 |
| | $F_6=-2.69+5.81.r_b-2.34.r_b^2$ | 1.46 | 0.72 |
| | $F_8(W_0)=0.37, F_8(moderat.)=0.68$ | | |

pH=pH ($H_2O$), N=kg n/ha, P=Bray-1 ppm, K=mmol($K^+$)kg$^{-1}$; $\rho_b$=bulk density Mgm$^{-3}$; D=rooting depth (cm); $F_8(W_0)$=no weeding; $F_8=(W_1)$=one weeding (moderate weeding in cassava); 1=cowpea data, 2=cowpea/soybean data, 3=cassava data.

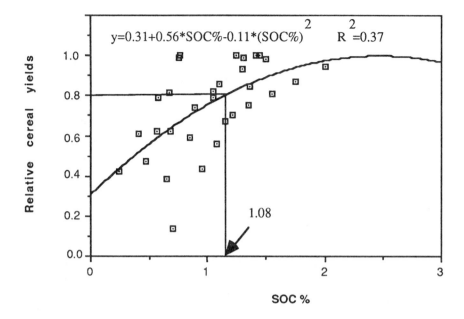

**Figure 1.** Effect of soil organic carbon on relative cereal yields.

nutrients for plant growth and increasing CEC, (ii) improving soil aggregation and root development, and (iii) facilitating soil and water conservation (Allison, 1973). Plants are able to grow without SOC, but results from the savannah region of West Africa showed that high soil productivity was not associated with low SOC (Pieri 1992 ).

The levels of SOC examined ranged from 0.3 to 2%, and the function relating SOC and productivity was a quadratic equation (Figure 1). The data in Figure 1 depicts that the effect of SOC on productivity is rather weak, being more pronounced as SOC drops below 1%. At a SOC level of 0.37%, cereal yield (maize and sorghum) yield is 50% of the potential yield. It is assumed that the relationship developed for cereals also can be used for other crops.

## B. Soil Acidity

Different crop species were evaluated for tolerance to pH and Al-saturation (Table 1). The experimental data on pH ranged from 3.9 to 6 and that of Al-saturation from 0 to 100%. Crop species are ranked in the following ascending order with regard to tolerance to Al-saturation: soybean > maize > groundnut > sweet potato > cowpea > rice > cassava. This ranking corresponds well with that of Wade et al. (1988) for crops in Indonesia except that in that study groundnut was slightly less resistant to Al-saturation than cowpea.

Maize and soybean were the least tolerant species to soil acidity, and both have a critical pH level of 5. Yields are very low for both species at pH 4. Soybean is also less tolerant to high Al-saturation than maize. Cowpea produced 55% of the potential yield at a pH as low as 4.4, and the critical pH was 4.7. Groundnut can produce satisfactory yield at moderate Al-saturation, but is unable to produce any yield at high Al-saturation. Sweet potato is moderately tolerant to Al-saturation. Cassava is the most adapted species to acid soil conditions as evidenced by the very low correlation between pH and relative yield. Cassava yield decreases by only 20% at an Al-saturation of about 70% and produces satisfactory yield even at an almost complete Al-saturation (Table 1).

## C. Soil Nutrients

### 1. Nitrogen

Soil N is a highly labile property and no single soil analysis is adequate to predict its supply to crop over the growing season. For this reason, the effect of N on crop productivity is not calculated using soil analysis. Instead, data on crop response to N-fertilizer were used. The N response also varies with cultivar and the expected yield level. It is, therefore, recommended to calculate the nitrogen response using site- and cultivar- specific data. For soils where such data are not available, the equation developed by IFDC for maize in the sub-humid regions of Africa can be used (IFDC 1985).

Sweet potato has only a moderate demand for N, and 20 kg N/ha produces 80% the maximum yield. This observation is in accord with general fertilizer recommendations for sweet potato of 35-45 kg N/ha (Onwueme 1978). Cassava responds much less to N than other species. Even without N application, cassava produces 83% of its optimum yield (Table 3).

### 2. Phosphorous

A problem in establishing functional relationship between available P and relative yield lies in variable methods of P extraction used by different laboratories. To overcome this limitation, P data were transformed to equivalent Bray-1 level by using site-specific functions. In Brazil, where the Mehrlich method is used, the data were transformed to Bray-1 by using the formula developed by TropSoil (1989) and in Indonesia data were transformed from Olsen-P by using paired analyses conducted by IRRI (1991).

The relationship between available-P and relative yield also follows the law of diminishing returns ,

$$Y = 1-B_1 e^{-B2*P}$$
(Eq. 3)

where, $B_1$ and $B_2$ are constants, and Bray-1 P is in ppm. The critical levels of available P for maize, grain legumes, and root crops are 7.6, 10.8, and 8.6 ppm,

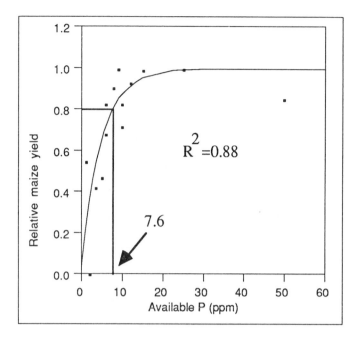

**Figure 2.** Relationship between available-P (ppm) and relative yield of maize.

respectively. Grain legumes are known to have a high demand for P and, therefore, have a high critical level. Figure 2 depicts the relationship between available P and relative maize yield. The shape of the curve for the root crops differs from that of other species. Cassava maintains its yield even at very low levels of available P.

## 3. Potassium

The model depicted in Equation 3 also is used to describe the relationship between exchangeable K and relative yield. The critical level of K for maize, rice, and grain legumes are 0.83, 0.64, and 0.7 mmol $(K^+)kg^{-1}$, respectively. The relationship between exchangeable K and relative yield of soybean/cowpea is presented in Figure 3 .

## D. Bulk Density

Bulk density is chosen as an indicator for soil physical properties because it is a key factor influencing productivity in the tropics (Stewart et al., 1991). It affects water infiltration, root growth and uptake of nutrients and water (Babolola and Lal, 1977). The functional relationship between relative maize yield and soil

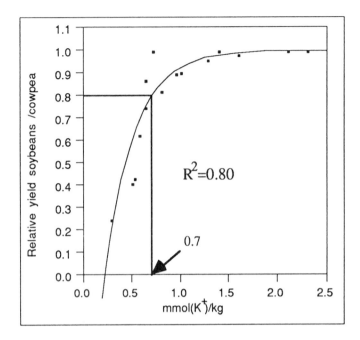

**Figure 3.** Relationship between exchangeable K and relative yield of soybean/cowpea.

bulk density is shown in Figure 4. The critical level of bulk density for maize, grain legumes, and cassava is 1.5, 1.39, and 1.5 Mg m$^{-3}$, respectively. Cowpea yields are less sensitive to high bulk densities than maize, probably because of its tap root system.

### E. Rooting Depth

Rooting depth is a crucial factor in soil productivity because it determines soil reserves of water and nutrients. Rooting depth is restricted by subsoil acidity, poor soil aeration and presence of hardpans. Accelerated soil erosion also reduces rooting depth. There is no direct method for measuring the effect of rooting depth on productivity. However, experimental data available from studies designed to evaluate the effects of factors limiting rooting depth can help establish functional relationships. These experiments are comprised of sub-liming, sub-tillage, and soil surface removal studies.

The relationship between rooting depth and productivity also is described according to the law of diminishing returns (Equation 3). The critical value of rooting depth for maize is 23 cm (Figure 5). Mean soil water holding capacity of soils in the tropics is about 1.3 mm water/cm soil (Lal, 1987). This implies that a soil depth of 23 cm has an available soil water holding capacity of 30 mm.

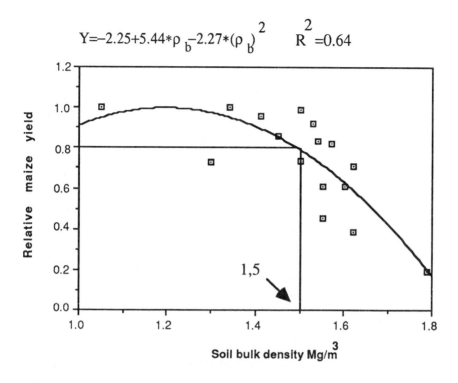

**Figure 4.** Relationship between bulk density and relative yield of maize.

For an evapotraspiration rate of 5 mm/day, this water is sufficient for about 6 days of crop growth. The critical rooting depth is 22 cm for cowpea.

## F. Weeds

Weed control   is essential for obtaining satisfactory yields. Heavy weed infestation, as often is the case in the tropics (Akobundu 1980), can offset the effect of other factors.  The yield suppressive effect of weeds is computed by dividing the yield in non-weeded treatments by that in weed-free treatments. Two weedings are normally sufficient in cereals and grain legumes, but cassava needs protection against weeds for the first three months. In the model, two weedings are assigned the value of one both in cereals and grain legumes.

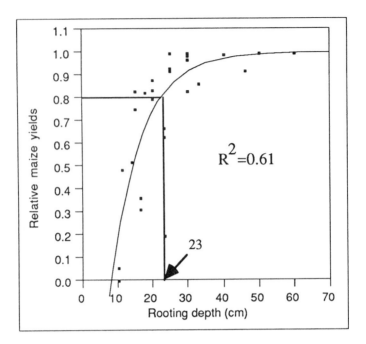

**Figure 5**. Relationship between rooting depth and relative productivity for maize.

## IV. Model Testing and Use

### A. Calibration

Data from long-term experiments conducted for more than 8 years were used to calibrate the model. To remove the effect of climatic variability on yield, yield trends were obtained by fitting a curve to data in each experiment. The model was then calibrated against the yield data obtained using the following procedure:

1. Choose a potential yield under best management practices. A potential yield of 10 Mg ha[-1] was chosen for maize in the sub-humid and humid climate and 7 Mg ha[-1] for sorghum in a semi-arid climate. A potential yield of 3.5 Mg ha[-1] was chosen for soybean.

2. Determine the nitrogen factor for the region. The N-factor can either be determined from independent studies in the same region or on the basis of the response of nitrogen in the first year of the experiment. The latter approach is used for calibrating this model.

3. Adjust $B_2$ factor for each location in order to fit the right yield level. The $B_2$ factor is adjusted on the basis of the first year data in long-term experiments, and it accounts for the difference between observed and the predicted yields.

4. Finally, different values for the B1 factor are tested and the value 2/3 produces the overall best fit. If B1 is set at 1, the model underestimates the predicted yield when the product of f-factors approaches zero. On the other hand, if B1 is 0.5, yields are overestimated when f-factors are low.

## B. Testing the Model

The model is tested using data from three long-term experiments conducted in Kasama in Zambia (Woode 1983), Saria in Burkina Faso (Pichot et al., 1981), and Yurimaguas in Peru (Sanchez et al. 1983). The soil data are input into Equation 5:

$$Y = \text{Potential yield} * (F_1 * F_2 * F_3 * F_4 * F_5 * F_6 * F_7)^{0.8} * B2 \quad \text{(Eq. 5)}$$

The B2 constant for Kasama, Yurimaguas, and Saria sites are 0.6, 0.32, and 0.6 respectively. When tested, the pH was used as an acidity indicator. The weed factor is not used when testing the model because weeds were controlled in all experiments used in this model. The yield was predicted for every year the soil properties were measured. In Saria, sorghum (*Sorghum bicolor*) was the crop grown, but the F-factors for maize were used instead because maize and sorghum have similar response to changes in soil properties. The experiments that were used to test the model must be considered almost completely independent of the data that were used to construct the model. The only common data are those for soil organic matter, but these experiments only represent two of the eight experiments that were used to develop the F-factor for soil organic matter.

The regression equation between measured yields (y) and predicted yields gives: $Y = 0.14 + 0.91 * X$, $r^2 = 0.91$. The perfect model would give an intercept of zero, inclination coefficient of 1, and $r^2$ of 1. Based on these characteristics, it is apparent that the model produces satisfactory results (Figure 6). Principal factors affecting productivity at these locations are acidity (Kasama, Yurimanas), SOC (Kasama and Saria), rooting depth (Yurimaguas), and fertilizer use (all sites).

The long-term experiment used to test the model for soybeans is an 8–year study conducted in Yurimaguas, Peru (Sanchez et al., 1983). The general model used to fit the data is the same as for maize except that parameters used have different values. The same B1 is used as for maize (0.8). Treatments included in this experiment were control and complete fertilization and liming. Soil parameters that improved over time in the complete fertilization and liming treatment were Al-saturation and rooting depth. However, SOC decreased with time. The model adequately describes the effect of changes in soil properties on yields with $r^2$ of 0.97 (Figure 7).

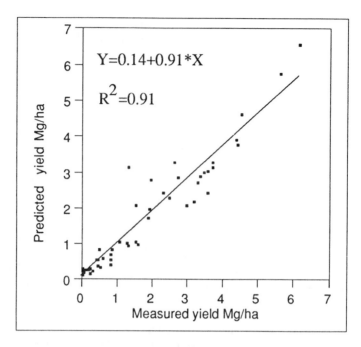

**Figure 6.** Relationship between measured and predicted yield for maize in Zambia and Peru and for sorghum in Burkina Faso

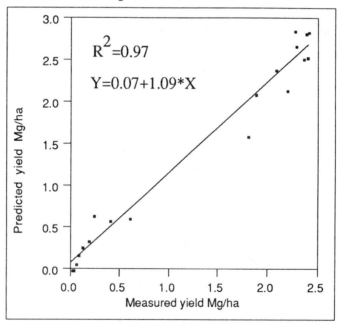

**Figure 7.** Relationship between measured yield and predicted yield for soybeans in Yurimaguas.

## C. Applications and Predictions of Long-Term Effects

The Tropical Soil Productivity Calculator is developed using data from Oxisols, Ultisols, and Alfisols in the tropics, and the model can, therefore, be used for these soils only. This model is usable for the following purposes:

1. Predicting short-term effects of liming and subliming, phosphorous and potassium applications, and tillage and subsoiling on productivity.

2. Predicting yield trends as a result of changes in soil management.

3. Evaluating land capability classification.

4. Assessing susceptibility of soil to accelerated erosion, by estimating changes in yield due to decrease in rooting depth.

If site-specific response functions are available for the relationship between growth factors and relative yield, it is appropriate to use such functions. Otherwise, functions can be developed from locally available data. These functions can then replace some of the f-functions proposed in the Tropical Soil Productivity Calculator. Nonetheless, the general model can still be used. If the response functions to N, P, and K are available, these functions are preferred over the functions between soil properties and relative yield reported in this study.

Using this model for long-term prediction requires some basic assumptions on how soil properties change over time. These basic assumptions include knowledge about changes in SOC, pH, N fertilization, P, K, bulk density, and rooting depth. The following methods are proposed to predict changes in these soil properties:

1. Changes in SOC for long-term predictions can be estimated from knowledge of cultivation history, residue management, and tillage practices. Nye and Greenland (1960) proposed that SOC can be calculated according to the model shown in Equation 7:

$$\text{Annual change in SOC} = \text{rate of addition of SOC} - k_c \text{SOC} \qquad \text{(Eq. 7)}$$

were $k_c$ is the decomposition constant. The Century and SCUAF models, which are based on the same principles as Equation 7, can also be used to predict changes in SOC.

2. Few, if any, models are available for predicting changes in pH and Al-saturation in soils of the tropics. Soil pH can, therefore, be estimated by considering soil cultivation history, nitrogen fertilizer use, liming and the soil-buffer capacity. The acidity factor can be set at one, if sufficient lime is applied or if Al-tolerant varieties are used.

3. The f-factors for N, P, and K can be estimated by making assumptions about future fertilizer use. With optimum fertilizer use, the f-factor can be set at one. Another option for predicting changes in P is using the model developed by Janssen et al. (1990).

4. Soil bulk density can be estimated by knowledge of cultivation history, tillage systems, soil texture, and intensity and frequency of farm traffic.

5. Rooting depth depends on several factors and may be increased as a consequence of decrease in subsoil acidity and bulk density and decreased due to soil erosion and increased subsoil compaction.

## D. Limitations

A principal criticism of this model lies in its general outline. In a multiplicative model, limited degree of substitution among factors is possible even though the maximum yield is determined by the most limiting factor. This kind of substitution does not normally occur in the real world (Paris, 1992). However, in this model true substitution is possible because not all factors included in the model are independent. For example, an increase in pH can increase P availability. Nitrogen supply and SOC also are interdependent (Janssen et al., 1990).

The relative yield used in the model also is subjective depending on the number and level of treatments used. The relative yield in the treatment with highest yield is always one, indicating that this growth factor is not limiting production. However, other treatment outside the range tested, may produce a higher yield, which can result in an overestimation of the f-factor.

The model may also have some problems in predicting yields for soils prone to severe crusting. Even though both bulk density and SOC are factors that influences soil crusting, the f-factors for these properties are not strong enough to account for a dramatic change in infiltration rate.

## V. Conclusions

1. The Tropical Soil Productivity Calculator is developed to predict yield in Oxisols, Ultisols, and Alfisols of the tropics.

2. Critical levels for soil properties vary among crop species.

3. Yield can be predicted by multiplying factors for SOC, soil acidity, N, P, K, bulk density, rooting depth, and weed infestation. The most limiting factor sets the limit for maximum productivity at any location.

4. The model can predict productivity in relation to fertilizer use, liming, residue management, tillage, vehicular traffic, and soil erosion.

5. The predicting capability of the model is high when tested with data from long-term experiments in the tropics.

6. The model can be used as a tool for predicting yield when analyses of some basic soil properties are available.

## References

Abruna, F., J. Rodrigues, J. Badillo-Feliciano, S. Silva, and J. Vicente-Chadlier 1978. Crop response to soil acidity factors in Ultisols and Oxisols in Puerto Rico-soybeans. *J. of Agric. of the Univ. of Puerto Rico* 62/63:90-112.

Acharya, C.L., S.K. Bishnoi, and H.S. Yaduvanshi. 1988. Effect of long term application of fertilisers, and organic amendments under continuous cropping on soil physical and chemical properties in Alfisols. *Indian J. of Agric. Sci.* 58:509-516.

Akobundu, I.O. 1980. Weed research at the International Institute of Tropical Agriculture and research needs in Africa. *Weed Science* 28:439-445.

Alegre J.C., D.K. Cassel, and D.E. Bandy 1986. Reclamation of an Ultisols damaged by mechanical land clearing. *Soil Sci. Soc. Am. J.* 50:1026-1031.

Alegre, J.C. and D.K Cassel. 1991. Conservation tillage in the humid tropics of Peru. p. 229-237. In: *International Soil Tillage Research Organisation, 12th International Conference.* Ibadan, Nigeria.

Allison, F.E. 1973. *Soil organic matter and its role in crop production.* Elsevier, Amsterdam, 637 pp.

Babolola O. and R. Lal 1977. Subsoil gravel horizon and maize root growth. I. Gravel concentration effects and bulk density effects. *Plant and Soil* 46:337-346.

Bandy, D.E. 1980. Lime experiment. Agronomic-Economic Research on soils of the tropics 1978-1979 Report. Soil Science Department, North Carolina State University Raleigh , N.C.

Bornemisca E. and A. Alvarado 1975. Lime response of corn and beans grown on typical Oxisols and Ultisols of Puerto Rico. p. 262-286 In: *Soil Management in Tropical America.* North Carolina State University, Raleigh.

Braide, F.G. 1991. The influence of machine traffic on growth of maize. p. 229-237. In: *International Soil Tillage Research Organization 12th International Conference.* Ibadan, Nigeria.

Chabalier, P.F. 1986. Evolution de la fertilité d'un sol ferralitique sous culture continue de mais en zone forestiére tropical. *L Agron. Trop.* 41:179-191.

Cox, F.R. and E. Uribe 1992a. Management and dynamics of potassium in a humid tropical Ultisols under a rice-cowpea rotation. *Agron. J.* 84:655-660.

Cox, F.R. and E. Uribe 1992b. Potassium in two humid tropical Ultisols under a corn and soybean cropping system. I. Management. *Agron. J.* 84:480-484.

Doll, J.D. 1977. Weeds: an economical problem in cassava. p. 65-70 In: T. Brekelbaum, A. Bellotti, and J.C. Lazanao (eds.), *Proc. cassava protection workshop.* CIAT. 7-12 November 1977. Cali, Colombia.

Food Agriculture Organization 1982. Weeds in tropical crops: review of abstracts. FAO Plant Production and Protection Paper 32 (suppl. 1), 63 pp.

Food Agriculture Organization. 1991. The state of food and agriculture. World and regional reviews. Agricultural policies and issues: lessons for the 1980's and prospects for the 1990's. Rome.

Friessen, D.K., A.S.R. Juo, and M.H. Miller. 1982. Residual value of lime and leaching of calcium in a kaolinitic Ultisols in the high rainfall tropics. *Soil Sci. Soc. Am. J.* 46:1146-1189.

Gill, D.W. and J.S. Andiningsih. 1986. The response of upland rice and soybeans to potassium fertilization , residue management and green manuring in Sitiung, West Sumatra. Pembr. Pen. Tanah. Dan Pupuk no 6:26-32.

Grove, T.L. 1978. Nitrogen fertility in Oxisols and Ultisols of the humid tropics. *Cornell Inter. Agric. Bulletin* 36:1-28.

International Fertilizer Development Centre. 1985. Fertiliser Research Program for Africa. The fate, sources, and management of nitrogen and phosphorus fertilisers in sub-Saharan Africa. Muscle Shoals, Alabama. 120 pp.

International Institute of Tropical Agriculture 1977. Annual Report 1976. Ibadan, Nigeria. 126 pp.

International Institute of Tropical Agriculture 1979. Annual Report 1978. Ibadan, Nigeria. 129 pp.

International Institute of Tropical Agriculture 1981. Annual Report 1980. Ibadan, Nigeria. 185 pp.

International Institute of Tropical Agriculture 1982. Annual Report 1981. Ibadan, Nigeria. 178 pp.

International Institute of Tropical Agriculture 1983. Annual Report 1981. Ibadan, Nigeria 217 pp.

International Institute of Tropical Agriculture 1987. Annual Report and Research Highlights 1986. Ibadan, Nigeria. 154 pp.

International Rice Research Institute 1991. Program Report for 1991. Los Banos, 322 pp.

Janssen, B.H., F.C.T. Guiking, D. van der Eikj, E.M.A. Smaling, J. Wolf, and H. van Reuler 1990. A system for quantitative evaluation of the fertility of tropical soils (QUEFTS). *Geoderma* 46:299-318.

Kamprath 1984. Crop response to lime of soil in the tropics. Soil Acidity and Liming. *Agronomy Monographs* 12:349-369.

Kang, B.T. and O.A. Osimane. 1979. Phosphorus response of maize grown on Alfisols of Southern Nigeria. *Agron. J.* 71:873-877.

Kayombo, B. 1991. Subsoil compaction effects on soil structure and yield of maize on an Oxisols. p: 247-252. In: *International Soil Tillage Research Organization 12th International Conference.* Ibadan, Nigeria.

Kayombo, B. and R. Lal 1986. Influence of traffic induced compaction on growth and yield of cassava. *J. Root Crops* 12:19-23.

Kayombo, B., R. Lal, G.C. Mrema, and H.E. Jensen 1991. Characterizing compaction effects on soil properties and crop growth in southern Nigeria. *Soil & Tillage Res.* 21:321-345.

Lal, R. 1976. Soil erosion problems on an Alfisols in western Nigeria and their control. IITA Monograph No. 1. Ibadan, Nigeria.

Lal, R. 1987. *Tropical ecology and physical edaphology.* John Wiley & Sons, New York. 732 pp.

Lal, R. 1992. Tropical Agriculture and sustainability of agricultural systems. A ten year watershed management project in Southwestern Nigeria. The Ohio State University. 303 pp.

Lathwell, D.J. 1979. Crop response to liming of Ultisols and Oxisols. *Cornell Inter. Agric. Bull.* 35:1-35.

Lins, I.D.G., F.R. Cox, and J.J. Nicholaides. 1985. Optimizing phosphorus fertilization rates for soybean grown on Oxisols and associated Entisols. *Soil Sci. Soc. of Am. J.* 49:1457-1460.

Manrique, L.A. 1987. Response of cassava to liming on a strongly acid Ultisols of Panama. *Commun. in Soil Sci. Plant Anal.* 18:115-130.

Mbagwu, J.S.C, R. Lal, and T.W. Scott. 1984a. Effects of desurfacing of Alfisols and Ultisols in Southern Nigeria: I. Crop performance. *Soil Sci. Soc. of Am. J.* 48:828-833.

Mbagwu, J.S.C., R. Lal, and T.W. Scott. 1984 b. Effects of artificial desurfacing on Alfisols and Ultisols in Southern Nigeria: II. Changes in soil physical properties. *Soil Sci. of. Am. J.* 48:834-838.

Mbagwu, J.S.C 1990. Maize (*Zea maize*) response to nitrogen fertiliser on an utilsol in Southern Nigeria under two tillage and mulch treatments. *J. Sci. Food Agric.* 52:365-376.

Nair, P.G., B. Mohankumar, M. Prabhakar, and S. Kabeerathumma. 1988. Response of cassava to graded doses of phosphorous in an acid laterite soil of high and low P-status. *J. Root Crops* 14:1-9.

North Carolina State University. 1976. Agronomic and economic research on tropical soils. Annual Report for 1975. Raleigh, NC. 312 pp

North Carolina State University. 1978. Agronomic and economic research on tropical soils. Annual Report for 1976-1977. Raleigh, NC. 267 pp.

North Carolina State University. 1980. Agronomic and economic research on tropical soils. Annual Report for 1978-1979. Raleigh, NC. 284 pp

Nye, P.H. and D.J. Greenland. 1960. The soil under shifting cultivation. Technical Communication No. 51. Commomwealth Bureau of Soils. Harpenden. 156 pp.

Martini, J.A., R.A. Kochann, O.J. Siqueira and C.M. Borkert 1974. Response to liming as relaaed to soil acidity, Al acidity and Mn toxicities and P in some Oxisols in Brasil. *Soil Sci. Soc. Amer. Proc.* 38:616-620.

Onofiok. O.E. 1989. Effect of soil compaction and irrigation interval on the growth and yield of cowpea on a Nigerian Ultisols. *Soil & Tillage Res.* 13:47-55.

Onwueme, I.C. 1978. *The Tropical Tuber Crops*. John Wiley & Sons. NY. 234 pp.

Osuji, G.E. 1986. Zero-tillage for erosion control and maize production in the humid tropics. *J. Environ. Management* 23:193-201.

Paris 1992. The return of Liebig's "law of the minimum". *Agron. J.* 84:1040-1046.

Parton, WJ., J.W.B. Stewart, D.S. and C.V. Cole 1988. Dynamics of C, N, P, and S in grassland soils: A model. *Biogeochemistry* 5:109-131.

Pichot K., M.P. Sedego, J.F. Polain, and R. Arrivets. 1981. Evolution de la fertilité d'un sol ferrugineux sous l'influence des fumures minérales et organiques. *L Agron. Trop.* 36:122-183.

Pieri, C. 1992. *Fertility of soils. A future for farming in the West African Savannah*. Springer-Verlag, Berlin. 348 pp.

Ritchie, K.D. 1979. Potassium fertility in Oxisols Ultisols of the humid tropics. *Cornell Inter. Agric. Bull.* 37:1-44.

Ritchie, J.T., U. Singh, D.C. Goodwin, and L. Hunt. 1989. A user guide to CERES maize-V. 2.10. International Fertilizer Development Center. Muscle Shoals, AL.

Rivera, E., F. Abruna, and J. Rodrigues. 1985. Crop response to soil acidity factors in Ultisols in Puerto Rico X1. Cassava. *J. of Agric. Univ. of Puerto Rico* LXIX:145-151.

Sanchez, P.A. 1977. *Properties and management of soil in the tropics*. John Wiley & Sons, NY. 619 pp.

Sanchez, P.A., J.H. Villachia, and D.E. Brady. 1983. Soil fertility dynamics after clearing a tropical rainforest in Peru. *Soil Sci. Soc. Am. J.* 47:1171-1178.

Sanchez, P.A. and T.J. Logan. 1992. Myths and science about the chemistry and fertility of soils in the tropics. p. 35-46. In: *Myths and Science of Soils in the Tropics*. SSSA special publication no. 49.

Siband, P. 1972. Etude de l'évolution des sols sous culture traditionelle en Haute-Casamance. Principaux résultats. *L Agron. Trop.* XXVII:574-591.

Singh, B. R. 1989. Evaluation of liming materials as ameliorant of acid soils in high rainfall areas of Zambia. *Norwegian J. of Agric. Sci.* 3:13-21.

Smyth T.J. and M.S. Cravo 1990. Critical phosphorous levels for corn and cowpea in a Brazilian Amazon Oxisols. *Agron. J.* 82:309-312.

Smyth, T.J. and M.S. Cravo. 1991. Continuos cropping experiment in Manaus: M-901. 1988-1989. p. 252-254.

Smyth T.J. and M.S. Cravo 1992. Aluminium and calcium constraints to continous crop production in a Brazilian Amazon Oxisol. *Agronomy J.* 84:843-850.

Spain, J.M., C.A. Francis, R.H. Howler, and F. Calvo. 1975. Differential species and varietal tolerance to soil acidity in tropical crops and pastures. p. 308-324. In: E. Bornemisza and A. Alvarado (eds.), *Soil Management in Tropical America*. N.C.S.U. Raleigh, NC.

Stewart, B.A., R. Lal, and S.A. El-Swaify. 1991. Sustaining the resource base of an expanding world agriculture, p. 125-144. In: R. Lal. and F.J. Pierce (eds.), *Soil Management for Sustainability*. Soil and Water Cons. Soc., Ankeny, IA.

Stoorvogel, J.J. and E.M.A. Smaling. 1990. Assessment of soil nutrient depletion in Sub-Saharan Africa: 1983-2000. Volume 1: Main report. The Winard Staring Centre. Wageningen, 137 pp.

Tveitnes S. and H. Svads. 1989. The effect of lime on maize and groundnut in the high rainfall areas of Zambia. *Norwegian J. of Agric. Sci.* 3:173-180.

TropSoils 1989. TropSoils technical report, 1986-1987. North Carolina State University, Raleigh, NC. 379 pp.

Wade, K.W., D.W. Gill, H. Subabangjo, M. Sudjadi, and P.A. Sanchez. 1988. Overcoming soil fertility constraints in a transmigration area of Indonesia. Trop. Soils Bull. 88-01. 60 pp.

Woode P.R. 1983. Long-term fertiliser experiments from Kasama, Zambia. M.Sc. thesis. University of Aberdeen. 123 pp.

World Bank 1993. World development report 1992. Development and environment. Washington, D.C..

Young, A. and P. Muraya. 1990. SCUAF Soil Changes under Agroforestry. A Predictive Model. Version 2, ICRAF. Nairobi, Kenya. 124 pp.

# Trends in World Agricultural Land Use: Potential and Constraints

### R. Lal

## I. Introduction

Land embodies one or several ecosystems but also can be a part of an ecosystem. Land use is a systematic manipulation of these ecosystems for human needs. The manipulation may lead to positive effects and to the enhancement of ecosystems' life support processes, or to negative effects resulting in land degradation and diminution of its productive capacity. Both land and ecosystems are, therefore, dynamic entities, and the magnitude of alterations in land's attributes depends on specific land use.

Agricultural land use includes arable (food crops), pastoral (animal grazing), silvicultural (forestry), and any combination of these, e.g., agro-pastoral, silvo-pastoral, or agro-silvo-pastoral land use for growing seasonal or annual domesticated plants, e.g., crops. The principal types of agricultural land uses are listed in Table 1.

In addition to a characterization by form of agriculture, agricultural land uses can also be classified on the basis of land-use intensity. This is expressed by a land-use factor ($L = C + F/C$, where C is the cropping period, and F is the fallow period). Shifting cultivation systems are listed in Table 2 in order of intensity. An extensive land use reflects subsistence agriculture. As the demographic

ISBN 1-56670-076-0

**Table 1.** Principal types of agricultural land use

Primitive agriculture
    Shifting (long fallow) agriculture
    Nomadic herding

Traditional agriculture
    Current fallow agriculture
    Continuing extensive mixed agriculture
    Labor-intensive, nonirrigated crop agriculture
    Labor-intensive, irrigated crop agriculture
    Labor-intensive, irrigated semicommercial crop agriculture
    Labor-intensive, nonirrigated semicommercial crop agriculture
    Low-intensive, semicommercial crop agriculture
    Large-scale, low-intensive, semicommercial agriculture

Market-oriented agriculture
    Mixed agriculture
    Intensive agriculture with fruit growing and/or market gardening dominant
    Large-scale, specialized agriculture with livestock breeding dominant
    Plantation agriculture
    Specialized, irrigated agriculture
    Specialized, large-scale grain crop agriculture
    Specialized, large-scale grazing (ranching)

Socialized agriculture
    Mixed agriculture
    Specialized fruit and/or vegetable agriculture
    Specialized industrial crop agriculture
    Specialized grain crop agriculture
    Specialized grazing
    Labor-intensive, nonirrigated crop agriculture
    Labor-intensive, irrigated crop agriculture

(Adapted from Vink, 1975.)


pressure increases, land scarcity leads to an intensive semicommercial or a commercial land use. If land is used more intensively due to increasing population density but without change in land management, severe land degradation may result. Crops can be efficiently grown only if weeds are removed by physical manipulation of the soil surface (e.g., plowing, harrowing, etc.) or by using growth regulators or herbicides.

    The nutrients harvested in crops must be replaced, in adequate amounts and in readily available form, if continuous high yields are desired. If not, land degradation is an inevitable consequence. Land degradation can be aggravated

**Table 2.** Shifting cultivation systems from most to least extensive

| Land use | Land-use factor[a] |
|---|---|
| 1. Nomadic herding: shifting cultivation (Phase 1) | >10 |
| 2. Bush fallowing or land rotation: shifting cultivation (Phase II) | 5-10 |
| 3. Rudimentary sedentary agriculture: shifting cultivation (Phase III) | 2-4 |
| 4. Compound farming and extensive subsistence: shifting cultivation (Phase IV) | <2 |

[a] Land-use factor (L) = C + F/C, where C = cropping period, F = fallow period.

(Data from Okigbo and Greenland, 1976; Benneh, 1972.)

if cropping intensity increases (the factor L in Table 2 decreases) without proper changes in land management, e.g., improved systems or off-farm input.

## II. Global Land Resources and Agricultural Land Use

Land resources of the world are finite, fragile, and nonrenewable. Agriculturally suitable land is only a small fraction of the total land area. The land area of the earth is $14.8 \times 10^9$ ha, which is 29% of the earth's surface. About 10% of the total world land area ($1.4 \times 10^9$ ha) is covered with ice. About $13.4 \times 10^9$ ha, therefore, is the available land area. In addition, 15% of the total land is too cold to grow crops, 17% is too dry, 18% is too steep, 9% is too rocky and stony with shallow soils, 4% is too wet, and 5% has other constraints (Buringh, 1989). This means that only 22% of all land or about $3.26 \times 10^9$ ha is potentially suitable for cultivation, and only about 50% of the potentially suitable land (about 11% of the total) is presently cultivated. Only 3% of the world's land area ($450 \times 10^6$ ha) has a high agriculturally productive capacity.

Only a fraction of the total land area is currently being used for cultivation. The cropland area in the world has increased 5.66 times since 1700 and has approximately tripled since 1850 (Table 3). The cultivated land area increased rapidly over the century ending in 1980. The current rate of expansion of cultivated land area is low, e.g., 0.94%/yr.

Although there is a potential to expand agricultural area, there are ecological risks involved. A principal risk is degradation of land resources due to excessive use, misuse, or inappropriate use. Land degradation implies diminution in land's productive capacity due to such processes as erosion, compaction, salinization, laterization, depletion of soil organic matter and plant nutrient reserve,

**Table 3.** Change in global land use from 1700 to 1980

| Year | Global cropland | |
| | Area change (ha $\times 10^6$) | Change (% yr$^{-1}$) |
|---|---|---|
| 1700 | 265 | - |
| 1850 | 537 | 0.68 |
| 1920 | 913 | 1.00 |
| 1950 | 1170 | 0.94 |
| 1980 | 1501 | 0.94 |
| 1988 | 1475 | - |

(Adapted from FAO, 1989; Richards, 1991; Schlesinger, 1984.)

acidification, or nutrient imbalance. While land area of high productive capacity is decreasing, world population is rapidly expanding, especially in regions with low reserves of cultivable land area. Consequently, there are five major concerns to be addressed:

- What is the actual and potential cultivable land area in the world and in different geographical regions?
- What is per capita land area?
- What are predominant land use systems?
- What are trends in land use for different geographical regions?
- What are the policy implications to ensure  sustainable land use?

## III. Potential Cultivable Land Area and Per Capita Land Base

The potential land base for rainfed agriculture for developing countries is shown in Table 4. Out of total land area of 6.4 x $10^9$ ha, 30.6% is suitable for agriculture and an additional 9.8% is marginal. About 60% of the land area is unsuitable for rainfed agriculture. Out of the total agriculturally suitable land area of 1.97 x $10^9$ ha, 41.4% is in South America, and 40.0% is in Africa. Asia, with a high population density, has only 17.3% of the potentially arable land base suitable for rainfed agriculture. Potential for expansion of agriculture, therefore, lies mostly in Africa and South America.

Like population, land resources in developing regions of the world are unequally distributed. There are 90 countries classified as developing regions. Because of the rapidly increasing population of these countries, per capita arable land area is decreasing. By the year 2000, the per capita arable land area is projected to be 0.15 ha in the Far East, 0.26 ha in the Near East, 0.39 ha in Asia, and 0.45 ha in Latin America. The average per capita land area of 0.37 ha in 1975 for the developing countries on the whole is projected to decrease to 0.25 by the year 2000 (FAO, 1981).

**Table 4.** Potential land base for rainfed agriculture

| Region | Area ($10^6$ ha) | | | |
| | Suitable | Marginal | Unsuitable | Total |
| --- | --- | --- | --- | --- |
| Africa | 789 | 231 | 1858 | 2878 |
| Southeast Asia | 294 | 226 | 378 | 898 |
| Southwest Asia | 48 | 16 | 613 | 677 |
| Central America | 24 | 15 | 182 | 221 |
| South America | 819 | 147 | 804 | 1770 |
| | 1874 | 635 | 3835 | 6444 |

(From FAO, 1984.)

The land area per agricultural worker in 1988 is estimated at 0.13 ha for centrally planned economies in Asia, 0.17 ha for the Far East, 0.40 ha for Africa, 0.73 ha for the Near East, and 1.3 ha for Latin America (FAO, 1988). While the arable land area per agricultural worker on the whole is 0.58 ha, that in the developed countries is 6.10 ha. Decline in arable land area per agricultural worker for the decade ending in 1988 was most drastic in Asia. For the decade ending in 1988, land area per agricultural worker declined by 74% in Asia (CPI), 83% in the Far East, 65% in the Near East, 73% in Africa, and 55% in Latin America. In comparison, decline in arable land per agricultural worker in developed countries was 32%. These calculations based on percentages can be misleading, however, if not considered in conjunction with the absolute land area available per worker.

Only 4.5% of the agricultural population of the world is from the developed countries. Asia, with an agricultural population of 1.6 billion, has 68% of world's agricultural-worker population. Therefore, productivity per agricultural worker is rather low in developing countries.

## IV. World Land Use

Statistics on cultivated land in different regions are shown in Table 5. Currently there are 1.474 x $10^9$ ha in cropland. Of this total, 3.3% is in Oceania, 9.4% in South America, 9.5% in Europe, 12.6% in Africa, 15.7% in the former USSR, 18.6% in North and Central America, and 30.9% in Asia.

In the 102-year period ending in 1984, there has been an increase of 72% in arable land (Buringh, 1989). This increase has accompanied decreases in areas of forests and other land reserves: on the basis of world land area, there was an increase of 4% in arable land and a decrease of 9% each in forest and other land uses.

Land-use statistics for 1973-88 are shown in Table 6. Worldwide, there were increases of 3.7% and 9.4% in areas under arable land and permanent crops, respectively. There were, however, decreases of 0.4% and 3.4% in areas under

**Table 5.** Cultivated land area in the world

| Region | Total land (ha × 10⁶⁾ | Cropland (ha × 10⁶⁾ | Cropland as a % of total land |
|---|---|---|---|
| Africa | 2965 | 184 | 6.20 |
| Asia | 2679 | 455 | 17.00 |
| North and Central America | 2139 | 274 | 12.80 |
| South America | 1753 | 139 | 7.93 |
| Europe | 473 | 140 | 29.60 |
| Former USSR | 2227 | 232 | 10.42 |
| Oceania | 843 | 49 | 5.82 |
| World | 13079 | 1475 | 11.28 |

(From Arnold et al., 1990.)

permanent pasture and forest/woodland, respectively. With the exception of Europe, the same pattern was observed in all major geographical regions. In Europe there were decreases of 1.9%, 3.5% and 5.3% in area under arable land, permanent crops, and pasture, respectively. In contrast, however, there was an increase of 2.3% in area under forest/woodland. The European land-use system over the 15 years ending in 1988 has shown an ecologically oriented trend.

The largest increase in arable land, about 22.6%, has occurred in South America. A high rate of tropical deforestation in the Amazon Basin is the major factor responsible for this drastic increase.

Projected changes in cultivated land area for developing regions of the world are shown in Table 7. The largest expansion in cultivated land area by the year 2025 is expected in Africa and South America. In comparison with the year 1975, there is a projected increase in cultivated land area of 14.5% in the Near East, 26.3% in Southeast Asia, 27.8% in Central America, 47.6% in Africa, and 79.8% in South America. Overall, an increase of 40.4% in the cultivated area in the developing world is expected between 1975 and 2025.

Even after this projected expansion in cultivated land area, there will be more land available for future cultivation in Central America (28 x 10⁶ ha), Africa (541 x 10⁶ ha), and South America (598 x 10⁶ ha). In contrast, the cultivated land area will exceed the potentially cultivable arable land in the Near East by 31 x 10⁶ ha. These additional areas brought under cultivation will be either marginal land or land with supplementary irrigation.

Expansion of cultivated land area by any means (deforestation, irrigation, or use of marginal land) is bound to have some adverse effects on the environment. These effects may be due to accelerated soil erosion, transport of excessive dissolved or suspended load in rivers, salinization, fertility depletion, and acidification. A major problem with converting forests and woodlands to arable

**Table 6.** Land use ($\times 10^6$ ha) in major regions of the world

| Region | Arable | | Permanent crop | | Permanent pasture | | Forest/woodland | |
|---|---|---|---|---|---|---|---|---|
| | 1988 | % change over 1973 | 1988 | % change over 1973 | 1988 | % change over 1973 | 1988 | % change over 1973 |
| World | 1373.4 | +3.7 | 102.0 | +9.4 | 3212.0 | -0.4 | 4049.0 | -3.4 |
| Africa | 167.9 | +7.8 | 18.7 | +12.7 | 729.9 | -0.6 | 683.3 | -6.1 |
| North and Central America | 267.1 | +2.3 | 6.8 | +11.5 | 368.2 | +2.9 | 686.9 | -1.4 |
| South America | 116.0 | +22.6 | 25.9 | +19.9 | 477.9 | +6.8 | 895.7 | -6.5 |
| Asia | 420.9 | +0.7 | 30.9 | +9.2 | 678.3 | -2.6 | 524.4 | -4.4 |
| Europe | 126.1 | -1.9 | 14.0 | -3.5 | 83.3 | -5.3 | 157.5[c] | +2.3 |
| Oceania | 47.7 | +1.1 | 1.0 | +10.6 | 439.4 | -4.5 | 156.3 | -17.0 |

(Calculated from FAO, 1989.)

**Table 7.** Projected change in cultivated land area ($\times$ $10^6$ ha)

| Region | Potentially cultivable area | Cultivated area 1975 | Cultivated area 2000 | Cultivated area 2025 |
|---|---|---|---|---|
| Africa | 789 | 168 | 204 | 248 |
| Near East | 48 | 69 | 74 | 79 |
| Southeast Asia | 297 | 274 | 308 | 346 |
| Central America | 74 | 36 | 41 | 46 |
| South America | 819 | 124 | 166 | 223 |
| Total | 2027 | 671 | 793 | 942 |

(From FAO, 1984.)

land is the emission of radiatively active gases into the atmosphere, with possible effects of global warming.

The environmental effects of conversion to arable land also depend on the farming system, degree of mechanization, cultivation intensity, and level of off-farm input used. The high demand for agricultural produce due to rapidly expanding population in Africa, Asia, and South America necessitates an increase in cultivated area planted with food crop staples, e.g. cereals and root/tubers. The data in Table 8 show the percentage change in cropland and yield for a 21-year period ending in 1985. Increase in crop yield per unit area has been an important factor for increase in food production. Crop yields of cereals increased by as much as 76 to 77% in Asia and Europe. There were, however, modest increases in cereal yield in Africa (13%), Oceania (25%), and the former USSR (35%). In addition to cereals, yield of root/tubers also increased an impressive 58% in Asia. In comparison, however, there were smaller increases of 13% in Oceania, 13% in the former USSR, 19% in Europe, 22% in Africa, and 23% in North and Central America. There was a decrease in yield of roots/tubers by 1% in South America for the period.

Yield increases in food crops are brought about by introduction of input-responsive and high-yielding varieties, irrigation, and intensive use of agrochemicals, including fertilizers and pesticides. While these science-based systems have been widely adopted in Asia, low-input and resource-based systems are still widely used in Africa. As a consequence, food production has seriously lagged behind the growth in population in sub-Saharan Africa.

## V. Irrigable Land Area

Expansion of agriculture in arid and semiarid regions is possible only through supplemental irrigation. In the 19th century, the rate of increase in irrigable land area was 5%/yr (Arnold et al., 1990). The highest rate or expansion of irrigable land area of 6.2%/yr was observed for the decade ending in 1959. The rate of expansion has drastically decreased since the early 1960s, to about 1.1%

**Table 8.** World increase in cropland, 1964-85

| | | % Change in yield | |
|---|---|---|---|
| Region | % Change in cropland | Cereals | Roots/tubers |
| Africa | 13.5 | 13 | 22 |
| Asia | 4.1 | 77 | 58 |
| North and Central America | 7.8 | 44 | 23 |
| Europe | 34.6 | 42 | -1 |
| Former USSR | 10.5 | 76 | 19 |
| Oceania | 1.3 | 35 | 13 |
| World | 23.5 | 25 | 13 |
| | 8.9 | 58 | 21 |

(From Arnold et al., 1990.)

**Table 9.** Prevalence of irrigation

| | ------------1980-------------- | | -------------2000------------- | |
|---|---|---|---|---|
| Region | Area equipped for irrigation ($\times$ 10$^6$ ha) | % of total arable land | Area equipped for irrigation ($\times$ 10$^6$ ha) | % of total arable land |
| 90 developing countries | 105.3 | 14 | 148 | 16 |
| Africa | 3.7 | 2 | 6 | 2 |
| Far East | 67.5 | 25 | 98 | 34 |
| Latin America | 13.4 | 7 | 19 | 7 |
| Near East | 20.7 | 23 | 25 | 27 |

(From FAO, 1981.)

annually, because economically feasible irrigable land has mostly been brought under irrigation.

Potentially irrigable land exists in several regions of developing countries (Table 9). For the two decades ending in the year 2000, irrigated land area in the 90 developing countries is expected to increase by 40.6%. The largest increase, 45.2%, is expected in the countries of the Far East, where more than one-third of all cropland will be irrigated by the year 2000. Irrigated agriculture is also expected to increase 41.8% in latin America. In Africa, although the area of irrigated arable land is expected to increase from 3.7 x 10$^6$ ha to 60 x 10$^6$ ha by the year 2000, the percentage of arable land under irrigation will remain almost the same. With Africa's perpetual food deficits and agrarian stagnation since the 1960s, the expansion of irrigated agriculture is an important strategy worthy of careful consideration. African water reserves are small, however, and suitable mainly for the development of small-scale irrigation schemes.

**Table 10.** Irrigated land area ($\times$ $10^6$ ha)

| Region | 1973 | 1978 | 1983 | 1988 | Area change in 15 yr | % change in 15 yr |
|---|---|---|---|---|---|---|
| World | 181.6 | 206.0 | 219.1 | 228.7 | 47.1 | +25.9 |
| Africa | 9.2 | 9.7 | 10.4 | 11.1 | 1.9 | +20.7 |
| North and Central America | 22.2 | 27.4 | 27.0 | 25.8 | 3.6 | +16.2 |
| South America | 6.2 | 7.1 | 8.0 | 8.8 | 2.6 | +41.9 |
| Asia | 117.3 | 129.5 | 137.3 | 142.8 | 25.5 | +21.7 |
| Europe | 12.3 | 14.0 | 15.4 | 17.3 | 5.0 | +36.2 |
| Oceania | 1.6 | 1.6 | 1.9 | 2.1 | 0.5 | +31.3 |

(Adapted from FAO, 1989.)

Even with these large projected increases overall, by the year 2000 the area of irrigated land as a percentage of all potentially irrigable land will be only 31.5% in Africa, 43.2% in Latin America, 45.0% in the Far East, and 86.2% in the Near East (FAO, 1984). It is apparent, therefore, that there is scope for further expansion of irrigated agriculture in Africa, Latin America, and the Far East.

Increases in irrigated land area in all regions of the world from 1973 to 1988 is shown in Table 10. For the world as a whole, irrigated land area increased by 47.1 x $10^6$ ha, or 25.9%. By far the largest increase occurred in Asia, with 25.5 x $10^6$ ha.

## VI. Rice Cultivation in the World

Rice is a food staple for a large section of the world population, especially in Asia. World rice production is estimated at 504 million metric tons (Mg) from a total area of 145.8 x $10^6$ ha (Table 11). Of the total land area used for rice cultivation, 89.9% lies in Asia. Among the principal rice-producing countries, the largest cultivated area of 28.5% is in India, followed by 22.2% in China. Outside of Asia, the principal rice-producing country is the United States, with 0.8% of the world area under rice cultivation.

The world average rice yield is about 3.5 Mg/ha. Yields are highest in the developed world, with 6.5 Mg/ha in the United States and 6.1 Mg/ha in Japan. Elsewhere in Asia, rice yield is 5.5 Mg/ha in China, 4.2 Mg/ha in Indonesia, and 2.7 Mg/ha in the Philippines. In the countries of South Asia, rice yield is generally low, i.e., 2.6 Mg/ha in India and 2.3 Mg/ha in Pakistan. There exists a tremendous scope to increase rice yield and decrease the area allocated to rice cultivation in these countries.

**Table 11.** Rice cultivation in the world

|  | Area ($\times 10^6$ ha) | Area as % of world total | Production ($\times 10^6$ Mg) |
|---|---|---|---|
| World regions |  |  |  |
| Africa | 5.5 | 3.8 | 10.5 |
| North and Central America | 1.8 | 1.2 | 9.3 |
| South America | 6.9 | 4.7 | 17.2 |
| Asia | 131.1 | 89.9 | 463.9 |
| Europe | 0.4 | 0.3 | 2.1 |
| Oceania | 0.1 | 0.1 | 0.8 |
| World total | 145.8 |  | 503.8 |
| Principal countries |  |  |  |
| China | 32.4 | 22.2 | 179.4 |
| India | 41.5 | 285 | 107.5 |
| Indonesia | 10.3 | 7.1 | 43.6 |
| Japan | 2.1 | 1.4 | 12.9 |
| Pakistan | 2.1 | 1.4 | 4.8 |
| Philippines | 3.5 | 2.4 | 9.5 |
| Thailand | 10.3 | 7.1 | 21.3 |
| Vietnam | 5.9 | 4.0 | 18.1 |
| USA | 1.1 | 0.8 | 7.1 |

(Adapted from FAO, 1989.)

## VII. Human Factors in Land Degradation

One of the reasons for confusion in the statistics involving land degradation is the nature of the problem itself. Land degradation is a social, managerial, and perceptual problem. Purely natural processes (e.g., leaching and geological erosion) occur without human intervention. However, compaction, accelerated erosion, impeded drainage and resulting salinization, fertilizer-induced soil acidification, and cropping-induced nutrient depletion are anthropogenic processes. The problem is perceptual and subjective because "lowering" of value is relative to actual or possible land uses, and relative to managerial skills and off-farm input. In many cases, loss in productivity due to degradation can be offset by adding chemical amendments or organic manures, improving drainage of waterlogged soils and leaching of salt-affected soil, decreasing bulk density of compacted soils, and managing accelerated erosion through appropriate use of preventive or ameliorative measures. Land degradation, therefore, becomes a social or managerial problem if society has no resources to alleviate land-related constraints to crop production.

Land degradation is also related to alternative land uses. For nomadic herding or hunting/gathering type of extensive shifting cultivation (Phase I in Table 2), replacement of forest by savannah with greater animal-carrying capacity is not a degradation. Nor is replacement of humid forest by pasture or arable land, provided that deforestation is done by appropriate methods of land clearing and soil management after clearing makes use of science-based technology.

Degradation is also related to the ability of land to support a crop or cropping system at the desired level of productivity. Anthropogenic land degradation may have a gradual effect on the productivity of a specific crop that cannot be offset even with improved systems of management, e.g., due to gully erosion, eluviation of clay and organic colloids from the root zone, or change in hydrology and plant available water-holding capacity of the soil. In these cases the value of the land has been reduced, because land cannot be used for cultivation of priority crops. In other words, "land capability" has been altered. Most degraded land is not totally lost for food production. Eroded cropland can be used for tree production or as grassland for pasture and livestock production. The land's capability has been transformed because rather than support good pasture, it can now support only thorny bushes, scrub vegetation, and low-grade pasture of low-carrying capacity.

The role of land management, however, is very critical. Land productivity and land capability can be altered. If land productivity changes due to soil-related constraints, it is important to alleviate those constraints. Production is also related to market. Market forces can regulate production and alter land capability because of change in demand. It is important to realize that land can be used to produce anything as long as it is economical to do so. Most farmers, however, cannot afford to invest heavily and take risks. Furthermore, higher yields do not always mean higher profit because inputs are expensive, and degraded land may require more inputs to produce the same yield than undegraded or less degraded land.

## VIII. Alternatives of Intensification and Extensification of Agricultural Land Use

Farming with traditional agricultural systems (Tables 1 and 2) produces low yields and requires more and more land area to support an increasing population. This is because traditional production systems use little or no input from outside the farm. Natural fallowing is used to improve soil fertility. Substituting off-farm inputs for fallowing means that more land is available for cultivation. External inputs become a viable alternative to extensive cultivation and lengthy fallow periods for fertility restoration. This means that more people can be fed from the same land area provided that external inputs can be judiciously used. Table 12 lists the world extent of different farming systems and their productivity. It is apparent that approximately 75% of all cropland is farmed using

**Table 12.** Cultivated land area and potential yields of different farming systems

| Farming system | Land area | | Potential yield (kg ha⁻¹) | Area of arable land needed (ha/capita) |
|---|---|---|---|---|
| | Actual ($\times 10^6$ ha) | % | | |
| Land rotation | 30 | 2 | <500 | 2.65 |
| Low traditional | 420 | 28 | 800 | 1.20 |
| Moderate traditional | 525 | 35 | 1200 | 0.60 |
| Improved traditional | 150 | 10 | 2000 | 0.17 |
| Moderate technological | 150 | 10 | 3000 | 0.11 |
| High technological | 150 | 10 | 5000 | 0.08 |
| Specialized technological | 750 | 5 | 7000 | 0.05 |

(Adapted from Buringh, 1989.)

traditional methods with low productivity. Improved technology is used in only 25% of farmland. Adopting technology can triple or quadruple yields and decrease the land area needed for achieving the same production. Presently, about 0.3 ha of arable land is used per capita. This can be reduced to one-half or one-third by use of science-based technology. For example, cultivated land area in India is 166 x $10^6$ ha (FAO, 1989). About 75% of this land is cultivated by moderate traditional methods, and another 10% to 15% by improved traditional methods. Productivity of these lands can be doubled or tripled by using improved technology.

The transitions from one mode of production to another is often easier said than done. In addition to biophysical constraints, socioeconomic factors limit adoption of improved technology. It is a very slow process.

## IX. Data Quality and Reliability

The most commonly used data source is the Food and Agricultural Organization of the United Nations (FAO) Yearbook, published annually. These data are compiled from reports submitted by each country. The accuracy of the data, therefore, depends on that of the data reported by each country. While agricultural statistics are up to date and reliable in most developed countries (North America, Western Europe, Australia, etc.), the data from most developing nations are questionable. The data estimates from developing nations may not be current or precise. Some of the data reported from these countries are guesstimates based on little if any validation. Most estimates from these regions may be five to ten years old. The accuracy of some estimates may be within 25 to 40% at best.

While the FAO statistics are a very useful data source, there is a need for an independent, reliable, and accurate assessment of world's land use in general and of the cultivated land area in particular. The cultivated land area is rapidly changing in the tropics, especially so in the equatorial rainforest ecosystem. Tropical rainforest is rapidly being converted to arable land, perennial crops, and pastures. However, the estimates of forest conversion and of allocation to different land use vary widely.

The data collation for cultivation and other land uses should be done by an international organization with a standard set of criteria and with methodology that is frequently verified with ground truth. Reliable survey data are required to improve the data bank, because such information has implications in planning for development of natural resources and in implementation of policies, if necessary. Thorough and systematic national, regional, and global surveys are needed using state-of-the-art technology, e.g., satellite pictures backed by ground truth validation.

## X. Conclusions

Estimates of potentially arable land area in the world indicate that it is sufficient to meet the present needs, and future requirements for a realistic global population of about 12 billion people (Buringh, 1989). However, land resources must be developed, used, and managed by science-based and conservation-effective techniques. Judicious management and protection of soil and water against degradation processes are essential for sustainable use of fragile and non-renewable resources.

An important task is to prepare an accurate inventory of the land resource base in relation to its productive capacity for principal soils in major ecoregions of the world. A reliable characterization of land resources and their capability is clearly needed for planning future strategies. Evaluation of the extent, kind, and severity of soil degradation caused by different processes is also needed. Restoration of degraded lands is another act of high priority. We must develop methods to regenerate productivity of degraded soils. Land-use planners should discourage cultivation of marginal lands and adoption of productivity-mining practices.

Expansion of agriculture to new areas should be done with careful planning, and only after a detailed inventory of natural resources has been done. This inventory must include assessment of soils, vegetation, climate, and water resources. There is also a need to assess socioeconomic factors, because land degradation is driven by a combination of biophysical and socioeconomic factors.

The need to bring new land under cultivation, a threat to natural ecosystems and an important factor in land and environmental degradation, can be drastically curtailed by intensive management of existing land through implementation of science-based practices of modern agriculture. Adoption of sustainable agricultural techniques should be done with the primary objective of enhancing per capita food production from existing arable land without jeopardizing its productivity potential and damaging the environment.

Intensification of agriculture is economically and ecologically feasible. However, sociopolitical and cultural factors require careful appraisal through long-term soil management experiments. There is a conspicuous paucity of such long-term experiments in the tropics.

Further degraded and marginal lands can also be brought under cultivation provided that long-term experiments are conducted on soil restoration. Most degraded land is not totally lost for production, and its productivity can be restored by judicious inputs and by change in management. The future lies in intensification of land use by transforming resource-based traditional agriculture to science-based modern agriculture, and data from well-planned and properly conducted long-term experiments can play a crucial role in achieving agricultural sustainability.

# References

Arnold, R.W., I. Szabolcs, and V.O. Targulian. 1990. *Global Soil Change*. International Institute for Applied Systems Analysis. Laxenberg, Austria. 110 pp.

Benneh, G. 1972. Systems of agriculture in tropical agriculture. *Econ. Geog.* 48:245-257.

Buringh, P. 1989. Availability of agricultural land for crop and livestock production. p. 70-85. In: D. Pimental and C.W. Hall (eds.), *Food and Natural Resources*. Academic Press, New York. 70-85.

Gupta, R.K. and I.P. Abrol. 1990. Salt-affected soils: Their reclamation and management for crop production. *Advances Soil Sci.* 11:223-288.

FAO. 1981. *Agriculture: Toward 2000*. FAO. Rome. Italy. 134 pp.

FAO. 1984. *Land, Food and People*. FAO. Rome, Italy. 96 pp.

FAO. 1989. *Yearbook*. Rome, Italy. 346 pp.

Lal, R. 1986. Conversion of tropical rainforest, agronomic potential and ecological consequences. *Advances Agronomy* 39:173-263.

Lal, R. 1991. Carbon emission and agricultural activities in the tropics. Environmental Protection Agency Report, Washington, D.C.

Lal, R. and B.A. Stewart. 1990. Soil degradation: A global threat. *Advances Soil Sci.* 11:13-17.

Okigbo, B.N. and D.J. Greenland. 1976. Intercropping systems in tropical Africa. p. 63-101. In: *Multiple Cropping*. American Sociey Agronomy. Madison, WI. 63-101

Richards, J. 1991. Land transformation. p. 163-178. In: B.L. Turner, W.C. Clarke, R.W. Kates, J.F. Richards, T.T. Matthews, and W.B. Meyers (eds.), *The Earth as Transformed by Human Action: Global and Regional Changes in Biosphere Over the Past 300 Years*. Cambridge University Press. United Kingdom.

Schlesinger, W.H. 1984. Soil organic matter: a source of atmospheric $CO_3$. p 111-127. In: G.M. Woodwell, (ed.), John Wiley & Sons, New York.

Vink, A.P.A. 1975. *Land Use in Advancing Agriculture*, Springer Verlag, New York. 394 pp.

World Resources Institute (WRI). 1990-91. *World Resources Report*. WRI. Washington, D.C.

# Need for Long-Term Experiments in Sustainable Use of Soil Resources

R. Lal and B.A. Stewart

## I. Introduction

Emphasis on agricultural sustainability in the 1980s and 1990s is indicative of an increasing awareness about the finite nature of arable land resources of the world, widespread problems of soil degradation, adverse effects of some agricultural activities on the environment, and excessive reliance on non-renewable resources. Most issues related to agricultural sustainability are related to soil quality, soil resilience, and the ability of soils to respond to inputs and science-based management. These issues on sustainable use of soil and water resources are global, cut across cultural and political boundaries, and require a coordinated and long-term effort to develop effective solutions.

## II. Assessment of Soil Quality from Long-Term Experiments (LTEs)

Soil quality refers to inherent attributes of soil and to characteristics and processes that determine the soil's capacity to produce economic goods and services and regulate the environment (Lal, 1993a; Lal and Miller, 1994). The response of soils to management and inputs also depends on soil quality.

ISBN 1-56670-076-0

Important attributes of soil quality are soil structure, porosity and pore size distribution, effective rooting depth, water retention and transmission properties, soil reaction, soil organic carbon content and biomass carbon, and plant-available nutrient reserves (Larson and Pierce, 1994; Parr et al., 1992; Lal, 1993b; NRC, 1993; Doran et al., 1994). Soil quality also depends on the balance between soil degradative and restorative processes (Hornick and Parr, 1987). In contrast, soil resilience refers to the ability of a soil to recover to the antecedent state following degradative pertubation or change in land use (Lal, 1993a; 1994). The rate and degree of recovery depends on the magnitude of pertubation, soil attributes, and management.

Understanding soil quality and soil resilience and their interaction with management is important to sustainable use of soil and water resources. It is important to establish the cause-effect relationship between management and soil quality. These relationships can only be established through long-term experiments (LTEs).

## III. Research and Development Priorities in LTEs

### A. Design of LTEs

Because of the long-term nature of LTEs, careful consideration should be given to types of experiments, principal issues to be addressed, goals, objectives, and methodology.

### 1. Types of LTEs

Two types of LTEs were identified. The first type involves critically important agricultural systems that are presently being used, but additional information is needed to improve or correct a problem in the system. The second type of LTEs deals with a truly new agricultural system, that is, a system that has not been used routinely by farmers, but might be used in the future.

### 2. Issues Relevant to LTEs

#### a. Why Are We Conducting LTEs?

LTEs are conducted to evaluate the use and conservation of soil resources in the production of food and raw materials for agribusiness: LTEs are excellent generators of process hypotheses. Basic principles are elucidated through good LTEs.

## b. Should LTEs be Process or Product Oriented?

Good LTEs should be both process and product oriented. Often, it takes years to obtain the result for either.

## c. Can LTEs be Conducted on Farmers' Fields?

The historical record in conducting LTEs, and in many cases, short-term experiments, in fields of farmers is not good. The central LTE should be conducted on a research station, but with supplementary short-term research conducted on farmers' fields.

## d. How Long is Long-Term?

The length of time required for a LTE depends on the issues to be addressed. LTEs are needed to assess long-term seasonal variability in productivity and environmental influences. There are *processes* and *mechanisms* involved which are not expressed in the short term. *Process rates* are dependent upon location in the hierarchy of activity, are relatively fast at a low position in the hierarchy, and become slower at higher positions in the hierarchy.

## e. Are LTEs Biased from a Scientific Standpoint?

LTEs can be biased in their design if only one or two disciplines are involved. LTEs should be conducted by a team of researchers having a broad base of knowledge. They should begin with identifying available data and determining if the problem is a "real problem" encountered by farmers or only a "perceived problem" identified by researchers. This step is often overlooked. The key issues are soils, water, and nutrients.

## f. Should Social Scientists and Economists be Included on the Research Team?

There are diverse views on this topic. Some argue that it is not possible to predict the social or economic situation 10 to 60 years in the future. Others believe that it is of little value to develop systems that would have too great an economic cost or be entirely socially unacceptable to the clientele. Involving social scientists and economists can make it easier to obtain policy changes which utilize results of LTEs.

**g. Are LTEs the Best Way to Put Together Predictive Tools?**

The final output of a well designed LTE should be a decision support system that can be used on a day-to-day basis. The preferred methodology is to: 1) capture existing knowledge; 2) focus on the problem; 3) find knowledge gaps in the data; and 4) develop the LTE to strengthen the data base.

**h. Why Is There Virtually No Funding for LTEs?**

Lack of funding for researchers at U.S. institutions is thought to be due to politics and local pressures. In general, the first allegiance of federal and state researchers is to solve domestic problems. Most citizens see little benefit, in fact they see negative benefit, in conducting LTEs in foreign lands. Apparently, the scientific community is not using the right approach.

**i. Are There Suitable Alternatives to LTEs?**

Some problems may be researched using data collected from identical agricultural systems operating at the same location on the landscape for different lengths of time.

## 3. Goals of LTEs

The principal goal of LTEs is to strengthen the data base needed for improving agroecosystem performance for present and future generations. Specifically, LTEs are designed with the following goals:
- Assess sustainability of production systems vis-a-vis short-term exploitation of natural resources.
- Enhance process level understanding, e.g., physical degradation and erosion, acidification and subsoil acidity, below-ground biological processes, and prediction of response to different input levels. Understanding of these processes requires minimum data sets.
- Facilitate evolution or development of improved systems through enhancement of water and nutrient use efficiencies, improvement of soil resilience, minimzation of adverse environmental effects, and increase in soil organic matter content and soil biodiversity.
- Develop innovative and revolutionary agricultural practices.
- Promote inter-disciplinary approach to address issues of agricultural sustainability.

4. Objectives of LTEs

LTEs should be designed with specific objectives such as:
  a. To produce products that can generate information to support decision making.
  b. To incorporate long-term processes into products that can generate information to help users (farmers, etc.) make better decisions.
The above two categories can be elaborated into the following specific objectives:
  1) To obtain data to identify process indicators.
  2) To obtain data for various kinds of models. No one model can include all factors.
  3) To validate models that mimic the full range of experience in the ecosystem.
  4) To organize new and existing data in a useful form to support decision making.
Collecting the appropriate data in LTEs allows the user to answer "What if" questions. For example, if climate change perturbations occur, we would hope to be able to predict the impacts. Hence, the environment must be appropriately characterized.

5. Experimental Methods

Experimental methods and design need careful planning. There are several issues that need to be addressed.

**a. Landscape or Watershed Level vs. Traditional Plot Approach**

The watershed or landscape approach enables spatial representation and interaction of all factors involved. This approach is better suited to technology transfer than to understanding basic soil processes. It is often difficult to replicate treatments on watershed scale. In comparison, field plots provide a targeted approach whereby multiple treatments can be easily replicated and data statistically analyzed. However, the problem of scaling for extrapolating results from plot to watershed or landscape level requires a careful consideration. Is it possible to develop a scale neutral design? If so, what is the appropriate plot size — 0.5 ha, 1 ha?

**b. Modeling before or after Experiments**

LTEs should be designed to couple models and experiments, and modelers and experimentalists. Whether modeling should be done before or after experimentation depends on the objectives. Modeling prior to experimentation sythesizes existing knowledge and is a useful strategy to identify knowledge gaps. Modeling after experimentation is a useful tool for extrapolation of results to other regions with similar soils and environments. Modeling, however, is not a substitute for experimentation.

### c. Analytical Techniques

Characterization of soil, water, crops, and environments should be done with new and state-of-the-art methodologies. The cause-effect relationship can only be established through systematic evaluation of soil physical, chemical and biological properties. Determination of spatial and temporal variations in these properties is critical to assessing sustainability. Parameters evaluated should facilitate scaling from field plot to watershed level and enable modeling. Data organization and accessibility for future use is an important consideration.

### d. Control

What are controls in LTEs? Does a control treatment reflect a specific production system? Is it important to involve farmers in defining a control treatment?

## B. Knowledge Gaps and Researchable Priorities

It is important to develop process-oriented research which can be transferred to other sites, and product-oriented research to provide decision support tools for application.

Review and analyses of the data from existing long-term experiments indicate knowledge gaps and researchable priorities with regard to the following:

- Identify factors and processes responsible for yield decline.
- Evaluate subtle soil changes affecting crop performance.
- Assess principles of water conservation and management.
- Determine water/nutrient interactions, and nutrient imbalance particularly in irrigated agriculture.
- Study root-rhizosphere and below-ground interactions.
- Monitor soil biological processes including the role of soil fauna.
- Evaluate nutrient behaviors influenced by soil amendments.
- Assess importance of soil organic matter.
- Synthesize available data into recommendations, and develop usable databases.
- Collate relevant plant data to complement soils data.

- Integrate livestock and trees with crops, including need for developing appropriate designs for evaluating these interactions.
- Develop scaling techniques to extrapolate data from field plots to landscape units or watershed.
- Assess impact of weather (and improved cultivar) on productivity.
- Develop indices of sustainability.

## C. Minimum Data Set

It is important to define the minimum data set (MDS) needed to assess sustainability, identify cause-effect relationships, and develop predictive models. Response variables must be clearly specified. The MDS can be defined on the basis of literature review or by conducting preliminary short-term experiments to identify the critical variables to be assessed. The MDS is required for soil properties, climate, economic variables, and plants and animals.

### 1. Soil Properties

- Organic matter content, pH, base saturation, N,P,S, and mineralizable nitrogen.
- Rooting depth, texture, aggregate stability, water transmission properties, infiltration capacity, water retention curves or "available $H_2O$."
- Well-described and preserved soil samples, detailed soil maps, landscape features, careful record of where samples should be collected.
- Frequency of measurements, depending on objectives and evaluation of processes, may involve short-term, detailed, and frequent measurement.
- Soil properties to be evaluated depend on questions and issues involved in sustainable management of soil resources.

### 2. Climate

The MDS recommended for climatic variables includes daily precipitation; rainfall intensity; minimum and maximum daily temperature; net solar radiation; wind speed at 2 m; vapor pressure; and soil temperature at 5, 10, 20, and 50 cm under bare soil conditions.

### 3. Plant Data

Total above-ground dry matter (over time) should be fractioned into product categories: food, feed, soil amendments. In addition, the following supporting MDS is needed:

- Mineral content of products
- Plant density, leaf area index
- Phenology
- Plant height
- Rooting depth
- Incidence of pests, diseases, nematodes
- Deficiency/toxicity symptoms

## 4. Animal Data

The MDS in animal-based and mixed-farming systems includes the following: species, sex, age, live weight and changes, grazing density/pressure, milk output, reproductive status, feed intake and feed refusal, manure and urine output.

## 5. Economic Variables

The MDS for economic evaluation includes:
- Opportunity and fixed costs of inputs in a farming setting including labor, production characteristics (major farming system), and cost of outputs (implies yield knowledge),
- The minimum data set to describe a farming system, and
- Total factor P productivity analysis including externalities.

## D. Networking and Role of U.S. Universities and Research Institutions

Effectiveness of LTEs can be vastly enhanced by networking. Objectives of networking are:
- Develop inter-institutional linkages,
- Exchange information,
- Adopt standardized methodology,
- Ensure continuity of new and on-going experiments, and
- Coordinate activities (e.g., training, workshops etc.) to avoid duplication.
  It is important that U.S. institutions are actively involved in these networks.
Principal benefits of participation by U.S. institutions are:
- Association with research programs involved in developing sustainable agricultural systems, biodiversity, global change, and environment.
- Availability of research products within a certain time period offers U.S. clientele benefits using decision support products.
- Opportunities for providing training facilities under diverse agroecosystems and socio-economic environments.

- Exchange of personnel for mutual enhancement of research and training opportunities. Deans from interested universities are encouraged to get together and develop a program for competitive funding for international research.

# References

Doran, J.W., D.C. Coleman, D.F. Bezdicek, and B.A. Stewart (eds.). 1994. *Defining Soil Quality for a Sustainable Environment*. SSSA Special Publication 35. Soil Science Society America. Madison, WI. 244 pp.

Hornick, S.B. and J.F. Parr. 1987. Restoring the productivity of marginal soils with organic amendments. *Am. J. Alternative Agric.* 2:64-68.

Lal, R. 1993a. Tillage effects on soil degradation, soil resilience, soil quality and sustainability. *Soil and Tillage Res.* 27:1-8.

Lal, R. 1993b. Sustainable land use systems and soil resilience. In: D.J. Greenland and I. Szabolcs (eds.), *Soil Resilience and Sustainable Land Use*. CAB International, Wallingford, United Kingdom.

Lal, R. and F.P. Miller. Soil quality and its management in humid subtropical and tropical environments. p. 1541-1550. In: *Proc. XVIII. Intl. Grassland Cong.*, Feb. 13-16, 1993, Palmerston North, New Zealand.

Lal, R. 1994a. Land Use and Soil Resilience. p. 246-261. In: *Proc. 15th World Congress of Soil Science*. Comm. 1D, Vol. 2A. International Society Soil Science, Vienna, Austria.

Lal, R. 1994b. Methods and guidelines for assessing sustainable use of soil and water resources in the tropics. SMSS Technical Monograph 21. USDA Soil Conservation Service. Washington, D.C. 78 pp.

Larson, W.E. and F.J. Pierce. 1994. The dynamic of soil quality as a measure of sustainable management. p. 37-51. In: J.W. Doran, D.C. Coleman, D.F. Bezdicek, and B.A. Stewart (eds.), *Defining Soil Quality for a Sustainable Environment*. SSSA Special Publication 35. Soil Science Society America, Madison, WI.

National Research Council (NRC). 1993. *Soil and Water Quality: An Agenda for Agriculture*. National Academy Press, Washington, D.C. 516 pp.

Parr, J.F., R.I. Papendick, S.B. Hornick, and R.E. Meyer. 1992. Soil quality: attributes and relationship to alternative and sustainable agriculture. *Am. J. Alternative Agric.* 7:5-11.

# Index

# Index